Advances in Conjugated Linoleic Acid Research
Volume I

Editors

Martin P. Yurawecz
Magdi M. Mossoba
Food and Drug Administration
Center for Food Safety and Applied Nutrition
Washington, DC

John K.G. Kramer
Food Research Center
Agriculture & Agri-Food Canada
Guelph, Ontario, Canada

Michael W. Pariza
Food Research Institute
Department of Food Microbiology and Toxicology
University of Wisconsin-Madison, Madison, Wisconsin

Gary J. Nelson
Western Human Nutrition Research Center/USDA
University of California at Davis
Davis, California

Champaign, Illinois

AOCS Mission Statement
To be a forum for the exchange of ideas, information, and experience among those with a professional interest in the science and technology of fats, oils, and related substances in ways that promote personal excellence and provide high standards of quality.

AOCS Books and Special Publications Committee

G. Nelson, chairperson, Western Regional Research Center, San Francisco, California
P. Bollheimer, Memphis, Tennessee
J. Derksen, Agrotechnological Research Institute, Wageningen, the Netherlands
N.A.M. Eskin, University of Manitoba, Winnipeg, Manitoba
J. Endres, Fort Wayne, Indiana
T. Foglia, USDA—ERRC, Wyndmoor, Pennsylvania
M. Gupta, Richardson, Texas
L. Johnson, Iowa State University, Ames, Iowa
H. Knapp, University of Iowa, Iowa City, Iowa
K. Liu, Hartz Seed Co., Stuttgart, Arkansas
J. Lynn, Congers, New York
M. Mathias, USDA-CSREES, Washington, D.C.
M. Mossoba, Food and Drug Administration, Washington, D.C.
F. Orthoefer, Monsanto Co., St. Louis, Missouri
E. Perkins, University of Illinois, Urbana, Illinois
J. Rattray, University of Guelph, Guelph, Ontario
A. Sinclair, Royal Melbourne Institute of Technology, Melbourne, Australia
G. Szajer, Akzo Chemicals, Dobbs Ferry, New York
B. Szuhaj, Central Soya Co., Inc., Fort Wayne, Indiana
L. Witting, State College, Pennsylvania
S. Yorston, Shur-Gain, Mississauga, Ontario

Copyright © 1999 by AOCS Press. All rights reserved. No part of this book may be reproduced or transmitted in any form or by any means without written permission of the publisher.

The paper used in this book is acid-free and falls within the guidelines established to ensure permanence and durability.

Library of Congress Cataloging-in-Publication Data

Advances in conjugated linoleic acid research/editors,
    Martin P. Yurawecz . . . [et al.].
      p. cm.
    Includes bibliographical references and index.
    ISBN 1-893997-02-2
    1. Linoleic acid—Metabolism. I. Yurawecz, Martin P.
QP752.L5A38   1999
612.3'97—dc21   99-32986
                                                                                                   CIP

Printed in the United States of America with vegetable oil-based inks.
03 02 01 00 99   5 4 3 2 1

Preface

This is the first book to be devoted solely to the subject of conjugated linoleic acid (CLA). It is the Editors' intent that it should document the state of knowledge about CLA as the 20th century draws to a close.

It has been known for more than 50 years that fatty acids with conjugated double bonds are present at varying levels in dairy products and other foods derived from ruminant animals. These conjugated fatty acids are common, though usually minor products of microbial lipid metabolism, which occur in the rumen of cattle, sheep, and other *Ruminantia*. Interest in conjugated fatty acids has increased substantially in recent years, following the observation that conjugated linoleic acid (CLA) is anti-carcinogenic in a number of rodent models. Since then, CLA has also been reported to induce a number of additional physiologic effects in a number of species, including the inhibition of atherosclerosis, enhancement of immunologic function, protection against the catabolic effects (cachexia) induced by immune stimulation, and nutrient repartitioning such that body fat is significantly reduced, whereas lean body mass is increased. These findings continue to drive CLA research.

As one should expect with any evolving scientific field, research to date on CLA has provided more questions than answers: How is it possible that CLA produces so many physiologic effects? What are the roles of the various CLA isomers in these effects? What are the precise biochemical mechanisms for these effects? What is the role of CLA metabolism, for example, elongation and desaturation of the various CLA isomers? Are there common mechanistic threads that tie these seemingly diverse effects together? Are the effects observed in animal models applicable to humans? How much of the various CLA isomers are in the diet? What are the best ways to analyze for CLA in foods/feeds and biological materials? These are some of the questions that are addressed (but not necessarily answered) in this book.

The nomenclature for the title compounds, "conjugated linoleic acid" and "CLA," varies from chapter to chapter. Sometimes the *cis*-9,*trans*-11 CLA isomer is referred to as "rumenic acid" or "RA." Conjugated linoleic acid is sometimes included in a larger class called "conjugated fatty acids" or "CFA." The term "octadecadienoates" is also sometimes used synonymously with CLA. The editors have not attempted to unify the nomenclature, preferring to leave this until consensus is achieved on the physiologic importance of the various CLA isomers.

Although this book is an up-to-date report of work that is still in progress, it is also clear that substantial progress has been made in some areas of CLA research. This is particularly so in the development of analytical methodologies. The reader is encouraged to utilize the newest methods as reported in this book, rather than older methods reported in the earlier literature.

At least 48 new peer-reviewed papers on CLA were published in 1998. If the reader wishes to keep track of CLA progress, an updated listing can be found on the internet at http://www.wisc.edu/fri/clarefs.htm.

Finally, the editors thank each of the contributors for their outstanding contribution, for their insight into the latest research findings, and the vision they have provided to make this book a truly valuable addition to the CLA literature.

Martin P. Yurawecz
Magdi M. Mossoba
John K.G. Kramer
Michael W. Pariza
Gary J. Nelson

Contents

Preface iii

Chapter 1 Conjugated Linoleic Acid: The Early Years 1
Peter W. Parodi

Chapter 2 The Biological Activities of Conjugated Linoleic Acid 12
Michael W. Pariza

Chapter 3 Preparation of Unlabeled and Isotope-Labeled 21
Conjugated Linoleic and Related Fatty Acid Isomers
Richard O. Adlof

Chapter 4 Commercial Production of Conjugated Linoleic Acid 39
Martin J.T. Reaney, Ya-Dong Liu, and Neil D. Westcott

Chapter 5 The Oxidation of Conjugated Linoleic Acid 55
Klaus Eulitz, Martin P. Yurawecz, and Yuoh Ku

Chapter 6 Methylation Procedures for Conjugated Linoleic Acid 64
Martin P. Yurawecz, John K.G. Kramer, and Yuoh Ku

Chapter 7 Separation of Conjugated Fatty Acid Isomers 83
John K.G. Kramer, Najibullah Sehat, Jan Fritsche,
Magdi M. Mossoba, Klaus Eulitz, Martin P. Yurawecz,
and Yuoh Ku

Chapter 8 Gas Chromatography/(Electron Impact) Mass 110
Spectrometry Analysis of Conjugated Linoleic
Acid (CLA) Using Different Derivatization Techniques
Volker Spitzer

Chapter 9 Identification of CLA Isomers in Food 126
and Biological Extracts by Mass Spectrometry
John A.G. Roach

Chapter 10 Confirmation of Conjugated Linoleic Acid Geometric 141
Isomers by Capillary Gas Chromatography-Fourier
Transform Infrared Spectroscopy
Magdi M. Mossoba, Martin P. Yurawecz, John K.G. Kramer,
Klaus D. Eulitz, Jan Fritsche, Najib Sehat, John A.G. Roach,
and Yuoh Ku

Chapter 11 Nuclear Magnetic Resonance Spectroscopic Analysis 152
of Conjugated Linoleic Acid Esters
Marcel S.F. Lie Ken Jie, M. K. Pasha, and M.S. Alam

Chapter 12 Identification and Quantification of Conjugated Linoleic 164
Acid Isomers in Fatty Acid Mixtures by ^{13}C NMR Spectroscopy
Adrienne L. Davis, Gerald P. McNeill, and David C. Caswell

Chapter 13	Biosynthesis of Conjugated Linoleic Acid and Its Incorporation into Meat and Milk in Ruminants *J. Mikko Griinari and Dale E. Bauman*	180
Chapter 14	Endogenous Synthesis of Rumenic Acid in Rodents and Humans *Donald L. Palmquist and Jamie E. Santora*	201
Chapter 15	Effect of Ionophores on Conjugated Linoleic Acid in Ruminal Cultures and in the Milk of Dairy Cows *V. Fellner, F.D. Sauer, and J.K.G. Kramer*	209
Chapter 16	Species-Dependent, Seasonal, and Dietary Variation of Conjugated Linoleic Acid in Milk *Gerhard Jahreis, Jan Fritsche, and Jana Kraft*	215
Chapter 17	Dietary Control of Immune-Induced Cachexia: Conjugated Linoleic Acid and Immunity *Mark E. Cook, D. DeVoney, B. Drake, M.W. Pariza, L. Whigham, and M. Yang*	226
Chapter 18	Incorporation of Conjugated Fatty Acid into Biological Matrices *Martin P. Yurawecz, John K.G. Kramer, Michael E.R. Dugan, Najibullah Sehat, Magdi M. Mossoba, Jun Jie Yin, and Yuoh Ku*	238
Chapter 19	Bone Metabolism and Dietary Conjugated Linoleic Acid *Bruce A. Watkins, Yong Li, and Mark F. Seifert*	253
Chapter 20	Conjugated Linoleic Acid (CLA) and the Risk of Breast Cancer *Flore Lavillonnière and Philippe Bougnoux*	276
Chapter 21	Conjugated Linoleic Acid (CLA) in Lipids of Fish Tissues *Robert G. Ackman*	283
Chapter 22	Conjugated Linoleic Acids in Human Milk *Mark A. McGuire, Michelle K. McGuire, Peter W. Parodi, and Robert G. Jensen*	296
Chapter 23	Influence of Dietary Conjugated Linoleic Acid on Lipid Metabolism in Relation to Its Anticarcinogenic Activity *Sebastiano Banni, Elisabetta Angioni, Gianfranca Carta, Viviana Casu, Monica Deiana, Maria Assunta Dessì, Leonardo Lucchi, Maria Paola Melis, Antonella Rosa, Silvana Vargiolu, and Francesco P. Corongiu*	307

Chapter 24	Conjugated Linoleic Acid Metabolites in Rats ....319 *J.L. Sébédio*	
Chapter 25	Effect of Conjugated Linoleic Acid on Polyunsaturated ....327 Fatty Acid Metabolism and Immune Function *M. Sugano, M. Yamasaki, K. Yamada, and Y.-S. Huang*	
Chapter 26	Regulation of Stearoyl-CoA Desaturase by ....340 Conjugated Linoleic Acid *James M. Ntambi, Youngjin Choi, and Young-Cheul Kim*	
Chapter 27	Conjugated Linoleic Acid for Altering Body Composition ....348 and Treating Obesity *Richard L. Atkinson*	
Chapter 28	Feeding CLA to Pigs: Effects on Feed Conversion, ...354 Carcass Composition, Meat Quality, and Palatability *Michael E.R. Dugan and Jennifer L. Aalhus*	
Chapter 29	Dietary Sources and Intakes of Conjugated Linoleic ....369 Acid Intake in Humans *Michelle K. McGuire, Mark A. McGuire, Kristin Ritzenthaler, and Terry D. Shultz*	
Chapter 30	Formation, Contents, and Estimation of Daily Intake of ....378 Conjugated Linoleic Acid Isomers and *trans*-Fatty Acids in Foods *J. Fritsche, R. Rickert, and H. Steinhart*	
Chapter 31	Conjugated Linoleic Acid and Experimental Atherosclerosis ....397 in Rabbits *David Kritchevsky*	
Chapter 32	Modulation of Diabetes by Conjugated Linoleic Acid ....404 *Martha A. Belury and John P. Vanden Heuvel*	
Chapter 33	Conjugated Linoleic Acid as a Nutraceutical: Observations in the Context of 15 Years of n-3 Polyunsaturated Fatty Acid Research ....412 *Howard R. Knapp*	
Chapter 34	Cancer Inhibition in Animals ....420 *Joseph A. Scimeca*	
Chapter 35	Intake of Dairy Products and Breast Cancer Risk ....444 *Paul Knekt and Ritva Järvinen*	
	Index 469	

Chapter 1

Conjugated Linoleic Acid: The Early Years

Peter W. Parodi

Human Nutrition Program, Dairy Research and Development Corporation, Melbourne, Australia

Introduction

During the early 1980s, Michael Pariza and his colleagues at the University of Wisconsin found that an isolate from grilled minced beef could inhibit carcinogenesis. The anticarcinogenic isolate was shown to consist of isomers of conjugated octadecadienoic acid in which the constituent double bonds are separated by a single carbon-to-carbon bond instead of a methylene group (1). The isomers were referred to collectively as conjugated linoleic acid (2) for which the acronym CLA is now used.

Since that time, an ever increasing number of studies using synthetically prepared CLA have shown that it can suppress cancer development at a number of sites in animal models and inhibit the growth of a large selection of human cancer cell lines *in vitro*. In addition, CLA possesses antiatherogenic, growth-promoting, lean body mass–enhancing properties; it can modulate food allergic reactions and normalize impaired glucose tolerance in noninsulin-dependent diabetes. These topics will be discussed in the following chapters. This chapter outlines the history of CLA detection in biological specimens and the determination of its structure.

Discovery of CLA in Milk Fat

In 1932, scientists from the National Institute for Research in Dairying at the University of Reading, collaborating with the Dunn Nutritional Laboratory, University of Cambridge, commenced a study of the seasonal variation in the vitamin content of milk. Vitamin A was estimated colorimetrically by measuring the intense blue color produced on treatment with antimony trichloride in chloroform solution—the Carr-Price reaction. However, they found that butterfat inhibited the blue color of the Carr-Price reaction. Further, it was established that inhibition was greater in summer butter, when cows were fed fresh pasture, than in the winter butter of stall-fed cows (3). In an associated study, this group noted that free fatty acids prepared from summer butter also differed from those prepared from winter butter in exhibiting much stronger spectrophotometric absorption at 230 μm as well as a much more rapid brown color formation when treated with antimony trichloride reagent (4).

Booth *et al.* (5) next reported that fatty acids obtained from butter fat by short-time saponification had definite absorption at 230 μm. On the other hand, the major unsaturated components of butter fat, oleic, and linoleic acids, treated in a similar manner, did not exhibit absorption at 230 μm. Another arm of this study showed that when the cow is turned out to pasture after winter, the fatty acids of milk fat showed a greatly increased absorption at 230 µm. Values were more than doubled during summer months when the cows had access to pasture. Although the nature of the fatty acid responsible for absorption at 230 µm was unknown at this time, this is the first study to record the seasonal variation in CLA content of milk fat.

In the same year, this group prepared mixed fatty acids from cod-liver oil, sardine oil, rapeseed oil, and linseed oil and showed that they exhibited little absorption at 230 µm. However, when the oils were fed to cows, the secreted mixed acids of milk fat possessed much greater absorption at 230 µm (6). This is the first study to demonstrate that the CLA content of milk fat could be increased by dietary manipulation.

What Causes Absorption at 230 µm?

Studies by Gillam *et al.* (7) and Edisbury *et al.* (8) showed that, although a selection of triglycerides of animal or vegetable origin lacked absorption in the ultraviolet region, the mixed acids derived from them by prolonged saponification often showed intense and highly selective absorption. Knowledge was extended when Dann and Moore (9) and Dann *et al.* (6) found that saponification of cod-liver oil was complete within a few minutes, and the resulting acids exhibited little selective absorption. However, prolonged heating with alcoholic potash greatly enhanced absorption at a number of wavelengths. Oxidation was ruled out as a cause of increased absorption. It was postulated that a molecular rearrangement occurs, probably with acids having two or more double bonds, with the production of an isomeric form.

On the other hand, Edisbury *et al.* (10) considered the most plausible explanation for the spectroscopic phenomenon to be cyclization to form polycyclic hydroaromatic compounds. Moore (11) then noted that the prolonged saponification of linseed oil produced two acids, one with intense absorption at 230 µm and the other, a solid, with an absorption maximum at 270 µm. This latter acid appeared similar to elaeostearic acid from tung oil. At this time, the absorption at 270 µm for elaeostearic acid (*cis*-9,*trans*-11,*trans*-13-octadecatrienoic acid) was considered to be due to the presence of three conjugated double bonds. Next, Moore (12) hydrogenated a sample of tung oil. This resulted in a change in the absorption maximum of the mixed acids from 270 to 230 µm. Moore (12) then concluded that absorption at 230 µm was the result of two conjugated double bonds.

Spectrophotometric Determination of Polyunsaturated Fatty Acids

It was now known that the wavelength of the absorption maxima depends on the number of double bonds present in the acids, and the intensity is proportional to the amount of the acid present. Thus alkali isomerization of polyunsaturated fatty acids, which lack absorption bands, to conjugated forms that have specific absorption bands allowed the detection and quantitative measurement of dienoic, trienoic, tetraenoic, pentaenoic, and hexaenoic acids as classes.

During the 1940s and early 1950s, a number of spectrophotometric methods, involving alkaline isomerism and the use of empirically determined constants, were developed for the determination of polyunsaturated fatty acids. These methods, which were performed rapidly and more accurately than precious techniques, such as ester fractionation analysis, are reviewed by Holman (13).

Polyunsaturated Fatty Acids of Milk Fat

Because of the nutritional significance of polyunsaturated fatty acids and their importance in the oxidation of milk fat, dairy scientists took advantage of the new spectrophotometric techniques to determine their composition. Riel (14) reviewed studies conducted in various countries up until 1963, which can be considered the period before gas–liquid chromatography (GLC). The review details the composition of monounsaturated, conjugated dienoic, trienoic, and tetraenoic fatty acids, together with nonconjugated polyunsaturated fatty acids with 2–6 double bonds.

Table 1.1 lists the range of conjugated dienoic acids for the major studies. The highest values occurred during the summer period. With the exception of the New Zealand study, the other studies were conducted in the Northern hemisphere where cows are stall-fed during winter. In these countries there was a two- to threefold increase in conjugated diene content when cows were turned out to fresh pasture.

Structure of the Conjugated Acid

Determination of Chain Length. Studies of milk fat fractions by Hilditch and Jasperson (23,24) and Mattsson (15) suggested that the conjugated unsaturation was associated with the C_{18} polyunsaturated fatty acids.

Infrared Absorption. Jackson *et al.* (25) examined the infrared spectra of conjugated isomers of linoleic acid. They found that *trans,trans*-conjugated linoleate was characterized by a strong absorption band at 988 cm^{-1}, whereas *cis,trans* (*trans,cis*)-conjugated linoleate was distinguished by a doublet at 948 and 982 cm^{-1}. No characteristic absorption was found for the *cis,cis*-conjugated isomer.

TABLE 1.1 Spectrophotometric Determination of Conjugated Diene Fatty Acids in Milk Fat

Country	Conjugated diene Range (wt%)	Reference
Sweden	0.6–3.7	15
New Zealand	0.7–1.4	16
Germany	0.5–2.3	17
U.S.	0.7–1.1	18
Holland	0.6–2.8	19
Canada	0.5–1.8	20
Sweden	0.4–1.6	21
France	0.4–1.9	22
Canada	0.2–2.0	14

Utilizing this infrared information, Lloyd Smith and colleagues at the University of California found that the C_{18}–C_{20} polyunsaturated fatty acid fraction of milk fat had strong infrared absorption bands at 948 and 982 cm^{-1}. This suggested *cis,trans-* or *trans,cis-*conjugated unsaturation. However, their spectra did not preclude the presence of small amounts of *trans,trans-*isomers (26). Conjugated *trans,trans-*diene was later detected in a concentrated unsaturated ester fraction of milk fat, further separated by urea-adduct stepwise crystallization with the use of ultraviolet and infrared spectroscopy (27). By using differential infrared spectroscopy Bartlet and Chapman (28) found that conjugated and isolated *trans* unsaturation were present in a constant ratio in milk fat. They used this characteristic as a basis for determining adulteration of milk fat.

Gas–Liquid Chromatography Studies

With the advent of GLC, this technique was soon utilized to determine the fatty acid composition of milk fat. Because milk fat contains >400 different fatty acids, it presents a complex GLC elution pattern. Magidman *et al.* (29) subjected methyl esters of milk fat fatty acids to distillation and silicic acid column chromatography to obtain fractions of less complexity for GLC analysis. With the aid of ultraviolet and infrared spectroscopy they were able to detect peaks representing octadecadienoic acids with *cis,trans-* (18:2 *ct*-conj.) or *trans,cis-* and *trans-trans-*conjugated unsaturation. The presence of the conjugated double bond system increased the retention time of the ester over that of a similar fatty acid with methylene interrupted double bonds. In a following report (30), the 18:2 *ct*-conj. and 18:2 *tt*-conj. acids were found to be present in a sample of milk fat at the 0.63 and 0.09% level, respectively.

Thin-Layer Chromatography Studies

The recently developed technique of silver ion absorption thin-layer chromatography (TLC) was used by Kuzdzal-Savoie (31) to separate the methyl esters of milk fat by the number and geometry of their double bonds. A band eluting between the *trans-*monounsaturated and *cis-* monounsaturated acids was thought to consist of conjugated dienes. A later study by Kuzdzal-Savoie *et al.* (32) used a combination of GLC, ultraviolet, and infrared spectroscopy, together with mass spectrometry to identify the presumed conjugated diene band as containing 18:2 *ct*- conj. or 18:2 *tc*-conj.

It is interesting to note that during silver ion absorption TLC, conjugated diene esters migrate with *cis-*monoenes when the common solvent system hexane:diethyl ether is used for development. However, when benzene or toluene is used as the solvent, they elute between the *cis-*monoenes and *trans-*monoenes (33,34). During studies with milk fat conjugated dienes, this author noted that they charred a brown color on silver nitrate–impregnated TLC plates after heating with sulfuric acid. On the other hand, other unsaturated acids produced a black color.

The Conjugated Fatty Acid Is cis-9, trans-11-Octadecadienoic Acid

Parodi (34) fractionated the methyl esters of milk fat by preparative GLC using a polyester phase. The effluent represented by the peak containing 18:2*ct* (*tc*)-conj. was collected and separated from co-eluting 18:3 and 20:1 by preparative silver ion absorption TLC. The band representing 18:2*ct* (*tc*)-conj. had a strong ultraviolet absorption maximum at 233 μm, indicating conjugated unsaturation. The infrared spectrum showed strong absorption at 949 and 982 cm^{-1}, designating conjugated, *cis,trans*- or *trans,cis*-unsaturation. GLC equivalent chainlengths (ECL) from both polar and nonpolar columns suggested that the isolated band was mainly *cis,trans*- or *trans,cis*-octadecadienoic acid.

The conjugated diene was subjected to reductive ozonolysis with the result that 90% of the cleavage products were represented by a C-7 aldehyde and a C-9 aldehyde-ester. This indicated that the original double bonds were at positions 9 and 11. Stereochemistry of the double bonds was determined by partial hydrazine reduction of the conjugated *cis,trans*-(*trans,cis*)-octadecadienoic acid. The resulting *cis*- and *trans*-monoene fractions were separated by preparative silver ion absorption TLC, then subjected to reductive ozonolysis. Cleavage products of the *cis*-monoene fraction consisted almost entirely of C-9 aldehyde and C-9 aldehyde-ester, indicating that the double bond at carbon 9 had the *cis*-configuration. The cleavage products of the *trans*-monoene fraction were mainly C-7 aldehyde and C-11 aldehyde-ester, showing that the double bond at carbon 11 had the *trans*-configuration. Thus, in milk fat, the conjugated dienoic acids are essentially *cis*-9,*trans*-11-octadecadienoic acid.

At various stages during the isolation and structural determination, small amounts of *trans,trans*-octadecadienoates were detected, and also octadecadienoates with double bonds at the 8,10- and 11,13-positions. Because of the labile nature of conjugated acids and the propensity of the *cis*-double bond to isomerize to the *trans*-configuration, it was considered that the minor components could have resulted from manipulative procedures.

cis-9, trans-11-Octadecadienoic Acid Acquires a Name

By reason of the importance now ascribed to synthetic CLA because of its many beneficial biological reactions, trivial names for the natural *cis*-9,*trans*-11-isomer have been suggested. McGuire *et al.* (35) proposed bovinic acid, but this name is considered too restrictive because the isomer is also produced in the rumen of a number of other species of commercial importance. For this reason, Parodi suggested rumenic acid, a name that is now gaining acceptance (36).

CLA in Milk Phospholipids

A study by Mattsson and Swartling (21) appears to be the first to measure conjugated fatty acids in milk phospholipids. They could not detect any absorption at 232 μm in phospholipids isolated from butter, collected during summer and winter, although the triglycerides from these butters contained 1% or more conjugated diene. The authors reported a high background absorption in the region of 230 µm. This may have obscured any absorption due to conjugation. A year later, Smith and Jack (37) reported that mixed phospholipids extracted from buttermilk powder, a rich source of phospholipid, contained 2.3% conjugated diene. This value was nearly twice that for triglycerides from the same source. Knowledge of conjugated dienes in phospholipids was extended when Hay and Morrison (38) presented the composition of the fatty acids from the sn-1 and sn-2 positions of phosphatidyl choline and phosphatidyl ethanolamine, prepared from spray-dried buttermilk powder. They reported that phosphatidyl ethanolamine contained 0.9 and 0.7% conjugated *cis,trans*-diene and *trans,trans*-diene, respectively. Phosphatidyl choline contained 0.65 and 0.35% of these isomers. The two conjugated isomers were nearly equally distributed between the sn-1 and sn-2 position. This distribution was in contrast to the *cis,cis*-, *cis,trans*-, and *trans,trans*-nonconjugated octatadecadienoates that exhibited preference for the sn-2 position.

Of interest here is the study of Christie (33), which showed that bile phosphatidyl cholines from cows and sheep contained 1.1 and 4.7% of *cis*-9,*trans*-11-18:2, respectively. In both cases, the conjugated diene was esterified exclusively at the sn-2 position. The question arises, was the *trans,trans*-conjugated diene reported by Hay and Morrison (38) in milk phospholipids, and by other early investigators in milk fat triglycerides, factual or artifactual? Conjugated double bonds are very labile, even at room temperature, and the *cis*-double bond can isomerize readily to the *trans*-form (39). No doubt heating during pasteurization and the subsequent processing of dairy products can result in isomerism. In addition, it is realized now that the method employed to prepare methyl ester derivatives for GLC analysis is critical. Methods used during the early years of CLA investigation produced considerable amounts of the *trans,trans*-isomer (40,41).

The Origin of CLA

During the 1930s, biological chemists began to realize that the cow could convert dietary nonconjugated fatty acids to a conjugated component. However, the mechanism for this transformation and the location at which it was effected were unknown. By 1950, it was established that there was a species-to-species difference in the unsaturated fatty acid content of body fat from pasture-fed animals. Linolenic acid is the predominant pasture fatty acid. When ruminant animals such as cows and sheep consume this acid, only trace amounts appear in body tissues or milk. On the other hand, the horse, a nonruminant, transfers a considerable proportion of dietary linolenate to its depot fat. Shorland (42) designated such fats as "heterolipoid" and "homolipoid," respectively.

Reiser (43) incubated rumen contents with an emulsion of linseed oil. This process reduced the linolenic acid content of the mixture with a corresponding increase in the linoleic acid content; the other acids remained unchanged. Later, Shorland *et al.* (44) showed that the hydrogenation process was more extensive than suggested by Reiser (43). When linolenic acid was incubated with sheep rumen contents, the process produced monoene as well as diene fatty acids and introduced *trans*-unsaturation.

The presence of conjugated diene was also established. Shorland et al. (45) next extended their study to include oleic and linoleic acid. In all cases, stearic acid was produced as well as trans fatty acids. A considerable quantity of conjugated diene was produced but only when linloeic acid was used as the substrate.

Kepler et al. (46) later showed that the conjugated diene was cis-9,trans-11-(or trans- 9,cis-11-, or both)octadecadienoic acid, and was produced as a stable first intermediate when the common rumen bacterium Butryivibrio fibrisolvens was incubated with linoleic acid. In a second step, the conjugated diene was hydrogenated to a mixture of trans-11-18:1 and trans-9-18:1. The enzyme responsible for the initial isomerism of linoleic acid to cis-9,trans-11-18:2 was isolated from B. fibrisolvens by Kepler and Tove (47) and identified as a linoleate cis-12,trans-11- isomerase.

CLA in the Depot Fat of Ruminants and Nonruminants

Observations by Reiser (43) and Shorland et al. (44) that rumen contents could hydrogenate polyunsaturated fatty acids to produce acids with trans-unsaturation provided a biological explanation for the presence of trans-fatty acids in the depot fat of sheep and oxen, noted as early as 1928 by Bertram (48). Because of its spectroscopic absorption at 230 μm, conjugated linoleic acid was used by Reiser (49) as a marker to monitor absorption and transfer of dietary fatty acids to the tissue lipids of rats.

In a study with pasture-fed horses, Shorland et al. (50) found that their tissue triglycerides and phospholipids contained conjugated diene. Next, this New Zealand group examined the trans-unsaturated fatty acid content of fat from a selection of ruminant and nonruminant animals (51). As expected, the depot fats from ruminants (Sambur, fallow deer, ox, and sheep) contained conjugated diene at around the 0.5% level. An unexpected finding was the presence of considerable conjugated diene in the body fat of the nonruminant marsupials, wallaby (3.5%), and quokka (2.9%). Depot fat trans-monounsaturated fatty acid content from these marsupials was also exceptionally high, 19 and 21%, respectively. The high conjugated diene and trans-monoene content is explained by their possession of a ruminant-like digestion. The stomach contents contain a rich microbial population similar to, but rather simpler than that of cows and sheep (52).

Hansen and Czochanska (53) tentatively identified cis-9,trans-11-18:2 in the depot fat of pasture-fed lambs on the basis of ECL from GLC analysis.

CLA in Vegetable Oils

A wide range of fatty acids with conjugated unsaturation occur in seed oils of various plant families (39). On the other hand, conjugated unsaturation is absent in fatty acids from the native form of common dietary vegetable oils. The small amount of CLA found in refined corn and peanut oil (54) no doubt results from heating, bleaching, and deodorization during the refining process (55).

Partial hydrogenation of vegetable oils containing linoleic acid results in the formation of isomers with conjugated unsaturation. Spectrophotometric methods were used to measure this conjugated unsaturation in various vegetable oil–based products. In the U.S., commercial shortenings and processed soybean oil contained from 0.3 to 0.6% conjugated diene (56). Different types of margarines had a range of 0.4 to 1.9% (57) and 0.6 to 0.8% (58). The conjugated diene content of Australian shortening ranged from 0.2 to 0.6%, whereas the level in various types of margarine was between 0.3 and 0.8% (59).

During this period, there appeared to be little interest in detecting the conjugated dienes by GLC or determining their structure. Currently the world-wide trend is to produce zero-*trans* margarines and shortenings. This will also result in the absence of CLA from these products.

The End of the Beginning

Imagine the biological chemists of the 1930s era being alive today. They would be astonished at the evolving literature, available analytical techniques, and the animal studies that have shown such a wide range of advantageous physiologic events, for the compound they struggled to isolate and identify. May CLA research continue to prosper.

References

1. Pariza, M.W. (1997) Conjugated Linoleic Acid, a Newly Recognised Nutrient, *Chem. Ind.*, 464–466.
2. Ha, Y.L., Grimm, N.K., and Pariza, M.W. (1987) Anticarcinogens from Fried Ground Beef: Heat-Altered Derivatives of Linoleic Acid, *Carconogenesis 8*, 1881–1887.
3. Booth, R.G., Kon, S.K., Dann, W.J., and Moore, T., (1933) A Study of Seasonal Variation in Butter-Fat. 1. Seasonal Variations in Carotene, Vitamin A and the Antimony Trichloride Reaction, *Biochem. J. 27*, 1189–1196.
4. Booth, R.G., Dann, W.J., Kon, S.K., and Moore, T. (1933) A New Variable Factor in Butter Fat, *Chem. Ind. 52*, 270.
5. Booth, R.G., Kon, S.K., Dann, W.J., and Moore, T. (1935) A Study of Seasonal Variation in Butter Fat. II. A Seasonal Spectroscopic Variation in the Fatty Acid Fraction, *Biochem J. 29*, 133–137.
6. Dann, W.J., Moore, T., Booth, R.G., Golding, J., and Kon, S.K. (1935) A New Spectroscopic Phenomenon in Fatty Acid Metabolism. The Conversion of "Pro-Absorptive" to "Absorptive" Acids in the Cow, *Biochem. J. 29*, 138–146.
7. Gillam, A.E., Heilbron, I.M., Hilditch, T.P., and Morton, R.A. (1931) Spectrographic Data of Natural Fats and Their Fatty Acids in Relation to Vitamin A, *Biochem. J. 25*, 30–38.
8. Edisbury, J.R., Morton, R.A., and Lovern, J.A. (1933) Absorption Spectra in Relation to the Constituents of Fish Oils, *Biochem. J. 27*, 1451–1460.
9. Dann, W.J., and Moore, T. (1933)The Absorption Spectra of the Mixed Fatty Acids from Cod-Liver Oil, *Biochem. J. 27*, 1166–1169.
10. Edisbury, J.R., Morton, R.A., and Lovern, J.A. (1935) The Absorption Spectra of Acids from Fish-Liver Oils, *Biochem. J. 29*, 899–908.
11. Moore, T. (1937) Spectroscopic Changes in Fatty Acids. 1. Changes in the Absorption Spectra of Various Fats Induced by Treatment with Potassium Hydroxide, *Biochem. J. 31*, 138–154.

12. Moore, T. (1939) Spectroscopic Changes in Fatty Acids. VI. General, *Biochem. J. 33,* 1635–1638.
13. Holman, R.T. (1957) Measurement of Polyunsaturated Fatty Acids, in *Methods of Biochemical Analysis,* (Glick, D., ed.) vol. 4, pp. 99–138, Interscience Publishers, New York.
14. Riel, R.R. (1963) Physico-Chemical Characteristics of Canadian Milk Fat. Unsaturated Fatty Acids, *J. Dairy Sci. 46,* 102–106.
15. Mattsson, S. (1949) Polyunsaturated Fatty Acids in Butter and Their Influence on the Oxidation of Butter, in *Proceedings of the XII International Dairy Congress,* vol. 2, pp. 308–323, Stockholm, Sweden.
16. McDowell, A.K.R. (1953) The Properties of New Zealand Butters and Butterfats. III. Seasonal Variations in the Nature of the Unsaturated Acids of Butterfat as Estimated by Spectrophotometric Methods, *J. Dairy Res. 20,* 101–107.
17. Lembke, A., and Kaufmann,W. (1954) Uber den Gehalt des Milchfettes an Konjuensauren, *Milchwissenschaft 9,* 113–114.
18. Smith, L.M., and Jack, E.L. (1954) The Unsaturated Fatty Acids of Milk Fat. II. Conjugated and Nonconjugated Constituents, *J. Dairy Sci. 37,* 390–398.
19. Stadhouders, J., and Mulder, H. (1955) The composition of Dutch Butterfat. 1. Seasonal Variations in the Unsaturated Fatty Acid Composition of Butter Fat, *Neth. Milk Dairy J. 9,* 182–193.
20. Wood, F.W., and Haab, W. (1957) Seasonal and Regional Variations in the Unsaturated Acids of Alberta Butterfat, *Can J. Anim. Sci. 37,* 1–7.
21. Mattsson, S., and Swartling, P. (1958) Note on the Spectrophotometric Determination of Polyunsaturated Fatty Acids in Butter, Milk and Dairy Research, Report No. 55, Alnarp, Sweden.
22. Kuzdzal-Savoie, S., and Kuzdzal, W. (1961) Les Acides Polyinsatures du Beurre, *Lait 41,* 606–617.
23. Hilditch, T.P., and Jasperson, H. (1941) Milk Fats from Cows Fed on Fresh Pasture and on Ensiled Green Fodder. I. Observations on the Component Fatty Acids, *J. Soc. Chem. Ind. 60,* 305–310.
24. Hilditch, T.P., and Jasperson, H. (1945) The Polyethenoid Acids of the C18 Series Present in Milk and Grass Fats, *J. Soc. Chem. Ind. 64,* 109–111.
25. Jackson, J.E., Paschke, R.F., Tolberg, W., Boyd, H.M., and Wheeler, D.H. (1952) Isomers of Linoleic Acid, Infrared and Ultraviolet Properties of Methyl Esters, *J. Am. Oil Chem. Soc. 29,* 229–234.
26. Smith, L.M., Freeman, N.K., and Jack, E.L. (1954) The Unsaturated Fatty Acids of Milk Fat. III. Geometrical Isomerism, *J. Dairy Sci. 37,* 399–406.
27. Scott, W.E., Herb, S.F., Magidman, P., and Riemenschneider, R.W. (1959) Unsaturated Fatty Acids of Butterfat, *J. Agric. Food Chem. 7,* 125–129.
28. Bartlet, J.C., and Chapman, D.G. (1961) Detection of Hydrogenated Fats by Measurement of *cis-trans* Conjugated Unsaturation, *J.Agric. Food Chem. 9,* 50–53.
29. Magidman, P., Herb, S.F., Barford, R.A., and Riemenschneider, R.W. (1962) Fatty Acids of Cows Milk. A. Techniques Employed in Supplementing Gas–Liquid Chromatography for Identification of Fatty Acids, *J. Am. Oil Chem. Soc. 39,* 137–142.
30. Herb, S.F., Magidman, P., Luddy, F.E., and Riemenschneider, R.W. (1962) Fatty Acids of Cows' Milk, B. Composition by Gas–Liquid Chromatography Aided by Other Methods of Fractionation, *J. Am. Oil Chem. Soc. 39,* 142–146.

31. Kuzdzal-Savoie, S. (1968) Fractionnement des Esters Methyliques du Beurre par Chromatographie sur Couche Mince de Silicagel Impregne de Nitrate d'Argent. 1. Etude Preliminaire, *Lait 48,* 121–130.
32. Kuzdzal-Savoie, S., Kuzdzal, W., and Langlois, D. (1969) Contribution a l'Etudie des Acids Gras Dienes du Beurre, *Fette Seifen Anstrichm. 71,* 326–330.
33. Christie, W.W. (1973) The Structures of Bile Phosphatidylcholines, *Biochim. Biophys. Acta 316,* 204–211.
34. Parodi, P.W. (1977) Conjugated Octadecadienoic Acids of Milk Fat, *J. Dairy Sci. 60,* 1550–1553.
35. McGuire, M.A., McGuire, M.K., McGuire, M.S., and Griinari, J.M. (1997) Bovinic Acid: The Natural CLA, in *Proceedings of the Cornell Nutrition Conference for Feed Manufacturers,* pp. 217–226, Rochester, New York, Cornell University, Ithaca, NY.
36. Kramer, J.K.G., Parodi, P.W., Jensen, R.G., Mossoba, M.M., Yurawecz, M.P., and Adlof, R.O. (1998) Rumenic Acid: A Proposed Common Name for the Major Conjugated Linoleic Acid Isomer Found in Natural Products, *Lipids 33,* 835.
37. Smith, L.M., and Jack, E.L. (1959) Isolation of Milk Phospholipids and Determination of Their Polyunsaturated Fatty Acids, *J. Dairy Sci. 42,* 767–778.
38. Hay, J.D., and Morrison, W.R. (1971) Polar Lipids in Bovine Milk. III. Isomeric *cis* and *trans* Monoenoic and Dienoic Fatty Acids and Alkyl and Alkenyl Ethers in Phosphatidyl Choline and Phosphatidyl Ethanolamine, *Biochim. Biophys, Acta 248,* 71–79.
39. Hopkins, C.Y. (1972) Fatty Acids with Conjugated Unsaturation, in *Topics in Lipid Chemistry,* (Gunstone, F.D., ed.) vol. 3, pp. 37–87, Elek Science, London.
40. Chin, S.F., Liu, W., Storkson, J.M., Ha, Y.L., and Pariza, M.W. (1992) Dietary Sources of Conjugated Dienoic Isomers of Linoleic Acid, a Newly Recognized Class of Anticarcinogens, *J. Food Comp. Anal. 5,* 185–197.
41. Kramer, J.K.G., Fellner, V., Dugan, M.E.R., Sauer, F.D., Mossoba, M.M., and Yurawecz, M.P. (1997) Evaluating Acid and Base Catalysts in the Methylation of Milk and Rumen Fatty Acids with Special Emphasis on Conjugated Dienes and Total *trans* Fatty Acids, *Lipids 32,* 1219–1228.
42. Shorland, F.B. (1950) Effect of the Dietary Fat on the Composition of the Depot Fats of Animals, *Nature 165,* 766.
43. Reiser, R. (1951) Hydrogenation of Polyunsaturated Fatty Acids by the Ruminant, *Fed. Proc. 10,* 236.
44. Shorland, F.B., Weenink, R.O., and Johns, A.T. (1955) Effect of the Rumen on Dietary Fat, *Nature 175,* 1129–1130.
45. Shorland, F.B.,Weenink, R.O., Johns, A.T., and McDonald, I.R.C. (1957) The Effect of Sheep-Rumen Contents on Unsaturated Fatty Acids, *Biochem. J. 67,* 328–333.
46. Kepler, C.R., Hirons, K.P., McNeill, J.J., and Tove, S.B. (1966) Intermediates and Products of the Biohydrogenation of Linoleic Acid by *Butyrivibrio fibrisolvens, J. Biol. Chem. 241,* 1350–1354.
47. Kepler, C.R., and Tove, S.B. (1967) Biohydrogenation of Unsaturated Fatty Acids. III. Purification and Properties of a Linoleate Δ^{12}-*cis,* Δ^{11}-*trans*-Isomerase from *Butyrivibrio fibrisolvens, J. Biol. Chem. 242,* 5686–5692.
48. Bertram, S.H. (1928) Die Vaccensaure, *Biochem. Z. 197,* 433–441.
49. Reiser, R. (1950) Conjugated Linoleic Acid in Rat Tissue Lipids After Ingestion as Free Acid and as Triglyceride, *Proc. Soc. Exp. Biol. Med. 74,* 666–669.

50. Shorland, F.B., Bruce, L.W., and Jessop, A.S. (1952) Studies on the Composition of Horse Oil. 2. The Component Fatty Acids of Lipids from Fatty Tissues, Muscle and Liver, *Biochem. J. 52,* 400–407.
51. Hartman, L., Shorland, F.B., and McDonald, I.R.C. (1955) The *trans*-Unsaturated Acid Contents of Fats of Ruminants and Non-ruminants, *Biochem. J. 61,* 603–607.
52. Moir, R.J., Somers, M., Sharman, G., and Waring, H. (1954) Ruminant-Like Digestion in a Marsupial, *Nature 173,* 269–270.
53. Hansen, R.P., and Czochanska, Z. (1976) Fatty Acid Composition of the Subcutaneous and Perinephric Fats of Lambs Grazed on Pastures in New Zealand, *N.Z. J. Sci. 19,* 413–419.
54. Ackman, R.G., Eaton, C.A., Sipos, J.C., and Crew, N.F. (1981) Origin of *cis*-9,*trans*-11- and *trans*-9,*trans*-11-octadecadienoic Acids in the Depot Fat of Primates Fed a Diet Rich in Lard and Corn Oil and Implications for the Human Diet, *Can. Inst. Food Sci. Technol. 14,* 103–107.
55. van den Bosch, G. (1973) Bleaching of Vegetable Oils. II. Conversions of Methyl Oleate and Linoleate, *J. Am. Oil. Chem. Soc. 50,* 487–493.
56. Scholfield, C.R., Davison,V.L., and Dutton, H.J. (1967) Analysis for Geometrical and Positional Isomers of Fatty Acids in Partially Hydrogenated Fats, *J. Am. Oil Chem. Soc. 44,* 648–651.
57. Carpenter, D.L., and Slover, H.T. (1973) Lipid Composition of Selected Margarines, *J. Am. Oil Chem. Soc. 50,* 372–376.
58. Smith, L.M., Dunkley, W.L., Franke, A., and Dairiki, T. (1978) Measurement of *trans* and Other Isomeric Unsaturated Fatty Acids in Butter and Margarine, *J. Am. Oil Chem. Soc. 55,* 257–261.
59. Parodi, P.W. (1976) Composition and Structure of Some Consumer-Available Edible Fats, *J. Am. Oil Chem. Soc. 53,* 530–534.

Chapter 2

The Biological Activities of Conjugated Linoleic Acid

Michael W. Pariza

> Food Research Institute, Department of Food Microbiology and Toxicology, University of Wisconsin-Madison, Madison, WI 53706

Introduction

In the first chapter of this book, Parodi elegantly describes the early years of research on conjugated linoleic acid (CLA), which culminated in the determination of $c9,t11$ CLA as the principal CLA isomer in dairy products (1) (Fig. 2.1). It is most fitting that this study should have been published just before our discovery of a "mutagenesis inhibitory" activity in extracts of grilled ground beef (2). At the time, of course, no one would have guessed that these two seemingly disparate observations might be related, much less central features in an evolving new branch of scientific inquiry.

Finding an antimutagen in a food extract was not in itself remarkable. Rather, what captured our attention was its apparent specificity in that the antimutagenic activity was evident under some, but not all conditions of test. I thought we might be dealing with a novel cytochrome P_{450} inhibitor, and thus ventured the speculation that ". . . it may also be found that the mutagenic inhibitory activity inhibits carcinogenesis" (2). We subsequently established that the antimutagen did indeed inhibit carcinogenesis (3) and then determined that the active anticarcinogenic principle was CLA (4). With this knowledge in hand, we began investigating the biochemical mechanism(s) whereby CLA inhibits cancer, beginning with the demonstration that CLA will act when administered orally (5). These studies have led to recognition of the multiple biological effects of CLA and would not have been possible without the concerted efforts of many scientific collaborators, most of whom are authors of chapters in this book. CLA research is now developing so rapidly that we have established a web site (**http://www.wisc.edu/fri/clarefs.htm**) to maintain a current listing of the scientific literature on CLA since 1987.

Effects of CLA on Carcinogenesis

The first investigator to take active interest in our CLA research was Clement Ip, with whom we published a collaborative study on the prevention of carcinogen-induced mammary carcinogenesis by CLA in rats (6). Ip and co-workers have continued this line of research with an elegant series of studies that are discussed in depth elsewhere in this book. A particularly noteworthy finding is that CLA suppresses proliferative activity in the rat mammary gland (7). This was the first published report indicating that CLA might inhibit carcinogenesis at least in part *via* specific effects on gene expression. More recently, they reported that dietary CLA affects the

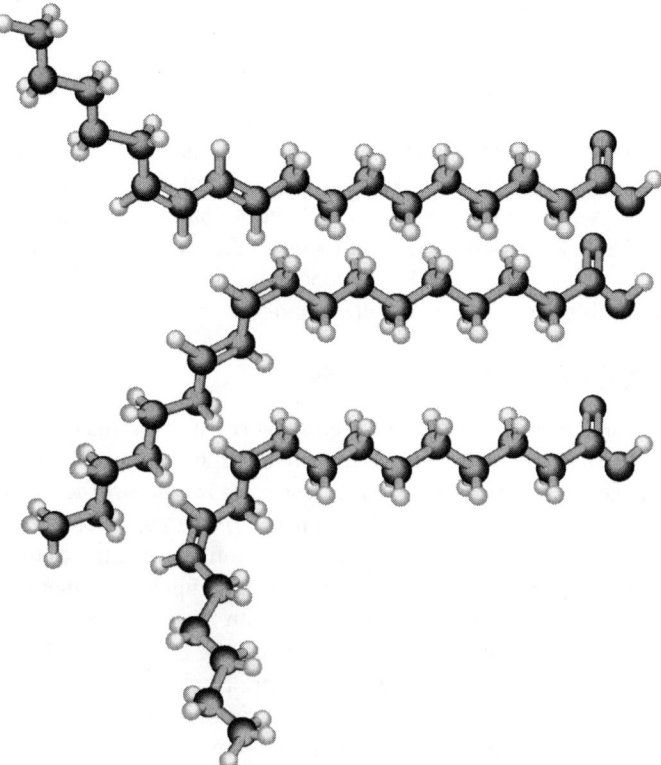

Fig. 2.1. Structures of trans-10,cis-12 CLA (upper panel), cis-9,trans-11 CLA (middle panel), and linoleic acid (lower panel). (*Source:* Ref. 39; Reprinted with permission from the *Journal of Chemical Education*, vol. 73, no. 12, 1996, pp. A302–A303; copyright © 1996, Division of Chemical Education, Inc.)

metabolism of retinol, a finding that is clearly relevant to anticarcinogenic effects; this is discussed further in Chapter 23 (Banni *et al.*). The effects of CLA on carcinogenesis are also reviewed in Chapter 34, whose author, Scimeca, has collaborated with us and others in this area for more than a decade.

Our work came "full circle" in the collaborative study with Dashwood's laboratory (8) in which it was shown that dietary CLA protected against colon neoplasia induced by 2-amino-3-methylimidazo[4,5-f]quinoline (IQ). IQ is one of the mutagens produced when beef is cooked. The findings indicated that CLA might interfere with the metabolism of IQ, possibly by inhibiting the arachidonic acid–dependent activation of heterocyclic amine carcinogens, which occurs in extrahepatic tissues, as well as the cytochrome P_{450}-mediated activation of these carcinogens, which occurs in the liver.

Dietary CLA has also been shown to inhibit the growth of human carcinoma cells injected into immune-deficient (SCID) mice (9). This work is noteworthy in that it indicates that CLA may eventually prove to be useful in cancer treatment as well as prevention. Of course, at this time, it is premature to extrapolate these promising findings in animal models to possible human application.

There is limited evidence for a direct association between CLA intake and cancer reduction in humans. Knekt et al. (10) found an inverse relationship between milk consumption and breast cancer risk in women. They speculated that some of this protective effect might be due to CLA in milk, although, as pointed out in Chapter 35 (Knekt et al.), this conclusion remains unproved in light of conflicting epidemiologic results. They conclude that future studies should focus on the relationship between blood levels of CLA and breast cancer risk. I fully concur, and would add that it is also crucial to assess blood levels of specific CLA isomers and their metabolites as a function of risk.

Effects of CLA on Atherosclerosis

Trans monounsaturated fatty acids have been shown to raise plasma low density lipoprotein (LDL) cholesterol levels, and may also increase triglyceride and lipoprotein(a) while lowering high density lipoprotein (HDL) cholesterol (11). Because the CLA preparations used routinely in animal studies consist primarily of the $c9,t11$ and $t10,c12$ isomers, both of which contain a double bond in the *trans* configuration, it was important to determine the effects of such CLA preparations on blood lipids and atherosclerosis.

We investigated this issue in collaboration with David Kritchevsky (12). Rabbits were fed a hypercholesterolemic diet with or without CLA. By 12 wk, total and LDL cholesterol and triglycerides were markedly lower in the blood of the CLA-fed rabbits. The experiment was terminated at 22 wk. Aortas from the CLA-fed rabbits exhibited less atherosclerosis than aortas from the control animals. Nicolosi et al. (13) reported that CLA also reduces plasma lipoproteins and early aortic atherosclerosis in the hamster model.

These effects of CLA on experimental atherosclerosis and the fascinating results from more recent studies are reported in detail in Chapter 31 (Kritchevsky).

Effects of CLA on the Immune System

Mark Cook and I began collaborating in the fall of 1990, after a chance meeting on a jogging trail. Within 6 months, we had obtained evidence indicating that feeding CLA to chicks provided partial protection against the catabolic effects of a subsequent exposure to endotoxin. The investigation was expanded to include rodents as well (14,15).

We also found that CLA-fed animals exhibited enhanced immune function (14,15), and further that CLA served as a growth factor for young rats (16). These observations are noteworthy in that the immune system competes for energy with other metabolic functions such as growth and tissue regeneration. For example, after major trauma, the immune system is actually suppressed (17). Presumably this occurs so that energy may be partitioned from immune function to tissue regeneration, a strategy that would favor enhanced healing at the expense of leaving the host susceptible to infection. Hence, to our knowledge, CLA is the only known dietary factor that both enhances the immune system while at the same time protecting against the catabolic effects of immune stimulation, a conclusion that suggests potential further application of CLA to improving health.

These findings may also be key in understanding the biochemical mechanisms whereby CLA produces many of its effects. When macrophages and other immune cells are stimulated, they produce *cytokines,* which are hormone-like mediators of immunity and inflammation. Two cytokines of central importance in this signaling process are tumor necrosis factor-α (TNF-α and interleukin-1 (IL-1). TNF-α and IL-1 induce a number of effects in immune cells including the inflammatory response. However, these cytokines also produce biochemical changes in other cells, for example, the induction of catabolism in skeletal muscle and changes in cell surface proteins. Virtually every cell in the body has receptors for TNF-α and IL-1, and many types of cells (e.g., nerve cells or adipocytes) can also produce these cytokines.

In recent years, it has become increasingly apparent that TNF-α and IL-1 are involved in a wide array of biochemical and pathological processes that include cachexia (18), atherosclerosis (19), carcinogenesis (20,21), and (paradoxically) obesity (22). Thus, effects of CLA on the synthesis and action of these cytokines may be a central feature of the biochemical pathways through which CLA exerts at least some of its biological action (23).

It is also noteworthy that both the synthesis and action of TNF-α and IL-1 are regulated by eicosanoids, in particular prostaglandin E_2 (PGE_2) (24). There is considerable evidence from many laboratories that CLA affects eicosanoid synthesis. This is discussed in several of the chapters that follow.

Effects of CLA on Lipid Metabolism and Body Composition

Dietary CLA has been shown to effect body composition change (reduction in body fat, enhancement of fat free mass) in a number of animal species including mice (25,26), rats (27,28), and pigs (23,29–32). The reader is referred to Chapter 28 (Dugan and Aalhus) for discussion of studies conducted at the Lacombe Research Centre in Canada relating to the effects of feeding CLA to pigs.

CLA appears to exert direct effects on adipocytes, which are the principal sites of fat storage, and skeletal muscle cells, which are the principal sites of fat combustion (25). Adding CLA to the culture medium of mouse 3T3-L1 adipocytes produced a dose-dependent reduction in lipoprotein lipase (LPL) activity and apparently induced lipolysis as well in this cell line. Skeletal muscle from mice fed CLA exhibited elevated carnitine palmitoyltransferase activity and enhanced whole-body protein accretion.

On the basis of these findings, we proposed that the physiologic mechanism of body fat reduction in mice by CLA involves inhibition of fat storage in adipocytes coupled with elevated β-oxidation in skeletal muscle, as well as an overall increase in skeletal muscle mass that may be mediated at least in part *via* the inhibition of immune-induced catabolism as discussed above (25,33). Moreover, these effects appear to correlate closely with CLA-induced effects on lipid metabolism (12,34), indicating that hepatocytes may also be affected by CLA.

Recently, evidence was obtained (33) indicating that dietary CLA (isomer composition: 41% *c*9,*t*11; 44% *t*10,*c*12) did not enhance adipocyte differentiation *in vivo*. As shown in Figure 2.2, when CLA was withdrawn from the diet of young mice previously fed CLA-supplemented diet for 4 wk, body fat accumulation lagged behind the controls (which had not been fed CLA) and did not return to control levels even after 8 wk. These findings indicate that CLA did not enhance, and may even have blocked, adipocyte differentiation during the time when CLA-supplemented diet was fed. These findings are in agreement with a recent report (35) involving the effects of CLA on the differentiation of 3T3-L1 cells. It will be important now to reconcile these data with the interesting findings presented in Chapter 32 (Belury and Vanden Heuvel) regarding the effects of CLA on adipocyte differentiation *in vitro*.

CLA Isomers

The apparent multifunctionality of CLA is both intriguing and perplexing (23). However, it should be kept in mind that the multiple biological effects seen in animal models to date have been produced by feeding mixtures of CLA isomers, in particular *c*9,*t*11 and *t*10,*c*12 CLA in approximately equal amounts. Emerging evidence indicates that these isomers induce different effects.

Evidence now indicates that the effects of CLA on lipid metabolism and body composition appear to be produced largely, if not solely, by *t*10,*c*12 CLA. Lee *et al.* (34) provided evidence that the *t*10,*c*12 isomer decreased the expression of hepatic stearoyl-CoA desaturase mRNA in mice. Park *et al.* (36) provided evidence indicating that *t*10,*c*12 CLA is responsible for body composition change in mice. Table 2.1, which is taken from Park *et al.* (36), provides support for this conclusion in that changes in body composition were correlated with the *t*10,*c*12 CLA isomer concentration in the test material. The reader is referred to the original publication (36) for more discussion and detail. In Chapter 13, Griinari and Bauman discuss evidence that the *t*10,*c*12 CLA isomer depresses milk fat synthesis in cows. Finally, it is of interest that the *t*10,*c*12 isomer appeared to be metabolized more rapidly than the *c*9,*t*11 isomer, particularly in skeletal muscle (33). Whether this is due to enhanced elongation/desaturation, enhanced β-oxidation, or both, is not yet known.

On the other hand, Ip has obtained evidence indicating that *c*9,*t*11 CLA and *t*10,*c*12 CLA are in essence equally effective in inhibiting carcinogenesis (personal communication). Belury and Vanden Heuvel (Chapter 32) indicate that *c*9,*t*11 CLA is a potent activator and high affinity ligand for peroxisome proliferator–activated receptor-γ (PPARγ).

Accordingly, the apparent multifunctionality of CLA is becoming more intriguing and less perplexing. The effects now appear to be produced by different isomers, alone or in combination, and likely are mediated through distinctly separate biochemical mechanisms (but see also Chapter 26 by Ntambi *et al.*).

The metabolism of the CLA isomers and the possible relationship of metabolism to biological activity are the focus of Chapter 24 (Sébédio) and are discussed in several

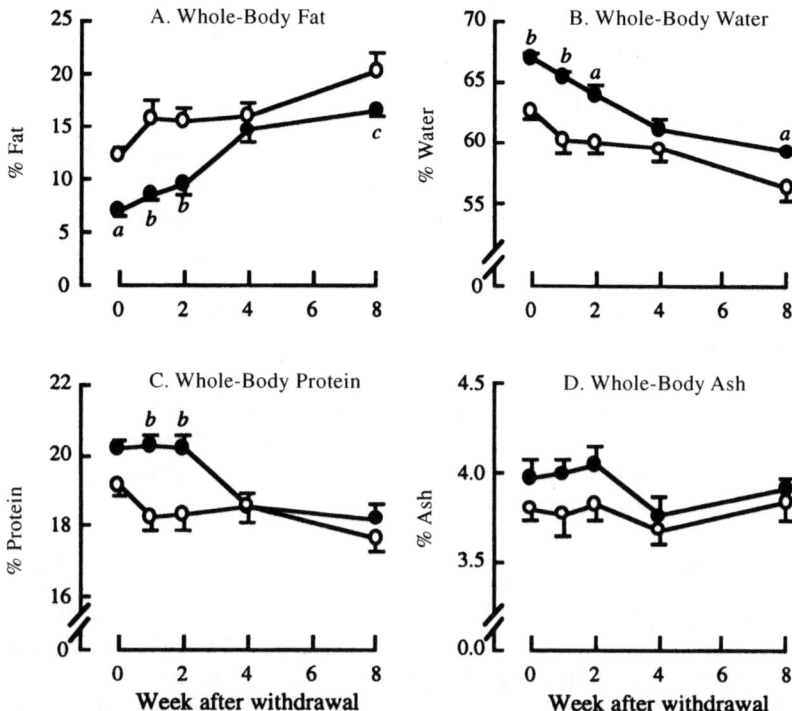

Fig. 2.2. Relative changes in whole-body fat, whole-body water, whole-body protein, and whole-body ash in mice after withdrawal of dietary CLA (●) vs. controls that were not fed CLA (○). Each data point represents the mean ± SEM, n = 7 or 8 animals. [a]$P < 0.05$, [b]$P < 0.01$, compared with control at same time point. [c]$P < 0.05$, compared with control at the same time point when cage effects, which were not significant, are ignored. See Park et al. (33) for experimental details.

TABLE 2.1 Evidence That the $t10,c12$ CLA Isomer Effects Body Composition Change in Mice[a]

	ECW (g)[b]	% Fat	% Water	% Protein	% Ash
Control	27.43 ± 1.21[c]	22.27 ± 1.80[c]	54.30 ± 1.35[c]	16.26 ± 0.49[c]	3.29 ± 0.13[c]
CLA-1[d]	24.28 ± 0.76[e]	6.69 ± 0.86[f]	65.59 ± 0.68[f]	19.04 ± 0.24[e]	3.78 ± 0.10[e]
CLA-2[g]	25.53 ± 0.59[ce]	13.08 ± 1.66[e]	60.99 ± 1.14[e]	18.09 ± 0.50[e]	3.54 ± 0.13[ce]
CLA-3[h]	23.44 ± 0.92[e]	6.80 ± 1.26[f]	65.35 ± 1.13[f]	19.33 ± 0.29[e]	3.83 ± 0.08[e]

[a]Source: Ref. 36. Female ICR mice were fed control diet or diet supplemented with 0.5% CLA-1, 0.3% CLA-2, or 0.25% CLA-3, for 4 wk. Reported body composition values are means ± SEM, n = 5–6. In each column, means with different letters are significantly different ($P < 0.05$). See Park et al. (36) for further experimental detail.
[b]ECW, empty carcass weight.
[d]CLA-1 preparation consisted of 41.1% $c9,t11$ plus 43.5% $t10,c12$.
[g]CLA-2 preparation consisted of 72.4% $c9,t11$ plus 13.0% $t10,c12$.

other chapters as well. It is clear that this issue is gaining in momentum as a new focal point of CLA research.

Dietary Sources of CLA

As discussed by Parodi in Chapter 1, the existence of fatty acids with conjugated double bonds has been known for more than 50 years. Various CLA isomers may form during oil processing, for example, in margarine manufacture (37). However, the principal dietary sources of CLA are dairy products and other foods derived from ruminant animals, e.g., cattle, sheep, and goats.

Dietary sources of CLA are reviewed in Chapter 29 (McGuire *et al.*) and also discussed in a number of other chapters. The production and occurrence of CLA in ruminants are reviewed and discussed extensively in several chapters, as well as enhancement of CLA in the human diet *via* modulating the diets of cattle and other species so as to increase the CLA content of foods derived from these species (see also Ref. 38). The commercial production of CLA for food and feed use is reviewed extensively in Chapter 4 (Reaney *et al.*).

CLA Analysis

As indicated above, emerging evidence indicates that specific biological effects may be due in some cases to the specific action of a single CLA isomer, in other cases to a more general effect that may be produced by two or more CLA isomers. Accordingly, it is imperative to have standardized methodologies for separating and quantifying individual CLA isomers in complex mixtures such as foods, biological materials, and manufactured products. In addition, as basic research reveals new information concerning the biological importance of specific CLA metabolites, it will be desirable to quantify the levels of such CLA metabolites in foods that are derived from animals that naturally produce CLA (i.e., ruminants) and those fed synthetic CLA (e.g., pigs). It is clear that methods developed in the 1980s are not suitable for these purposes. Considerable improvements in CLA analytical methodologies have occurred in recent years, as discussed in several of the following chapters.

References

1. Parodi, P.W. (1977) Conjugated Octadecadienoic Acids in Milk Fat, *J. Dairy Sci. 60,* 1550–1553.
2. Pariza, M.W., Ashoor, S.H., Chu, F.S., and Lund, D.B. (1979) Effects of Temperature and Time on Mutagen Formation in Pan-Fried Hamburger, *Cancer Lett. 7,* 63–69.
3. Pariza, M.W., and Hargraves, W.A. (1985) A Beef-Derived Mutagenesis Modulator Inhibits Initiation of Mouse Epidermal Tumors by 7,12-Dimethylbenz[a]anthracene, *Carcinogenesis 6,* 591–593.
4. Ha, Y.L., Grimm, N.K., and Pariza, M.W. (1987) Anticarcinogens from Fried Ground Beef: Heat-Altered Derivatives of Linoleic Acid, *Carcinogenesis 8,* 1881–1887.
5. Ha, Y.L., Storkson, J., and Pariza, M.W. (1990) Inhibition of Benzo[a]pyrene-Induced Mouse Forestomach Neoplasia by Conjugated Dienoic Derivatives of Linoleic Acid, *Cancer Res. 50,* 1097–1101.

6. Ip, C., Chin, S.F., Scimeca, J.A., and Pariza, M.W. (1991) Mammary Cancer Prevention by Conjugated Dieonic Derivatives of Linoleic Acid, *Cancer Res. 51*, 6118–6124.
7. Ip, C., Singh, M., Thompson, H.J., and Scimeca, J.A. (1994) Conjugated Linoleic Acid Suppresses Mammary Carcinogenesis and Proliferative Activity of the Mammary Gland in the Rat, *Cancer Res. 54*, 1212–1215.
8. Liew, C., Schut, H.A.J., Chin, S.F., Pariza, M.W., and Dashwood, R.H. (1995) Protection of Conjugated Linoleic Acids Against 2-Amino-3-Methylimidazo[4,5-f]quinoline-Induced Colon Carcinogenesis in the F344 Rat: A Study of Inhibitory Mechanisms, *Carcinogenesis 16*, 3037–3043.
9. Cesano, A., Visonneau, S., Scimeca, J.A., Kritchevsky, D., and Santoli, D. (1998) Opposite Effects of Linoleic Acid and Conjugated Linoleic Acid on Human Prostatic Cancer in SCID Mice, *Anticancer Res. 18*, 1429–1434.
10. Knekt, P., Jarvinen, R., Seppanen, R., Pukkala, E., and Aromaa, A. (1996) Intake of Dairy Products and the Risk of Breast Cancer, *Br. J. Cancer 73*, 687–691.
11. Katan, M.B., and Zock, P.L. (1995) *Trans* Fatty Acids and Their Effects on Lipoproteins in Humans, *Annu. Rev. Nutr. 15*, 473–493.
12. Lee, K.N., Kritchevsky, D., and Pariza, M.W. (1994) Conjugated Linoleic Acid and Atherosclerosis in Rabbits, *Atherosclerosis 108*, 19–25
13. Nicolosi, R.J., Rogers, E.J., Kritchevsky, D., Scimeca, J.A., and Huth, P.J. (1997) Dietary Conjugated Linoleic Acid Reduces Plasma Lipoproteins and Early Atherosclerosis in Hypercholesterolemic Hamsters, *Artery 22*, 266–277.
14. Cook, M.E., Miller, C.C., Park, Y., and Pariza, M.W. (1993) Immune Modulation by Altered Nutrient Metabolism: Nutritional Control of Immune-Induced Growth Depression, *Poult. Sci. 72*, 1301–1305.
15. Miller, C.C., Park, Y., Pariza, M.W., and Cook, M.E. (1994) Feeding Conjugated Linoleic Acid to Animals Partially Overcomes Catabolic Responses Due to Endotoxin Injection, *Biochem. Biophys. Res. Commun. 198*, 1107–1112.
16. Chin, S.F., Storkson, J., Albright, K.J., Cook, M.E., and Pariza, M.W. (1994) Conjugated Linoleic Acid Is a Growth Factor for Rats as Shown by Enhanced Weight Gain and Improved Feed Efficiency, *J. Nutr. 124*, 2344–2349.
17. Hensler, T., Hecker, H., Heeg, K., Heidecke, C.-L., Bartels, H., Barthlen, W., Wagner, H., Siewert, J.-R., and Holzmann, B. (1997) Distinct Mechanisms of Immunosuppression as a Consequence of Major Surgery, *Infect. Immun. 65*, 2283–2291.
18. Freeman, L.M., and Roubenoff, R. (1994) The Nutrition Implications of Cardiac Cachexia, *Nutr. Rev. 52*, 340–347.
19. Ross, R. (1993) The Pathogenesis of Atherosclerosis: A Perspective for the 1990s, *Nature 362*, 801–809.
20. Okahara, H., Yagita, H., Miyake, K., and Okumura, K. (1994) Involvement of Very Late Activation Antigen 4 (VLA-4) and Vascular Cell Adhesion Molecule 1 (VCAM-1) in Tumor Necrosis Factor α Enhancement of Experimental Metastasis, *Cancer Res. 54*, 3233–3236.
21. Suganuma, M., Okabe, S., Sueoka, E., Iida, N., Komori, A., Kim, S.-J., and Fugiki, H. (1996) A New Process of Cancer Prevention Mediated Through Inhibition of Tumor Necrosis Factor α Expression, *Cancer Res. 56*, 3711–3715.
22. Hotamisligil, G.S., and Spiegelman, B.M. (1994) TNFα: A Key Component of Obesity-Diabetes Link, *Diabetes 43*, 1271–1278.
23. Pariza, M.W. (1997) Conjugated Linoleic Acid, a Newly Recognized Nutrient, *Chem. Ind. 12*, 464–466.

24. Lewis, G.P. (1983) Immunoregulatory Activity of Metabolites of Arachidonic Acid and Their Role in Inflammation, *Br. Med. Bull. 39,* 243–248.
25. Park, Y., Albright, K.J., Storkson, J.M., Cook, M.E., and Pariza, M.W. (1997) Effect of Conjugated Linoleic Acid on Body Composition in Mice, *Lipids 32,* 853–858.
26. West, D.B., DeLany, J.P., Camet, P.M., Blohm, F., Truett, A.A., and Scimeca, J. (1998) Effects of Conjugated Linoleic Acid on Body Fat and Energy Metabolism in the Mouse, *Am. J. Physiol.-Reg. I. 44,* R667–R672.
27. Houseknecht, K.L., Vanden Heuvel, J.P., Moya-Camarena, S.Y., Portocarrero, C.P., Peck, L.W., Nickel, K.P., and Belury, M.A. (1998) Dietary Conjugated Linoleic Acid Normalizes Impaired Glucose Tolerance in the Zucker Diabetic Fatty *fa/fa* Rat, *Biochem. Biophys. Res. Commun. 244,* 678–682.
28. Sisk, M., Azain, M.J., Hausman, D.B., and Jewell, D.E. (1998) Effect of Conjugated Linoleic Acid on Fat Pad Weights and Cellularity in Sprague-Dawley and Zucker Rats, *FASEB J. 12,* A536.
29. Dugan, M.E.R., Aalhus, J.L., Schaefer, A.L., and Kramer, J.K.G. (1997) The Effects of Conjugated Linoleic Acid on Fat to Lean Repartitioning and Feed Conversion in Pigs, *Can. J. Anim. Sci. 77,* 723–725.
30. Cook, M.E., Jerome, D.L., Crenshaw, T.D., Buege, D.R., Pariza, M.W., Albright, K.J., Schmidt, S.P., Scimeca, J.A., Lofgren, P.A., and Hentges, E.J. (1998) Feeding Conjugated Linoleic Acid Improves Feed Efficiency and Reduces Whole Body Fat in Pigs, *FASEB J. 12,* A836.
31. Dunshea, F.R., Ostrowska, E., Muralitharan, M., Cross, R., Bauman, D.E., Pariza, M.W., and Skarie, C. (1998) Dietary Conjugated Linoleic Acid Decreases Back Fat in Finisher Gilts, *J. Anim. Sci. 76 (Suppl. 1),* 131.
32. Thiel, R.L., Sparks, J.C., Wiegand, B.R., Parrish, F.C., Jr., and Ewan, R.C. (1998) Conjugated Linoleic Acid Improves Performance and Body Composition in Swine, *Midwest Animal Science Meetings 1998 Abstracts,* Abstract #127 (p. 61), Midwestern Section, American Society of Animal Science, Savoy, IL.
33. Park, Y., Albright, K.J., Storkson, J.M., Liu, W., Cook, M.E., and Pariza, M.W. (1999) Changes in Body Composition During Feeding and Withdrawal of Dietary Conjugated Linoleic Acid, *Lipids 34,* 243–248.
34. Lee, K.N., Pariza, M.W., and Ntambi, J.M. (1998) Conjugated Linoleic Acid Decreases Hepatic Stearoyl-CoA Desaturase mRNA Expression, *Biochem. Biophys. Res. Commun. 248,* 817–821.
35. Brodie, A.E., Manning, V.A., Ferguson, K.R., Jewell, D.E., and Hu, C.Y. (1999) Conjugated Linoleic Acid Inhibits Differentiation of Pre- and Post-Confluent 3T3-L1 Preadipocytes but Inhibits Cell Proliferation Only in Preconfluent Cells, *J. Nutr. 129,* 602–606.
36. Park, Y., Albright, K.J., Storkson, J.M., Liu, W., and Pariza, M.W. (1999) Evidence That the *trans*-10,*cis*-12 Isomer of Conjugated Linoleic Acid Induces Body Composition Changes in Mice, *Lipids 34,* 235–241.
37. Carpenter, D.L., and Slover, H.T. (1973) Lipid Composition of Selected Margarines, *J. Am. Oil Chem. Soc. 50,* 372–376.
38. Dhiman, T.R., Helmink, E.D., McMahon, D.J., Fife, R.L., and Pariza, M.W. (1999) Conjugated Linoleic Acid Content of Milk and Cheese from Cows Fed Extruded Oilseeds, *J. Dairy Sci. 82,* 412–419.
39. Steinhart, C. (1996) Conjugated Linoleic Acid—The Good News About Animal Fat. *J. Chem. Edu. 73,* A302–A303.

Chapter 3

Preparation of Unlabeled and Isotope-Labeled Conjugated Linoleic and Related Fatty Acid Isomers

Richard O. Adlof

Food Quality and Safety Research, National Center for Agricultural Utilization Research, USDA, Agricultural Research Service, Peoria, IL 61604

Introduction

Conjugated linoleic acid (CLA; 9-*cis*,11-*trans*-octadecadienoic acid; 9-*cis*,11-*trans*-18:2) has been associated with the reduction of chemically-induced cancers in mice and rats and the suppression of atherosclerosis in rats (1,2). Commercially available CLA diet supplements often contain conjugated fatty acids other than the expected 9-*cis*,11-*trans*- and 10-*trans*,12-*cis*-isomer pair (3). The other fatty acids consist of the 8,10- and 11,13-isomers, all primarily *cis*/*trans* or *trans*/*cis* in configuration, and ~2–5% of the *cis*/*cis* and *trans*/*trans* isomers. The number of isomers is dependent on the severity of the alkali isomerization conditions used to prepare the CLA (4). Although the 9-*cis*,11-*trans*-isomer is considered to be the active constituent (5), contributions by other isomers have not been ruled out. The term "CLA" is thus not limited only to the 9,11-isomer, but encompasses a number of conjugated dienoic fatty acids. The perceived biological activity may be due to one or more of the CLA isomers available in commercial preparations. To study the effects, metabolism, and interaction of the CLA isomers, however, pure samples are required. The health benefits and promotion of CLA as a diet supplement may be traced to the mid-1980s, whereas the synthesis and identification of CLA isomers dates to the 1940s in the U.S., to the German work of the 1930s (see Ref. 6 for examples of early chemical syntheses), and to the work of Mangold with hydroxy fatty acids in the 1890s (7).

The pre-1970 history/evolution of conjugated fatty acids (from the 19th century to ca. 1970) has been examined in an excellent, very detailed review by Hopkins (8). Hopkins covers the isolation of conjugated dienoic and more highly unsaturated fatty acids from seed oil sources and their chemical synthesis. He also includes the isolation/synthesis of oxygenated diene acids, acetylenic acids, and hydroxy-acetylenic acids. In this chapter, we will concentrate on post-1970 methods for isolation/synthesis of conjugated fatty acids, with special emphasis on biosynthetic (enzymatic) methods, and on the preparation of conjugated dienoic fatty acids labeled with stable and radioactive isotopes. We will also review the synthesis/isolation of trienoic and more highly unsaturated fatty acids containing one or more sites of conjugation, and of oxygenated and acetylenic fatty acids.

The synthesis part of the chapter is divided into two sections (nonlabeled and labeled fatty acids); each section is further divided into chemical and biochemical methods, the latter method including the extraction/isolation of the desired isomer from seed oils. Only an overview of the many synthetic methods is presented; a comprehensive listing of synthesized CLA isomers, polyunsaturated fatty acid (PUFA) isomers, and isotopically labeled CLA and PUFA isomers (including those commercially available) is provided in Tables 3.1–3.3, respectively. Table 3.3 also includes a listing of commercially available precursors (hydroxy acids/chemical fragments) useful for preparation of labeled CLA isomers. Yields are provided whenever possible, but separation of CLA isomers is difficult even with the powerful high-performance liquid chromatography (HPLC) and LC columns available to the researcher of today. Recent work by Sehat and co-workers (9) has demonstrated even recently published CLA percentage composition data to be suspect. In addition, Christie (10), Kramer (11), and others have recently shown that *trans* isomers and/or oxygenated adducts are formed during the acid-catalyzed (BF_3 or HCl in methanol) conversion of CLA fatty acids (FA) or CLA-containing triglycerides (TG) to fatty acid methyl esters (FAME). Final FA composition data may show elevated levels of *trans* isomers or other impurities (and lower yields) due to the FAME preparation step rather than to the isolation method/synthesis employed.

Preparation of CLA isomers by dehydration of an unsaturated, hydroxy-containing fatty acid (12-hydroxy-9-*cis*-octadecenoic acid; ricinoleic acid, for example) or by the alkali, acid, or light-catalyzed isomerization of a methylene-interrupted fatty acid precursor (9-*cis*,12-*cis*-octadecadienoic acid; linoleic acid, for example) usually yields a mixture of isomers (for a review, see Ref. 12), with the number of isomers dependent on the severity (time/temperature) of the reaction conditions (see below). Single CLA isomers are often not readily separable from these mixtures by conventional methods, such as urea adduct formation, crystallization, or liquid chromatography. Gupta and Kummerow (13) used crystallization to isolate 8-*trans*,10-*trans*-18:2 (in 15–20% yield) from the mixture of CLA isomers obtained by the bromination/debromination of oleic acid (9-*cis*-18:1). Crystallization has also been used to fractionate the *trans*,*trans*-, *cis*,*trans*- and *cis*,*cis*-18:2 isomers (14). More recent methods rely on reversed-phase and/or silver ion–HPLC to isolate individual isomers (see Chapter 10). Improved separations of CLA isomers by Ag-HPLC have recently been demonstrated (9). Many of these recently developed methods of purification/isolation could be applied to improve yields and purities of CLA isomers prepared using earlier (pre-1970) methods.

Biosynthesis utilizing specific enzymes isolated from algal or bacterial sources to convert nonconjugated fatty acid precursors to CLA isomers has been employed primarily for preparation of nonlabeled 9-*cis*,11-*trans*-18:2 from 9-*cis*,12-*cis*-18:2 (see below). Yields are often poor. In the preparation of 5-*cis*,8-*cis*,12-*trans*,14-*cis*-20:4 by incubation of arachidonic acid (AA) with the coraline red alga *Bussiella orbigniana*, 2 kg of algae incubated with 40 mg AA yielded 5 mg of product (15). An *in vivo* source (a rat) has also been used (15) to prepare small quantities of highly unsaturated, conjugated fatty acids. A number of unsaturated, hydroxy-containing fatty acids have been prepared by incubation of fatty acid precursors with soybean lipoxygenase, hydration, or oxidation enzymes from seed oil, algal, or bacterial sources (12).

Biosynthesis is the preferred method for the preparation of carbon-13 (^{13}C-) and carbon-14 (^{14}C-) uniformly labeled FA and, to a lesser extent, their ^{3}H- and ^{2}H-labeled analogs. At this writing, only the ^{14}C-labeled 9-*cis*,11-*trans*-18:2 isomer is available commercially. Realistically, 9-*cis*,11-*trans*-18:2 labeled with stable isotopes such as deuterium (^{2}H) or (^{13}C), or with radioactive isotopes such as tritium (^{3}H) or (^{14}C), could be prepared by a similar conversion of the appropriately labeled 9-*cis*,12-*cis*-18:2 precursor.

A number of algal and bacterial sources are also available for preparation of labeled fatty acids. Starting materials are usually ^{x}CO$_2$ (x = 13 or 14), [1-^{14}C]acetate, or ^{x}H$_2$O (x = 2 or 3); their incubation yields isotopically labeled FA. This procedure has been used to prepare a variety of labeled FA; isolation of the desired fatty acid from the fatty acid pool is often difficult and care must be exercised in the choice of culture used. Although biosynthesis has been used to prepare FA uniformly or totally labeled with isotopes, specifically labeled FA may be metabolized to yield other FA, prostaglandins, thromboxins, prostacyclines, and other prostanoids specifically labeled with both radioactive and stable isomers. (A good review of this topic may be found in Ref. 16.) Stereochemistry can be controlled by careful choice of the enzymes used to desaturate and/or elongate the FA.

Multistep syntheses using Wittig or acetylenic coupling reactions have been developed to prepare FA, with the geometry and location of the double bonds carefully controlled and from which any by-products may be readily removed. For preparation of deuterium-labeled fats, for example, the number and location of the deuterium atoms in the fatty acid can be controlled by careful selection of starting materials. A drawback to this approach is the time required for the synthesis/purification of the intermediates and the often low overall yields (5–15%) of the desired product.

A final, often overlooked resource for methods of conjugated diene and ene-yne preparation may be found in the field of insect pheromones. Many pheromone structures contain conjugated aldehydes, alcohols, acetates, or esters that are readily convertible to fatty acids or whose syntheses may be adapted to the preparation of CLA isomers (17,18).

Care must also be taken to minimize oxidation during isolation of conjugated FA. The oxidative susceptibility of CLA has been demonstrated to lie between arachidonic and docosahexaenoic acid (19). Conjugated trienoic fatty acids are more readily oxidized than CLA and have been shown to decompose during their separation with silica thin-layer chromatography (TLC) plates or LC columns of silica gel impregnated with silver nitrate (20). Gas chromatography (GC) analysis of conjugated trienoic FAME is frequently unsatisfactory due to decomposition and/or isomerization (21). The extreme sensitivity of conjugated tetraenoic FA to air and light often requires that the final coupling reactions and purifications be done in a glove box under red light and an argon atmosphere (22). Millar (23) synthesized methyl 2-*trans*,4-*cis*,6-*cis*-10:3, one of the sex-specific compounds from the stink bug (*Thyanta pallidovirens*) and noted that the compound rearranged under GC conditions *via* a 1,7 sigmatropic rearrangement.

Nonlabeled CLA Isomers

Chemical Synthesis

Dehydration of Hydroxy Fatty Acids. See Ref. 24 for a review. The dehydration of ricinoleic (12-hydroxy-9-*cis*-octadecenoic) or ricinelaidic (12-hydroxy-9-*trans*-octadecenoic) acid has been used to prepare 9-*cis*,11-*trans*-18:2 [activated alumina as catalyst, 235°C (25)] and 9-*trans*,11-*trans*-18:2 [potassium acid sulfate and other dehydration catalysts, 200°C, 1 h, under vacuum (26)], respectively. Isolated product yields were low (15–35%). Lie Ken Jie and Wong (27) prepared methyl 9-*trans*, 11-*trans*-18:2 from methyl 9,12-dihydroxy-10-*trans*-octadecenoate in 72% yield using chlorotrimethylsilane and sodium iodide in acetonitrile.

Gunstone and Said (28) reported improved yields of 9-*cis*,11-*trans*-18:2 from methyl ricinoleate if the hydroxy group is first converted to its mesyl ester (methanesulfonate). When the mesyl ester is heated for 1 h at 100°C with sodium methoxide in DMSO, with DBU (1,5- diazobicyclo(5.4.0)undec-5-ene) or with DBN (1,5-diazobicyclo(4.3.0)non-5-ene) as base, the 9-*cis*,11-*trans*-18:2 conjugated ester is the major product in yields of 66, 72, and 72%, respectively. Berdeaux *et al.* (29) adapted this procedure (DBU as base) to prepare 50- to 60-g batches of 9-*cis*,11-*trans*-18:2 from methyl ricinoleate in overall yields of >70%. The methyl ricinoleate was isolated from castor oil FAME by countercurrent distribution (30).

Yurawecz and co-workers (31) prepared mixtures of CLA isomers by acid catalysis of allylic hydroxy oleate. The allylic hydroperoxides of methyl oleate (9-*cis*-18:1; with peroxy groups in the 8- or 12-position) were reduced to hydroxy groups with sodium borohydride (80% overall yield). Treatment with HCl/methanol or BF_3/methanol generated a mixture of CLA isomers (9-*cis*,11-*trans*-18:2, 9-*trans*,11-*trans*-18:2; 7-*trans*,9-*cis*-18:2, 7-*trans*,9-*trans*-18:2) in total yields of 72 and 78%, respectively. By choosing the appropriate monoene precursor, this method was used to prepare mixtures of CLA isomers for use as analytical standards.

Bromination/Dehydrobromination. Bromination of an unsaturated fatty acid followed by dehydrobromination or debromination has been used to prepare a number of conjugated *trans*,*trans*- and *cis*,*trans*-18:2 positional isomers.

Schmidt and Lehmann (32) prepared mixtures of 7,9-, 8,10-, 9,11- and 10,12-18:2 isomers by allylic bromination/dehydrobromination of oleic (9-*cis*-18:1) and elaidic (9-*trans*-18:1) acid. The same authors also used the allylic bromination/dehydrobromination of brassidic acid (11-*trans*-22:1) to prepare a mixture of conjugated 22:2 isomers, including the 13,15-22:2 isomer. Purification of the isomer mixture by crystallization from petroleum ether (PE) to constant melting point yielded a mixture of conjugated *trans*,*trans*-22:2 positional isomers.

Gupta and Kummerow (13) developed a multistep synthesis to prepare 8-*trans*,10-*trans*-18:2, but in <10% overall yield. The synthesis involved the allylic bromination (NBS in carbon tetrachloride in the presence of benzoyl peroxide) of ethyl oleate, further bromination of the double bond (by addition of free bromine), debromination using zinc in ethanol, fractional distillation, saponification (alcoholic potassium hydroxide), isomerization [ultraviolet (UV) irradiation in PE in presence of iodine], and final purification by repeated crystallizations from PE.

Isomerization. (a) *Alkali isomerization.* Isomerization of linoleic (9-*cis*,12-*cis*-18:2) acid or the methyl ester to yield the CLA isomer(s) has been accomplished using different basic catalysts, solvent systems, and at different temperatures. Procedures include the following: (i) ethylene glycol, sodium hydroxide, 200°C (32,33); (ii) ethylene glycol, potassium hydroxide, 180°C [(34), modified conditions (5)]; and (iii) potassium *t*-butoxide in *t*-butanol at 60, 90, and 140°C (35). Nugteren (15) modified procedure (ii) (145°C) to prepare 8-*cis*,10-*trans*-20:2 and 9-*trans*,11-*cis*-20:2 from 8-*cis*,11-*cis*-20:2, and to prepare 11-*cis*,13-*trans*-20:2 and 12-*trans*,14-*cis*-20:2 from 11-*cis*,14-*cis*-20:2.

(b) *Photoisomerization.* Seki *et al.* (36) used a 100-W high-pressure Hg lamp to isomerize methyl linoleate in the presence of iodine (petroleum ether as solvent). Under optimum conditions, 80% of the linoleate was converted to roughly equal amounts of the 9,11- and the 10,12-18:2 isomers (the ratio of *cis/trans* to *trans/trans* isomers in the final mixture was ~3:7). Seki *et al.* (37) also applied the methodology to the isomerization of methyl linolenate (9-*cis*,12-*cis*,15-*cis*-18:3). Conjugated diene-trienoates (10,12,15-18:3 and 9,12,14-18:3) or mainly trienoates (10,12,14-18:2) could be generated by adjustment of the iodide concentration and/or irradiation times.

Reduction of Olefinic or Acetylenic Bonds. A number of chemical syntheses are available for the preparation of conjugated dienes, ene-ynes, diynes, and more highly unsaturated acetylenic/olefinic fatty acids (38–41).

(a) *Polyunsaturated fatty acids.* Hydrazine reduction of α-eleostearic acid (9-*cis*,11-*trans*,13-*trans*-octadecatrienoic acid) resulted in selective reduction of the 9- and 13- double bonds and formation of a ~30% conjugated diene mixture of 9-*cis*,11-*trans*-18:2 and 11-*trans*,13-*trans*-18:2 (42).

(b) *Mono-acetylenic fatty acids.* Santalbic acid (also called ximenynic acid; 11-*trans*-octadec-9-ynoic acid) is a useful precursor to CLA and can be isolated from *Santalum album* seed oil (43), prepared from ricinoleic acid [12-hydroxy-9-*cis*-octadecenoic acid; (29)] or synthesized (44). Lindlar catalyst and H_2 gas (45), zinc in isopropyl alcohol (46), or iron in isopropyl alcohol/water (47) have been used to reduce the acetylenic bond to the *cis*-bond. Yields are usually 70–80%.

(c) *Diacetylenic fatty acids.* Chemically synthesized conjugated diacetylenic acids (6,8-/7,9-/8,10-/9,11-18:2) have been reduced to the corresponding *cis/cis* dienes by using H_2 gas and an iron catalyst in isopropyl alcohol/water (47) or by use of Lindlar catalyst (48). Although H_2 uptake during the Lindlar-catalyzed reduction of methylene-interrupted polyacetylenes ceases at the polyolefin stage, the Lindlar-catalyzed reduction of conjugated systems does not, and it must be monitored carefully to prevent overreduction of the substrate (49).

Biochemical Methods

1. *Extraction from Seed Oil Samples.* The topic is reviewed in Hopkins (8), and selected examples are provided in Tables 3.1–3.3. Continued improvements in separation technologies have allowed the isolation of specific fatty acids from seed oils. Limitations to this procedure include the availability of oils such as that obtained from *Santalum album* seeds (see below), the often low concentration of the desired FA, and, as noted before, difficulties in the isolation and purification of the material.

(a) *Santalum album* seeds were crushed with a mortar and pestle and the oil was extracted with hexane. The solvent was removed and the residue chromatographed through silica gel (diethyl ether/hexane, 10:90, vol/vol). The triglycerides (TG) were converted to fatty acid methyl esters by room temperature alkali-catalyzed transesterification, and the methyl santalbate (72% of FAME) was isolated (97.9% pure) by preparative reversed-phase HPLC. *S. album* seeds (5.83 g) yielded 1.86 g oil (72.8% santalbic acid by GC) from which 1.14 g of highly purified (>98%) material was isolated (49).

(b) Methyl 10-*trans*,12-*trans*-18:2 was isolated from the seed oil of *Chilopsis linearis* in which it constitutes ~10% of total fatty acids. Unfortunately, fractionation of 24 g of seed oil fatty acids by fractional crystallization yielded only 0.08 g of highly pure conjugated diene (50).

(c) Takagi and Itabashi (51) converted selected seed oil TG to FAME and utilized a reversed-phase HPLC system to isolate conjugated trienoates. Methyl 9,11,13-18:3 isomers were separated from tung oil (67% *c,t,t*; 11% *t,t,t*), *Momordica charantia* (56% *c,t,t*), *Punica. granatum* (83% *c,t,c*) and *Catalpa ovata* (42% *t,t,c*). Methyl 8-*trans*,10-*trans*,12-*cis*-18:3 was isolated from *Calendula officinalis*.

2. *Enzymatic Methods.* Although biosynthetic methods have been used to prepare a large number of methylene-interrupted fatty acids, the use of algal or bacterial enzymes (*in vivo* or *in vitro*) to produce CLA isomers is limited at this time to preparation of the 9-*cis*,11-*trans*-18:2 isomer and a number of unsaturated mono- and dihydroxy fatty acids. The hydroxy fatty acids (see above) can be converted to CLA isomers by incubation with selected bacterial strains or with isolated bacterial enzymes (see Ref. 12 for review). A limitation to these procedures is the often low conversion of the substrate and the difficulty of isolating specific fatty acids from the resultant FA mixture. Examples of enzymatic conversion of FA (*in vivo* or *in vitro*) include the following: (i) Preparation of the 9-*cis*,11-*trans*-18:2 isomer from linoleic acid using the enzyme linoleate isomerase isolated from the rumen bacterium *Butyrivibrio fibrisolvens* (52 and 5). (ii) Methyl 9,12-dihydroxy-10-*trans*-octadecenoate [precursor to methyl 9-*trans*,11-*trans*-18:2 (see above)] has been obtained by the incubation of oleic acid with bacterial strain PR3 (53). (iii) Nugteren (15) fed an essential fatty acid–deficient rat 100 mg of 8-*cis*,12-*trans*,14-*cis*-eicosatrienoic acid per day for

2 wk. The animal was terminated, the organs from the chest and abdomen were removed, and the organ lipids were extracted. Silica LC and argentation-TLC were used to isolate 5-*cis*,8-*cis*,12-*trans*,14-*cis*-eicosatetraenoic acid (as FAME, 80% pure).

Isotopically Labeled, Conjugated Fatty Acids

Nonconjugated FA labeled with stable isotopes such as deuterium (^{2}H) or carbon-13 (^{13}C) or with radioactive isotopes such as tritium (^{3}H), carbon-11 (^{11}C) or carbon-14 (^{14}C) have found a prominent role in the study (*in vivo* and *in vitro*) of lipid incorporation, metabolism, oxidation, and interaction. A variety of isotopically labeled FA have been prepared chemically using well-documented syntheses and biologically by incubating algal or bacterial cultures with labeled FA, water, carbon dioxide, or acetate. The same synthesis may even be used to incorporate different isotopes. As noted above, the acetylenic bond in santalbic acid (11-*trans*-octadec-9-ynoic acid) may be reduced with Lindlar catalyst and ^{2}H$_2$ (rather than H$_2$) gas to an olefinic bond in the preparation of 9-*cis*,11-*trans*-18:2-9,10-^{2}H$_2$. The Lindlar-catalyzed reduction of fatty acids containing one or more acetylenic bonds is considered one of the easiest ways to prepare unsaturated FA labeled with tritium (^{3}H) atoms on the double bond(s).

At this time, published preparations of isotopically labeled isomers of conjugated FA are limited to FA labeled with deuterium and, in one or two examples, ^{14}C (See Table 3.3). Most current research utilizing labeled CLA isomers is being done with the use of deuterium-labeled FA. Deuterium-labeled FA isomers can be safely used in humans to trace FA metabolism and interaction(s). Deuterium-labeled fatty acids have thus found applications in a variety of studies, including lipid metabolism, membrane structure elucidation, and mechanisms of lipid oxidation. Several different FA labeled with differing numbers of deuterium atoms (a mixture of a di-, a tetra- and a hexadeuterated FA, for example) may be fed and tracked simultaneously. Direct comparison of several fats in a single system is a great advantage when comparing FA interactions, rates of metabolism, or rates of tissue incorporation. Several syntheses have been developed to yield CLA isomers that are deuterium atom–labeled, in sufficient yield and of high enough isotopic and chemical purity for use in humans (54–56).

A single example of a conjugated dienoic fatty acid labeled with carbon-14 (^{14}C) is available (Table 3.3), and that only through a commercial supplier. Although no dienoic CLA isomers labeled with carbon-13 (^{13}C), tritium (^{3}H), or carbon-11 (^{11}C) are available at this writing, precursor isomers may be prepared by conventional techniques used to prepare stable and radioactive, nonconjugated FA isomers (see Ref. 54 for review). Isomerization of ^{13}C-, ^{3}H-, or ^{11}C-labeled nonconjugated FA could provide a ready source of individual FA or conjugated FA standard mixtures. Hydroxy-FA precursors labeled with tritium atoms (hydroxy-5,8,10,14-eicosatrienoic acid, 12(*S*)-[5,6,8,9,11,12,14,15-^{3}H], for example) are commercially available, as are ^{2}H- and ^{13}C-labeled aldehydes and phosphonium salt precursors for the Wittig coupling reaction (see Table 3.3).

Chemical Synthesis

Wittig/Acetylenic Coupling. The deuterium atoms are incorporated by reduction of an acetylenic and/or olefinic alcohol or halide (chloride) with Wilkinson's catalyst and $^2H_2$ gas. Wilkinson's catalyst is stereoselective and yields little hydrogen-deuterium exchange. The saturated, deuterium-labeled fragment is used in the synthesis of monounsaturated fatty acids (MUFA) and PUFA *via* a series of acetylenic coupling, Grignard, or Wittig coupling reactions. By careful selection of starting materials, the number and location of the deuterium atoms in the fatty acid can be controlled. A drawback to this approach is the time required for the synthesis/purification of the intermediates and the often low overall yields (5–15%) of the desired product. The example provided below (56) is for preparation of methyl 9-*cis*,11-*trans*- and 9-*trans*,11-*trans*-octadecadienoate-17,17,18,18-$^2H_4$, two of the isomers of conjugated linoleic acid.

"A multi-step synthesis was used to prepare the *cis*-9,*trans*-11- and *trans*-9,*trans*-11-isomers (in a ratio of 46/54) of conjugated linoleic acid (*cis*-9,*trans*-11-octadecadienoic acid) labeled with deuterium atoms on the 17- and 18-carbon atoms (17,17,18,18-d_4). The methyl *cis/trans*-9, *trans*-11-octadecadienoate-17,17,18,18-d4 isomer pair was obtained from the Wittig coupling of *trans*-2-nonenyltriphenylphosphonium bromide (8,8,9,9-d_4) and methyl 9-oxononanoate. To prepare the phosphonium bromide, 5-hexyn-1-ol was reduced with deuterium gas/Wilkinson's catalyst to yield 1-hexanol-5,5,6,6-d_4. The alcohol was converted to the iodide with phosphorous pentoxide/phosphoric acid/potassium iodide. Coupling of the iodide with 2-propyn-1-ol *via* lithium amide in liquid ammonia gave 2-nonyn-1-ol-d_4. The acetylenic alcohol was reduced with lithium metal in liquid ammonia to yield the *trans*-2-nonen-1-ol-d_4. The alcohol was converted to the bromide (using triphenylphosphine dibromide) and then converted to the phosphonium salt. The aldehyde ester was prepared by the reductive ozonization of methyl *cis*-9-octadecenoate. The two conjugated linoleic acid isomers, formed during the final Wittig coupling reaction, were readily separated by a combination of reversed-phase and silver resin chromatography. Isotopic and chemical purities were >95% for each geometric isomer. Overall yield (both isomers) from the 8-step synthesis was 12%."

The addition of an alcohol during the Wittig coupling reaction has been shown to increase the percentage of the *trans*-isomer in the generated double bond; this increase has been demonstrated in the synthesis of monoenoic or methylene-interrupted dienoic fatty acids. During the preparation of *trans*-10,*cis*-12-octadecadienoate-17,17,18,18-$^2H_4$ (formation of the double bond in the 10-position by the Wittig coupling reaction), however, a comparison of the percentage of *trans* formed by the addition of methanol vs. the reaction without methanol showed little (2–3%) increase in the percentage of *trans* from the addition of methanol (unpublished results).

Reduction of Acetylenic to Olefinic FA. (a) Methyl santalbate (methyl 11-*trans*-octadecen-9-ynoate), obtained from *S. album* (L.) seed, was reduced with Lindlar catalyst, quinoline, and deuterium gas to produce, in yields of 65–75%, the gram quantities of methyl 9-*cis*,11-*trans*-octadecadienoate-9,10-d_2 (CLA-d_2) required for metabolism and oxidation studies. Unlike mono-acetylenic and methylene-interrupted polyacetylenic fatty acid methyl esters, the conjugated system was reduced with no noticeable break in the rate of deuterium uptake. The quantity of poison (quinoline) present did influence the amount of CLA-d_2 produced, but the production of overreduced fatty acid methyl esters (perhaps because of the conjugated system) could not be prevented. Fractionation of the reaction mixture by silver resin chromatography resulted in the isolation of >99% chemically pure CLA-d_2 (as FAME) in yields of 60–70% (49).

(b) Goerger and Hudson (22) prepared parinaric acid (9-*trans*,11-*trans*,13-*trans*,15-*trans*-18:4) specifically labeled with deuterium atoms at the vinyl positions by using the Wittig reaction to couple a diene phosphorane with an α,β-unsaturated aldehyde ester. The deuterium-labeled *trans* double bonds of each component were formed by the stereoselective reduction of a substituted propynoic ester with lithium aluminum deuteride in tetrahydrofuran at −78°C.

Photoisomerization. Cawood *et al.* (57) demonstrated that UV irradiation of [1-^{14}C]linoleic acid in the presence of human albumin or gamma globulin could produce, in low yields, a mixture of [1-^{14}C]-labeled 9-*cis*,11'-*cis*-; 9-*cis*,11-*trans*-; 10-*cis*,12-*cis*-; and 10-*trans*,12-*cis*-18:2.

Conversion of Unlabeled FA to Labeled FA Ester. BF_3/C^3H_3OH has been used to esterify unsaturated FA, placing the label in the methyl group of the FAME (58). Conjugated fatty acids are prone to isomerization in the presence of acidic catalysts such as BF_3 or HCl (11), but basic catalysts such as sodium methoxide can be employed (59). Although this procedure for incorporating an isotopic label is less useful than methods that place the label on the fatty acid backbone, the FA labeled in this way could still be used in the study of FA hydrogenation, isomerization, or oxidation. This method may also be used to prepare [^{13}C-] and [^{14}C-] labeled FA.

Biosynthesis/Conversion of Fatty Acids Labeled with Stable and/or Radioisotopes

Biosynthesis is the preferred method for the preparation of ^{13}C- and ^{14}C- uniformly labeled FA and, to a lesser extent, their ^{3}H- and ^{2}H-labeled analogs. A number of algal and bacterial sources are available. Starting materials may be $^xCO_2$ (x = 13 or 14), [1-^{14}C]acetate or ^xH_2O (x = 2 or 3); their incubation yields isotopically labeled FA. Although these procedures can be used to prepare a variety of labeled FA, isolation of the desired fatty acid from the fatty acid pool is often difficult and care must be exercised in the choice of culture used. These methods have been used extensively to prepare FA labeled with radioisotopes.

(a) Crombie and Holloway (60) used a homogenate of marigold (*Calendula officinalis*) seeds to form conjugated trienoic acids (from a variety of labeled precursors). Although sodium [1-^{14}C]acetate ($CH_3^{14}COOH$) was poorly converted to [1-^{14}C] calendic acid (8-*trans*,10-*trans*,12-*cis*-18:3) in germinating seeds, [1-^{14}C]acetate incubated with the homogenate of 15-d-old marigold seeds gave good incorporation (~0.80%). [1-^{14}C] and ^{3}H-labeled oleic acids were readily converted to [1-^{14}C]- and

3H-labeled calendic acids at 0.52 and 0.32% label incorporation, respectively. The labeled FA were purified by crystallization (7–9 times) from pentane until constant count was achieved. Approximately 15% of the ^{14}C-label in the calendic acid was distributed from the 1-position to the FA chain by degradation of the labeled calendic acid with loss of labeled acetate and resynthesis of calendic acid from the acetate.

(b) Kepler and co-workers (see above also) prepared ^{14}C- and ^{3}H-labeled 9-*cis*,11-*trans*-18:2 from linoleic acid using the enzyme linoleate isomerase isolated from the rumen bacterium *Butyrivibrio fibrisolvens*. The ^{14}C-isomer was prepared by incubating 0.2 mL of the enzyme preparation with 2 microCurie (0.12 mg) of [1-^{14}C]linoleic acid and 4 mg of bovine serum albumin in 1.0 mL of 0.1 mol/L potassium phosphate buffer, pH 7. The conjugated isomer was isolated by preparative silver nitrate (25% Ag) TLC (61). The ^{2}H-labeled CLA (with a single ^{2}H atom on the C-13 carbon) was also prepared by incubation of linoleic acid for 23 h with the enzyme in the presence of 99% ^{2}H$_2$O (62).

The preparation of CLA-containing triacylglycerols (TG) or phospholipids (PL), for example, by use of enzymes is a practical alternative to purely chemical methods (63). Isolated enzyme systems or bacterial cultures may be used to convert specifically labeled fatty acids to labeled CLA isomers (see above), or the FA can be metabolized to yield prostaglandins, thromboxins, prostacyclines, and other prostanoids specifically labeled with both radioactive and stable isomers. A method for preparing synthetic TG based on CLA has been patented (64).

Miscellaneous

The chemical synthesis of CLA isomers usually involves a combination of synthetic steps and purification methodologies. Ha *et al.* (1), for example, isolated 10-*trans*, 12-*cis*-18:2 from alkali-isomerized linoleic acid by crystallization (13). Sehat and co-workers (9) prepared the 10-*trans*,12-*trans*- and 10-*cis*,12-*cis*-isomers from the 10-*trans*,12-*cis*-isomer by isomerization (light and iodine; purification by reversed-phase HPLC), and Lie Ken Jie (27) isolated the three isomers of 9,11-18:2 (*cis,cis*-; *cis,trans*-; *trans,trans*-) from ricinoleic or ricinelaidic acid [purification by argentation-HPLC (65)]. Although many of the pre-1970 preparations of conjugated fatty acids yielded mixtures of difficult-to-separate isomers, the development of Ag-HPLC and other improved HPLC columns may now allow the separation of formerly "inseparable" isomers.

Although the synthesis/isolation of individual isomers of CLA has proceeded slowly, interest in these isomers as a diet supplement in Europe and the U.S. continues to grow; the interest is not limited to only CLA isomers. The nutritional characteristics of α-eleostearic acid (9-*cis*,11-*trans*,13-*trans*-18:3) in rats have been studied (66), and two types of hydroxy acids [9-hydroxy-10-*trans*,12-*cis*-18:2 (a) and 13-hydroxy-9-*cis*,11-*trans*-18:2 (b)] have been found to be cytotoxic against cancer cells in mice (67). The hydroxy acids were extracted from rice bran, with 10 kg of bran yielding 100 mg (a) and 60 mg (b). The low recoveries of these isomers and of others listed in Tables 3.2 and 3.3 should fuel the search for improved methods of synthesis. The methods of synthesis and commercially available isomers presented in this chapter will therefore certainly continue to increase in number, both in response to the demands of the diet supplement and food additive industries and as researchers attempt to better understand the effects exhibited by conjugated fats.

Acknowledgment

Ms Erin Walter, who once again helped bring order out of chaos, is gratefully acknowledged.

TABLE 3.1 CLA Isomers

Fatty acid	Mixture?	Purity[a]	Yield (%)	Method[b]	Ref./Note
2-cis,4-cis-10:2	No	a	—	B, F	48
2-cis,4-trans-10:2	No	a	—	B, F	48
2-trans,4-cis-10:2 (stillingic acid)	No	a	—	F	48
2-trans,4-trans-10:2	No	a	—	F	48
4-trans,6-trans-10:2	—	—	—	—	8[c]
3-trans,5-cis-13:2 (megatomic acid)	—	—	—	—	8[c]
3-trans,5-trans-13:2	—	—	—	—	8[c]
2,4-16:2 (all 4 isomers)	—	—	—	—	8[c]
8-cis,10-cis-16:2	No	a	73	B,F	68
6-cis,8-cis-18:2	No	a	7	B,F	69
6-cis,8-cis-18:2	Yes	—	76	B	47
7-cis,9-cis-18:2	Yes	—	65	B	47
8-cis,10-cis-18:2	Yes	—	57	B	47
8-trans,10-trans-18:2	No	a	<10	E,F	13
9-cis,11-cis-18:2	No	e	—	B	47
	No	a	75	B,C	12
9-cis,11-trans-18:2	Yes	—	86	B,C	47
	No	a	82	B,C	12
	Yes	c	66	A	29
9-cis,11-trans-18:2[d] (Mangold's acid)	No	d	—	A	28
9-trans,11-cis-18:2[e]	No	a	15	A	12
9-trans,11-trans-18:2	Yes	—	30–35	A	26
9-trans,11-trans-18:2	No	a	75	A	12
10-cis,12-cis-18:2	No	a	10	B,F	69
10-trans,12-trans-18:2 (Mikusch's acid)	No	c	9	F	69
				C	8[c]
8-cis,10-trans-20:2 and 9-trans,11-cis-20:2	Yes	—	40	E	15
11-cis,13-trans-20:2 and 12-trans,14-cis-20:2	Yes	—	40	E	15
13-trans,15-trans-22:2	Yes	—	—	F	32

[a]Purity: a = 95–100%; b = 90–95%; c = 80–90%; d = 70–80%; e = <70%.
[b]Method: A, from hydroxy fatty acid; B, chemical reduction of acetylenic precursor; C, from seed oil fatty acid precursor; D, biosynthesis; E, isomerization (acid, alkali, photo-); F, chemical synthesis.
[c]And references cited therein.
[d]21% 9-cis,11-cis-18:2.
[e]Side-product from preparation of 9-trans,11-trans-18:2.

TABLE 3.2 Miscellaneous Polyunsaturated Fatty Acids Containing Conjugation

Fatty acid	Purity[a]	Yield (%)	Method[b]	Ref./Note
2-trans,4-trans,6-trans-16:3	b	<40	F	70
2-trans,4-trans,6-trans,8-trans-16:4	c	<20	F	70
6-cis,8-trans,10-trans-18:3	—	—	D	71
7-cis/trans,9-cis,12-cis-18:3	—	—	—	72
8-cis,10-trans,12-cis-18:3 (jacarandic acid)	—	—	F	6
8-trans,10-trans,12-cis-18:3 (calendic acid)	—	—	F	6[c]
	—	—	D	71
8-trans,10-trans,12-trans-18:3	—	75	E	73
9-cis,11-cis,13-trans-18:3	—	—	—	8[c]
9-cis,11-trans,13-cis-18:3 (punicic acid)	—	—	—	6[c]
9-cis,11-trans,13-trans-18:3 (α-eleostearic acid)	—	30–35	A	26
9-cis,11-trans,15-cis-18:3	—	—	D	61
9-trans,11-trans,13-cis-18:3 (catalpic acid)	—	—	—	6[c]
9-trans,11-trans,13-trans-18:3 (β-eleostearic acid)	—	—	—	8[c]
9-cis,12-yne,14-yne-18:3	—	10	F	75
10-trans,12-trans,14-trans-18:3	—	—	—	8[c]
9-trans,11-trans,13-trans,15-trans-18:4 (β-parinaric acid)	—	—	—	8[c]
9-cis,12-yne,14-yne,16-cis-18:4	—	5–7	F	75
9-cis,12-yne,14-yne,16-trans-18:4	—	7–9	F	75
9-cis,12-yne,14-yne,16-yne-18:4	—	10	F	75
8-cis,10-trans,14-cis-20:3	—	—	—	15
8-cis,12-trans,14-cis-20:3	a	<15	B,F	74
9-trans,11-cis,14-cis-20:3	—	—	—	15
5-cis,8-cis,11-cis,13-trans-20:4	a	<15	F	76
5-cis,8-cis,12-trans,14-cis-20:4	c	—	D	15
6-cis,8-trans,10-trans,12-cis-20:4	—	29	D	77
5-cis,8-cis,10-trans,12-trans,14-cis-20:5	—	—	—	77
5-cis,8-cis,11-cis,13-trans,15-trans-20:5[d]	a	<15	F	76
4-cis,7-cis,9-trans,11-trans-,13-cis,16-cis,19-cis-22:7	—	—	D	78

[a]Purity: a = 95–100%; b = 90–95%; c = 80–90%.
[b]Method: A, from alcoholic precursor; C, from seed oil fatty acid precursor; D, biosynthesis; E, isomerization (acid, alkali, photo-); F, chemical synthesis.
[c]And references cited therein.
[d]Prepared in dark.

TABLE 3.3 Conjugated Fatty Acids Labeled with Stable or Radioactive Isotopes

Fatty acid	Location[a,b]	Method[c]	Isotopic purity[d]	Reference
Deuterium-labeled				
11-cis,13-cis-16:2	15,15,16,16,16	A	a	79
9-cis,11-trans-18:2	13	D	c	61
	9,10	B	b	49
	17,17,18,18	B	b	56
9-trans,11-trans-18:2	17,17,18,18	B	b	56
10-cis,12-cis-18:2	15,15,16,16	C	c	55
10-trans,12-cis-18:2	15,15,16,16	C	c	55
10-trans,12-trans-18:2	15,15,16,16	C	c	57
9-trans,11-trans,13-trans,15-trans-18:4	9,10,11,12,13,14,15,16	E	—	22
12-hydroxy-9-cis-18:2	12,12	F	—	80
12-hydroxy-5-cis,8-trans,10-trans-17:3	5,6,8,9	B,F	—	81
12-hydroxy-5-cis,8-cis,10-trans,14-cis-20:4	5,6,8,9	B,F	—	81
trans-4-hydroxy-2-nonenal-dimethyl acetal	(5,5,6,6,7,7,8,8,9,9,9)	F	—	CDN[e]
³H-labeled				
9-hydroxy-10-trans,12-cis-18:2	9	D	—	62
5-hydroxy-6-trans,8-cis,11-cis,14-cis-20:4 (5-HETE)	(5,6,8,9,11,12,14,15)	—	—	NEN[f]
12-hydroxy-5-cis,8-cis,10-trans,14-cis-20:4 [12(S)HETE]	(5,6,8,9,11,12,14,15)	—	—	NEN
15-hydroxy-5-cis,8-cis,11-cis,13-trans-20:4 (15-HETE)	(5,6,8,9,11,12,14,15)	—	—	NEN
¹⁴C-labeled				
9-cis,11-trans-18:2	(1)	—	—	ARC[g]
6-cis,8-trans,10-trans,12-cis-18:4	1	D	b	82
5-cis,8-cis,10-trans,12-trans,14-cis-20:5	1	D	—	83
9-cis,12-yne,14-yne,16-yne-18:4	18	F	—	75
	9	F	—	75

[a] Location of label.
[b] (a) - ( ) Indicates commercial availability.
[c] Methods of preparation: A, from alcoholic precursor; B, chemical reduction of acetylenic precursor; C, from seed oil fatty acid; precursor; D, biosynthesis; E, isomerization (acid, alkali, photo-); F, chemical synthesis.
[d] isotopic purity: a = 95–100%; b = 90–95%; c = 80–90%.
[e] CDN: C/D/N Isotopes, Quebec, Canada; (http://www.cdniso.com).
[f] NEN: New England Nuclear, Boston, MA; (http://www.nenlifesci.com).
[g] ARC: American Radiolabeled Chemicals, St. Louis, MO; (http://www.arc-inc.com).

References

1. Ha, Y.L., Grimm, N.K., and Pariza, M.W. (1989) Newly Recognized Anticarcinogenic Fatty Acids: Identification and Quantification in Natural and Processed Foods, *J. Agric. Food Chem. 37*, 75–81.
2. Belury, M.A. (1995) Conjugated Dienoic Linoleate: A Polyunsaturated Fatty Acid with Unique Chemoprotective Properties, *Nutr. Rev. 53*, 83–89.
3. Christie, W.W., Dobson, G., and Gunstone, F. (1997) Isomers in Commercial Samples of Conjugated Linoleic Acid, *J. Am. Oil Chem. Soc. 74*, 1231.
4. Ackman, R.G. (1998) Laboratory Preparation of Conjugated Linoleic Acids, *J. Am. Oil Chem. Soc. 75*, 1227.
5. Chin, S.F., Liu, W. Storkson, J.M. Ha, Y.L., and Pariza, M.W. (1992) Dietary Sources of Conjugated Dienoic Isomers of Linoleic Acid, a Newly Recognized Class of Anticarcinagens, *J. Food Comp. Anal. 5*, 185–197.
6. Bergel'son, L.D., and Shemyakin, M.M. (1968) Synthesis of Naturally Occurring Unsaturated Fatty Acids by Sterically Controlled Carbonyl Olefination, *N. Methods Prep. Org. Chem. 5*, 155–175.
7. Mangold, C. (1894) Einige Beitrage zur Kenntniss der Ricinusol-, Ricinelaidin- und Ricinstearolsäure. *Monatsch. Chem. 15*, 307–315.
8. Hopkins, C.Y. (1972) Fatty Acids with Conjugated Unsaturation, in *Topics in Lipid Chemistry*, (F.D. Gunstone, ed.) pp. 37–87, Elek Science, London.
9. Sehat, N., Yurawecz, M.P., Roach, J.A.G., Mossoba, M.M., Kramer, J.K.G., and Ku, Y. (1998) Silver Ion–High-Performance Liquid Chromatographic Separation and Identification of Conjugated Linoleic Acid Isomers, *Lipids 33*, 217–221.
10. Christie, W.W. (1997) The Analysis of Conjugated Fatty Acids, *Lipid Technol.*, May, 73–75.
11. Kramer, J.K.G., Fellner, V., Dugan, M.E.R., Sauer, F.D., Mossoba, M.M., and Yurawecz, M.P. (1997) Evaluating Acid and Base Catalysts in the Methylation of Milk and Rumen Fatty Acids with Special Emphasis on Conjugated Dienes and Total *trans* Fatty Acids, *Lipids 32*, 1219–1228.
12. Lie Ken Jie, M.S.F., Chueng, Y.K., Pasha, M.K., and Syed-Rahmatullah, M.S.K. (1997) Fatty Acids, Fatty Acid Analogues and Their Derivatives, *Nat. Prod. Rep. 14*, 136–189.
13. Gupta, S.C., and Kummerow, F.A. (1960) Preparation of 8t,10t-Octadecadienoic Acid. *J. Am. Oil. Chem. Soc. 37*, 32–34.
14. Scholfield, C.R., and Koritala, S. (1970) A Simple Method for Preparation of Methyl *trans*-10,*cis*-12 Octadecadienoate. *J. Am. Oil Chem. Soc. 47*, 303.
15. Nugteren, D.H. (1970) Inhibition of Prostaglandin Biosynthesis by 8-*cis*,12-*trans*,14-*cis*-Eicosatrienoic Acid and 5-*cis*,8-*cis*,12-*trans*,14-*cis*-Eicosatetraenoic Acid, *Biochim. Biophys. Acta 210*, 171–176.
16. Crombie, L. (1996) Synthesis in the Isotopic Labelling of Plant Fatty Acids: Their Use in Biosynthesis, in *Synthesis in Lipid Chemistry* (Tyman, J.H.P., ed.), pp. 34–56, Royal Society of Chemistry, UK.
17. Gunstone, F.D. (1998) Movements Toward Tailor-Made Fats, *Prog. Lipid Res. 37*, 277–305.
18. Camps, F., and Guerro, A. (1998) Synthesis of Long-Chain Compounds, in *Lipid Synthesis and Manufacture* (Gunstone, F., ed.) Ch. 3, pp. 94–126, Sheffield Academic Press, England.

19. Zhang, A., and Chen, Z.Y. (1997) Oxidative Stability of Conjugated Linoleic Acids Relative to Other Polyunsaturated Fatty Acids, *J. Am. Oil Chem. Soc. 74*, 1611–1613.
20. Joh, Y.-G., Kim, S.J., and Christie, W.W. (1995) The Structure of Triacylglycerols, Containing Punicic Acid, in the Seed Oil of *Trichosanthes kirilowii*, *J. Am. Oil Chem. Soc. 72*, 1037–1042.
21. Tulloch, A.P. (1979) Analysis of the Conjugated Trienoic Acid Containing Oil From *Faveoli trilobata* by ^{12}C Nuclear Magnetic Resonance Spectroscopy, *Lipids 14*, 996–1002.
22. Goerger, M.M., and Hudson, B.S. (1988) Synthesis of all-*trans*-Parinaric Acid-d_8 Specifically Deuterated at All Vinyl Positions, *J. Org. Chem. 53*, 3148–3153.
23. Millar, J.G. (1965) Methyl (2*E*,4*Z*,6*Z*)-deca-2,4,6-Trienoate, a Thermally Unstable, Sex-Specific Compound from the Stink Bug *Thyanta pallidovirens*, *Tetrahedron Lett. 38*, 7971–7972.
24. Ucciani, E. (1979) Structure et Synthese des Acides Gras Polyinsatures, *Rev. Fr. Corps Gras*, 397–404.
25. Grummitt, O., and Marsh, D. (1953) Alternate Methods for Dehydrating Castor Oil, *J. Am. Oil Chem. Soc. 30*, 21–25.
26. Schneider, W.J., Gast, L.E., and Teeter, H.M. (1964) A Convenient Laboratory Method for Preparing *trans,trans*-9,11-Octadecadienoic Acid, *J. Am. Oil Chem. Soc. 47*, 605–606.
27. Lie Ken Jie, M.S.F, and Wong, K.P. (1992) Dehydration Reactions Involving Methyl 9,12-Dihydroxy-10-*trans*-octadecenoate, *Chem. Phys. Lipids 62*, 177–183.
28. Gunstone, F.D., and Said, A.I. (1971) Fatty Acids, Part 29: Methyl 12-Mesyloxyoleate as a Source of Cyclopropane Esters and of Conjugated Octadecadienoates, *Chem. Phys. Lipids 7*, 121–134.
29. Berdeaux, O., Christie, W.W., Gunstone, F.D., and Sébédio, J.L. (1997) Large Scale Synthesis of Methyl *cis*-9,*trans*-11-Octadecadienoate from Methyl Ricinoleate, *J. Am. Oil Chem. Soc. 74*, 1011–1015.
30. Tassignon, P., de Waard, P., de Rijk, T., Tournois, H., de Wit, D., and de Buyck, L. (1994) An Efficient Countercurrent Distribution Method for the Large-Scale Isolation of Dimorphecolic Acid Methyl Ester, *Chem. Phys. Lipids 71*, 187–196.
31. Yurawecz, M. P., Hood, J.K., Roach, J.A.G., Mossoba, M.M., Daniels, D.D., Ku, Y., Pariza, M.W., and Chin, S.F. (1994) Conversion of Allylic Hydroxy Oleate to Conjugated Linoleic Acid and Methoxy Oleate by Acid-Catalyzed Methylation Procedures, *J. Am. Oil Chem. Soc. 71*, 1149–1155.
32. Schmidt, H., and Lehmann, A. (1950) 195. Darstellung einer neuen 9,11-Oktadekadiensaure und der 13,15-Dokosadiensaure aus den einfach ungesattigten Fettsauren. *Helv. Chim. Acta 195*, 1494–1502.
33. Nichols, P.L., Jr., Herb, S.F., and Riemenschneider, R.W. (1951) Isomers of Conjugated Fatty Acids. I. Alkali-Isomerized Linoleic Acid, *J. Am. Chem. Soc. 73*, 247–252.
34. Association of Official Analytical Chemists (1990) Acids (Polyunsaturated) in Oils and Fats Spectrophotometric Method Final Action, AOAC Official Methods of Analysis. 957.13, 960–963.
35. Mounts, T.L., Dutton, H.J., and Glover, D. (1970) Conjugation of Polyunsaturated Acids, *Lipids 5*, 997–1005.
36. Seki, K., Kaneko, R., Kobayashi, K. (1998) Photoconjugation of Methyl Linolenate in the Presence of Iodine as a Sensitizer, *Yukagaku 38*, 949–954.
37. Seki, K., Kaneko, R., Kobayashi, K. (1998) Photoconjugation of Methyl Linolenate in the Presence of Iodine as a Sensitizer, *Yukagaku 38*, 955–958.

38. Lie Ken Jie, M.S.F., Chueng, Y.K., Chau, S.H., and Yan, B.F.Y. (1991) C13-NMR Spectra of Positional Isomers of Long-Chain Conjugated Diacetylenic Fatty Esters, *Chem. Phys. Lipids 60,* 179–188.
39. Mhaskar, S.Y., and Lakshminarayana, G. (1990) Synthesis of a Novel Acetylenic Ester Methyl (E)-5-Octadecen-7,9-diynoate, *Syn. Commun. 20,* 2001–2009.
40. Sprecher, H. (1978) The Organic Synthesis of Unsaturated Fatty Acids. *Prog. Chem. Fats Other Lipids 15,* 219–254.
41. Singh, A. and J.M. Schnur. (1986) A General Method for the Synthesis of Diacetylenic Acids, *Syn. Commun. 16,* 847–852.
42. Mikolajczak, K.L., and Bagby, M.O. (1965) Partial Reduction of α-Eleostearic Acid with Hydrazine, *J. Am. Oil Chem. Soc. 42,* 43–45.
43. Lie Ken Jie, M.S.F., Pasha, M.K., and Ahmad, F. (1996) Ultrasound-Assisted Synthesis of Santalbic Acid and a Study of Triacylglycerol Species in *Santalum album* (Linn.) Seed Oil, *Lipids 31,* 1083–1089.
44. Crombie, L., and Jacklin, A.G. (1957) Lipids. Part V. Total Synthesis of Ximenynic Acid, Homoricinstearolic Acid, and Two Fatty Hydroxy-Acids with Allenic Side-Branches, *J. Chem. Soc.,* 1622–1646.
45. Smith, G.N., Taj, M., and Braganza, J.M. (1991) On the Identification of a Conjugated Diene Component of Duodenal Bile as 9Z,11E-Octadecadienoic Acid, *Free Radic. Biol. Med. 10,* 13–21.
46. Lie Ken Jie, M.S.F., Pasha, M.K., and Alam, M.S. (1997) Synthesis and Nuclear Magnetic Resonance Properties of All Geometrical Isomers of Conjugated Linoleic Acids, *Lipids 32,* 1041–1044.
47. Morris, S.G., Magidman, P., and Herb, S.F. (1972) Hydrogenation of Triple Bonds to Double Bonds in Conjugated Methyl Octadecadiynoate and Methyl Santalbate, *J. Am. Oil Chem. Soc. 49,* 505–507.
48. Crombie, L. (1955) Amides of Vegetable Origin. Part V. Stereochemistry of Conjugated Dienes, *J. Chem. Soc.,* 1007–1025.
49. Adlof, R. The Lindlar-Catalyzed Reduction of Methyl Santalbate: A Facile Preparation of Methyl 9-*cis*,11-*trans*-Octadecadienoate-9,10-d$_2$, *J. Am. Oil Chem Soc. 76,* 301–304.
50. Hopkins, C.Y, and Chisholm, M.J. (1964) Isolation of a Natural Isomer of Linoleic Acid from a Seed Oil, *J. Am. Oil Chem. Soc. 41,* 42–44.
51. Takagi, T., and Itabashi, Y. (1981) Occurrence of Mixtures of Geometrical Isomers of Conjugated Octadecatrienoic Acids in Some Seed Oils: Analysis by Open-Tubular Gas–Liquid Chromatography and High-Performance Liquid Chromatography, *Lipids 16,* 546–551.
52. Kepler, C.R., and Tove, S.B. (1969) Linoleate *cis*-12,*trans*-11 Isomerase, *Methods Enzymol. 14,* 105–110.
53. Knothe, G., Bagby, M.O., Peterson, R.E., and Hou,C.T.. (1992) 7,10-Dihydroxy-8(*E*)-Octadecenoic Acid: Stereochemistry and a Novel Derivative, 7,10-Dihydroxyoctadecanoic Acid, *J. Am. Oil Chem. Soc. 69,* 367–371.
54. Adlof, R.O.(1998) Isotopically-Labelled Fatty Acids, in *Lipid Synthesis and Manufacture* (F. Gunstone, ed.) Ch. 2, pp. 46–93, Sheffield Academic Press, England.
55. Adlof, R.O., Walter, E.L., and Emken, E.A. (1997) Synthesis of Five Conjugated Linoleic Acid Isomers Labelled with Deuterium Atoms, in *Synthesis and Applications of Isotopically Labelled Compounds,* (Heys, J.R., and Melillo, D.G., eds.) John Wiley and Sons, New York, NY, pp. 387–390.

56. Adlof, R. (1997) Preparation of Methyl *cis*-9, *trans*-11- and *trans*-9, *trans*-11-Octadecadienoate-17,17,18,18-d_4, Two of the Isomers of Conjugated Linoleic Acid, *Chem. Phys. Lipids 88*, 107–112.
57. Cawood, P., Wickens, D.G., Iversen, S.A., Braganza, J.M., and Dormandy, T.L. (1983) The Nature of Diene Conjugation in Human Serum, Bile and Duodenal Juice, *FEBS Lett. 162*, 239–243.
58. Mounts, T. L., and Dutton, H. J. (1967) Methyl Esters of Unsaturated Fatty Acids Labelled with Tritium in the Methyl Group, *J. Label. Compd. Radiopharm. 3*, 343–345.
59. Glass, R. L. (1971) Alcoholysis, Saponification, and the Preparation of Fatty Acid Methyl Esters, *Lipids 6*, 919–925.
60. Crombie, L., and Holloway, S.J. (1985) The Biosynthesis of Calendic Acid, Octadeca-(8*E*,10*E*,12*Z*)-trienoic Acid by Developing Marigold Seeds: Origins of (*EEZ*) and (*ZEZ*) Conjugated Triene Acids in Higher Plants, *J. Chem. Soc. Perkins Trans. I*, 2425–2434.
61. Kepler, C.R., and Tove, S.B. (1967) Biohydration of Unsaturated Fatty Acids III: Purification and Properties of a Linoleate 12-*cis*,11-*trans*-Isomerase from *Butyrivibrio fibrisolvens*, *J. Biol. Chem. 242*, 5686–5692.
62. Kepler, C.R., Hirons, K.P., McNeill, J.J., and Tove, S.B. (1966) Intermediates and Products of the Biohydrogenation of Linoleic Acid by *Butyrivibrio fibrisolvens*, *J. Biol. Chem. 241*, 1350–1354.
63. Arcos, J.A., Oturo, C., and Hill, C.G. (1998) Rapid Enzymatic Production of Acylglycerols from Conjugated Linoleic Acid and Glycerol in a Solvent-Free System, *Biotech. Lett. 20*, 617–621.
64. Timmerman, F., Gauf, R., Gierke, J., Von Kries, R., Adams, W., and Sander, A. (1998) Synthetic Triglycerides Based on Conjugated Linoleic Acid, Their Manufacture and Use, German DE 19,718,245 (Cl. C07C69/58) 30 Jul 1998.
65. Scholfield, C.R. (1980) Argentation High-Performance Liquid Chromatography of Methyl Esters, *J. Am. Oil Chem. Soc. 57*, 331–334.
66. Dhar, P., and Bhattacharyya, D.K. (1998) Nutritional Characteristics of Oil Containing Conjugated Octadecatrienoic Fatty Acid, *Ann. Nutr. Metab. 42*, 290–296.
67. Hayashi, Y., Nishikawa, Y., Mori, H., Tamura, H., Matsushita, Y., and Matsui, T. (1998) Antitumor Activity of (10*E*,12*Z*)-9-Hydroxy-10,12-octadecadienoic Acid from Rice Bran, *J. Ferm. Bioeng. 86*, 149–153.
68. Gunstone, F.D., and Sykes, P.J. (1962) Fatty Acids, Part VII: the Synthesis of Hexadeca-8,10-dienoic, Octadeca-7,11-dienoic, and Eicosa-7,13-dienoic Acid, *J. Chem. Soc.*, 3055–3058.
69. Gunstone, F.D., and Lie Ken Jie, M. (1970) Fatty Acids. Part 24. The Synthesis of Ten Octadecadiynoic Acids and the Related *cis*,*cis*- and *trans*,*trans*-Octadecadienoic Acids, *Chem. Phys. Lipids 4*, 1–14.
70. Hussain, M.G., and Gunstone, F.D. (1977) Polyunsaturated Acids Part 1.The Synthesis of Hexadeca *trans*-2, *trans*-4, *trans*-6 trienoic (16:3) and Hexadeca-*trans*-2, *trans*-4, *trans*-6, *trans*- 8 tetraenoic (16:4) Acid, *Bangladesh J. Sci. Ind. Res. 12*, 215–218.
71. Wise, M.L., Hamberg, M., and Gerwick, W.H. (1994) Biosynthesis of Conjugated Triene-Containing Fatty Acids by a Novel Isomerase from the Red Marine Alga *Ptilota filicina*, *Biochemistry 33*, 15223–15232.
72. Stoffel, W., Schiefer, H.G., and Ditzer, R. (1966) Metabolism of Unsaturated Fatty Acids. Viii. Chemical Synthesis and Metabolism of *cis*-7, *cis*-9, *cis*-12- and *trans*-7, *cis*-9,*cis*-12-Octadecatrienoic-1-^{14}C Acids, *Z. Physiol. Chem. 345*, 52–60.

73. Chisolm, M.J., and Hopkins, C.Y. (1960) Conjugated Fatty Acids of *Tragopogon* and *Callendula* Seed Oils, *Can. J. Chem. 38*, 2500–2507.
74. Beerthuis, R.K., Nugteren, D.H., Pabon, H.J.J., Steenhoek, A., and van Dorp, D.A. (1971) Synthesis of a Series of Polyunsaturated Fatty Acids, Their Potencies as Essential Fatty Acids and as Precursors of Prostaglandins, *Recueil De Travaux Chimiques De Pays-Bas 90*, 943–960.
75. Fallis, A.G., Hearn, M.T.W., Jones, E.R.H., Thaller, V., and Turner, J.L. (1973) Synthesis of Polyacetylenic C18 and C16 Esters with 9-ene-12,14-diyne Unsaturation, and Their Labelling, *J. Chem. Soc. Perkin Trans. I, 743*–749.
76. Labelle, M., Falgueyret, J.-P., Riendeau, D., and Rokach, J. (1990) Synthesis of Two Analogues of Arachidonic Acid and Their Reactions with 12-Lipoxygenase, *Tetrahedron 46*, 6301–6310.
77. Burgess, J.R., de la Rosa, R.I., Jacobs R.S., and Butler, A. (1991) A New Eicosapentaenoic Acid Formed from Arachidonic Acid in the Coralline Red Algae *Bossiella orbigniana*, *Lipids 26*, 162–165.
78. Mikhailova, M.V., Bemis, D.L., Wise, M.L., Gerwick, W.H., Norris, J.N., and Jacobs, R.S. (1995) Structure and Biosynthesis of Novel Conjugated Polyene Fatty Acids from the Marine Green Alga *Anadyomene stellata*, *Lipids 30*, 583–589.
79. Barrot, M., Fabrias, G., and Camps, F. (1994) Synthesis of {16,16,16-^{2}H$_3$}11-Hexadecynoic Acid and {15,15,16,16,16,^{2}H$_5$}(Z,Z)-11,13-Hexadecadienoic Acid and Their Use as Tracers in a Key Step of the Sex Pheromone Biosynthesis of the Processionary Moth, *Tetrahedron 50*, 9789–9796.
80. Tulloch, A.P. (1979) Synthesis of Methyl [18-^{2}H$_3$],[17-^{2}H$_2$],[16-^{2}H$_2$],[14-^{2}H$_2$] and [12-^{2}H$_2$] *cis*-9-Octadecenoates, *Chem. Phys. Lipids 23*, 69–76.
81. Russell, S. W., and Pabon, H.J.J. (1982) Synthesis of (*R,S*)-(5Z,8E,10E)-12-Hydroxyheptadeca-5,8,10-trienoic Acid and of (*R,S*) and (*S*)-(5Z,8Z,10E,14Z)-12-Hydroxyeicosa-5,8,10,14-tetraenoic Acid and Their Racemic 5,6,8,9-Tetradeuterioisomers, *J. Chem. Soc. Perkin Trans. I, 545*–552.
82. Hamberg, M. (1992) Metabolism of 6,9,12-Octadecatrienoic Acid in the Red Alga *Lithothamnion corallioides:* Mechanism and Formation of a Conjugated Tetraene Fatty Acid, *Biochem. Biophys. Res. Commun. 188*, 1220–1227.
83. Gerwick, W.H., Asen, P., and Hamberg, M. (1993) Biosynthesis of 13*R*-Hydroxyarachidonic Acid, an Unusual Oxylipin from the Red Alga *Lithothamnion corallioides*, *Phytochemistry 34*, 1029–1033.

Chapter 4
Commercial Production of Conjugated Linoleic Acid

Martin J.T. Reaney, Ya-Dong Liu, and Neil D. Westcott
  Agriculture and Agri-Food Canada, Saskatoon, SK, Canada S7N OX2

Introduction

Industrial conjugated linoleic acid (CLA) is a poorly defined blend of compounds (1). Early commercial syntheses focused on maximizing total CLA content. Many early products were rich in CLA but contained a number of positional isomers. Market demand has now shifted toward a product that contains two predominant isomers, specifically *cis,trans*-9,11-octadecadienoic acid and *trans,cis*-10,12-octadecadienoic acid. It is not surprising that alkali isomerization produced some undesirable positional isomers of CLA. In 1970, Mounts and Dutton (2) showed unequivocally that when potassium *t*-butoxide was used, at least four positional isomers of CLA were produced. It was not until 1997, after the use of CLA as a dietary supplement began, that Christie *et al.* (1) elegantly demonstrated that commercial CLA was a blend of positional isomers. In response to this discovery, new commercial CLA products have been introduced that have comparatively high levels of the preferred isomers. In spite of the improvements, all currently available commercial CLA products contain some level of the less desirable isomers and other components that may or may not be desirable.

Commercial processes for the synthesis of any compound of economic value normally constitute proprietary information and the commercial methods of CLA production are no exception. The process by which each brand of commercial CLA is synthesized is not known by the authors of this review. Therefore, this review is directed at the patent literature on CLA synthesis, major problems encountered in CLA synthesis, and analysis of CLA from commercial suppliers.

CLA Production Raw Materials

The raw material for CLA production must be a material that is rich in linoleic acid. This product could be in the form of triglycerides, fatty acids, or fatty acid esters. The concentration of CLA in the final product is directly dependent on the level of linoleic acid in the starting material. The highest level of linoleic acid available from botanical sources is not available in commercial products. Extraction and refining equipment would be required to obtain oils with the highest linoleic acid levels. Table 4.1 lists the commercial and noncommercial sources of oil and fatty acids that are known to be rich in linoleic acid and their availability as triglycerides and fatty acids.

If commercial CLA is to be synthesized from a fatty acid, it must be recognized that commercial fatty acids are generally not intended for use in the production of

TABLE 4.1 Linoleic and Linolenic Acid Contents of Some Vegetable Oils[a]

Oil source	% Linoleic acid	% Linolenic acid	Commercial oil available	Commercial fatty acid available
Corn	57	0	yes	
Cottonseed	53	0	yes	
Cucumber	72	0	no	
Grapeseed	70	0	yes	
Linola flaxseed	72	3	yes	
Poppy	77	0	yes	
Safflower	75	0	yes	
Sunflower	64	0	yes	yes
Soybean	51	8	yes	yes
Squash (pumpkin)	60	0	yes	
Walnut	62	12	yes	

[a]Source: Ref. 4.

CLA. Commercial fatty acids are usually produced by the reaction of water (steam) and triglyceride oil at high temperatures in a continuous reaction (Reaction 4.1). This reaction is accelerated through the use of a solid phase acid catalyst, which is readily separated from the fatty acid and glycerol products after hydrolysis (3). The disadvantage of this hydrolysis process is that the reaction is reversible, and products generated by this process contain appreciable amounts of mono- and diglycerides, which may have undesirable side reactions in CLA synthesis.

Fatty acids may also be produced by the hydrolysis of triglycerides in a pressurized reactor at 200°C without the addition of a catalyst (3). This reaction may be catalyzed at lower temperatures using zinc oxide in a batch reactor (3). The product of these batch reactions also contains substantial amounts of mono- and diglycerides.

Hydrolysis of triglycerides is possible using water and a strong base to produce soaps. This reaction proceeds to completion and can be conducted at the modest temperatures required to maintain the reaction mixture as a fluid. More than 3 mol of potassium hydroxide or sodium hydroxide are required to hydrolyze 1 mol of triglyceride oil. Because the caustic alkali cannot catalyze the reverse reaction, this process can produce soaps that are virtually free of glycerides in a single step (3). The soaps from alkali hydrolysis of triglycerides are readily converted to fatty acids by acidification with the addition of citric acid or strong mineral acids, including HCl, H_2SO_4, or H_3PO_4. Regardless of the method chosen for production of fatty acids, the acids should be dried under vacuum after washing with brine or by a combination of other acceptable methods (3).

There are commercially available fatty acids suitable for use in CLA production. For example, Henkel sells a series of fat products including those shown in Table 4.2 (5). However, none of the products listed in Table 4.2 would be preferred as starting materials for CLA production for reasons that will be discussed.

Reaction 4.1. Alkali or acid hydrolysis of glycerides.

TABLE 4.2 Henkel Products[a]

Product	14:0	16:0	18:0	16:1	18:1	18:2	18:3
				(%)			
Soy	0.5	16	4	1	25.5	48	5
Linoleic	0.5	3.5	0.5	Tr	19.5	65.5	10.5
Methyl	0.5	3.5	0.5	Tr	19.5	65.5	10.5

[a]Source: Ref. 5.

Enrichment of Linoleic Acid

A commercial interest may wish to produce CLA at concentrations greater than can be obtained by modifying high linoleic acid plant oils. Several methods exist that will improve the starting material by increasing the concentration of linoleic acid, but only a few methods are used in industrial settings. Industrial separation of fatty acids has been reviewed by Fritz and Johnson (6,7). A limited discussion of these methods will be presented.

Crystallization is used to separate saturated fats and oleic acid from linoleic acid. If a highly concentrated product is required, the linoleic acid may be crystallized once or repeatedly as the last step in purification. Crystallization is a mild procedure but usually requires the use of a solvent (8) such as acetone or methanol. The use of low boiling point and flammable solvents raises concerns over plant safety, government regulations on manufacturing, and market acceptance of the product. Furthermore, the removal of oleic acid by crystallization in solvent is possible only by lowering the temperature of the liquor to below −40°C (8). To crystallize linoleic acid, the temperature must be reduced to −75°C.

Dry or solvent free crystallization is also possible, but these methods often require the addition of crystal modifiers, which become incorporated into the product (8).

Losses during crystallization can be very high because the crystals entrain large amounts of fatty acid. However, these losses may be reduced by physically pressing the crystals to remove the entrained solution (9). Linoleic acid–rich products of dry crystallization would be preferred starting materials for CLA production over solvent-crystallized products, but the losses incurred in dry crystallization may prohibit this method of manufacture. Crystal modifiers may be selected so that they do not adversely affect the quality or acceptance of the final product.

Specific fatty acids may be concentrated by sequentially removing contaminating fatty acids as urea adducts and forming the urea adduct of the desired fatty acid. This process requires dissolving the fatty acids or esters in urea and hot methanol (or other alcohol) and cooling to effect adduct formation. The adduct is filtered from the liquor and if conditions are carefully controlled, the adducts can be used to sequentially crystallize saturates, monounsaturates, diunsaturates, and triunsaturates. A urea adduct rich in linoleic acid could be produced by first removing adducts of saturates and monounsaturates from a suitable oil and then forming the desired adduct. Once formed, the adduct may then be decomposed by the addition of water to the solid phase. Enriched linoleic acid could be recovered by solvent extraction of the urea:water solution with a nonpolar solvent such as hexane. All problems associated with crystallization in solvent mentioned previously also occur in the formation of urea adducts with the exception of the requirement for very low temperatures. Typically, urea adducts form between room temperature and 0°C (8).

Fatty acids may also be enriched by the use of various absorption media. Molecular sieves can separate saturated fatty acids from unsaturated fatty acids dissolved in acetone (10). Oleic acid and linoleic acid dissolved in blends of solvents including acetonitrile, tetrahydrofuran, water, and formamide may be separated with the use of cross-linked polystyrene polymers such as Amberlite XAD-2 or XAD-4.

Selective extraction methods using two-phase solvent systems may also be used to enrich fatty acids. Solvent systems such as dimethyl formamide(DMF), hexane, and ethylene glycol can form a two-phase system that effectively partitions sunflower oil triglycerides rich in linoleic acid from those depleted in linoleic acid (11). Triglycerides partitioned in this way may contain up to 84.7% linoleic acid. This method would not likely be used because the magnitude of the losses is usually unacceptable in industry.

Approaches to CLA Production

CLA has been produced by the reaction of soaps with strong alkali bases in alcohol, ethylene glycol, and glycerol (12,13,14; Reaction 4.2). The CLA product is generated by acidification of the soap solution with a strong acid (sulfuric or hydrochloric acid) and repeatedly washing the product with brine or an aqueous $CaCl_2$ solution.

CLA has been synthesized from fatty acid and soap blends using SO_2 in the presence of a substoichiometric amount of soap-forming base (15). This reaction produced predominantly the all-*trans* configuration of CLA.

Reaction 4.2.
Isomerization of cis,cis-9,12-octadecadienoyl soaps.

Of these methods, alkali isomerization of soaps is the least expensive process for bulk preparation of CLA isomers; however, the use of either monohydric or polyhydric alcohols in alkali isomerization of CLA can be problematic. Lower alcohols are readily removed from the CLA product, but they require that the production facility be constructed to support the use of flammable solvents. Higher molecular weight alcohols and polyhydric alcohols are considerably more difficult to remove from the product, and residual levels of these alcohols (e.g., ethylene glycol) may not be acceptable in the CLA product.

Water may be substituted for the alcohols in the production of CLA by alkali isomerization of soaps (16,17). When water is used in this reaction, it is necessary to perform the reaction in a pressure vessel, whether in a batch (16) or continuous mode of operation (17). The process for synthesis of CLA from soaps dissolved in water still requires a complex series of reaction steps. Bradley and Richardson (18) were able to produce CLA directly from triglycerides by mixing sodium hydroxide, water, and oil in a pressure vessel. Their method eliminated the need to synthesize fatty acids followed by soap formation before the isomerization reaction. However, they reported that they were able to produce an oil with only 40% CLA. Quantitative conversion of the linoleic acid in soybean oil to CLA would have produced a fatty acid mixture with ~51% CLA.

Reaction Kinetics and Production of Positional Isomers

The kinetics of conversion of linoleyl soaps to conjugated linoleyl soaps have been well described by first-order reaction kinetics (19). Total conjugation is readily measured by simple methods such as increases of ultraviolet (UV) absorbance at 231.5 nm. In industry, it is necessary to allow the reaction to proceed until most of the linolyl soaps are conjugated, but desirable to stop the reaction as soon as possible after that point. The reaction is between 97 and 98.5% complete after 5 and 6 half-lives, respectively. There is little advantage in continuing the reaction longer than this time and, as will be discussed below, undesired reactions may occur with longer reaction times. The reaction constant is readily determined early in the reaction when changes in the level of conjugation are large. The task of determining the rate constant in a large reactor is complicated by the mass of the reactor and its contents. A large batch reactor often requires several hours to reach the optimum heat of reaction at which time the reaction may be almost complete. Similarly, cooling the contents of a large reactor as a means of stopping a reaction is usually impractical. A reaction that nears 99% completion in 2 h in a laboratory has a half-life of <20 min. Control of a batch reaction in a commercial operation may utilize analytical data from periodic sampling, but the analytical method must be very rapid to be an effective tool for decision making.

The authors have found that the half-life of the reaction may vary by ~10% per 1°C. It follows that precise reactor temperature control is essential to standardize quality control. A reaction planned to continue for 6 half-lives could vary from 5 to 7 half-lives if the reactor temperature control is ± 2°C.

Just as the formation of total conjugated linoleyl soaps from linoleyl soaps is governed by first-order kinetics, the isomerization reaction that leads to the production of positional isomers (Reaction 4.3) has a similar mechanism and probably has similar kinetics. Using this assumption, we have modeled the sequential conversion of linoleyl soaps to a mixture of *cis,trans*-9,11-octadecadienoyl and *trans,cis*-10,12-octadecadienoyl soaps (Reaction 4.2) and finally, the conversion of these two soaps to 8,10-octadecadienoyl and 11,13-octadecadienoyl soaps (Reaction 4.3). We expect a

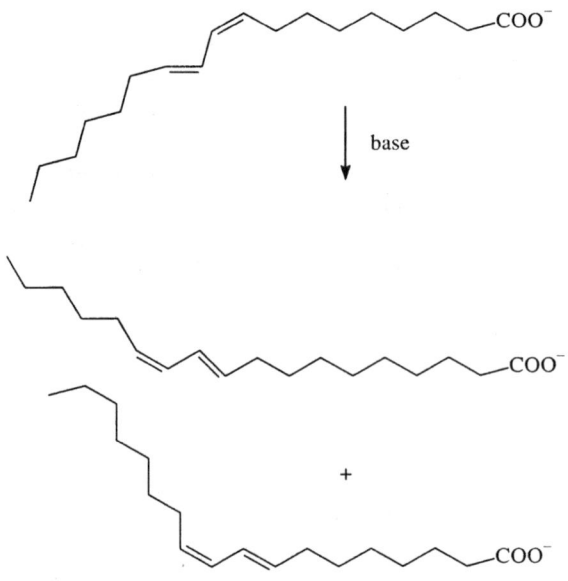

Reaction 4.3. Isomerization of *cis,trans*-9,11-octadecadienoyl soap to *trans,cis*-8,10-octadecadienoyl and *trans,cis*-10,12-octadecadienoyl soaps.

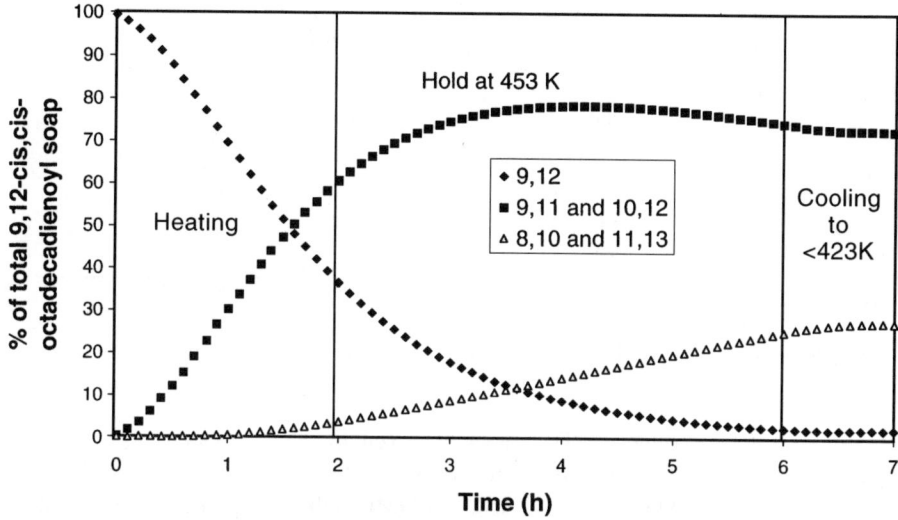

Fig. 4.1. Model of conversion of linoleic acid to a mixture of conjugated fatty acids in a commercial batch reactor. The model depicts disappearance of linoleic acid and the formation of 9,11 and 10,12 CLA isomers followed by the formation of 8,10 and 11,13 isomers as expected in a sequential reaction. The reaction kinetics are based on actual experimental data. Details of the reaction mixture (reactor contents) of fatty acids, solvent and catalyst are proprietary. The temperature regime in the reactor was based on heating rapidly from 293 to 453K. Approximately 2 h is required to heat the rector from 433 to 453K, at which the reaction rate is significant. The reactor was held at reaction temperature for 4 h and subsequently cooled to below 433K in 1 h.

considerable interconversion of 9,11 and 10,12 soaps and accumulation of other isomers by a similar mechanism, but these reactions are not modeled here.

In our model, the reaction mixture is heated from 423K to 453K over a 2-h period and then held at 453K for 4 h before cooling to 423K in just 1 h (Fig. 4.1). The model and the reaction conditions were selected to demonstrate how isomers arise and do not demonstrate the best conditions under which to control the production of positional isomers. In the model and in our experience, isomer formation can be suppressed by limiting the duration of the reaction. In our experience, if the market required a CLA mixture that was essentially free of undesirable isomers (<1%), the reaction would have to be limited to <70% conversion. The choice of reaction medium, catalyst, and temperature all affect the quantity of undesirable isomers. The optimum conditions of reaction are the proprietary information of each CLA manufacturer.

Solvents Used in Production of CLA by Alkali Catalysts

Research reports describe the use of at least eight solvent systems for the production of CLA using alkali catalysts (Table 4.3). The choice of solvent greatly affects the reaction

TABLE 4.3 Summary of Solvents Used in Conjugated Linoleic Acid Production[a]

Solvent (reference)	Mol wt.	B.P. (760)	B.P. (Vac)	Temp (°C)	Time (h)
Ethylene glycol (20)	62	198	93[13d]	180	2
Glycerol (20)	92	290	182[20]	180	0.75
Propylene glycol (21)	76	189	97[21]	170	2.5
t-Butyl alcohol (2)	74	82	Low	90	4
Water (18)	18	100	Low	225	2.5
DMSO (22)	78	189	83[17]	30	1.5
DMF (22)	73	153	76[39]	—	—

[a]Abbreviations: DMSO, dimethyl sulfoxide; fp, flash point; DMF.
[b]Color is as described in references.
[c]Phase separation = Yes if a two-phase system is formed after acidification of the soap.
[d]Superscript numbers refer to vacuum in mm Hg.

conditions of CLA production. The choice of solvent by the manufacturer is determined by a number of considerations. Many markets will not accept low levels of ethylene glycol, t-butanol, dimethyl sulfoxide (DMSO), or DMF in the final product. This limitation could restrict the choice of solvents to only glycerol, propylene glycol, water, and ethanol. The reaction in water and ethanol proceeds only above the boiling temperature of these solvents; therefore, a pressure reactor would be required to operate using these solvents. Glycerol is expensive but it could be recovered from a commercial operation that produces its own fatty acids. The quality of glycerol necessary to produce high quality CLA has not been investigated, but refining this glycerol stream to remove the salt might prove difficult in a small operation. Recovery would have to be very efficient because fatty acid production generates only 10% of the weight of the oil as glycerol.

Catalyst Selection

Numerous catalysts have been used in the production of CLA. We have found that hydroxides of lithium, sodium, and potassium are all capable of generating CLA in various solvents. Because fatty acids neutralize the catalyst, it is necessary to add at least 1 mol of catalyst for every mole of fatty acid in the reaction to ensure that soap is generated. We have found that on a molar basis, potassium hydroxide has proven to be a more effective catalyst than sodium hydroxide, with lithium hydroxide the least effective and not suited for industrial CLA production. On a weight basis, sodium and potassium hydroxide have similar efficiency of conversion. Although sodium hydroxide is much less expensive than potassium hydroxide, the disposal costs for the waste neutralized alkali should also be considered. Potassium salts are easily utilized as fertilizer and can be applied to fields, whereas sodium salts cannot be disposed of in a similar fashion.

The effective form of the catalyst is not necessarily determined by the nature of the catalyst added but rather by the solvent used. When water is used as the solvent and sodium ethoxide is the catalyst, the effective form of the catalyst is likely the hydroxide ion. If t-butanol is used as the solvent and sodium methoxide is added to cat-

% Reaction	Phase separation[c]	Color[b]	Toxicity	Other
100	Yes	Poor	Yes	
100	Yes	Good	None	Viscous
99.1	Yes	Good	Minimal	
98.5	No	Unknown	Yes	High pressure
40	Yes	Good	None	High pressure
78	No	Unknown	Yes	fp 95
—	No	Unknown	Yes	fp 67

alyze the reaction at 90°C, the reaction mixture will quickly release methanol vapors and the *t*-butoxide ion will become the effective catalyst.

Water consumes alkoxide catalysts; in industrial production, maintaining alkoxide catalysts in a water-free environment is difficult. Water is produced by the neutralization of fatty acids with an alkali hydroxide. Because many alkoxide catalysts contain some alkali hydroxide, it is not uncommon for the catalyst to be consumed by this reaction.

Reaction Vessels

Alkali isomerization of linoleyl soaps requires a containment vessel that is tolerant of both heat and caustic. When a low boiling point solvent such as ethanol or water is used, the vessel must also be capable of maintaining the reaction under pressure. There are a limited number of materials that will meet these criteria. Polytetrafluoroethylene and other fluoropolymers are capable of withstanding both the heat of the reaction and the caustic environment, but they cannot withstand pressure and are poor heat conductors. Fluoropolymer-coated parts and nickel and nickel alloys such as Monel may be used in the construction of reaction vessels for production of CLA. The high cost of these materials rules out their use in construction of large batch reactors. Furthermore, none of these materials has sufficient strength for use in pressure reactors if a reaction in water or alcohol is planned. The preferred choice for reactor construction is nickel-plated steel, which has the desired strength, heat transfer, and chemical properties for conducting reactions in strong caustic solutions. A coated vessel of this design requires regular inspection because a flaw in the coating could lead to vessel failure.

Microbial Production of CLA

Pariza and Yang (25) have recently described the microbial production of *cis*, *trans*-9,11-octadecadienoic acid from linoleic acid using cultures of *Lactobacillus* sp. In their patented method, early stationary phase *Lactobacillus* cultures were incubated with linoleic acid dissolved in propylene glycol. A total CLA level of 7 mg/g cells was produced, which was >96% *cis,trans*-9,11-octadecadienoic acid. This type of conversion may lead to improved CLA products in the future.

Synthesis of CLA by Dehydration of Ricinoleic Acid (12-Hydroxy-*cis*-9-octadecadienoic Acid)

The most attractive method for production of pure *cis,trans*-9,11-octadecadienoic acid is through the dehydration of ricinoleic acid. Synthesis from this relatively inexpensive starting material has proven elusive because it is difficult to control the formation of dehydration products (26). Synthesis of *cis,trans*-9,11-octadecadienoic from ricinoleic acid has been reported (27); although it is an efficient reaction, it uses expensive elimination reagents such as 1,8-diazobicyclo-(5,4,0)-undecene. For most applications, the high cost of the elimination reagent increases the production cost beyond the level at which commercial production of CLA is economically viable.

The Quality of Commercial CLA Products

The fate of other fatty acids and minor components during processing has not been investigated. The conditions used to conjugate linoleic acid have little or no effect on either monounsaturated or saturated fatty acids; however, any polyunsaturated fatty acids may be conjugated. The products of the reaction of alkali catalysts on these fatty acids is more complex than that discussed for linoleic acid (Reaction 4.4), and it will not be discussed except to note that these reactions may produce undesirable products.

From our observations, glycerol does not form undesirable compounds under the conditions of alkali isomerization of linoleic acid. However, we have found that a number of commercial fatty acids and CLA preparations contain appreciable levels of monoglycerides. Monoglycerides themselves are not toxic, but it is possible that toxic compounds (such as ethylene glycol) may be incorporated into the final product through alcoholysis of the monoglyceride by the low volatility alcohols used as a reaction medium (Reaction 4.5). Nuclear magnetic resonance (NMR) and liquid chromatography analysis of CLA samples from all sources indicated that although CLA was the predominant compound, CLA esters and other unknown compounds may also have been formed.

The minor components of vegetable oil such as tocopherol, sterol, or squalene are stable to heat and strong alkali. However, tocopherol and other components present in vegetable oil react readily with oxygen in the presence of metals (28). Tocopherols stabilize vegetable oils against oxidation and tocopherol loss through processing; otherwise, high metal content may lead to a decreased shelf life of the CLA product. Three commercial CLA products were analyzed for metal contents using Inductively Coupled Plasma Spectometry (ICP). The total metal contents are given in Table 4.4.

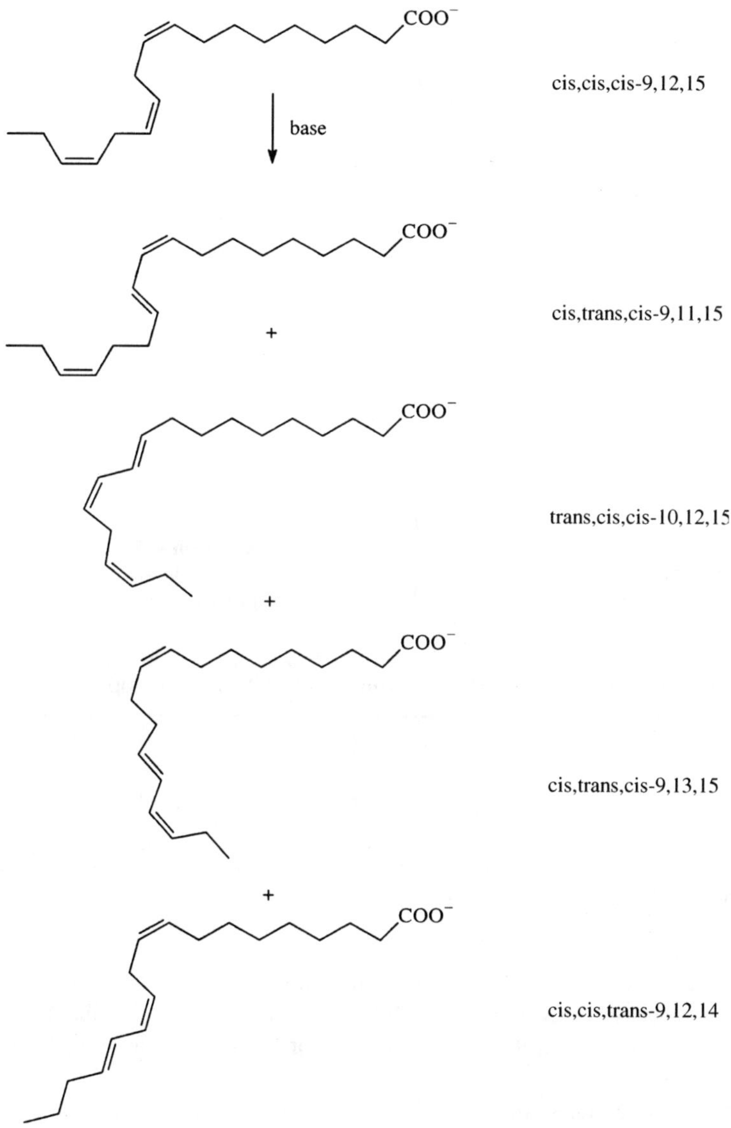
COO⁻
cis,cis,cis-9,12,15
base
COO⁻
cis,trans,cis-9,11,15
+
COO⁻
trans,cis,cis-10,12,15
+
COO⁻
cis,trans,cis-9,13,15
+
COO⁻
cis,cis,trans-9,12,14

Reaction 4.4. Isomerization of *cis,cis,cis*-9, 12, 15-octadectrienoyl soap to isomers.

The CLA samples tested were free of potentially toxic metals at the levels tested, with the exception of a trace level of barium in product C. Products A and B have very low total metal contents, whereas product C has appreciable levels of calcium. The soaps probably arise from a $CaCl_2$ wash in a late stage of processing.

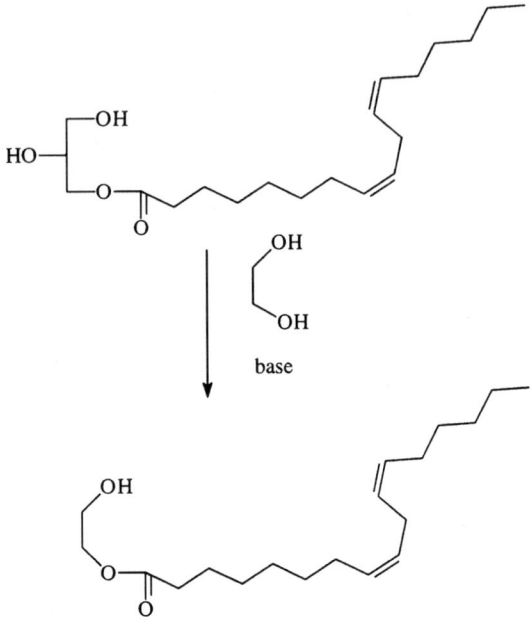

Reaction 4.5. Alcoholysis of monoglyceride with ethylene glycol.

TABLE 4.4 Metal Contents of Three Commercial CLA Products (ppm)[a]

	Na	K	Fe	Cr	B	P	Ca	Ba	Mg	Li	Total
A	3	3	1	0	5	2	5	0	1	0	20
B	2	28	0	0	3	8	1	0	1	0	43
C	4	6	11	8	3	3	917	1	2	1	956

[a]Not detected: Si, Pb, Cu, Sn, Al, Ni, Ag, Ti, Zn, Mo, V, Sb, Be.

Product C also contains iron and chromium, suggesting the use of a stainless steel reactor for processing. Even these small levels of metals may contribute to rapid oxidation of product C, and it probably has a shortened shelf life compared with the other products.

Three commercial samples of CLA were subjected to a series of tests that are reported in Table 4.5. Viscosity was measured using ASTM test D445 (29). Large differences in viscosity were not found, but it was presumed that increased viscosity might be related to increased oxidation, if it had occurred. The color of the three samples is reported, but it is only a subjective statement regarding the apparent quality of the three materials. Included in the color analysis is the absorbance maximum observed for CLA dissolved in hexane (1:1000 CLA/hexane) using a Cary UV/Vis spectrometer. The UV spectra indicated that conjugated dienes were the major source of absorption of the oil samples.

TABLE 4.5 Summary of Analyses of Three Commercial Conjugated Linoleic Acid Products[a]

	Viscosity cSt (40°C)	Color	¹H-NMR (400 MHz)	Size exclusion	RP-LC	GC-FID	Isomer mix
A	28.9	Clear $\lambda = 231.5$ Å $= 0.720$	Glycerides	Glycerides	Polar contam.	Contam.	Good 95%
B	27.9	Light yellow $\lambda = 231.5$ Å $= 0.667$	No glycerides	No glycerides	Polar contam.	OK	Poor 70%
C	29.7	Brown $\lambda = 231.5$ Å $= 0.676$	No glycerides	No glycerides	Nonpolar contam.	OK	Good 90%

[a]Abbreviations: NMR, nuclear magnetic resonance; RP-LC, reversed-phase liquid chromatography; GC-FID, gas chromatography-flame ionization detection.

Proton and ^{13}C NMR spectra were obtained from the three commercial samples. A series of unknown spectral components were observed in product A at 3.68, 3.98, 4.08, 4.15, 5.05, and 5.17 ppm. Comparison of the unknown peaks with the spectra of monoglycerides indicated that most of the observed peaks correlated with those present in the spectra of monoglycerides. An exact assignment of the spectra was not attempted because many forms of monoglycerides are possible. Observation of the spectral region of the olefinic protons revealed that the three samples were predominantly *cis,trans* or *trans,cis* fatty acids (30). We have found that a convenient measure of the CLA content of an oil can be obtained by comparing the integrated values of the protons at 6.0 and 6.3 ppm with the integrated value of the α-methylene group (adjacent to the carbonyl) at 2.3 ppm. The olefinic protons were chosen because they do not overlap with oleic acid olefinic protons or olefinic protons from conjugated linoleic acid with *trans,trans* configurations. The ratio of *cis,trans* olefin to α-methylene protons is a useful measure of CLA purity.

Carbon-13 olefinic carbons were observed at 100 MHz. The carbon-13 spectrum clearly demonstrates the formation of positional CLA isomers. Because pure standards were not available, it was not possible to unequivocally assign the spectra of 8,10- or 11,13-CLA, but it is clear that these isomers are predominantly *cis,trans* or *trans,cis* fatty compounds.

Size exclusion chromatography was performed on a Waters GPC-StyragelHR 0.5 column (7.8 × 300 mm) at 35°C using tetrahydrofuran as a solvent flowing at 1.0 mL/min and detecting compounds with both UV absorbance and evaporative light scattering detection (ELSD, 40°C, 4.2 L/min gas flow). The goal of this analysis was to observe polymerization of the CLA products or the occurrence of glycerides. Under these conditions, 18-carbon fatty acids had a retention time of 6.6 min and monoglycerides had a retention time of 5.9 min. All three samples had two peaks observed by

ELSD, one at the position expected for 18-carbon fatty acids and the other as expected for their respective monoglycerides. Product A had the largest peak at the position expected for monoglycerides. Observations made using UV absorbance at 233 nm reflected the observations made with the ELSD.

Reversed-phase liquid chromatography was performed on a 150 mm (3.0 mm Waters Symmetry C-8 column (Waters, Milford, MA) at 30%C and a flow rate of 1.0 mL/min. The solvent phase was acetonitrile/tetrahydrofuran/0.1% aqueous phosphoric acid (50.4:21.6:28 by vol). Under these operating conditions, most of the UV absorbance occurred as a peak at 3.83 min for all samples. Chromatograms of all samples had some small peaks (presumably the more polar compounds) eluting before the major peak. Product C presented a small but significant UV absorbing peak that eluted after the peak at 3.83 min.

Rapid Analytical Methods

Industrial CLA syntheses must be controlled to both maximize the content of preferred isomers such as 9,11-*cis*,*trans*-octadecadienoic acid and minimize the formation of undesirable isomers. For the analysis to be useful, the results of the analysis should be available on-line or as quickly as possible. The authors are not aware of any existing on-line tests for the quality of CLA preparations, but several methods may have promise for use off-line.

The potential off-line analytical methods include UV, Fourier transform infrared spectroscopy (FTIR), proton NMR, ^{13}C NMR, gas chromatography (GC), and capillary electrophoresis. With the exception of capillary electrophoresis and UV absorbance, none of these methods can effectively analyze soap solutions; therefore, for most analytical methods, neutralization of soaps would be necessary.

A reaction medium that contains 40% soaps by weight can usually be dissolved in ethanol. We have found that 100 mg of reaction mixture will dissolve totally in 10 mL of 95% ethanol when glycerol, water, or ethylene glycol is the reaction solvent and alkali hydroxides are used as the catalyst. It is then possible to dilute the reaction mixture solution 1000-fold to determine the UV absorbance at 231.5 nm. When there is no other interfering UV absorbance, this method is an excellent indication of the total conjugated double bonds. This method is also sensitive to the presence of conjugated linolenic acids derived from linolenic acid, which are indicated by a UV absorbance at 268 nm. None of the samples observed showed three conjugated double bonds.

To obtain more detailed information regarding the composition of fatty acids requires additional analytical methods some of which require extensive and time-consuming sample preparation. For these methods, we have found that it is possible to rapidly prepare a free fatty acid fraction. Alkali soaps from most reaction mixtures are readily dissolved in a mixture of hexane and ethanol (1:1) or ethanol alone. When the soaps are neutralized by the addition of hydrochloric acid and water, a two-phase system is evolved. The fatty acids remain in the nonpolar phase, whereas the polar solvent used in the reaction medium dissolves in the water. The solution of dissolved fatty acids can be injected directly onto a GC column specifically designed for separation of free fatty acids, such as the DB-FFAP column, or a suitable nonpolar GC column, such as the HP-5 column. Analysis by chromatography without derivitization affords the potential for rapid analytical feedback. We have found that a 30 m, DB-FFAP column (0.32 mm i.d., 0.25 μm film) gave baseline resolution of most fatty acids without derivitization (program 50°C for 1 min, 50–200°C at 25°C/min, 20–240°C at 5°C/min, hold 10 min, flow He carrier 3 mL/min). Similarly, the nonpolar HP-5 column (0.32 mm i.d., 0.25 μm film) also gave good resolution of underivatized fatty acids with some tailing (program 50°C for 1 min, 50–150°C at 25°C/min, 150–290°C at 10°C/min, hold 6 min, flow He carrier 2 mL/min). These columns also partially separate isomers of CLA. It was possible to observe the formation of the all-*trans* isomers, but detailed analysis of positional isomers was not possible without additional effort.

Conclusions

Market forces have generated an environment that has pushed industry to consistently improve the commercial CLA product. We have observed that although the content of desirable isomers in commercial CLA products has improved, there is still a demand for highly enriched or pure *cis,trans*-9,11-octadecadienoic acid products. The kinetic control of CLA synthesis will allow the development of CLA products that are virtually free of isomers other than 9,11 and 10,12. Kinetic control of reactions requires exceedingly rapid analytical techniques that can be applied inexpensively and on-line or virtually on-line.

References

1. Christie, W.W., Dobson, G., and Gunstone, F.D. (1997) Isomers in Commercial Samples of Conjugated Linoleic Acid, *J. Am. Oil Chem. Soc. 74,* 1231.
2. Mounts, T.L., and Dutton, H.J. (1970) Conjugation of Polyunsaturated Fatty Acids, *Lipids 5,* 997–1005.
3. Johnson, R.W., and Daniels, R.W. (1996) Carboxylic Acids (Manufacture), in *Kirk Othmer Encyclopedia of Chemical Technology,* vol. 5, 4th edn., pp. 168–177, John Wiley and Sons, New York.
4. Padley, F.B., Gunstone, F.D., and Harwood, J.L. (1994) Occurrence and Characteristics of Oils and Fats, in *The Lipid Handbook,* (Gunstone, F.D., Harwood, J.L., Padley, F.B., eds.) 2nd edn., pp. 47–224, Chapman and Hall, London.
5. Henkel Corporation (1991) Specifications and Characteristics of Emery Oleochemicals, Emery Group, Technical Bulletin A-473, Cincinnati, OH.
6. Fritz, E., and Johnson, R.W. (1988) *Fatty Acids in Industry,* Marcel Dekker, New York.
7. Johnson, R.W. (1990) *World Conference on Oleochemicals into the 21st Century,* (Applewhite, T.H., ed.) p. 189, American Oil Chemists' Society, Champaign, IL.
8. Gunstone, F.D., Kates, M., and Harwood, J.L. (1994) Separation and Isolation Procedures, in *The Lipid Handbook,* (Gunstone, F.D., Harwood, J.L., Padley, F.B., eds.) 2nd edn., pp. 47–224, Chapman and Hall, London.

9. Keulemans, C., Counter Current Dry Fractional Crystallization, European Patent Application 90201235.0 (1990).
10. Cleary, M.T., Kulprathipanja, S., and Neuzil, R.W., Process for Separating Fatty Acids, U.S. Patent 4,524,029 (1985).
11. Cleary, M.T., and Kulprathipanja, S., Process for Separating a Monoethanoid Fatty Acid, U.S. Patent 4,353,838 (1982).
12. Kirshner, H. (1946) Conjugated Unsaturated Fats, U.S. Patent 2,389,260, cited in *Chemical Abstracts* 1682-1.
13. Burr, G.O. (1941) Producing Conjugation in Unconjugated Fatty Polyene Compounds, U.S. Patent 2,242,230, cited in *Chemical Abstracts* 5732-5.
14. Crawley, J.D., Procedure for the Preparation of Substances Containing Conjugated DoubleBonds, U.S. Patent 2,343,644 (1944).
15. Struve, A. Process for the Conjugation of the Double Bonds of Polyunsaturated Fatty Acids and Fatty Acid Mixtures, U.S. Patent 4,381,264 (1983).
16. Bradley. T.F., Alkali-Induced Isomerization of Drying Oils and Fatty Acids, U.S. Patent 2,350,583 (1944).
17. Krajca, K.E., Flow Process for Conjugating Unconjugated Unsaturation of Fatty Acids, U.S. Patent 4,164,505 (1979).
18. Bradley. T.F., and Richardson, D. (1942) Drying Oils and Resins. Alkali-Induced Isomerization of Drying Oils and Fatty Acids, *Ind. Eng. Chem. 34*, 237–242.
19. Nichols, P.L., Jr., Riemenschneider, R.W., and Herb, S.F. (1950) Kinetics of Alkali Isomerization of Linoleic, Linolenic and Arachidonic Acids, *J. Am. Oil Chem. Soc. 27*, 329–336.
20. Brice, B.A., Swain, M.L., Herb, S.F., Nichols, P.L., Jr., and Reimenschneider, R.W. (1949) Standardization of Spectrophotometric Methods for Determination of Polyunsaturated Fatty Acids Using Pure Natural Acids, *J. Am. Oil Chem. Soc. 29*, 279–287.
21. Iwata, T., Kamegai,T., Sato,Y., Wantanabe, K., and Kasai, M., Method for Producing Conjugated Linoleic Acid, Canadian Patent 2,219,601 (1997).
22. Unknown source as cited in Iwata, T., Kamegai, T., Sato,Y., Wantanabe, K., and Kasai, M., Method for Producing Conjugated Linoleic Acid, Canadian Patent 2,219,601 (1997).
23. Richardson, T.A., Industrial Plastics: Theory and Application. South West Publishing Co., Cincinnati, 1983.
24. Skinner, E.N., and G. Smith (1987) Nickel Alloys, in *McGraw-Hill Encyclopedia of Science and Technology*. 6th Edition, McGraw-Hill Book Company, New York, Volume II, pp. 656–658.
25. Pariza, M.W., and Yang, X.-Y. Method of Producing Conjugated Fatty Acids, U.S. Patent 5,856,149 (1999).
26. Solomon, D.H. (1977) *The Chemistry of Natural Film Formers,* 2nd edn., pp. 33–74, Robert E. Krieger Publishing, Huntington, NY.
27. Zabolotskij, D.A., Demin, P.M., and Mylagkova,G.I., Synthesis of 9Z,11E-Octadecadienoic Acid Methyl Ester, Russian Patent 2,021,252 (1994).
28. Merck Index (1996) 12th edn., Merck and Co., Inc., Whitehouse Station, NJ.
29. American Standard Test Method (1991) Standard Test fo for the Kinematic Viscosity of Transparent and Opaque Liquids (and the Calculation of Dynamic Viscosity). Method D 455. Section 5, Vol. 05.01. ASTM, 1916 Race St., Philadelphia, PA, 19103–1178.
30. Lie Ken Jie, M.S.F, Pasha, M.K., and Alam, M.S. (1977) Synthesis and Nuclear Magnetic Resonance Properties of All Geometric Isomers of Conjugated Linoleic Acids, *Lipids 32*, 1041–1044.

Chapter 5

The Oxidation of Conjugated Linoleic Acid

Klaus Eulitz, Martin P. Yurawecz, and Yuoh Ku

Center for Food Safety and Applied Nutrition, U.S. Food and Drug Administration, Washington, DC 20204

Introduction

Ha *et al.* (1) and Ip *et al.* (2) reported that CLA had strong antioxidative effects and proposed this as a possible explanation for the anticarcinogenic and antiatherosclerotic properties. However, other investigations have shown that CLA functions more as a prooxidant (3,4), and its effectiveness depends on its chemical form, i.e., as free fatty acids, methyl esters, or as a constituent in triacylglycerols or phospholipids (3–5). In biological systems, CLA is incorporated into phospholipids (2,6,7) and may provide protective biological effects.

These studies, although dealing with oxidation, did not identify any oxidation products of CLA or any other compound in the mixture. Ha *et al.* (1) used a phosphate buffer/ethanol mixture with free CLA and measured the formation of peroxides as an indicator of oxidation. Ip *et al.* (2) used the thiobarbituric acid-reactive substances assay, which is known to measure only selective oxidation products. Van den Berg *et al.* (3) investigated model phospholipid membranes consisting of 16:0,18:2n-6-phosphatidylcholine. They added CLA as the free fatty acid and measured the absorbance at 233 nm to determine the formation of conjugated dienes (3). The loss of CLA during open-air oxidations was examined using gas chromatography/mass spectrometry (GC/MS), but no oxidation products were identified. Chen *et al.* (4) prepared mixtures of CLA (as free acid, methyl ester, or triacylglycerol) with canola oil in hexane and measured the decrease of oxygen in the headspace after heating the solutions at 90°C for 1–2 d.

Zhang and Chen (5) showed that the oxidation rate of CLA as the free fatty acid was similar to that of free docosahexaenoic acid in air at 90°C and considerably greater than those of free linoleic acid (LA), linolenic acid (LNA), and arachidonic acid (AA). They analyzed the remaining fatty acids by GC after esterification with BF_3 (5). However, CLA in the triacylglycerol form was more stable, although it still oxidized more readily than LA, LNA, or AA. These findings are in contrast to those of Holman and Elmer (8) who found similar oxidation rates for CLA and LA in the free fatty acid form during autoxidation at 37°C by measuring the oxygen uptake. Furthermore, Allen *et al.* (9) showed that CLA autoxidizes significantly more slowly than LA when the peroxide value is used as an indicator for the extent of oxidation.

Oxidation of CLA by Singlet Oxygen

The oxidation of 1,3-dienes has been thoroughly examined. There are three principal pathways for the oxidation of dienes by singlet oxygen as shown using model substances such as 2,4-hexadienes (10,11). With sterically unhindered dienes such as 2,4-hexadiene, the main oxidation products are cyclic endoperoxides (11) as shown in Figure 5.1, left side. Depending on the geometrical configuration of the double bond system, the cyclic endoperoxides were formed in total yields of 80, 60, and 50% from the *E,E*-, the *E,Z*-, and the *Z,Z*-2,4-hexadiene, respectively. This 4+2-addition is often described to be of the Diels-Alder-reaction type, which involves a simple concerted reaction.

With more hindered dienes such as 2,5-dimethylhexa-2,4-diene (10) and indenes (12), or substituted dienes such as 1,4-di-*tert*-butoxy-1,3-dienes (13,14), the major oxidation products are dioxetanes (see Fig. 5.1, right side) formed by a 2+2-cycloaddtion. These dioxetanes are unstable at room temperature and cleave to the corresponding aldehydes. The third principal pathway leads to hydroperoxides (10–12) as known from the oxidation of nonconjugated double bonds (15) (Fig. 5.1, center). A fourth reaction occurring with hindered and nonhindered dienes, but causing no oxidation, is the rapid isomerization of geometrical isomers that favors the formation of the *E,E*-isomers over *E,Z*-isomers and generally decreases the amount of *Z,Z*-isomers present (11).

Several studies have been conducted to elucidate the mechanism of the initial attack of singlet oxygen on a conjugated double bond system and to identify intermediates that occur (10–14,16). Manring and Foote (10) proposed the initial formation of a perepoxide as shown in Figure 5.2. This perepoxide can directly form hydroperoxides, thus following the ene-reaction typical for nonconjugated double bonds. In the case of conjugated double bonds, the formation of a zwitterionic or biradical structure is preferred because the positive charge or the radical character can be distributed over the entire double bond system. From these intermediates, the formation of both the endoperoxides (2+4-addition) as well as the dioxetanes (2+2-addition)

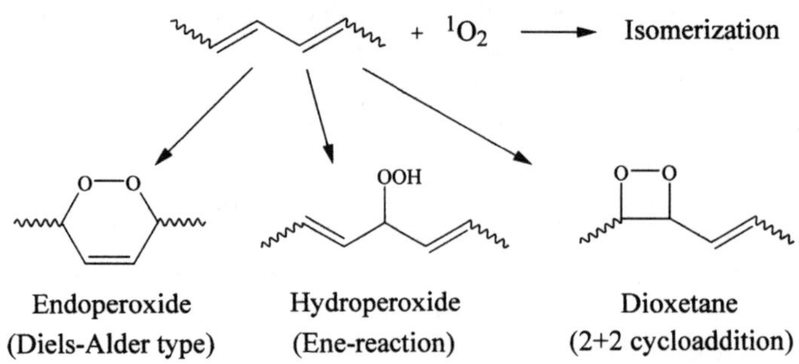

Fig. 5.1. Principal oxidation scheme of 1,3-dienes with singlet oxygen.

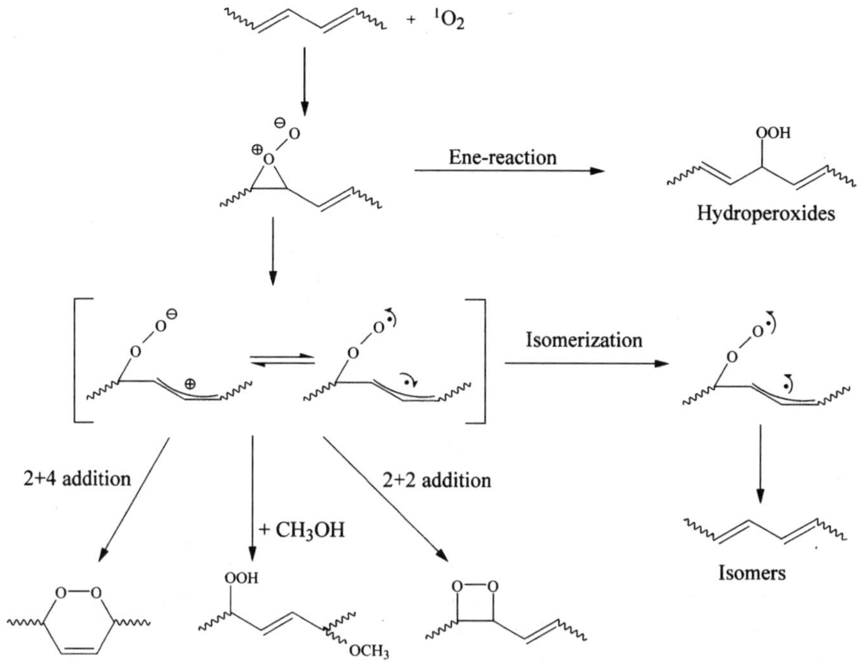

Fig. 5.2. Oxidation mechanism proposed by Manring and Foote (10) for the singlet oxygen oxidation of conjugated dienes.

can be easily rationalized. The main evidence for this zwitterionic/biradical mechanism is the occurrence of isomerization, depending on the free rotation of the former double bond and the occurrence of methoxy-compounds when the oxidation is carried out in methanol, indicating the formation of a very polar intermediate (10–14).

As expected, the peroxide value of CLA methyl ester after photo-oxidation with methylene blue was much lower than that of methyl linoleate (17). On the other hand, methylene blue was bleached much more rapidly by CLA methyl ester than by methyl linoleate. Gunstone and Wijesundra (18) converted the 8E,10E-CLA into the cyclic peroxide in a 5-d photo-oxidation. The photo-oxidation of 9E,11E-CLA methyl ester using methylene blue as sensitizer resulted in the formation of the cyclic peroxide after 16 h with a yield of >80% (19). Further oxidation of the purified cyclic peroxide with singlet oxygen did not occur even after 5 d, but the cyclic peroxide was decomposed by Fe^{2+} and heat (e.g., during GC analysis) to the corresponding furan fatty acid (9,12-epoxy-9,11-octadecadienoic acid methyl ester, F9,12) (19). This decomposition can be described briefly as a dehydration, although the proposed mechanism includes several steps showing biradicals and zwitterions leading to a 4-hydroxy-2-en-1-one substructure, which then condensates with ring closure to the furan (19).

The major oxidation products of the oxidation of CLA with singlet oxygen are cyclic peroxides yielding furan fatty acids upon decomposition, dioxetanes yielding aldehydes upon decomposition, but no hydroperoxides.

Autoxidation of CLA

The open air oxidation of CLA results in a product pattern very similar to that of singlet oxygen oxidation of CLA (20) when GC/flame ionization detection (FID), GC/MS and GC/Fourier transform infrared spectroscopy (FTIR) were used to identify the oxidation products. In addition to saturated and unsaturated aldehydes and aldehyde esters, α,β-unsaturated lactones and furan fatty acids were also found. The aldehydes and aldehyde esters were probably formed from dioxetanes analogous to the singlet oxygen oxidation. The lactones can be explained by an intramolecular reaction of the aldehyde group with the carboxy group. The occurrence of furan fatty acids suggests the presence of cyclic peroxides, which, as in the photo-oxidation, were converted to furan fatty acids by impurities during the oxidation or during GC analysis as described above (19). Interestingly, only small amounts of the furan fatty acids were found, although the formation of the cyclic peroxide has been shown to be the dominant oxidation pathway with singlet oxygen (19). Similar to singlet oxygen oxidation, no hydroperoxides were found. The oxidation rate of CLA was temperature dependent, and the half-lives of CLA were 14 d at ambient temperature and 7 h at 75°C (20). However, CLA in solution is much more stable to oxidation (21). No oxidation was detected for CLA in apolar solvents, in methanol, or in a 90:10 (vol/vol) mixture of methanol and water. Mixtures with higher amounts of water in methanol resulted in an increase in the formation of furan fatty acid, depending on the water content (21).

The furan fatty ester formed is in equilibrium with an open-chain dioxo-structure under acidic conditions (19,22). The furan fatty ester completely converted to the open-chain dioxo compound when treated with dilute sulfuric acid in methanol. Interestingly, the same diketo compounds were also found in the absence of water with BF_3/absolute methanol or anhydrous HCl in absolute methanol, whereas no methoxy compounds were detected under these conditions (22).

Oxidation of 9,12-Epoxy-9,11-octadecadienoic acid (F9,12)

Furan fatty acids (F-acids) are among the secondary oxidation products of CLA. They were first found in *Exocarpus* seed oil (23) and subsequently in fish (24), bovine liver (25), and plants (26). In these F-acids, the 3- and 4-positions of the furanyl moiety may also contain one or two methyl groups. The tetra-alkyl substituted F-acids show considerable antioxidant activity, whereas the trialkyl substituted derivatives show only about half the antioxidant potential of the tetra-alkylated F-acids (27). The dialkyl substituted F-acids possess no antioxidant activity (27). The tetra-alkylated F-acids in particular are capable of inhibiting free-radical oxidation as well as singlet oxygen–induced oxidation (27).

F-acids can be oxidized by autoxidation (28,29), by ultrasonically stimulated peroxides (30), and by potassium peroxomonosulfate (31). The oxidation of CLA by soybean lipoxygenase-1 was also reported (32). However, further investigation showed that CLA is not a substrate for soybean lipoxygenase-1 (33,34). CLA is actually oxidized by hydroperoxides of linoleic acid after the linoleic acid is oxidized by the lipoxygenase-1. Interestingly, the extent of oxidation of CLA bound at the *sn*-2 position of a synthetic phosphatidylcholine depends on the fatty acid bound in the *sn*-1 position (34). On the basis of these results, F-acids may play a strategic role in antioxidative processes in plant cells (34).

The oxidation of 2,5-dialkylated F-acids (see Fig. 5.3) with potassium peroxomonosulfate as well as the ultrasonically stimulated oxidation with magnesium monoperoxyphthalate and *m*-chloroperoxybenzoic acid yielded pure dioxo-ene-compounds (30,31) in the *cis*-configuration, whereas the use of pyridinium chlorochromate led to *trans* dioxoenes (30). Oxidation of F-acids by lipid hydroperoxides in the presence of lipoxygenase-1 gave primarily the *cis*-dioxoenes, which were isomerized in the course of the further oxidation to the more stable *trans*-dioxoenes (33). The unsubstituted dioxoenes formed are stable and can be isolated, in contrast to the dialkyl-substituted derivatives, which are unstable and react to form other oxidation products (24,25).

The dioxoenes are known to be strong inhibitors of blood platelet aggregation (35). They are also inhibitors of bacterial urease (29). Rosenblat *et al.* (29) suggested that this inhibition is caused by the reaction of the oxo-groups with thiol groups at the

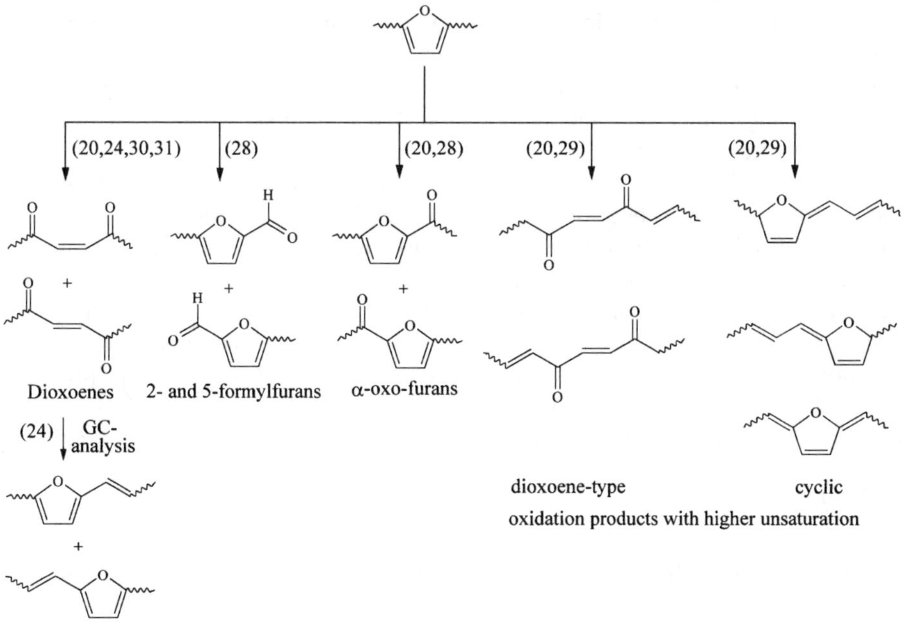

Fig. 5.3. Principal oxidation scheme of furan fatty acids.

active center of the enzymes. This hypothesis is supported by the fact that the addition of free cysteine to the reaction mixture reversed the inhibitory effects by reducing the thiol groups of the enzyme. The biologically active oxidation products were converted to more stable but less active compounds during thin-layer chromatography.

Sehat et al. (28) showed that F9,12 is oxidized slightly faster than CLA itself under similar conditions. This might explain the relatively small amount of F-acids found during autoxidation experiments with CLA. They also found other oxidation products such as α-oxo-F-acids and 2- and 5-formyl-furans (see Fig. 5.3), but were unable to propose a mechanism for the formation of these compounds. Several F-acids with additional double bonds have been found (20,24,29, shown in Fig. 5.3), but no mechanism for their formation has been suggested. Ishii et al. (24) attributed these compounds to conversions taking place during GC analysis.

Conclusion

Both the singlet oxygen oxidation and the autoxidation of CLA seem to yield identical products (shown in Fig. 5.4). Three possible pathways are known for the oxidation of conjugated dienes. These include the formation of hydroperoxides through the ene-reaction, the formation of aldehydes from dioxetanes resulting from 1,2-addition, and the formation of cyclic peroxides resulting from 1,4-addition. In the case of CLA, no hydroperoxides were found during photo-oxidation or autoxidation.

The aldehydes and aldehyde esters expected from the dioxetane pathway have been found in both photo-oxidation and autoxidation mixtures together with lactones formed from the aldehyde esters. The presence of cyclic peroxides has been established only for photo-oxidations, but their presence is supported by the formation of F-acids during both photo- and autoxidation. F-acids are known to result from the conversion of cyclic peroxides. The rate of the formation of cyclic peroxides or dioxetanes should depend on the steric conditions of the oxidized material. However, these rates have not yet been quantified for the different geometric isomers of CLA. The F-acids are themselves subject to oxidation, leading to the formation of highly biologically active dioxoene-compounds. The F-acids are oxidized slightly more rapidly under similar oxidation conditions. Other compounds such as F-acids with additional double bonds, α-oxo-F-acids, or 2- and 5-formyl-F-acids have been found in oxidation experiments with F-acids. Whether these products are actually products formed during prolonged oxidation or are artifacts arising from GC analysis is a question that requires additional research

References

1. Ha, Y.L., Storkson, J., and Pariza, M.W. (1990) Inhibition of Benzo(α)pyrene-Induced Mouse Forestomach Neoplasia by Conjugated Dienoic Derivatives of Linoleic Acid, *Cancer Res. 50,* 1097–1101.
2. Ip, C., Chin, S.F., Scimeca, J.A., and Pariza, M.W. (1991) Mammary Cancer Prevention by Conjugated Dienoic Derivatives of Linoleic Acid, *Cancer Res. 51,* 6118–6124.

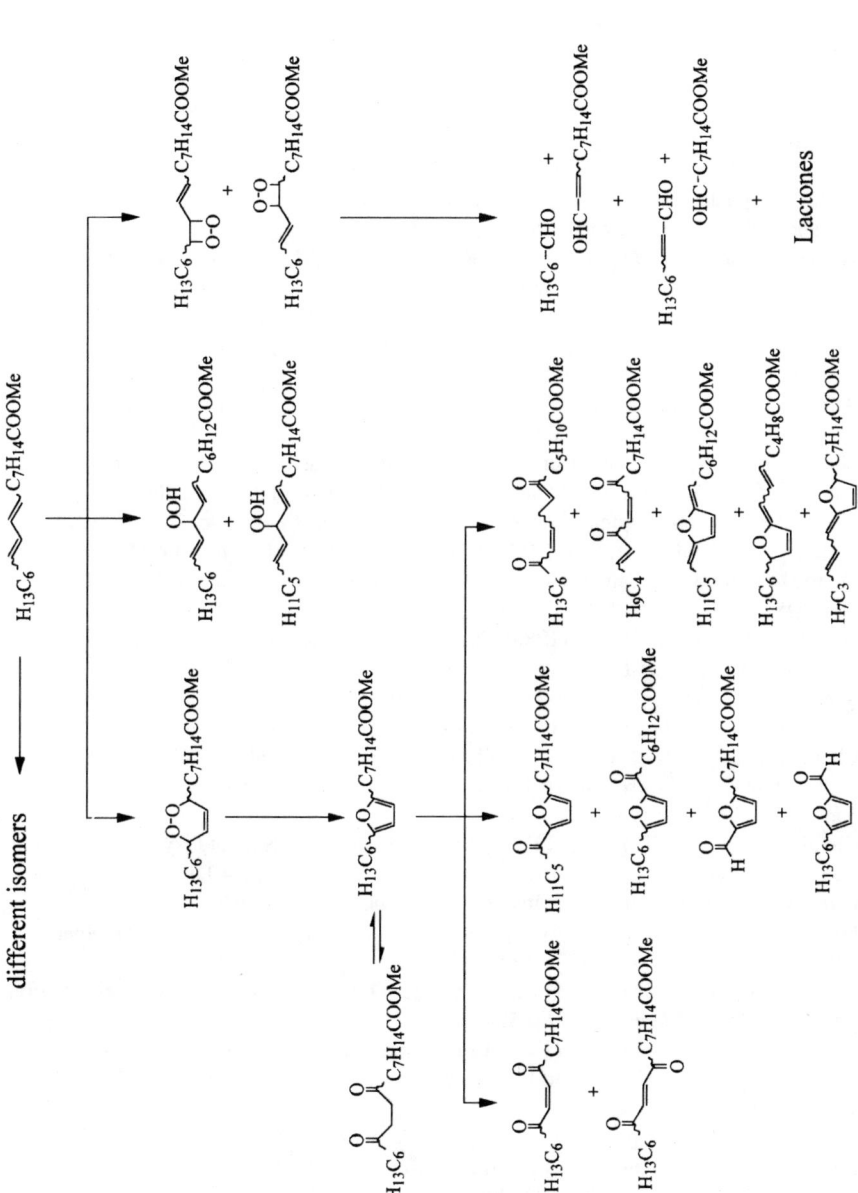

Fig. 5.4. Overview of the oxidation of conjugated linoleic acid (CLA).

3. van den Berg, J.J.M., Cook, N.E., and Tripple, D.L. (1995) Reinvestigation of the Antioxidant Properties of Conjugated Linoleic Acid, *Lipids 30,* 599–605.
4. Chen, Z.Y., Chan, P.T., Kwan, K.Y., and Zhang, A. (1997) Reassessment of the Antioxidant Activity of Conjugated Linoleic Acid, *J. Am. Oil Chem. Soc. 74,* 749–753.
5. Zhang, A., and Chen, Z.Y. (1997) Oxidative Stability of Conjugated Linoleic Acids Relative to Other Polyunsaturated Fatty Acids, *J. Am. Oil Chem. Soc. 74,* 1611–1613.
6. Sugano, M., Tsujita, A., Yamasaki, M., Yamada, K., Ikeda, I., and Kritchevsky, D. (1997) Lymphatic Recovery, Tissue Distribution, and Metabolic Effects of Conjugated Linoleic Acid in Rats, *Nutr. Biochem. 8,* 38–43.
7. Kramer, J.K.G., Sehat, N., Dugan, M.E.R., Mossoba, M.M., Yurawecz, M.P., Roach, J.A.G., Eulitz, K., Aalhus, J.L., Schaefer, A.L., and Ku. Y. (1998) Distributions of Conjugated Linoleic Acid (CLA) Isomers in Tissue Lipid Classes of Pigs Fed a Commercial CLA Mixture Determined by Gas Chromatography and Silver-Ion High-Performance Liquid Chromatography, *Lipids 33,* 549–558.
8. Holman, R.T., and Elmer, O.C. (1947) The Rates of Oxidation of Unsaturated Fatty Acids and Esters, *J. Am. Oil Chem. Soc. 24,* 127–129.
9. Allen, R.R., Jackson, A., and Kummerow, F.A. (1949) Factors Which Affect the Stability of Highly Unsaturated Fatty Acids. I. Differences in the Oxidation of Conjugated and Nonconjugated Linoleic Acid, *J. Am. Oil Chem. Soc. 26,* 395–399.
10. Manring, L.E., and Foote, C.S. (1983) Chemistry of Singlet Oxygen. 44. Mechanism of Photooxidation of 2,5-Dimethylhexa-2,4-diene and 2-methyl-2-penten, *J. Am. Chem. Soc. 105,* 4710–4717.
11. O'Shea, K.E., and Foote C.S. (1988) Chemistry of Singlet Oxygen. 51. Zwitterionic Intermediates from 2,4-Hexadienes, *J. Am. Chem. Soc. 110,* 7167–7170.
12. Manring, L.E., Kanner, R.C., and Foote, C.S. (1983) Chemistry of Singlet Oxygen. 43. Quenching by Conjugated Olefins, *J. Am. Chem. Soc. 105,* 4707–4710.
13. Clennan, E.L., and L'Esperance, R.P. (1985) The Unusual Reactions of Singlet Oxygen with Isomeric 1,4-di-*tert*-Butoxy-1,3-butadienes. A 2s+2a Cycloaddition, *J. Am. Chem. Soc. 107,* 5178–5182.
14. Clennan, E.L., and L'Esperance, R.P. (1985) Mechanism of Singlet Oxygen Addition to Conjugated Butadienes. Solvent Effects on the Formation of a 1,4-Diradical. The 1,4-Diradical/1,4-Zwitterion Dichotomy, *J. Org. Chem. 50,* 5424–5426.
15. Foote, C.S. (1971) Mechanism of Addition of Singlet Oxygen to Olefins and Other Substrates, *Pure Appl. Chem. 27,* 635–645.
16. Kearns, D.R. (1969) Selection Rules for Singlet-Oxygen Reactions. Concerted Addition Reactions, *J. Am. Chem. Soc. 91,* 6554–6563.
17. Jiang, J., and Kamal-Eldin, A. (1998) Comparing Methylene Blue-Photosensitized Oxidation of Methyl-Conjugated Linoleate and Methyl Linoleate, *J. Agric. Food Chem. 46,* 923–927.
18. Gunstone, F.D., and Wijesundra, R.C. (1979) Fatty Acids, Part 54: Some Reactions of Long-Chain Oxygenated Acids with Special Reference to Those Furnishing Furanoid Acids, *Chem. Phys. Lipids 24,* 193–208.
19. Bascetta, E., Gunstone F.D., and Scrimgeour, C.M. (1984) Synthesis, Characterisation, and Transformation of a Lipid Cyclic Peroxide, *J. Chem. Soc. Perkin Trans. I,* 2199–2205.
20. Yurawecz, M.P., Sehat, N., Mossoba, M.M., Roach, J.A.G., and Ku, Y. (1997)

Oxidation Products of Conjugated Linoleic Acid and Furan Fatty Acid, in *New Techniques and Applications in Lipid Analysis,* (McDonald, R.E., and Mossoba, M.M., eds.) pp.183–215, AOCS Press, Champaign, IL.
21. Yurawecz, M.P., Hood, J.K., Mossoba, M.M., Roach, J.A.G., and Ku, Y. (1995) Furan Fatty Acids Determined as Oxidation Products of Conjugated Octadecadienoic Acid, *Lipids 30,* 595–598.
22. Lie Ken Jie, M.S.F., and Sinha, S. (1981) Fatty Acids, Part 21: Ring Opening Reactions of Synthetic and Natural Furanoid Fatty Esters, *Chem. Phys. Lipids 28,* 99–109.
23. Morris, L.J., Marshall, M.O., and Kelly, W. (1966) A Unique Furanoid Fatty Acid from *Exocarpus* Seed Oil, *Tetrahedron Lett. 36,* 4249–4253.
24. Ishii, K., Okajima, H., Koyamatsu, T., Okada, Y., and Watanabe, H. (1988) The Composition of Furan Fatty Acids in the Crayfish, *Lipids 23,* 694–700.
25. Schödel, R., and Spiteller, G. (1986) Über das Vorkommen von F-Säuren in Rinderleber und deren enzymatischen Abbau bei Gewebeverletzungen (The Occurrence of F-Acids in Cattle Liver and Their Enzymatic Degradation During Tissue Damage), *Liebigs Ann. Chem. 5,* 459–462.
26. Hannemann, K., Puchta, V., Simon, E., Ziegler, H., Ziegler, G., and Spiteller, G. (1989) The Common Occurrence of Furan Fatty Acids in Plants, *Lipids 24,* 296–298.
27. Okada, Y., Okajima, H., Konishi, H., Terauchi, M., Ishii, K., Liu, I.-M., and Watanabe, H. (1990) Antioxidant Effect of Naturally Occurring Furan Fatty Acids on Oxidation of Linoleic Acid in Aqueous Dispersion, *J. Am. Oil Chem. Soc. 67,* 858–862.
28. Sehat, N., Yurawecz, M.P., Roach, J.A.G., Mossoba, M.M., Eulitz, K., Mazzola, E.P., and Ku, Y. (1998) Autoxidation of the Furan Fatty Acid Ester, Methyl 9,12-Epoxyoctadeca-9,11-dienoate, *J. Am. Oil Chem. Soc. 75,* 1313–1319.
29. Rosenblat, G., Tabak, M., Lie Ken Jie, M.S.F., and Neeman, I. (1993) Inhibition of Bacterial Urease by Autoxidation of Furan C-18 Fatty Acid Methyl Ester Products, *J. Am. Oil Chem. Soc 70,* 501–505.
30. Lie Ken Jie, M.S.F., Pasha, M.K., and Lam, C.K. (1997) Ultrasonically Stimulated Oxidation Reactions of 2,5-Disubstituted C18 Furanoid Fatty Ester, *Chem. Phys. Lipids 85,* 101–106.
31. Lie Ken Jie, M.S.F., and Pasha, M.K. (1998) Epoxidation Reactions of Unsaturated Fatty Esters with Potassium Peroxomonosulfate, *Lipids 33,* 633–637.
32. Boyer, R.F., Litts, D., Kostishak, J., Wijesundra R.C., and Gunstone, F.D. (1979) The Action of Lipoxygenase-1 on Furan Derivatives, *Chem. Phys. Lipids 25,* 237–246.
33. Batna, A., and Spiteller, G. (1994) Oxidation of Furan Fatty Acids by Soybean Lipoxygenase-1 in the Presence of Linoleic Acid, *Chem. Phys. Lipids 70,* 179–185.
34. Batna, A., and Spiteller, G. (1994) Effects of Soybean Lipoxygenase-1 on Phosphatidylcholines Containing Furan Fatty Acids, *Lipids 29,* 397–403.
35. Graff, G., Gellerman, G.J., Sand, D.M., and Schlenk, H. (1984) Inhibition of Blood Platelet Aggregation by Dioxo-Ene Compounds, *Biochim. Biophys. Acta 799,* 143–150.

Chapter 6
Methylation Procedures for Conjugated Linoleic Acid

Martin P. Yurawecz[a], John K.G. Kramer[b], and Yuoh Ku[a]

[a]Food and Drug Administration, Center for Food Safety and Applied Nutrition, Washington, DC 20204
[b]Southern Crop Protection, Food Research Center, Agriculture & Agri-Food Canada, Guelph, ON, Canada N1G 2W1

Introduction

Conjugated linoleic acid or CLA is an imprecise term for one or a mixture of octadecadienoic (18:2) fatty acids containing two conjugated double bonds. When CLA is synthetically derived by alkali isomerization of linoleic acid [(9Z,12Z) octadeca-9,12-dienoic acid, 9Z,12Z-18:2], the conjugated double bond position can vary from 8,10-18:2 to 11,13-18:2 (1–3). In natural products, conjugated double bond positions have been shown to occur from 6,8-18:2 to 12,14-18:2 (4–6). Each one of these positional isomers may occur as one or more of four possible geometric isomers, i.e., *trans,trans*; *cis,trans*; *trans,cis*; and *cis,cis*. In practice, all of these octadecadienoic acids with conjugated double bonds are included under the term CLA. Figure 6.1 shows structures of some of the major naturally occurring and synthetic CLA isomers.

CLA, like other fatty acids, are included in food extracts, dietary supplements, and other products in a variety of chemical forms, including, but not limited to free fatty acids (FFA), fatty acid methyl esters (FAME), mono-, di-, and triacylglycerols (MAG, DAG, TAG), phospholipid esters [choline (PC), ethanolamine (PE), serine (PS), inositol (PI), diphosphatidylglycerol (DPG)], sphingolipids [sphingomyelin (SM)], and sterol esters. Among the fatty acids, CLA is more labile than most; it is similar in stability to arachidonic acid and docosahexaenoic acids (7,8). Furthermore, CLA may be easily isomerized from *cis/trans* to *trans,trans* isomers (9–11), destroyed as a result of exposure to acid catalysts during extraction and methylation (11,12), or formed from endogenous precursors during methylation (13). In the past, the use of poorly chosen methylation and separation for CLA analysis has led to questionable results.

In general, the analysis for CLA isomers requires their conversion to derivatives that can be separated from the other fatty acids by using either gas chromatography (GC), high-performance liquid chromatography (HPLC), or silver-ion HPLC (Ag$^+$-HPLC). In this discussion, we will restrict ourselves to the preparation of CLA derivatives for the analysis of CLA from synthetic mixtures, foods, and biological tissues. No single hydrolysis or esterification procedure applies to all of these cases. In addition, the chemistry of acid-catalyzed *cis/trans* isomerization will be discussed as will the implications of reported results of previously published CLA analyses.

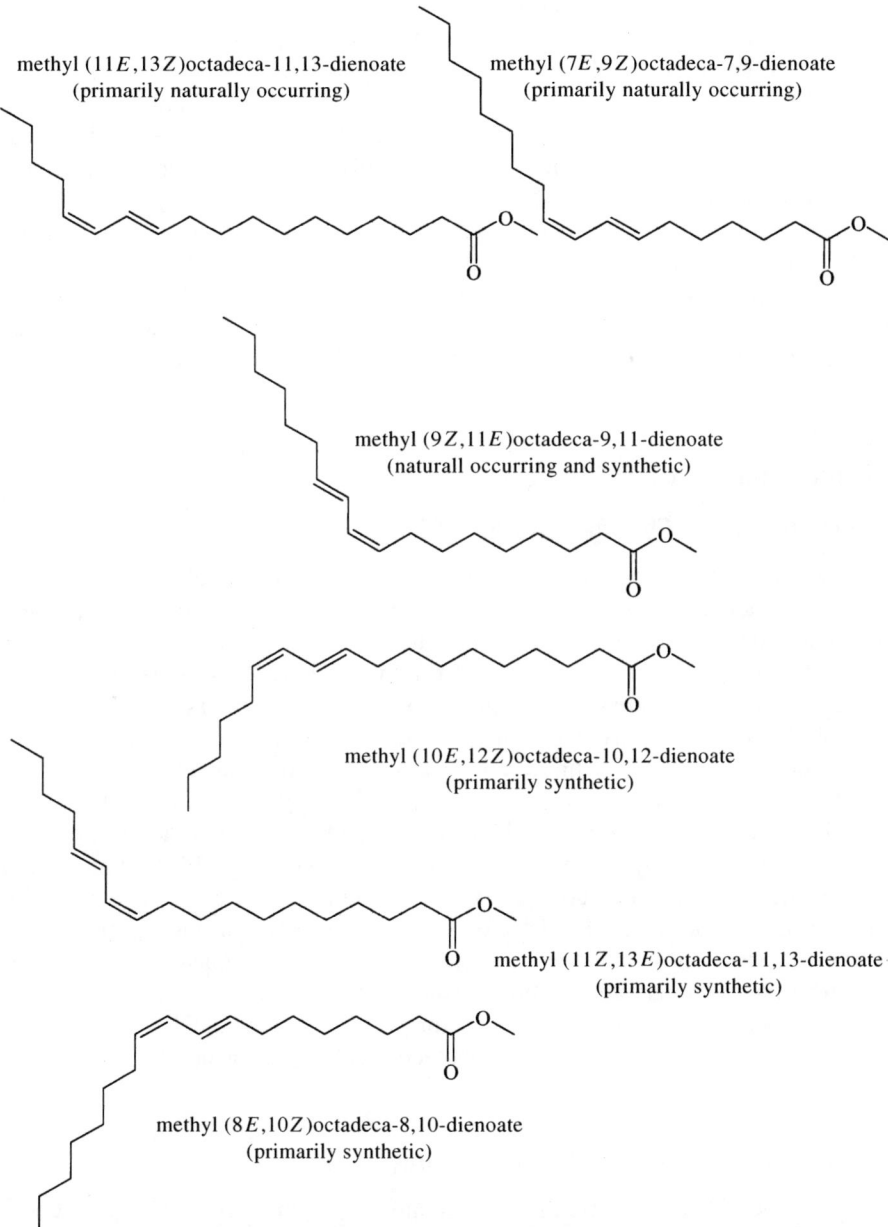

Fig. 6.1. Fatty acid methyl ester (FAME) structures of major conjugated linoleic acid (CLA) isomers that have been synthesized in mixtures and/or occur naturally. The (9Z,11E) isomer is the only one that occurs as a major product in both natural products and synthesized mixtures.

Types of Materials Containing CLA

CLA is present in a variety of matrices. A complex array of CLA isomers consisting mainly (~70–80%) of (9Z,11E)octadeca-9,11-dienoic acid (9Z,11E-18:2) are incorporated as esters in TAG and phospholipids that occur naturally in milk, cheese, and beef (5,14–16). Other isomers, e.g., (7E,9Z)octadeca-7,9-dienoic acid (7E,9Z-18:2), are also commonly present in milk, cheese, and beef products at ~3–16% (5). In addition, CLA is found in tissues of humans (5,17–19) and in animals intentionally fed products containing CLA (20–27).

CLA is sold by specialty chemical companies as a pure compound(s) or isomeric mixture(s). Dietary supplements containing CLA were shown to consist of two types containing approximately equal amounts of two (9Z,11E-18:2) and (10E,12Z-18:2) or four (8E,10Z-18:2), (9Z,11E-18:2), (10E,12Z-18:2), and (11Z,13E-18:2) positional isomers (28). Other matrices of analytical interest include supplements that have CLA as one of several added components (e.g., power bars and dry powders).

Direct Analysis of Total CLA

Supercritical Fluid Chromatography (SFC)

There are several options available for the determination of total CLA content. A simple procedure might include dilution of a test portion of a nutritional supplement in hexane and direct injection of an aliquot onto a chromatographic column that uses CO_2 as the mobile phase at supercritical fluid conditions. An example of such an analysis is shown in Figure 6.2. The figure shows a chromatogram of a commercially available CLA capsule separated on a 10 m SB cyano 50 column using supercritical CO_2 as the eluant (29–31). Palmitic acid (16:0) elutes before stearic acid (18:0), followed by monoenoic fatty acids (18:1) and linoleic acid (9Z,12Z-18:2). CLA eluted at 22.5 min. TAG eluted between 35 and 40 min on this column. The amounts present were quantitated by flame ionization detection, and the results for total CLA were consistent with results obtained by GC techniques. These findings were confirmed using an infrared (IR) detector (Magna 550 FTIR spectrometer, Nicolet Instrument, Madison, WI) that was also interfaced to the instrument (model 501 supercritical fluid chromatograph, Dionex, Sunnyvale, CA).

The supercritical fluid method allows for total quantitation of CLA as FFA without the need to prepare FAME, but the method totally lacks the ability to determine individual CLA isomers.

High-Performance Liquid Chromatography

HPLC techniques using reversed-phase columns have been successful in quantitating total CLA and CLA metabolites as their FFA (32–34) or FAME (4,24,35,36). This method partially separates CLA isomers into two peaks, *cis/trans* and *trans,trans* isomers (35), and separates CLA metabolites based on chain length and number of double

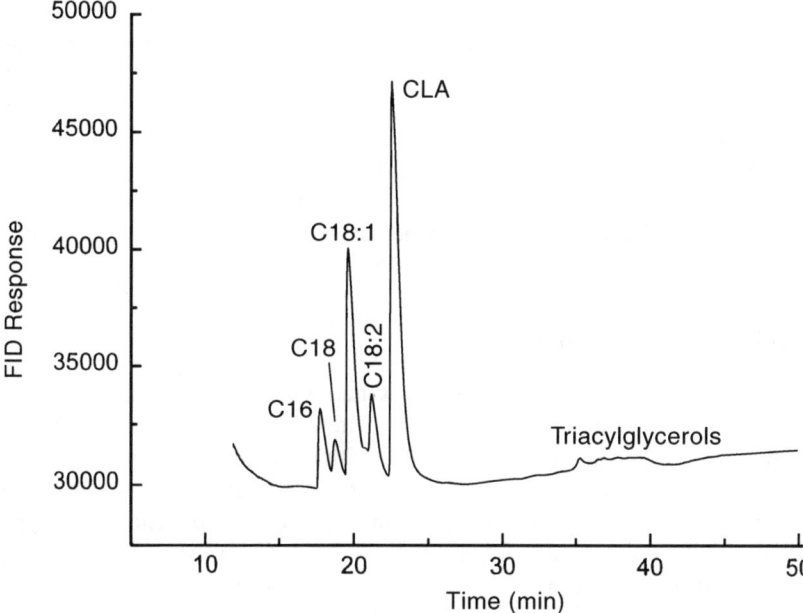

Fig. 6.2. Supercritical fluid chromatogram with flame ionization detection (FID) of hexadecanoic (16:0), octadecanoic (18:0), octadecenoic (18:1), methylene interrupted octadecadienoic (18:2), conjugated octadecadienoic (CLA) acids, and triacylglycerols (TAG). Pressure (atm) program: 110, hold 10 min; to 125 at 10 atm/min; to 140 at 0.5 atm/min; to 350 at 10 atm/min. Temperature program: 65°C, hold 26 min; to 100°C at 10°C/min; hold 15 min.

bonds (24,33,36). Reversed-phase HPLC is particularly useful as a preparative means of obtaining enriched fractions of CLA and their metabolites (4,24,32).

Neither SFC nor reversed-phase HPLC was effective in resolving CLA isomers. Therefore, extraction followed by derivatization techniques must be used to obtain more information about these mixtures.

Preparation of Lipid Extracts

The criteria for obtaining total CLA from different matrices are that the extraction must be quantitative and the structural integrity of the CLA isomers must be preserved. It is not the intent of this review to describe in detail each of the extraction methods, but simply to list a few techniques used successfully in CLA studies. The interested reader is referred to books on lipid methodology such as Kates (37) and Christie (38) for greater detail.

The composition of most commercial synthetic mixtures is generally complex and unknown. Therefore, alkali hydrolysis in ethanol is recommended to convert all lipids to FFA, which are subsequently extracted from the ethanol phase with diethyl ether/hexane (28,38). Milk lipids are effectively extracted using chloroform/methanol (39), isopropanol/hexane (40), or freeze-dried and subsequently extracted with chloroform/methanol (41). Cheese was successfully extracted with chloroform/methanol (32), isopropanol/hexane (4), or diethyl ether/hexane (6). Prior acid digestion of cheese (42,43) should be avoided because of possible *cis* to *trans* isomerization of the CLA isomers. Meat or tissues were first pulverized at dry ice (26,44) or liquid N_2 (27) temperatures to avoid lipolysis of lipids, followed by chloroform/methanol extraction. This method proved particularly applicable to the pulverization and extraction of cancer tissues that were "leathery" (unpublished results).

Methylation Procedures

The advantages of methylating CLA are that the FAME derivatives allow for GC and Ag^+-HPLC identification and quantitation of individual CLA isomers. The FAME including the CLA can be quantitated by GC with inclusion of internal reference material such as heptadecanoic (17:0) or tricosanoic (23:0) acids into the lipid mixture before methylation. The methylation procedure chosen will depend on the type and chemical composition of material to be methylated. A variety of matrices that contain CLA will be discussed below.

It is not always the case that FAME are the requisite or best derivative for an analysis. Detection of CLA using mass spectrometry (MS) generally employs 4,4'-dimethyloxazoline (DMOX) or other derivatives that are described in the MS chapters in this book.

Examples of methylation procedures are presented here for (i) dietary supplements, (ii) natural products, and (iii) reference materials. There are no standardized procedures for biological extracts and composited food products containing CLA. However, the same principles apply that are employed for the methylations described below.

Hydrolysis Followed by Trimethylsilyl(TMS)-Diazomethane Methylation

Dietary supplements are considered to be unknowns. As mentioned above, they may contain a combination of FFA and TAG. They may be FFA in an oil-based mixture or in an aqueous mixture. They do not necessarily contain what is stated on the label. To convert all of the CLA to the FFA form, a small representative test portion is weighed, hydrolyzed in base, and the FFA is isolated before methylation. The following procedure was adapted from methods described previously (3,38,45): A known amount (~25 mg) of product was weighed into a test tube with a teflon-lined screw cap. The material was dissolved in 2 mL 1 N KOH in ethanol (95%) and hydrolyzed overnight in the dark at room temperature. For quantitative analyses, 1 mg of tricosanoic acid (23:0) was added as internal standard. After hydrolysis, 5 mL of H_2O and 1 mL of 6 N HCL were added, and the free fatty acids were extracted three times with 5 mL diethyl ether/petroleum ether (1:1). The combined extracts were washed with H_2O, dried over anhydrous Na_2SO_4, and the solvents removed under a stream of argon.

The procedure that follows is applicable only to FFA. Methylation of a small amount of isolated FFA containing CLA was adapted from a previously described method (3,45): A free fatty acid extract (5–25 mg) is weighed into a screw-capped test tube with a teflon-lined screw cap. One milliliter of 20% methanol/benzene and 0.5 mL of 10% TMS-diazomethane in hexane solution (available from TCI, America, or Aldrich Inc.) are added. The reaction is allowed to run for 30 min with occasional gentle shaking or constant stirring. Five drops of glacial acetic acid are added, one at a time, with gentle swirling to remove excess TMS-diazomethane. If the yellow color persists after 5 min, one more drop of glacial acetic acid can be added. Glacial acetic acid is added dropwise until the yellow color is gone. If color does not disappear after 10 drops, proceed to the next step. Add 5 mL H_2O and 5 mL petroleum ether, shake for 30 s. Dry the petroleum ether extract over Na_2SO_4. Use petroleum ether solution directly for GC or HPLC analysis.

Methylation of Ester Lipids Using Sodium Methoxide in Methanol. Natural products generally contain CLA in the form of neutral lipids or phospholipids. It is not necessary to hydrolyze the lipids after extraction. The following procedure is applicable to a total lipid mixture containing TAG, cholesterol esters, and phospholipids. Cholesterol esters may require a longer reaction time than TAG; FFA and the *N*-acyl lipids of SM will not be methylated. If the lipid classes are separated by thin-layer chromatography (TLC) or HPLC before methylation, then all of the individual ester lipids can be methylated as follows: ester lipids using sodium methoxide, FFA using TMS-diazomethane, and SM using an acid-catalyzed methylation procedure at 80°C for at least 1 h. The acid-catalyzed methylation will result in significant isomerized CLA products, which is unavoidable, but SM are not known to contain much CLA. The procedure is as follows: Weigh 5–30 mg of test portion into a screw-capped test tube. Dissolve test portion in a small amount (<1 mL) of benzene or toluene, add 2 mL of 0.5 N NaOMe/methanol (available from Supelco Inc., Bellefonte, PA), and store overnight in the dark, or heat for 15 min at 50°C. Add H_2O to make a 95:5 mixture of methanol/H_2O which expels hexane. Add 2–3 mL hexane and mix well. Allow layers to separate and remove hexane layer containing the FAME. Dry hexane layer over Na_2SO_4 and use directly for GC or HPLC analysis.

Methylation of Lipids Using Anhydrous HCl/Methanol or BF_3/Methanol at Low Temperatures. The use of anhydrous HCl/methanol or BF_3/methanol is a universal methylation procedure because these catalysts will methylate all ester lipids, FFA, and *N*-acyl lipids present in SM. The usual acid-catalyzed conditions are 80–100°C for 1 h. Under these conditions, *cis/trans* CLA isomers will undergo extensive isomerization to *trans,trans* CLA isomers (9–11,46), form methoxy artifacts (11,13), and produce artifact CLA isomers from hydroxy fatty acid precursors if present in the lipid mixture (12,13); see theoretical discussion below.

If acid-catalyzed methylation procedures are carried out at lower temperatures, for instance at 60°C using 4% HCl/methanol (15), or at room temperature using BF_3/methanol (46), isomerization and artifact formation are greatly reduced. It was estimated that only ~5% of the *cis/trans* CLA isomers were converted to *trans,trans* CLA isomers (15). However, there is concern that methylation of the phospholipids at lower temperatures is not complete (11).

An acid-catalyzed methylation procedure is given for reference as follows: Weigh 5–30 mg of test portion into a screw-capped test tube. Dissolve test portion in a small amount (<1 mL) of benzene or toluene, add 2 mL of anhydrous 4% HCl/methanol, and heat for 20 min at 60°C. Add H_2O to make a 95:5 mixture of methanol/H_2O which expels hexane. Add 2–3 mL hexane and mix well, allow layers to separate, and remove hexane layer containing the FAME. Dry hexane layer over Na_2SO_4 and use directly for GC or HPLC analysis. HCl/methanol is prepared by passing dry HCl gas into anhydrous methanol. Anhydrous methanol is obtained by distillation over Grignard reaction.

For BF_3-catalyzed methylations, freshly obtained BF_3/methanol (commercially available from several sources) is added to the reaction, which is kept at room temperature for 30 min and mixed occasionally by using a vortex.

Methylation with BF_3/Methanol, a Useful Reference Mixture

Borontrifluoride (BF_3) is a Lewis acid. When the BF_3/methanol procedure was scaled up to methylate 1 g of CLA FFA (47), we were able to obtain a usable quantity of CLA FAME for oxidation studies and reference material for GC and HPLC. The results of the analyses obtained by GC for the scaled-up procedure are shown in Figure 6.3. The major peaks are those of 9Z,11E-18:2 and 9Z,11Z-18:2. A lesser quantity of 9E,11E-18:2 is present. The group of peaks eluting after 9E,11E-18:2 corresponds to allylic methoxyoctadecenoates such as methyl 11-methoxy(9E)octadecen-9-oate. In this particular reaction, these peaks accounted for <1% of the total FAME. Additional material concerning the identification of these peaks is described in the IR chapter of this book. This procedure is useful for synthesizing CLA FAME from CLA FFA. The authors do not recommend, and in many instances do not condone, the use of the BF_3 procedure for analysis for the reasons expressed below.

A variation of the procedure for methylation of marine oils was used (48). A high percentage of CLA FFA in reference standards can be converted to FAME by the following procedure: An accurately weighed quantity (~25 mg) of CLA FFA is added to a screw-capped test tube. Fresh BF_3/methanol (2 mL, 14%) is added. The tube is purged with argon, capped, and heated at 100°C for 5 min. Isooctane (1 mL) is added to the tube which is capped and vortexed for 15 s. Saturated NaCl/H_2O (5 mL) is added to the tube, and the tube is recapped and vortexed for 15 s. The isooctane is transferred to another tube with a Pasteur pipette. The BF_3 is reextracted with 1 mL of isooctane. The isooctane extracts are combined in one tube and dried over Na_2SO_4 before chromatographic examination. The isooctane solution is analyzed directly by GC or HPLC.

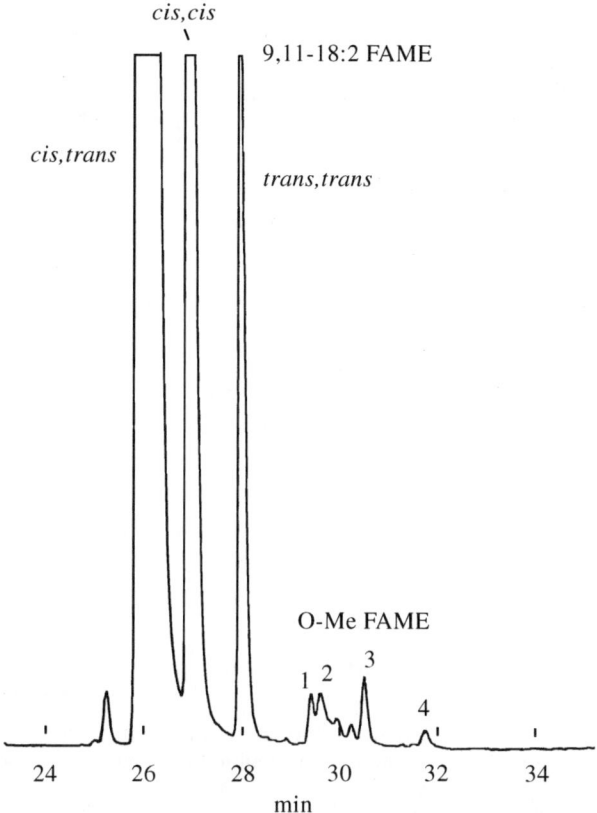

Fig. 6.3. Partial gas chromatography/flame ionization detection (GC/FID) chromatogram (24–32 min) of a mixture of *cis/trans* 9,11 conjugated linoleic acid (CLA) isomers produced by reaction with BF_3/methanol. Peaks 1 to 4 are allylic methoxy fatty acid methyl esters (FAME). A 50 m CP-Sil-88 fused silica capillary column (Chrompack, Raritan, NJ) was used, programmed from 75°C (2 min.) to 175°C at 4°C/min.

Theory of Acid-Catalyzed Methylation

The occurrence of methoxyoctadecenoates is an indication of the imperfections of this reaction. Other nuances of this reaction are illustrated in the figures that follow. Figure 6.4 shows that the reaction of CLA with H^+ (or with a Lewis acid such as BF_3) produces a series of products that undergo the addition of methanol in 1,2 or 1,4 fashion with or without retention of the *cis/trans* geometry. This reaction may be reversible, thus partially explaining the *trans,trans* isomer formation. Figure 6.5 shows a simplified explanation of how naturally occurring precursors to CLA may be formed

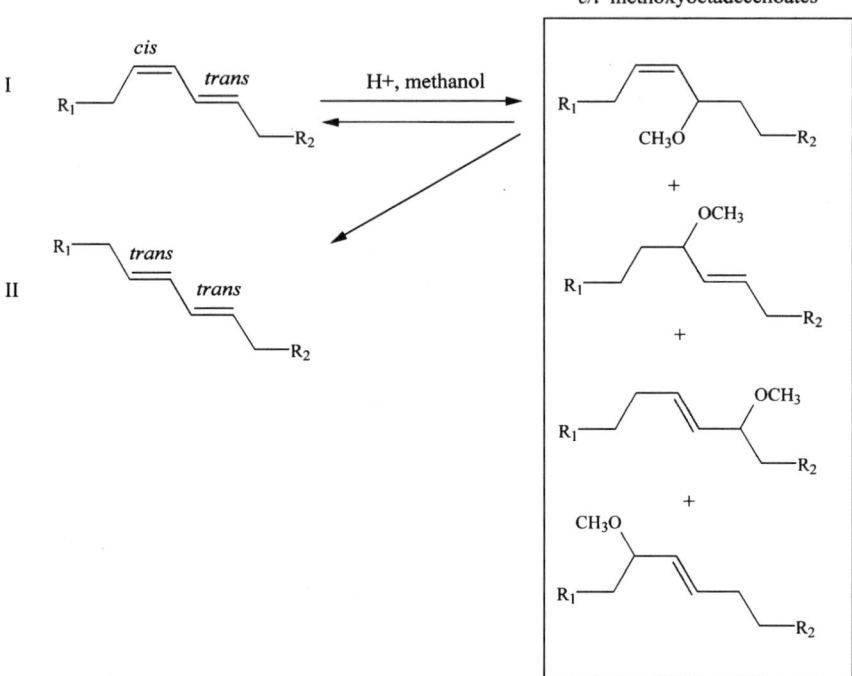

Fig. 6.4. In the presence of H+/methanol compound (I), a *cis/trans* conjugated linoleic acid (CLA) isomer, adds methanol across the double bond system to produce *cis/trans* methoxyoctadecenoates. The reaction may be reversible and this may partially explain the presence of *trans,trans* compounds (II), commonly detected in this reaction.

in an acid-catalyzed reaction. Monoenoic fatty acid moieties are oxidized, with or without a change in *cis/trans* geometry or carbon position, to the hydroperoxides that are indicated in Figure 6.5. Reduction of the hydroperoxides occurs naturally. Figure 6.6 shows the reactions of these naturally occurring (13) precursors in the presence of an acid catalyst in methanol with the conversion of the allylic hydroxymonoenes to methoxyoctadecenoates or CLA isomers.

Two additional figures are presented to emphasize the point that acid methylation procedures must not to be used for CLA analyses. Figure 6.7 shows a sample of olive oil that was examined by two methylation procedures: (A) a test portion that was acid (BF_3)-catalyzed and (B) a test portion that was base (tetramethylguanidine or TMG)-catalyzed (13). Enhanced CLA response in both the *cis/trans* and *trans/trans* region of the chromatogram is apparent only with BF_3. The amount of CLA produced as an artifact exceeded 0.2%. The methoxy artifacts can arise either from CLA or from other precursors. We are skeptical of reports indicating that CLA is produced by cooking beef or by aging cheese when the method used to make such a determination is based on an acid-catalyzed procedure. It is more likely that

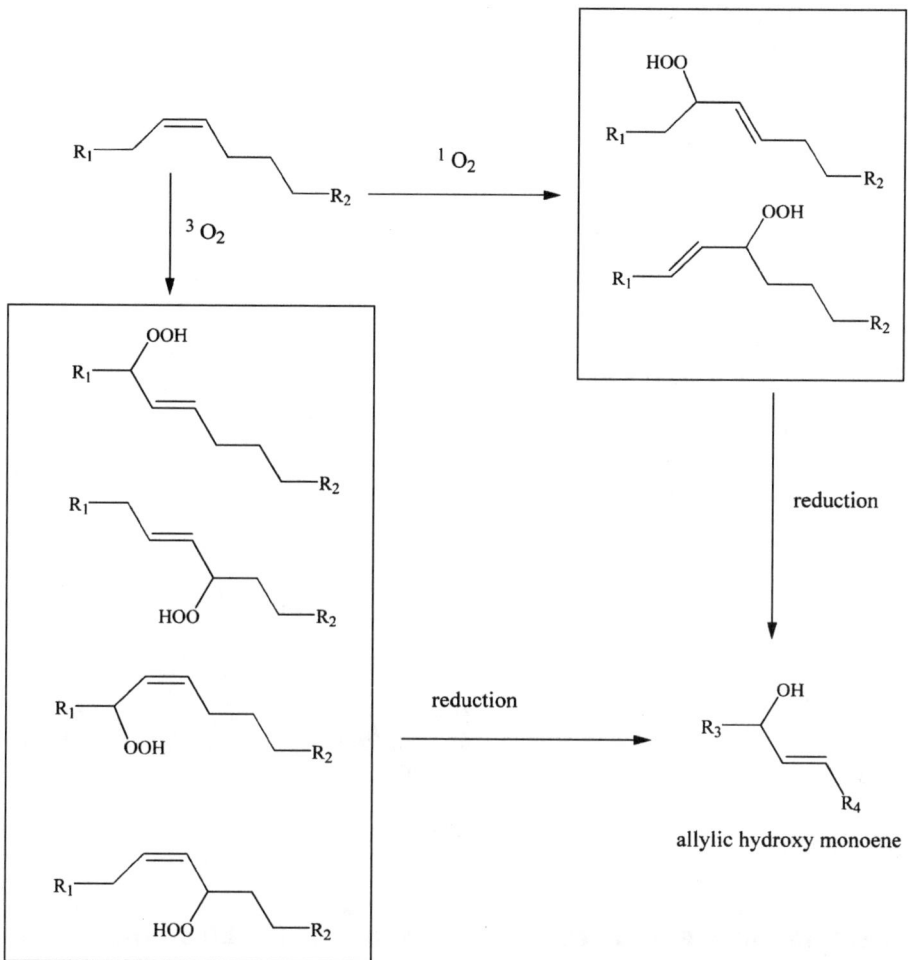

Fig. 6.5. Monoenes autoxidize to produce allylic hydroperoxides with or without a change in double bond *cis/trans* configuration. When oxidized by singlet oxygen ($^1O_2$) monoenes shift double bond position and *cis/trans* configuration to form hydroperoxides allylic to double bonds. In either case, an allylic hydroxy monoene is formed after reduction.

precursors in these products are increasing *via* mechanisms shown in Figures 6.6 and 6.7. Finally, Figure 6.8 shows a series of Ag$^+$-HPLC chromatograms that indicate that the major CLA isomer occurring in natural products (i.e., cow milk, cheese, beef, human adipose, and human milk) is rumenic acid, 9Z,11E-18:2 (5). These results differ substantially from results of an earlier report that suggested that *trans,trans* isomers (32) were the major CLA components present in cheese.

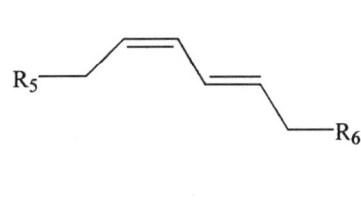

Fig. 6.6. The allylic hydroxy monoene formed, as shown in Figure 6.5, becomes a precursor to methoxyoctadecenoates and conjugated linoleic acid (CLA) that are formed as artifacts of acid-catalyzed procedures.

Methylation Methods Used in the Analysis of CLA Isomers in the Past 10 Years

Over the past 10 years, CLA analysis has undergone many changes and improvements. The summary in Table 6.1 is given so that the reader can interpret seeming inconsistencies in CLA analyses that appear to be based on differences in methodology. Furthermore, researchers may use this as a guide in choosing methods appropriate for their specific needs.

In the early studies, acid-catalyzed methylations (HCl or BF_3) at high temperatures (generally 80–100°C for 1 h) were used, which resulted in excessive formation of *trans,trans* CLA isomers (Table 6.1, first group). By using these methods, the major CLA isomers in cheese (32,42) and some tissue lipids (25) were found to be *trans,trans* isomers. In addition to isomerization, significant amounts of allylic methoxy artifacts were formed (11). Even though isomerization was recognized as a problem during acid-catalysis as early as 1966 (9), and well documented in later reports

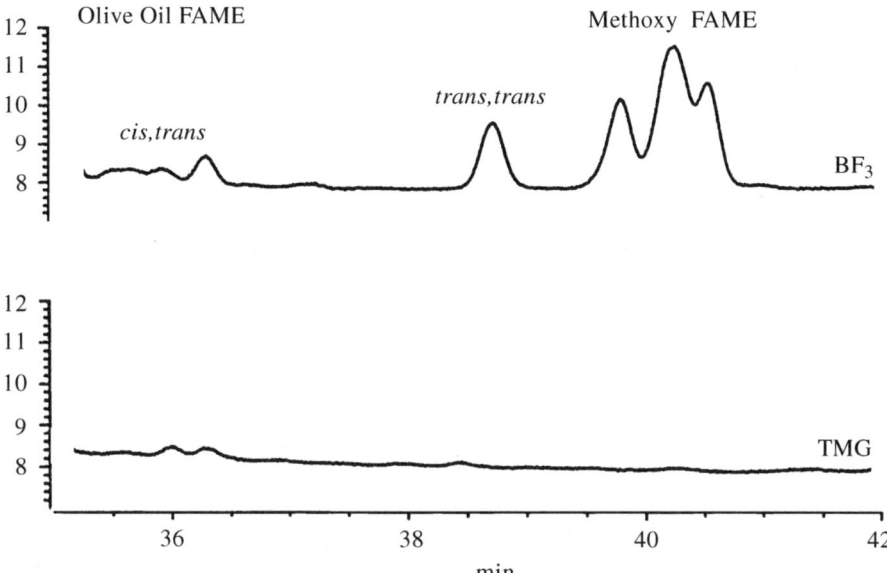

Fig. 6.7. Partial gas chromatography/flame ionization detection (GC/FID) chromatograms (35–42 min) of an olive oil that was methylated using either BF_3 or tetramethylguanidine (TMG).

(10,11,46), high-temperature acid-catalyzed methylations are still being used (25,50,52). The extent of isomer formation under acid-catalyzed conditions is not consistent within subsequent methylation runs even in the same laboratory (unpublished results) because the factors involved in isomerization, such as moisture, may differ among tests. Therefore, the CLA isomer distribution data from these studies are highly questionable. At best these data can be used only as indicators of total CLA content by summing all CLA isomers.

Isomerization was drastically reduced by lowering the temperature and time of methylation (Table 6.1, groups 2 and 3). The 4% HCl-catalyzed methylation for 20 min at 60°C was estimated at ~5% isomerized product (15). The BF_3-catalyzed reaction for 30 min at room temperature showed only a 1% content of *trans,trans* isomers in the standard CLA mixture (46). Under these mild methylation conditions, test samples consisting mainly of TAG (such as the fat of milk, cheese, beef, and adipose tissue) were completely methylated, but not the phospholipids (11). Therefore, these mild methylation procedures are a problem when determining the CLA content in tissues of animals fed CLA because of the high content of phospholipids. The two-stage method of NaOH hydrolysis followed by BF_3- or HCl-catalyzed methylation (Table 6.1, group 3) appears to have the advantage of methylating ester lipids, FFA, and *N*-acyl (SM) lipids with a minimum of isomerization (11). However, the completion of phospholipid and SM methylation should be checked by TLC (11).

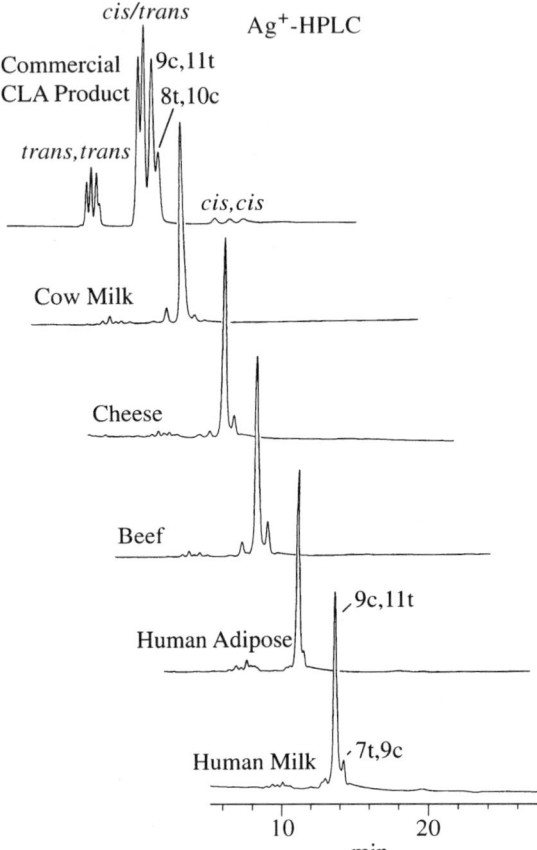

Fig. 6.8. Silver-ion high-performance liquid chromatography (Ag+-HPLC) chromatograms of a commercial conjugated linoleic acid (CLA) mixture containing 12 CLA isomers, cow milk, cheese, beef, human adipose tissue, and human milk. In the natural products indicated here, the rumenic acid fatty acid methyl ester (FAME) is always the major CLA isomer found.

TABLE 6.1 Comparison of Selected References of CLA Analysis in the past 10 Years Based on the Methylation Procedure Used. Extraction Conditions and GC Columns Used Are Included[a]

Authors, (year),(reference)	Extraction	Methylation catalyst/condition[b]	GC column/length
Ha et al.(1989) (32)	C/M	(i) NaOH; (ii) BF_3	60 m Supelcowax-10
Ha et al.(1990) (49)	C/M	HCl	60 m Supelcowax-10
Ip et al. (1991) (20)	C/M	Hcl	60 m Supelcowax-10
Shantha et al. (1992) (42)	Acid digestion/ether: hexane	BF_3	60 m Supelcowax-10
Chen et al. (1997) (50)	No extraction required	BF_3 90%C 45 min	100 m SP 2560
Sugano et al. (1997) (25)	Folch et al. (51)	BF_3	60 m Supelcowax-10
Salamin et al. (1998) (52)	Dichloromethane/M	Acid methylation	60 m SP 2380

TABLE 6.1 *(continued)*

Authors, (year),(reference)	Extraction	Methylation catalyst/condition[b]	GC column/length
Chin et al. (1992) (15)	Bligh and Dyer (53)	4% HCl 60°C 20 min	60 m Supelcowax-10
Chin et al. (1994) (54)	Folch et al. (51)	4% HCl 60°C 20 min	60 m Supelcowax-10
Ip et al. (1996) (21)	Bligh and Dyer (53)	4% HCl 60°C 20 min	60 m Supelcowax-10
Ip et al. (1997) (23)	Bligh and Dyer (53)	4% HCl 60°C 20 min	60 m Supelcowax-10
Stanton et al. (1997) (55)	Fat by centrifugation	4% HCl 60°C 20 min	60 m Supelcowax-10
Park and Pariza (1998) (56)	C/M	4% HCl 60°C 20 min	60 m Supelcowax-10
Werner et al.(1992) (46)	Folch et al. (51)	(i)NaOH; (ii)BF_3 RT 30 min	60 m Supelcowax-10
Lin et al. (1995) (16)	Bligh and Dyer (53)	(i)NaOH; (ii)BF_3 RT 30 min	60 m Supelcowax-10
Jiang et al. (1996) (40)	Isopropanol/hexane	(i)NaOH; (ii)BF_3 RT 30 min	50 m CP Sil 88
Jiang et al. (1997) (57)	Isopropanol/hexane	(i) KOH; (ii) BF_3 RT	50 m CP Sil 88
McGuire et al. (1997) (58)	Bligh and Dyer (53)	(i)NaOH; (ii)BF_3 RT 30 min	60 m Supelcowax-10
Lin et al. (1998) (59)	Bligh and Dyer (53)	(i)NaOH; (ii)BF_3 RT 30 min	60 m Supelcowax-10
Sébédio et al. (1997) (24)	Folch et al. (51)	(i) NaOH; (ii) HCl	50 m BPX 70
Shantha et al. (1993) (10)	Bligh and Dyer (53)	TMG; $NaOCH_3$; BF_3	60 m Supelcowax-10
Shantha et al. (1994) (60)	Isopropanol/hexane	TMG	60 m Supelcowax-10
Shantha et al. (1995) (43)	Acid digestion/ether: hexane	TMG	60 m Supelcowax-10
Belury et al. (1997) (22)	Bligh and Dyer (53)	TMG	30 m Omegawax 320
Shantha et al. (1997) (61)	Isopropanol/hexane	TMG	60 m Supelcowax-10
Fogerty et al. (1988) (14)	Bligh and Dyer (53)	$NaOCH_3$	25 m BP20-0.5
Jahreis et al. (1997) (41)	C/M	$NaOCH_3$	100 m SP 2560
Kramer et al. (1997) (11)	C/M	$NaOCH_3$ or HCl	100 m SP 2560
Precht et al. (1997) (62)	Butter fat	$NaOCH_3$	100 m CP Sil 88
Griinari et al. (1998) (63)	Isopropanol/hexane	$NaOCH_3$	100 m CP Sil 88
Kelly et al. (1998) (64)	Isopropanol/hexane	$NaOCH_3$	60 m Supelcowax-10
Kramer et al. (1998) (26)	Cold pulverization; C/M	$NaOCH_3$	100 m CP Sil 88
Lavillonnière et al. (1998) (4)	Isopropanol/hexane	$NaOCH_3$	50 m BPX 70
Li and Watkins (1998) (27)	Cold pulverization; sep pack	$NaOCH_3$	30 m DB 23; DB 225
Sehat et al. (1998) (6)	K oxalate/ether:hexane	$NaOCH_3$	100 m CP Sil 88

[a]Abbreviations: CLA, conjugated linoleic acid; GC, gas chromatography; C/M, chloroform/methanol; RT, room temperature; TMG, tetramethylguanidine.
[b]Some methylations were undertaken in two steps (i) and (ii)

The use of tetramethylguanidine (TMG) as a base catalyst in the methylation of CLA-containing lipids was extensively investigated by Shantha *et al.* (10) (Table 6.1, group 4), and appears promising. Although TAG are generally completely methylated, FFA and phospholipids may not be (11). Further investigation will be required to address these issues.

Alkali-catalyzed methylation has been proven to be the most accurate method for CLA analysis (Table 6.1, group 5). It does not isomerize conjugated double bonds or form allylic methoxy derivatives (11,65). However, FFA and N-acyl lipids, such as SM, are not methylated under these conditions. If both of these lipid constituents are minor, their contribution may be ignored. Nevertheless, the best solution is to separate all of the lipid components by TLC or HPLC before methylating each lipid class independently. Ester lipids can then be methylated using $NaOCH_3$, and FFA can be methylated using TMS-diazomethane, leaving N-acyl lipids (such as SM) to be methylated using HCl/methanol at 80°C for 1 h. The last-mentioned will cause some isomerization of CLA, which is unavoidable in this case. This approach was used recently to determine the distribution of CLA isomers in tissues of pigs fed a commercial CLA mixture (26).

References

1. Mounts, T.L., Dutton, H.J., and Glover, D. (1970) Conjugation of Polyunsaturated Acids, *Lipids 5*, 997–1005.
2. Christie, W.W., Dobson, G., and Gunstone, F.D. (1997) Isomers in Commercial Samples of Conjugated Linoleic Acid, *Lipids 32*, 1231.
3. Sehat, N., Yurawecz, M.P., Roach, J.A.G., Mossoba, M.M., Kramer, J.K.G., and Ku, Y. (1998) Silver-Ion High-Performance Liquid Chromatographic Separation and Identification of Conjugated Linoleic Acid Isomers, *Lipids 33*, 217–221.
4. Lavillonnière, F., Martin, J.C., Bougnoux, P., and Sébédio, J.-L. (1998) Analysis of Conjugated Linoleic Acid Isomers and Content in French Cheeses, *J. Am. Oil Chem. Soc. 75*, 343–352.
5. Yurawecz, M.P., Roach, J.A.G., Sehat, N., Mossoba, M.M., Kramer, J.K.G., Fritsche, J., Steinhart, H., and Ku, Y. (1998) A New Conjugated Linoleic Acid Isomer, 7 *trans*,9 *cis*-Octadecadienoic Acid, in Cow Milk, Cheese, Beef and Human Milk and Adipose Tissue, *Lipids 33*, 803–809.
6. Sehat, N., Kramer, J.K.G., Mossoba, M.M., Yurawecz, M.P., Roach, J.A.G., Eulitz, K., Morehouse, K.M., and Ku, Y. (1998) Identification of Conjugated Linoleic Acid Isomers in Cheese by Gas Chromatography, Silver-Ion High-Performance Liquid Chromatography and Mass Spectral Reconstructed Ion Profiles. Comparison of Chromatographic Elution Sequences, *Lipids 33*, 963–971.
7. van den Berg, J.J.M., Cook, N.E., and Tribble, D.L. (1995) Reinvestigation of the Antioxidant Properties of Conjugated Linoleic Acid, *Lipids 30*, 599–605.
8. Zhang, A., and Chen, Z.Y. (1997) Oxidative Stability of Conjugated Linoleic Acids Relative to Other Polyunsaturated Fatty Acids, *J. Am. Oil Chem. Soc. 74*, 1611–1613.
9. Kepler, C.R., Hirons, K.P., McNeill, J.J., and Tove, S.B. (1966) Intermediates and Products of the Biohydrogenation of Linoleic Acid by *Butyrivibrio fibrisolvens*, *J. Biol. Chem. 241*, 1350–1354.
10. Shantha, N.C., Decker, E.A., and Hennig, B. (1993) Comparison of Methylation Methods for the Quantitation of Conjugated Linoleic Acid Isomers, *J. Assoc. Off. Anal. Chem. Int. 76*, 644–649.
11. Kramer, J.K.G., Fellner, V., Dugan, M.E.R., Sauer, F.D., Mossoba, M.M., and Yurawecz, M.P. (1997) Evaluating Acid and Base Catalysts in the Methylation of Milk and Rumen Fatty Acids with Special Emphasis on Conjugated Dienes and Total *trans* Fatty Acids, *Lipids 32*, 1219–1228.

12. Yurawecz, M.P., Sehat, N., Mossoba, M.M., Roach, J.A.G., and Ku, Y. (1997) Oxidation Products of Conjugated Linoleic Acid and Furan Fatty Acids, in *New Techniques and Applications in Lipid Analysis,* (McDonald, R.E., and Mossoba, M.M., eds.) pp. 183–215, AOCS Press, Champaign, IL.
13. Yurawecz, M.P., Hood, J.K., Roach, J.A.G., Mossoba, M.M., Daniels, D.H., Ku, Y, Pariza, M.W., and Chin, S.F. (1994) Conversion of Allylic Hydroxy Oleate to Conjugated Linoleic Acid and Methoxy Oleate by Acid-Catalyzed Methylation Procedures, *J. Am. Oil Chem. Soc. 71,* 1149–1155.
14. Fogerty, A.C., Ford, G.L., and Svoronos, D. (1988) Octadeca-9,11-dienoic Acid in Foodstuffs and in the Lipids of Human Blood and Breast Milk, *Nutr. Rep. Int. 38,* 937–944.
15. Chin, S.F.,Liu, W., Storkson, J.M., Ha, Y.L., and Pariza, M.W. (1992) Dietary Sources of Conjugated Dienoic Isomers of Linoleic Acid, a Newly Recognized Class of Anticarcinogens, *J. Food Comp. Anal. 5,* 185–197.
16. Lin, H., Boylston, T.D., Chang, M.J., Luedecke, L.O., and Shultz, T.D. (1995) Survey of the Conjugated Linoleic Acid Contents of Dairy Products, *J. Dairy Sci. 78,* 2358–2365.
17. Ackman, R.G., Eaton, C.A., Sipos, J.C., and Crewe, N.F. (1981) Origin of *cis*-9, *trans*-11- and *trans*-9, *trans*-11-Octadecadienoic Acids in the Depot Fat of Primates Fed a Diet Rich in Lard and Corn Oil and Implications for the Human Diet, *Can. Inst. Food Sci. Technol. J. 14,* 103–107.
18. Britton, M., Fong, C., Wickens, D., and Yudkin, J. (1992) Diet as a Source of Phospholipid Esterified 9,11-Octadecadienoic Acid in Humans, *Clin. Sci. 83,* 97–101.
19. Fritsche, J., Mossoba, M.M., Yurawecz, M.P., Roach, J.A.G., Sehat, N., Ku, Y., and Steinhart, H. (1997) Conjugated Linoleic Acid (CLA) Isomers in Human Adipose Tissue, *Z. Lebensm. Unters. Forsch. A 205,* 415–418.
20. Ip, C., Chin, S.F., Scimeca, J.A., and Pariza, M.W. (1991) Mammary Cancer Prevention by Conjugated Dienoic Derivative of Linoleic Acid, *Can. Res. 51,* 6118–6124.
21. Ip, C., Briggs, S.P., Haegele, A.D., Thompson, H.J., Storkson, J., and Scimeca, J.A. (1996) The Efficacy of Conjugated Linoleic Acid in Mammary Cancer Prevention Is Independent of the Level or Type of Fat in the Diet, *Carcinogenesis 17,* 1045–1050.
22. Belury, M.A., and Kempa-Steczko, A. (1997) Conjugated Linoleic Acid Modulates Hepatic Lipid Composition in Mice, *Lipids 32,* 199–204.
23. Ip, C., Jiang, C., Thompson, H.J., and Scimeca, J.A. (1997) Retention of Conjugated Linoleic Acid in the Mammary Gland is Associated with Tumor Inhibition During the Post-Initiation Phase of Carcinogenesis, *Carcinogenesis 18,* 755–759.
24. Sébédio, J.L., Juanéda, P., Dobson, G., Ramilison, I., Martin, J.C., Chardigny, J.M., and Christie, W.W. (1997) Metabolites of Conjugated Isomers of Linoleic Acid (CLA) in the Rat, *Biochim. Biophys. Acta 1345,* 5–10.
25. Sugano, M., Tsujita, A., Yamasaki, M., Yamada, K., Ikeda, I., and Kritchevsky, D. (1997) Lymphatic Recovery, Tissue Distribution, and Metabolic Effects of Conjugated Linoleic Acid in Rats, *J. Nutr. Biochem. 8,* 38–43.
26. Kramer, J.K.G., Sehat, N., Dugan, M.E.R., Mossoba, M.M., Yurawecz, M.P., Roach, J.A.G., Eulitz, K., Aalhus, J.L., Schaefer, A.L., and Ku, Y. (1998) Distributions of Conjugated Linoleic Acid (CLA) Isomers in Tissue Lipid Classes of Pigs Fed a Commercial CLA Mixture Determined by Gas Chromatography and Silver-Ion High-Performance Liquid Chromatography, *Lipids 33,* 549–558.

27. Li, Y., and Watkins, B.A. (1998) Conjugated Linoleic Acids Alter Bone Fatty Acid Composition and Reduce *ex vivo* Prostaglandin E_2 Biosynthesis in Rats Fed n-6 or n-3 Fatty Acids, *Lipids 33*, 417–425.
28. Yurawecz, M.P., Sehat, N., Mossoba, M.M., Roach, J.A.G., Kramer, J.K.G., and Ku, Y. (1999) Variations in Isomer Distribution in Commercially Available Conjugated Linoleic Acid, *Fett/Lipid 101* (in press).
29. Calvey, E.M., Yurawecz M.P., and Mossoba M.M. (1997) Supercritical Fluid Chromatography/Fourier Transform Infrared Spectroscopy of Commercial Conjugated Linoleic Acid, *Proceedings of the Ninteenth International Symposium on Capillary Chromatography and Electrophoresis,* Wintergreen, Virginia, May 18–22, No. 97, pp. 396–397.
30. Calvey, E.M. (1998) Supercritical Fluid Chromatography and Extraction: Application to Food Analysis, in *Final Program and Book of Abstracts, The Eighth International Symposium on Supercritical Fluid Chromatography and Extraction,* July 12–16, St. Louis, MO.
31. Calvey, E.M., McDonald, R.E., Page, S.W., Mossoba, M.M., and Taylor, L.T. (1991) Evaluation of SFC/FT-IR for Examination of Hydrogenated Soybean Oil, *J. Agric. Food Chem. 39,* 542–548.
32. Ha, Y.L., Grimm, N.K., and Pariza, M.W. (1989) Newly Recognized Anticarcinogenic Fatty Acids: Identification and Quantification in Natural and Processed Cheeses, *J. Agric. Food Chem. 37,* 75–81.
33. Banni, S., Carta, G., Contini, M.S., Angioni, E., Deiana, M., Dessì, M.A., Melis, M.P., and Corongiu, F.P. (1996) Characterization of Conjugated Diene Fatty Acids in Milk, Dairy Products, and Lamb Tissues, *J. Nutr. Biochem. 7,* 150–155.
34. Banni, S., Angioni, E., Contini, M.S., Carta, G., Casu, V., Iengo, G.A., Melis, M.P., Deiana, M., Dessì, M.A., and Corongiu, F.P. (1998) Conjugated Linoleic Acid and Oxidative Stress, *J. Am. Oil Chem. Soc. 75,* 261–267.
35. Smith, G.N., Taj, M., and Braganza, J.M. (1991) On the Identification of a Conjugated Diene Component of Duodenal Bile as 9Z,11*E*-Octadecadienoic Acid, *Free Radic. Biol. Med. 10,* 13–21.
36. Banni, S., Day, B.W., Evans, R.W., Corongiu, F.P., and Lombardi, B. (1995) Detection of Conjugated Diene Isomers of Linoleic Acid in Liver Lipids of Rats Fed a Choline-Devoid Diet Indicates That the Diet Does Not Cause Lipoperoxidation, *J. Nutr. Biochem. 6,* 281–289.
37. Kates, M. (1972) Techniques of Lipidology, in *Laboratory Techniques in Biochemistry and Molecular Biology,* (Work, T.S., and Work, E., eds.) North-Holland Publishing Co., Amsterdam, The Netherlands.
38. Christie, W.W. (1982) *Lipid Analysis*, 2nd edn., Pergamon Press, Oxford, U.K.
39. Jensen, R.G. (1989) *The Lipids of Human Milk,* p. 30, CRC Press, Boca Raton, FL.
40. Jiang, J., Bjoerck, L., Fondén, R., and Emanuelson, M. (1996) Occurrence of Conjugated *cis*-9, *trans*-11-Octadecadienoic Acid in Bovine Milk: Effects of Feed and Dietary Regimen, *J. Dairy Sci. 79,* 438–445.
41. Jahreis, G., Fritsche, J., and Steinhart, H. (1997) Conjugated Linoleic Acid in Milk Fat: High Variation Depending on Production System, *Nutr. Res. 17,* 1479–1484.
42. Shantha, N.C., Decker, E.A., and Ustunol, Z. (1992) Conjugated Linoleic Acid Concentration in Processed Cheese, *J. Am. Oil Chem. Soc. 69,* 425–428.

43. Shantha, N.C., Ram, L.N., O'Leary, J., Hicks, C.L., and Decker, E.A. (1995) Conjugated Linoleic Acid Concentrations in Dairy Products as Affected by Processing and Storage, *J. Food Sci. 60*, 695–697,720.
44. Kramer, J.K.G., and Hulan, H.W. (1978) A Comparison of Procedures to Determine Free Fatty Acids in Rat Heart, *J. Lipid Res. 19*, 103–106.
45. Hashimoto, N., Aoyama, T., and Shioiri, T. (1981) New Methods and Reagents in Organic Synthesis. 14. A Simple Efficient Preparation of Methyl Esters with Trimethylsilyldiazomethane ($TMSCHN_2$) and Its Application to Gas Chromatographic Analysis of Fatty Acids, *Chem. Pharm. Bull. 29*,1475–1478.
46. Werner, S.A., Luedecke, L.O., and Shultz, T.D. (1992) Determination of Conjugated Linoleic Acid Content and Isomer Distribution in Three Cheddar-Type Cheeses: Effects of Cheese Cultures, Processing, and Aging, *J. Agric. Food Chem. 40*, 1817–1821.
47. Association of Official Analytical Chemists (1995) AOAC Official Method 969.33, Fatty Acids in Oils and Fats, sec. 41.1.28, *Official Methods of Analysis of AOAC International*, vol. II, (Cunniff, P., ed.) 16th edn., AOAC International, Alexandria, VA.
48. Official Methods and Recommended Practices of the American Oil Chemists' Society, method Ce 1b-89, Champaign, IL 1973 (revised 1990).
49. Ha, Y.L., Storkson, J., and Pariza, M.W. (1990) Inhibition of Benzo(a)pyrene-Induced Mouse Forestomach Neoplasia by Conjugated Dienoic Derivatives of Linoleic Acid, *Can. Res. 50*, 1097–1101.
50. Chen, Z.Y., Chan, P.T., Kwan, K.Y., and Zhang, A. (1997) Reassessment of the Antioxidant Activity of Conjugated Linoleic Acids, *J. Am. Oil Chem. Soc. 74*, 749–753.
51. Folch, J., Lees, M., and Sloane-Stanley, G.H. (1957) A Simple Method for the Isolation and Purification of Total Lipids from Animal Tissues, *J. Biol. Chem. 226*, 497–509.
52. Salminen, I., Mutanen, M., Jauhiainen, M., and Aro, A. (1998) Dietary *trans* Fatty Acids Increase Conjugated Linoleic Acid Levels in Human Serum, *J. Nutr. Biochem. 9*, 93–98.
53. Bligh, E.G., and Dyer, W.J. (1959) A Rapid Method of Total Lipid Extraction and Purification, *Can. J. Biochem. Physiol. 37*, 911–917.
54. Chin, S.F., Storkson, J.M., Liu, W., Albright, K.J., and Pariza, M.W. (1994) Conjugated Linoleic Acid (9,11- and 10,12-Octadecadienoic Acid) Is Produced in Conventional but Not Germ-Free Rats Fed Linoleic Acid, *J. Nutr. 124*, 694–701
55. Stanton, C., Lawless, F., Kjellmer, G., Harrington, D., Devery, R., Connolly, J.F., and Murphy, J. (1997) Dietary Influences on Bovine Milk *cis*-9,*trans*-11-Conjugated Linoleic Acid Content, *J. Food Sci. 62*, 1083–1086.
56. Park, Y., and Pariza, M.W. (1998) Evidence That Commercial Calf and Horse Sera Can Contain Substantial Amounts of *trans*-10,*cis*-12 Conjugated Linoleic Acid, *Lipids 33*, 817–819.
57. Jiang, J., Björck, L., and Fondén, R. (1997) Conjugated Linoleic Acid in Swedish Dairy Products with Special Reference to the Manufacture of Hard Cheeses, *Int. Dairy J. 7*, 863–867.
58. McGuire, M.K., Park, Y., Behre, R.A., Harrison, L.Y., Shultz, T.D., and McGuire, M.A. (1997) Conjugated Linoleic Acid Concentrations of Human Milk and Infant Formula, *Nutr. Res. 17*, 1277–1283.
59. Lin, H., Boylston, T.D., Luedecke, L.O., and Shultz, T.D. (1998) Factors Affecting the Conjugated Linoleic Acid Content of Cheddar Cheese, *J. Agric. Food Chem. 46*, 801–807.

60. Shantha, N.C., Crum, A.D., and Decker, E.A. (1994) Evaluation of Conjugated Linoleic Acid Concentrations in Cooked Beef, *J. Agric. Food Chem. 42*, 1757–1760.
61. Shantha, N.C., Moody, W.G., and Tabeidi, Z. (1997) Conjugated Linoleic Acid Concentration in Semimembranosus Muscle of Grass- and Grain-Fed and Zeranol-Implanted Beef Cattle, *J. Muscle Foods 8*, 105–110.
62. Precht, D., and Molkentin, J. (1997) Effect of Feeding on Conjugated *cis*Δ9, *trans*Δ11-Octadecadienoic Acid and Other Isomers of Linoleic Acid in Bovine Milk Fats, *Nahrung 41*, 330–335.
63. Griinari, J.M., Dwyer, D.A., McGuire, M.A., Bauman, D.E., Palmquist, D.L., and Nurmela, K.V.V. (1998) *Trans*-Octadecenoic Acids and Milk Fat Depression in Lactating Dairy Cows, *J. Dairy Sci. 81*, 1251–1261.
64. Kelly, M.L., Berry, J.R., Dwyer, D.A., Griinari, J.M., Chouinard, P.Y., Van Amburgh, M.E., and Bauman, D.E. (1998) Dietary Fatty Acid Sources Affect Conjugated Linoleic Acid Concentrations in Milk from Lactating Dairy Cows, *J. Nutr. 128*, 881–885.
65. Koritala, S., and Rohwedder, W.K. (1972) Formation of an Artifact During Methylation of Conjugated Fatty Acids, *Lipids 7*, 274.

Chapter 7

Separation of Conjugated Fatty Acid Isomers

John K.G. Kramer[a], Najibullah Sehat[b], Jan Fritsche[b], Magdi M. Mossoba[b], Klaus Eulitz[b], Martin P. Yurawecz[b], and Yuoh Ku[b]

[a]Southern Crop Protection, Food Research Center, Agriculture and Agri-Food Canada, Guelph, ON, Canada N1G 2W1
[b]U.S. Food and Drug Administration, Washington, DC 20204

Introduction

Conjugated fatty acid (CFA) is a general term that by now has come to include many natural and synthetic octadecadienoic (18:2) acids with a conjugated double bond system. The main CFA isomer in milk fat is 9*cis*,11*trans*-octadecadienoic acid (1) which has since been found to be the main CFA isomer in all dairy products and meats from ruminants. Recently, it was proposed that this acid be given the name rumenic acid (2). Even though CFA in dairy products and meats from ruminants is the main emphasis in this review, conjugated fatty acids have also been found in a variety of oilseeds [see review by Hopkins and Chisholm (3)].

We use the abbreviation CFA throughout this chapter to refer to any octadecadienoic acid (18:2) containing a conjugated double bond system. The CFA range from 6,8-18:2 to 12,14-18:2, which have been reported in natural dairy products or in alkali-isomerized linoleic acid (9*cis*,12*cis*-18:2) (4). It is unfortunate that the term conjugated linoleic acid (CLA) was coined because some of these CFA are not derived from linoleic acid either biologically in the rumen or chemically by alkali-isomerization of linoleic acid. We will further restrict our use of CFA to the straight-chain C-18 fatty acids; conjugated fatty acids of different chain length will be clearly specified. A mixture of CFA can therefore be very complex and a challenge for analysts. For every positional CFA, there are four possible geometric pairs of isomers, i.e., *cis,trans*; *trans,cis*; *trans,trans* and *cis,cis*. CFA isomers will be abbreviated as 9*cis*,11*trans*-18:2 or 9,11-18:2, and so on.

In natural products, these CFA occur together with many other fatty acids with different chain lengths, and number, position and geometry of double bonds, plus branched-chain and additional oxygen- or nitrogen-containing functional groups. In milk fat, >400 different fatty acids exist (5). This complex fatty acid mixture plays an amazing number of roles; some of these fatty acids serve as energy for the offspring, and some as a source of essential fatty acids. Others may provide health benefits protecting against cancer, atherosclerosis, and diabetes, or provide immunity and partition fat to lean muscle; see appropriate chapters in this book for references.

The aim of this chapter is to trace the development over the past 60 years that led to our ability to identify and resolve this complex CFA mixture. We will restrict ourselves to what we believe were major steps toward the present state-of-the-art separation techniques. Further developments in the methodology of CFA continue to be reported such as the application of nuclear magnetic resonance (NMR) to the identification of CFA isomers (see Chapter 12). In light of the fact that different CFA isomers may prove to have different biochemical or physiologic responses (6), it is even more critical to be able to separate and identify various CFA and their metabolites with reliability and confidence.

Characteristic Identifiers of CFA

The study of CFA began in the 1930s with the observation that milk fats showed ultraviolet (UV) absorption at 230 nm, which did not increase after brief saponification (7). However, prolonged saponification of animal fats and vegetable oils increased this absorption. In 1951, Nichols *et al.* (8), with the aid of UV absorption, undertook an elaborate fractional crystallization procedure to purify two conjugated octadecadienoic acids from alkali-isomerized linoleic acid, each having different melting points. Based on theoretical considerations, these authors predicted the products of linoleic acid (9*cis*,12*cis*-18:2) to be 9*cis*,11*trans*-18:2 and 10*trans*,12*cis*-18:2 with small amounts of the corresponding *trans,trans* but no *cis,cis* isomers. The structure of the conjugated fatty acids in milk fat, however, was suspected to be a more complex mixture, consisting of geometric and positional isomers of different chain lengths.

The second characteristic of CFA reported in the early 1950s, showed a unique doublet in the infrared (IR) region at 948 and 982 cm^{-1} (9). This doublet could be clearly distinguished from an isolated *trans* double bonds at 968 cm^{-1} and a conjugated *trans,trans* double bond system at 988 cm^{-1} (Fig. 7.1). The IR doublet was first thought to be attributable to a mixture of *cis,cis* and *trans,trans* isomers. However, repeated purification, with the use of several separation techniques, failed to further separate this isolated fraction from isomerized linoleic acid, and the relative intensity of these two IR bands remained the same. Therefore, the two IR absorption peaks were ascribed to the *cis/trans* conjugated system.

In 1963, Riel (10) summarized the results of several reports that had appeared from different countries in which the conjugated dienoic acid content of milk fat was

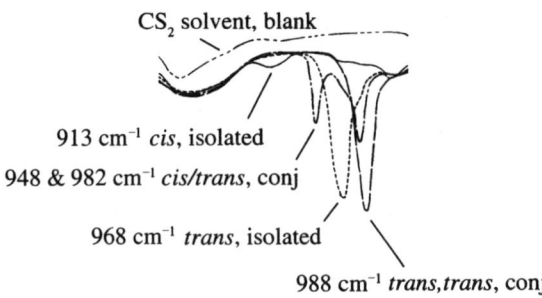

Fig.7.1. Characteristic infrared absorption differences for the =C–H deformation vibration for isolated *trans* double bonds, and conjugated *trans,trans* and *cis/trans* double bonds first reported by Jackson *et al.* (9). [Reproduced by permission of the *Journal of the American Oil Chemists' Society* and redrawn from the original publication.]

investigated on the basis of spectroscopic observations. The range for the conjugated dienoic acids was found to be 0.24–2.81% of the milk fat. However, the identity of the CFA in milk remained unknown, although they were suspected to be C-18 fatty acids. With the introduction of chromatographic techniques in the 1950s, the era of determining the fatty acid composition by UV detection came to an end. However, the characteristic UV and IR absorptions of CFA are still used today as reliable indicators of their presence [see Chapter 10 (Mossoba *et al.*) for recent developments of the IR technique].

Structural Determination of CFA by Chemical and Chromatographic Methods

During the 1960s, lipid research was revolutionized by the application of thin-layer chromatography (TLC), silver nitrate TLC, and gas chromatography (GC). Separations that seemed impossible were achieved, and they became tools to complement the chemical methods developed to undertake structural studies.

The application of these new chromatographic methods was most elegantly demonstrated in the identification of the 9,11-18:2 as an intermediate in the biohydrogenation of linoleic acid (9*cis*,12*cis*-18:2) to *trans* monoenes by the rumen bacteria *Butyrivibrio fibrisolvens* (11). These authors isolated the fatty acid intermediates by preparative GC or silver-ion thin-layer chromatography (Ag$^+$-TLC), cleaved the double bond by reductive ozonolysis, and analyzed the aldehyde and aldehyde-esters by GC. They were able to establish with certainty the double bond position of the CFA as Δ9 and Δ11, with supporting TLC, GC, UV, IR, MS and NMR data, but they were not able to discern their geometric configuration. On the basis of previous theoretical considerations, a 9*cis*,11*trans*-18:2 was more probable than the opposite 9*trans*,11*cis*-18:2 isomer. Two *trans* monoenes were also identified in the reaction mixture as 9*trans*-18:1 and 11*trans*-18:1. These authors clearly showed that these CFA could be separated by GC on a packed column with a polar stationary phase (10% diethylene glycol succinate, DEGS). The methyl ester of *cis/trans* 9,11-18:2 eluted after linoleic acid, and was followed in turn by 9*trans*,11*trans*-18:2 (Fig. 7.2). Interestingly, these authors already noted that acid extraction procedures and acid-catalyzed

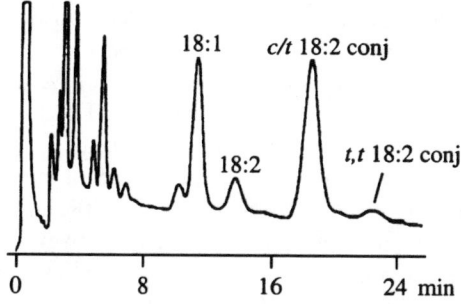

Fig. 7.2. Gas chromatographic separation of the methyl esters of the biohydrogenation by *Butyrivibrio fibrisolvens* of linoleic acid using packed diethylene glycol sucinate (DEGS) columns reported by Kepler *et al.* (11). [Reproduced by permission of the *Journal of Biological Chemistry* and redrawn from the original publication.]

methylations using BF_3 resulted in partial isomerization of the CFA to *trans,trans* fatty acids. To avoid this isomerization, the authors used chloroform/methanol to extract the lipids and methylated the CFA using diazomethane. Furthermore, the authors investigated the dehydration mixture of ricinoleic acid (12-hydroxy-9*cis*-18:1) and identified three CFA in the mixture: 8,10-18:2, 9,11-18:2, and 10,12-18:2.

In 1973, Christie (12) reported the isolation of phosphatidylcholine containing two CFA (18:2 and 18:3) from the bile of cows and sheep. Phosphatidylcholine was isolated by silicic acid column chromatography and preparative TLC, and methylated using sodium methoxide. The CFA were separated by Ag^+-TLC using either hexane/diethyl ether (9:1) or toluene. In the former system, the 18:2 CFA migrated with the *cis*-monoenes; in the latter, the 18:2 CFA migrated ahead of the *cis*-monoenes. The isolated CFA exhibited strong UV absorption at 233 nm; on hydrogenation, they yielded only stearic acid (18:0). Partial hydrazine reduction, isolation of the resultant *cis* and *trans* monoenes, followed by oxidative fission of each monoene, and GC analysis of the mono- and dibasic acid products established the structure of the 18:2 CFA as 9*cis*,11*trans*-18:2. This method clearly differentiated 9*cis*,11*trans*-18:2 from 9*trans*,11*cis*-18:2. The latter geometric isomer, if present, occurred only in very small amounts.

In 1977, Parodi (1) isolated CFA from milk fat after BF_3-methylation and the use of Ag^+- TLC; he used benzene and toluene as developing solvents. He also observed, as stated above (12), that the CFA migrated between the *cis*- and *trans*-monoenes using Ag^+-TLC. The isolated CFA showed the characteristic IR and UV spectrum. This CFA coeluted with linolenic acid by GC using a DEGS or EGSS-X packed column, just after linolenic acid using a BDS packed column, and after 18:0 using a wall-coated open tubular column coated with Apiezon L. The double bond position of the CFA was established as $\Delta 9$ and $\Delta 11$ by reductive ozonolysis. Partial hydrazine reduction of the CFA, followed by Ag^+-TLC separation of the resultant monoenes and subsequent reductive ozonolysis and analysis of the products (aldehydes and aldehyde-esters), established the major CFA in milk fat as 9*cis*,11*trans*-18:2. Small amounts of 8,10-18:2 and 11,13-18:2 were also observed in milk fat lipids, which were attributed to double bond migration during preparative GC, as suggested by Emken *et al.* (13). In many subsequent studies in which milk and dairy products were analyzed, the authors based the assignment of CFA as 9*cis*,11*trans*- 18:2 on this report.

This elaborate chemical/chromatographic procedure was extended successfully to the analysis of two minor CFA isomers, 9*cis*,11*trans*-18:2 and 9*trans*,11*trans*-18:2, in the depot fat of monkeys fed a lard/corn oil mixture (14). However, several steps used in this procedure, such as preparatory GC (13), preparative Ag^+-TLC, and excessive handling, were believed to cause either rearrangement, isomerization, or oxidation. The availability of new spectroscopic techniques, such as GC-MS of fatty acid derivatives that measure directly the double bond position, GC-FTIR, and NMR, that the chemical/chromatographic methods are not used today. However, several chemical methods still serve as useful complementary techniques in structural determinations today, particularly partial hydrazine reduction and Ag^+-TLC.

Separation of CFA Isomers Using Capillary GC Columns

The resolution of CFA present in natural products using packed GC columns is severely limited because the amounts of CFA are generally very small (<1% of total FAME), and the CFA isomers coelute with linolenic or eicosenoic (20:1) acids using many of the polar GC phases. On the other hand, nonpolar columns may be useful for preparative GC because CFA elute after 18:1 (1).

Separation of Synthetic CFA Isomers

The separations of CFA using GC stainless-steel capillary columns were much more promising compared to packed GC columns. In 1971, Scholfield and Dutton (15) reported an extensive comparison of many synthetic unsaturated fatty acids and CFA, establishing their equivalent chain length (ECL) by GC using 61-m stainless-steel capillary columns coated with four different polar GC coatings. Similar results were obtained using a 50-m Silar 10C capillary column (16); see summary in Table 7.1. Their results indicated three general observations. First, the elution order of a given positional CFA isomer is as follows: *cis/trans* followed by *cis,cis* followed by *trans,trans*. Second, *cis,trans* eluted before *trans,cis* for a given pair of positional *cis/trans* CFA isomers. Third, the elution time increased by moving the double bond from Δ9 to Δ10 for an identical geometric pair of *cis/trans* or *cis,cis* isomers; for the *trans,trans* CFA isomers, however, there was no consistent pattern using any of the four GC coatings (Table 7.1). The elution order for the *trans,trans*-18:2 isomers was reported as 11,13-18:2, 9,11-18:2, and 10,12-18:2 (16). Similar results confirmed the first two observations stated above were obtained using a 30-m SP2380 capillary column with (17) or without deuterium labeling (18). A nonpolar capillary column

TABLE 7.1 Comparison of Equivalent Chain Length (ECL) of Conjugated Fatty Acids (CFA) Determined by Different Gas Chromatography (GC) Capillary Columns

CFA isomers[a]	GC stationary phases				
	PPE[b]	DEGS[b]	XF1150[b]	TCPE[b]	Silar 10C[c]
9c,11t	19.04	20.10	20.10	20.60	20.72
9t,11c	—	—	20.15	20.65	—
10c,12t	19.07	—	—	—	—
10t,12c	19.16	20.30	20.24	20.77	20.8
9c,11c	19.31	20.32	—	20.87	—
10c,12c	19.32	—	—	—	—
11t,13t	—	—	—	—	21.19
10t,12t	19.61	20.72	20.57	21.16	21.23
9t,11t	19.60	20.70	20.59	21.19	21.22

[a]CFA, conjugated fatty acid isomers; *c, cis; t, trans*; DEGS, diethylene glycol succinate; PPE, polyphenyl ether; TCPE, tetracyanoethylated pentaerythritol.
[b]*Source:* Scholfield and Dutton (15).
[c]*Source:* Scholfield (16).

(OV-1) gave a similar elution pattern of the three geometric CFA isomers (19). Regarding the elution order of the *trans,trans* CFA isomers, recent studies indicate that there is a consistent decrease in elution order of the *trans,trans* CFA isomers from 12,14- to 7,9-18:2 isomers (4).

Separation of CFA Isomers from Biological Matrices

The identification of two CFA in the depot fat of monkeys fed a lard/corn oil mixture was the first report to identify CFA in biological matrices using a 46-m stainless-steel open-tubular column coated with either Silar 7CP or Silar 5 CP (14). On both of these GC phases, the 9*cis*,11*trans*-18:2 and 9*trans*,11*trans*-18:2 eluted in the linolenic acid to 20:1 region.

Separation of CFA Isomers Using the 60-m Supelcowax-10 Column

In 1987, Pariza's group reported the presence of a lipid fraction in fried ground beef, which was shown to have anticarcinogenic activity. They identified the active component as CFA (20). This lipid fraction was isolated by reversed-phase semi-preparatory HPLC and exhibited a strong UV absorption at 233 nm. The mixture was methylated with BF_3 and was shown to consist of five peaks when analyzed by GC using a 60-m Supelcowax-10 fused silica capillary column (Supelco Inc., Bellefonte, PA). All peaks had a molecular ion at *m/z* 294. This lipid fraction was shown to be identical to a synthetic mixture of CFA produced by alkali-isomerization of linoleic acid.

In a subsequent paper, Pariza's group reported the separation and identity of the CFA isolated from a cheese sample and the synthetic mixture obtained by alkali-isomerized linoleic acid in their laboratory (21). They separated the CFA as methyl esters (prepared by BF_3) using a 60-m Supelcowax-10 fused silica capillary column (Fig. 7.3). Seven CFA isomers were separated and identified in cheese lipids by comparison of their equivalent chain length (ECL) data to published ECL results (Table 7.1), and by GC-chemical ionization mass spectrometry (GC-CIMS). Some of the CFA isomers do not appear to be correctly identified and should be reinvestigated using appropriate GC-MS derivatives. It would appear that peak 1 is 9*cis*,11*trans*-18:2, peak 2 is a mixture of 9*trans*,11*cis*-18:2 and 10*cis*,12*trans*-18:2, and peak 4 is not 11*cis*,13*cis*. In addition, peak 7 is a major peak because of extensive *cis* to *trans* isomerization resulting from the BF_3-catalyzed methylation procedure used. In response to this isomerization problem, the temperature of the BF_3-(22) or HCl-catalyzed (23) methylation procedures was reduced, and base-catalyzed methylation methods were introduced by others (24,25) (see further discussion in Chapter 6).

The choice of GC column depends totally on the purpose of the study. If only the major CFA isomer in dairy products or biological matrices is to be determined, then the 60-m Supelcowax-10 fused silica capillary column, or an equivalent capillary column, is perfectly satisfactory. The advantages are that the identity of the different CFA is easily recognizable by comparison with a standard mixture, and CFA isomers elute in a region of the GC chromatogram that is fairly free of interfering

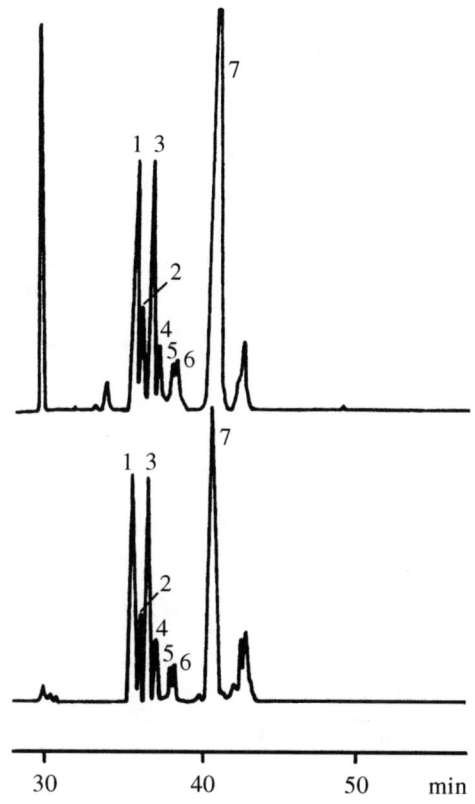

Fig. 7.3. Gas chromatographic separation of the methyl esters of a lipid fraction from cheese (upper panel) and the products of alkali-isomerization of linoleic acid (lower panel) using a 60-m Supelcowax-10 fused silica capillary column reported by Ha et al. (21). The CFA isomers were identified as follows: 1) c-9,t-11/t-9,c-11; 2) c-10,t-12; 3) t-10,c-12; 4) c-11,c-13; 5) c-9,c-11; 6) c-10,c-12; 7) t-9,t-11/t-10,t-12. [Reproduced by permission of the authors and the *Journal of Agriculture and Food Chemistry*, and redrawn from the original publication.]

fatty acids present in biological matrices. It is therefore not surprising that this GC column has been and continues to be used today (Table 7.2). Several other 30- and 50-m capillary columns of similar polarity and resolution have been used (see Table 7.2). However, if a detailed analysis of the minor CLA isomers is desired, or if different CFA isomers are to be investigated, one of the longer GC columns should be considered.

Separations Using 100-m Capillary Columns

A 100-m fused silica capillary column, such as SP 2560 (Supelco), CP Sil 88 (Chrompack Inc., Middelburg, The Netherlands), or any other 100-m polar capillary column, is highly recommended for the analysis of CFA isomers. Up to 10 CFA isomers are resolved in a typical commercial CFA mixture (Fig. 7.4, left side) (53), which is dominated by four positional isomers as follows: 8,10-, 9,11-, 10,12-, and 11,13-18:2 known to be present in this mixture from Ag^+-HPLC results (Fig. 7.4, right side) (58). However, even the 100-m column resolves neither 9*cis*,11*trans*-18:2 and 8*trans*,10*cis*-18:2 (a partial split or a tailing shoulder may occur) nor three (10,12-, 9,11-, and 8,10-18:2) of the four *trans,trans* CFA isomers (53) of this commercial

TABLE 7.2 Comparison of Extraction Methods, Methylation Conditions and GC Columns Used in the Analysis of CFA in the past 10 Years[a]

Authors, (year), (reference)	Extraction	Methylation catalyst/condition[b]	GC column/length
Fogerty et al. (1988) (26)	Bligh and Dyer (27)	NaOCH$_3$	25 m BP20-0.5
Ha et al.(1989) (21)	C/M	(i) NaOH; (ii) BF$_3$	60 m Supelcowax-10
Ha et al.(1990) (28)	C/M	HCl	60 m Supelcowax-10
Ip et al. (1991) (29)	C/M	HCl	60 m Supelcowax-10
Chin et al. (1992) (30)	Bligh and Dyer (27)	4% HCl 60°C 20 min	60 m Supelcowax-10
Shantha et al. (1992) (31)	Acid digestion/ether:hexane	BF$_3$	60 m Supelcowax-10
Werner et al. (1992) (22)	Folch et al. (32)	(i)NaOH; (ii)BF$_3$ RT 30 min	60 m Supelcowax-10
Shantha et al. (1993) (24)	Bligh and Dyer (27)	TMG; NaOCH$_3$; BF$_3$	60 m Supelcowax-10
Chin et al. (1994) (33)	Folch et al. (32)	4% HCl 60°C 20 min	60 m Supelcowax-10
Shantha et al. (1994) (34)	Isopropanol/hexane	TMG	60 m Supelcowax-10
Lin et al. (1995) (35)	Bligh and Dyer (27)	(i)NaOH; (ii)BF$_3$ RT 30 min	60 m Supelcowax-10
Shantha et al. (1995) (36)	Acid digestion/ether:hexane	TMG	60 m Supelcowax-10
Ip et al. (1996) (37)	Bligh and Dyer (27)	4% HCl 60°C 20 min	60 m Supelcowax-10
Jiang et al. (1996) (38)	Isopropanol/hexane	(i)NaOH; (ii)BF$_3$ RT 30 min	50 m CP Sil 88
Belury et al. (1997) (39)	Bligh and Dyer (27)	TMG	30 m Omegawax 320
Chen et al. (1997) (40)	No extraction required	BF$_3$ 90%C 45 min	100 m SP 2560
Ip et al. (1997) (41)	Bligh and Dyer (27)	4% HCl 60°C 20 min	60 m Supelcowax-10
Jahreis et al. (1997) (42)	C/M	NaOCH$_3$	100 m SP 2560
Jiang et al. (1997) (43)	Isopropanol/hexane	(i) KOH; (ii) BF$_3$ RT	50 m CP Sil 88
Kramer et al. (1997) (25)	C/M	NaOCH$_3$ or HCl	100 m SP 2560
McGuire et al. (1997) (44)	Bligh and Dyer (27)	(i)NaOH; (ii)BF$_3$ RT 30 min	60 m Supelcowax-10
Precht et al. (1997) (45)	Butter fat	—	100 m CP Sil 88
Sébédio et al. (1997) (46)	Folch et al. (32)	(i) NaOH; (ii) HCl	50 m BPX 70
Shantha et al. (1997) (47)	Isopropanol/hexane	TMG	60 m Supelcowax-10
Stanton et al. (1997) (48)	Fat by centrifugation	4% HCl 60°C 20 min	60 m Supelcowax-10
Sugano et al. (1997) (49)	Folch et al. (32)	BF$_3$	60 m Supelcowax-10
Griinari et al. (1998) (50)	Isopropanol/hexane	NaOCH$_3$	100 m CP Sil 88
Jensen et al. (1998) (51)	Folch et al. (32)	Not specified	60 m Supelcowax-10
Kelly et al. (1998) (52)	Isopropanol/hexane	NaOCH$_3$	60 m Supelcowax-10
Kramer et al. (1998) (53)	Cold pulverization; C/M	NaOCH$_3$	100 m CP Sil 88
Lavillonnière et al. (1998) (54)	Isopropanol/hexane	NaOCH$_3$	50 m BPX 70
Li and Watkins (1998) (55)	Cold pulverization; sep pack	NaOCH$_3$	30 m DB 23; DB 225

TABLE 7.2 *(continued)*

Authors, (year), (reference)	Extraction	Methylation catalyst/condition[b]	GC column/length
Lin et al. (1998) (56)	Bligh and Dyer (27)	(i)NaOH; (ii)BF$_3$ RT 30 min	60 m Supelcowax-10
Park and Pariza (1998) (23)	C/M	4% HCl 60°C 20 min	60 m Supelcowax-10
Salamin et al. (1998) (57)	Dichloromethane/M	Acid methylation	60 m SP 2380
Sehat et al. (1998) (58)	K oxalate/ether:hexane	NaOCH$_3$	100 m CP Sil 88

[a]Abbreviations: GC, gas chromatography; CFA, conjugated fatty acid; C/M, chloroform/methanol; RT, room temperature; TMG, tetramethylguanidine.
[b]Some methylations were undertaken in two steps (i) and (ii). In one reference, the methylation catalyst was not specified.

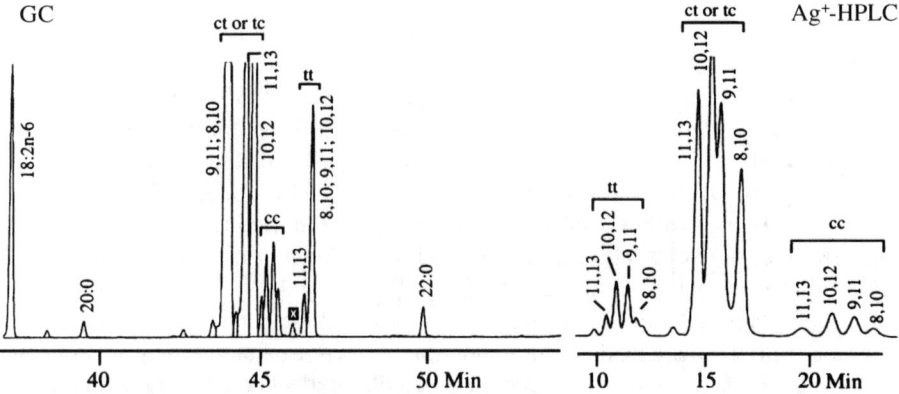

Fig. 7.4. Partial gas chromatography (GC) and silver-ion high-performance liquid chromatography (Ag$^+$-HPLC) profiles of the methyl esters of a commercial conjugated fatty acid (CFA) mixture. For the GC separation, a 100-m SP2560 fused silica capillary column was used with flame ionization detection (FID), and a single 25-cm ChromSpher 5 Lipids column was used for the Ag$^+$-HPLC separation and detection at 233 nm (53). [Reproduced by permission of *Lipids* and redrawn from the original publication.]

CFA mixture. The identities of all of the CFA isomers were confirmed by GC-EIMS as their 4,4'-dimethyloxazoline (DMOX) derivatives, by GC-FTIR, and by comparison to Ag$^+$-HPLC separations (4,53,58).

Recently, an attempt was made to resolve all eight possible *cis/trans* CFA isomers of 8,10-, 9,11-, 10,12-, and 11,13-18:2. A commercial mixture of the four positional CFA isomers was isomerized by I$_2$ (59), and the resultant methyl esters of the CFA were analyzed by GC using a 100-m CP Sil 88 fused silica capillary column (60) (Fig. 7.5B). All eight *cis/trans* CFA isomers were resolved and the contents of all *trans,trans* CFA isomers was greatly increased. The identity of each of the *cis/trans* peaks was established by comparison with the original CFA mixture (Fig. 7.5A).

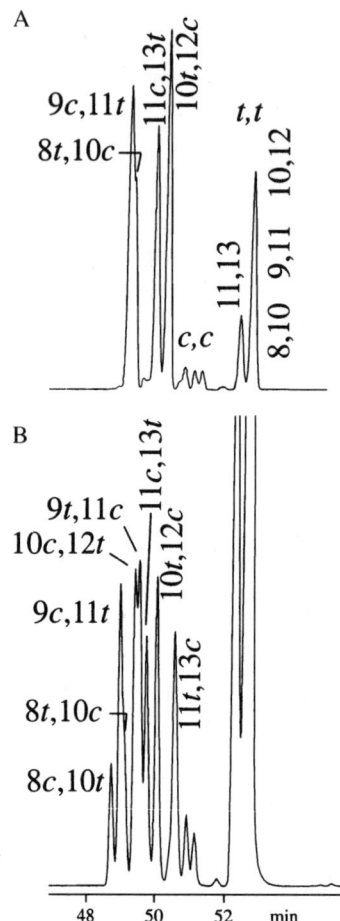

Fig. 7.5. Partial gas chromatography (GC) profile of (A) the methyl esters of a commercial conjugated fatty acid (CFA) mixture, and (B) the I_2-isomerized product of the CFA methyl ester mixture using a 100-m CP Sil 88 fused silica capillary column and flame ionization detection (FID) (60). [Reproduced by permission of *Lipids* and redrawn from the original publication.]

Portions of reference 9*cis*,11*trans*-18:2 and 10*trans*,12*cis*-18:2 isomers obtained from Matreya Inc., (Pleasant Gap, PA) were similarly isomerized by I_2 (59) or *p*-toluenesulfinic acid (61) and analyzed by GC under the same conditions, or co-injected with the isomerized eight *cis/trans* isomer mixture. The elution order found confirmed the order previously predicted by Sehat *et al.* (4) and shown in Table 7.3. The elution order of the *cis/trans* CFA isomers increases as the Δ *cis* value increases; for identical Δ *cis* values, the lower Δ *trans* value elutes first. There is an overlap of *cis/trans* CFA eluting after 10*trans*,12*cis*-18:2 and the *cis,cis* CFA isomers, namely, 11*trans*,13*cis*-18:2 coeluted with 9*cis*,11*cis*-18:2 (Fig. 7.5B).

It should be noted, that this separation (Fig. 7.5B) occurs under optimum conditions in which the relative concentration of all of the *cis/trans* CFA isomers is essentially similar. However, natural products consist of one major CFA isomer, which obscures many of the minor isomers, as seen in the GC separations of cheese CFA (4,54). In such cases, reconstructed MS ion chromatograms (4,62), selected MS ion

TABLE 7.3 Gas Chromatographic Elution Order of Positional and Geometric Conjugated Fatty Acid (CFA) Methyl Ester Isomers on a 100-m CP-Sil 88 Capillary Column[a]

cis/trans-18:2[b]	cis,cis-18:2[c]	trans,trans-18:2[d]
7c,9t	(7c,9c)[e]	12t,14t
(6t,8c)	8c,10c	11t,13t
8c,10t	9c,11c	10t,12t
7t,9c	10c,12c	9t,11t
9c,11t	11c,13c	8t,10t
8t,10c	12c,14c	7t,9t
10c,12t		
9t,11c		
11c,13t		
10t,12c		
12c,14t		

[a]The elution order was as follows: all of the cis/trans, followed by all of the cis,cis, followed by all of the trans,trans CFA positional isomers. Several cis/trans CFA isomers overlapped with some cis,cis geometric isomers.
[b]The observed elution time of cis/trans CFA isomers increased as the Δ value of the cis double bond increased in the molecule. For a pair of cis/trans isomers in which the cis double bond has the same Δ cis value, the isomer with the lower Δ trans value eluted first. Therefore, it followed that for the same positional isomer, the cis,trans eluted before the trans,cis geometric isomer.
[c]The observed elution time of cis,cis CFA isomers increased with increased Δ values.
[d]The observed elution time of trans,trans CFA isomers increased with decreased Δ values.
[e]CFA isomers shown in parentheses were predicted.

scans (63), a combination of chemical, chromatographic, and MS methods (54), or the method of co-injection (60,62) are necessary to identify the different CFA isomers (see also Chapter 9). We have successfully used these techniques in the identification of CFA isomers in milk fat and cheese samples (4,62). In addition to the major 9cis,11trans-18:2 identified in milk fat by Parodi (1), six more CFA isomers were reported by Ha et al. in cheese (21), six more by Lavillonnière et al. (54), three more by Yurawecz et al. (62), and seven more by Sehat et al. (4). The sum of the CFA isomers in cheese is misleading because misidentifications were made. The total number of CFA isomers in cheese determined to date by GC and Ag$^+$-HPLC (discussed below) is 20; they are listed in Tables 7.3 and 7.4.

Separation of CFA by Silver-Ion Thin-Layer Chromatography (Ag$^+$-TLC)

The cis/trans CFA can be separated as their methyl esters by Ag$^+$-TLC and the use of the developing solvents hexane/diethyl ether (9:1), benzene, or toluene (12). Using the first solvent system, the cis/trans CFA migrate together with the cis-monoenes, and, with either of the last two developing solvents, between the cis- and trans-monoenes (1,12). However, care should be taken not to separate geometric CFA isomers

TABLE 7.4 Elution Order of Positional and Geometric Conjugated Fatty Acid (CFA) Methyl Ester Isomers by Silver-Ion High-Performance Liquid Chromatography (Ag⁺-HPLC)[a]

trans,trans-18:2[b]	cis/trans-18:2[b,c]	cis,cis-18:2[b]
(13t,15t)[d]	(13,15 c/t)	(13c,15c)
12t,14t	12,14 c/t	12c,14c
11t,13t	11t,13c	11c,13c
10t,12t	11c13t	10c,12c
9t,11t	10,12 c/t	9c,11c
8t,10t	9c,11t	8c,10c
7t,9t	9t,11c	7c,9c
(6t,8t)	8,10 c/t	
	7t,9c	

[a]The elution order was as follows: all of the *trans,trans*, followed by all of the *cis/trans*, followed by all of the *cis,cis* CFA positional isomers.
[b]The observed elution order of each group of geometric CFA isomers increased as the Δ values decreased.
[c]For the same positional isomer, the *cis,trans* and the *trans,cis* geometric isomers were resolved for odd pairs of Δ values, but not for the odd Δ values (given as c/t). The elution order was found to be opposite for the 11,13-18:2 compared with the 9,11-18:2 geometric isomers. The elution order for 7,9-18:2 has not been determined.
[d]CFA isomers shown in parentheses were not confirmed.

unintentionally. Ackman *et al.* (14) showed that the *cis/trans* (R_f 0.61) separated from the *trans,trans* CFA isomers (R_f 0.65) by Ag⁺-TLC. In general, this method is potentially useful to collect larger amounts of CFA and possible metabolites from biological matrices.

Separation of CFA by Silver-Ion High-Performance Liquid Chromatography (Ag⁺- HPLC)

Commercial Ag⁺-HPLC columns are presently available from Chrompack. These columns are based on the discovery by Christie that a silica matrix containing chemically bonded phenylsulfonic acid (Nucleosil) can exchange their protons for silver ions that will not elute with organic solvents (see review by Dobson *et al.* 64). Because ~30% of the original silanol groups in these Ag⁺-HPLC columns are free, these columns show a partial normal-phase character as well as the expected binding of the π-electron of the double bonds to the silver ion (65). The combined effects explain many of the separations observed using this Ag⁺-HPLC column, such as the separation of mixtures of triacylglycerols containing saturated fatty acids (65), and positional and geometric isomers of monounsaturated (66–68), diunsaturated (17), triunsaturated (69), and deuterium-labeled unsaturated fatty acids (70), or mixtures thereof (18).

Separation of CFA by Ag⁺-HPLC

In 1997, Adlof (17) reported the synthesis of 9*cis*,11*trans*-18:2 and 9*trans*,11*cis*-18:2, labeled with deuterium in carbon positions 17,17,18,18-d_4. These two isomers separated by GC (indicated above), but not by Ag$^+$-HPLC, even with the use of two columns in series. Adlof used 0.35% acetonitrile in hexane as the mobile phase. Sehat *et al.* (58) reported that Ag$^+$-HPLC separated all of the positional and geometric isomers of CFA as their methyl esters using 0.1% acetonitrile in hexane as the mobile phase, operated isocratically, and using UV detection at 233 nm. The original separation was performed using one Ag$^+$-HPLC column (Fig. 7.4 right side). Sehat *et al.* now tested up to six Ag$^+$-HPLC columns in series with successive improvements in resolution (71). The separation presented in Figure 7.6A is that of a commercial CFA mixture using three Ag$^+$-HPLC columns in series, which was regarded as adequate to resolve most CFA isomers at moderate pressures of the columns (71). In general, commercial samples of CFA were shown to contain four major positional CFA isomers (8,10-, 9,11-, 10,12-, and 11,13-18:2), all of which were identified and characterized by GC-EIMS, GC-DD-FTIR, and by comparison to standards (58,71).

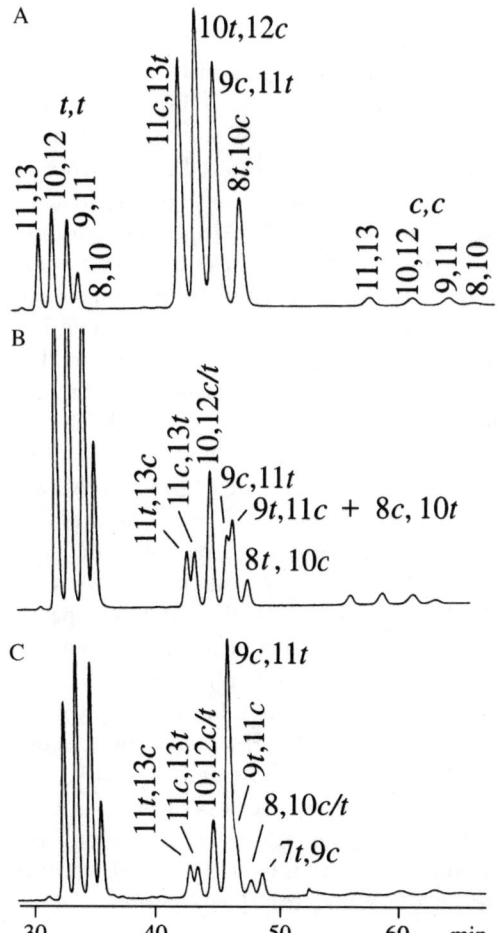

Fig. 7.6. Partial silver-ion high-performance liquid chromatography (Ag$^+$-HPLC) profile of (A) the methyl esters of a commercial conjugated fatty acid (CFA) mixture, (B) the I$_2$-isomerized product of the CFA methyl ester mixture, and (C) the I$_2$-isomerized product of the CFA methyl ester mixture co-injected with cheese total lipid fatty acid methyl esters (60). Detection at 233 nm. [Reproduced by permission of *Lipids* and redrawn from the original publication.]

Separation of Isomerized CFA Isomers

Separation of the I_2-isomerized commercial CFA methyl ester mixture (see above), which contained all eight possible *cis/trans* isomers is shown in Figure 7.6B (60). As expected, the content of all *trans,trans* isomers was greatly increased during I_2-isomerization. Unlike the GC separation in which all of the eight *cis/trans* isomers were resolved (Fig. 7.5B), only the 8,10-, 9,11-, and 11,13-18:2, but not 10,12-18:2 *cis/trans* CFA were separated using three Ag^+- HPLC columns in series. Not even six Ag^+-HPLC columns in series resolved the 10,12-18:2 geometric pair of CFA isomers (unpublished data). This phenomenon may be due to the structure of CFA and appears to be similar to the alternating melting and boiling point patterns of fatty acids (72). It was of interest to note that the *cis/trans* geometric isomer pairs of 11,13-18:2 and 9,11- 18:2 eluted in the reverse order. The main 11,13-18:2 isomers in cheese (11*trans*,13*cis*-18:2) eluted before the 11*cis*,13*trans* isomer present in commercial preparations (4,71). The identity of the 9*cis*,11*trans*-18:2 peak was confirmed by co-injecting the isomerized FAME mixture with the methyl esters prepared from cheese lipids in which 9*cis*,11*trans*-18:2 is the major CFA isomer (Fig. 7.6C). The 9*cis*,11*trans*-18:2 isomer clearly eluted before the 9*trans*,11*cis*-18:2 isomer. The 8*cis*,10*trans*-18:2 coeluted with 9*trans*,11*cis*-18:2 as determined by quantitative measurements (60). Coincidentally, the co-injected mixture showed the presence of an additional peak due to 7*trans*,9*cis*-18:2 identified in cheese lipids (4,62). In addition, samples of reference 9*cis*,11*trans*-18:2 and 10*trans*,12*cis*-18:2 (Matreya Inc.) were isomerized with I_2, and used as a standard to identify peaks due to the 9,11- and 10,12-18:2 *cis/trans* isomers. These results confirm the predicted elution pattern of CFA isomers reported previously (4); see the elution order presented in Table 7.4.

The two reports by Adlof's group (17,18), in which the same Ag^+-HPLC columns were used to separate CFA isomers, require some explanation. The first report indicated no separation of the deuterated (d_4) 9*cis*,11*trans*-18:2 and 9*trans*,11*cis*-18:2 isomers despite the use of two Ag^+- HPLC columns in series (17). In the second case, 9*cis*,11*trans*-18:2 was reported to elute after 9*trans*,11*cis*-18:2 (18), which is inconsistent with the data reported in Table 7.4. There is no explanation for this observation. The only apparent difference in these reports is the concentration of acetonitrile used as the mobile phase [0.1% (58) vs. either 0.35% (17) or 0.5% (18)], which may have caused this inversion. To test this principle, the I_2-isomerized commercial CFA mixture was analyzed by using increasing amounts of acetonitrile from 0.05 to 0.4% with hexane used as the mobile phase (Fig. 7.7). The resolution of all of the *cis/trans* CFA isomers was optimum at 0.1% acetonitrile in hexane and a flow rate of 1 mL/min as reported previously (58). Increasing the concentration of acetonitrile decreased the retention time and sharpened the peaks, but the resolution was progressively lost. At no time was an inversion of the two 9,11-18:2 geometric isomer pairs observed. At 0.05% acetonitrile in hexane, some CFA isomers were baseline resolved, but the retention time was also doubled, and the 9,11-18:2 geometric pair did not resolve.

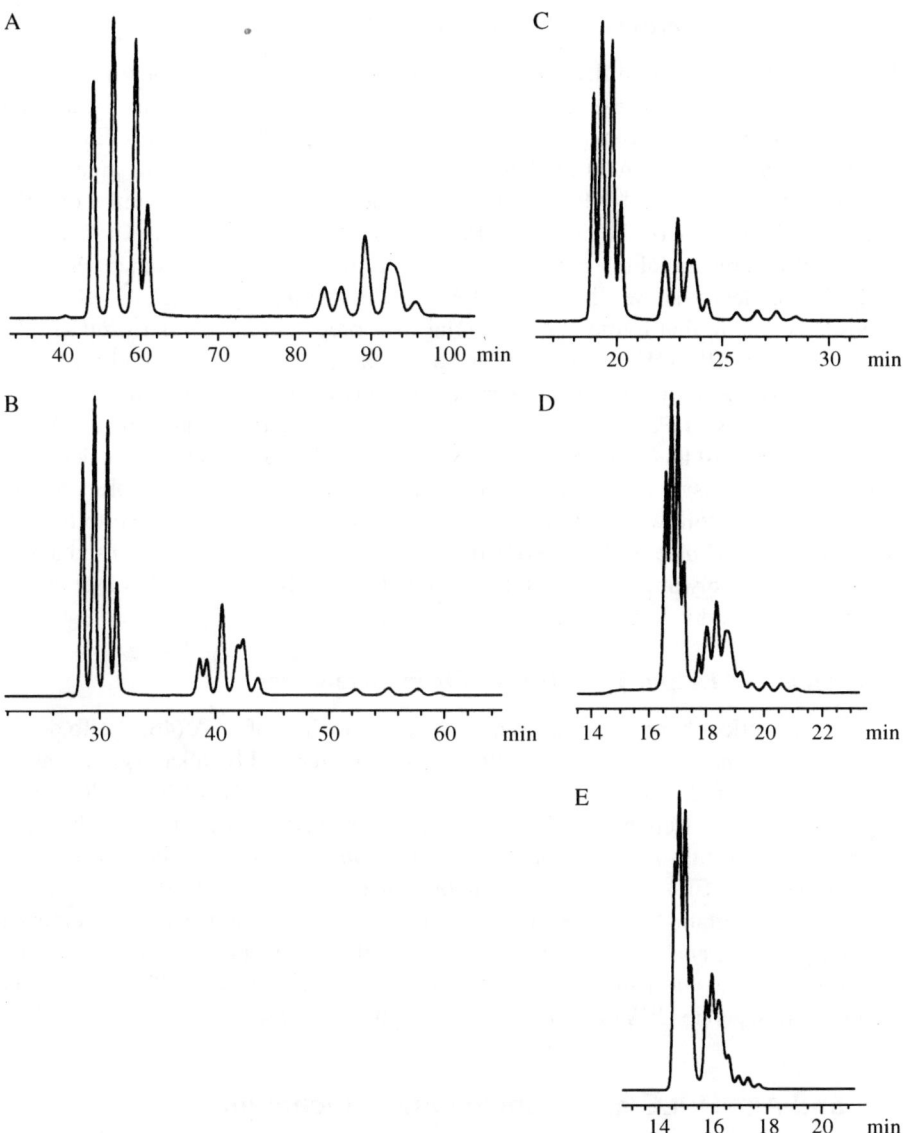

Fig. 7.7. Partial silver-ion high-performance liquid chromatography (Ag^+-HPLC) profiles of the I_2-isomerized product of a commercial conjugated fatty acid (CFA) methyl ester mixture, using increasing amounts of acetonitrile (ACN) in hexane as the mobile phase and a flow rate of 1 mL/min: (A) 0.05% ACN; (B) 0.1% ACN; (C) 0.2% ACN; (D) 0.3% ACN; and (E) 0.4% ACN. Detection at 233 nm (unpublished data).

Separation of CFA Present in Natural Products

This method was successfully used to separate and identify numerous minor new CFA isomers in milk, cheese, beef, and human milk and adipose tissue (4,62). The 9cis,11trans-18:2 was the major CFA isomer present at ~80% of total CFA (4,62). The resolution of the CFA in the dairy products (cheese and milk) was greatly improved by using three Ag^+-HPLC columns in series; compare the resolution of CFA isomers in Figure 7.4 (right side) with Figure 7.8. Co-injection of milk CFA methyl esters with a commercial mixture containing the four *cis/trans* CFA isomers (8,10- to 11,13-18:2), clearly shows that milk CFA contain 11*trans*,13*cis*-18:2, which is the opposite isomer to that found in commercial CFA preparations (11*cis*,13*trans*-18:2). In addition, Ag^+-HPLC helped identify and quantitate the isomeric distribution of CFA isomers in tissues of pigs fed a commercial CFA mixture (see Chapter 18).

All of the Ag^+-HPLC results of natural products were based on using a UV detector at 233 nm. At this wavelength, unsaturated fatty acids are generally not detected, unless they are present at very high concentration in a sample. For example, one can see a small peak following the *cis/trans* isomers in Figure 7.8 that is due to the absorption of methyl oleate. The use of a dual UV wavelength detector was used to detect CFA selectively at 233 nm and unsaturated fatty acids at 205 nm to observe any interfering peaks due to the latter.

Separation of CFA and Their Longer Chain Metabolites

Biological matrices have been shown to contain metabolites of CFA (46,73). To determine their elution order, conjugated fatty acids were prepared by alkali-isomerization of 11*cis*,14*cis*-20:2. The two C-20 conjugated dienes (11*cis*,13*trans*-20:2 and 12*trans*,14*cis*-20:2) were resolved from the corresponding C-18 CFA (71), as shown in Figure 7.9. The results indicated that the C-20 *cis/trans* CFA isomers eluted just before the C-18 *cis/trans* CFA isomers, but still in the general *cis/trans* CFA region of the Ag^+-HPLC chromatogram. Therefore, if C-20 and C-22 CFA metabolites are not removed before Ag^+-HPLC analysis, there will be overlap of these isomers, particularly when the total lipids of biological matrices are investigated. Reversed-phase HPLC is a useful technique to separate CFA of different chain length (see below).

GC and Ag^+-HPLC as Complementary Techniques

In the total analysis of CFA, the application of both GC using very long capillary columns and Ag^+-HPLC is mandatory. Some comparisons of the advantages and disadvantages of each method follow: GC, using flame ionization detection (FID), measures all fatty acids from which the total content of CFA in a sample can be determined. Care should be taken to ensure that other fatty acids are not eluting in the CFA region, or coeluting with CFA isomers, which will lead to misidentification. For example, in the analysis of cheese (4) one peak was marked as "11*t*,13*c*?" in Figure 7.1 of reference (4). This peak turned out not to be a CFA as demonstrated by monitoring

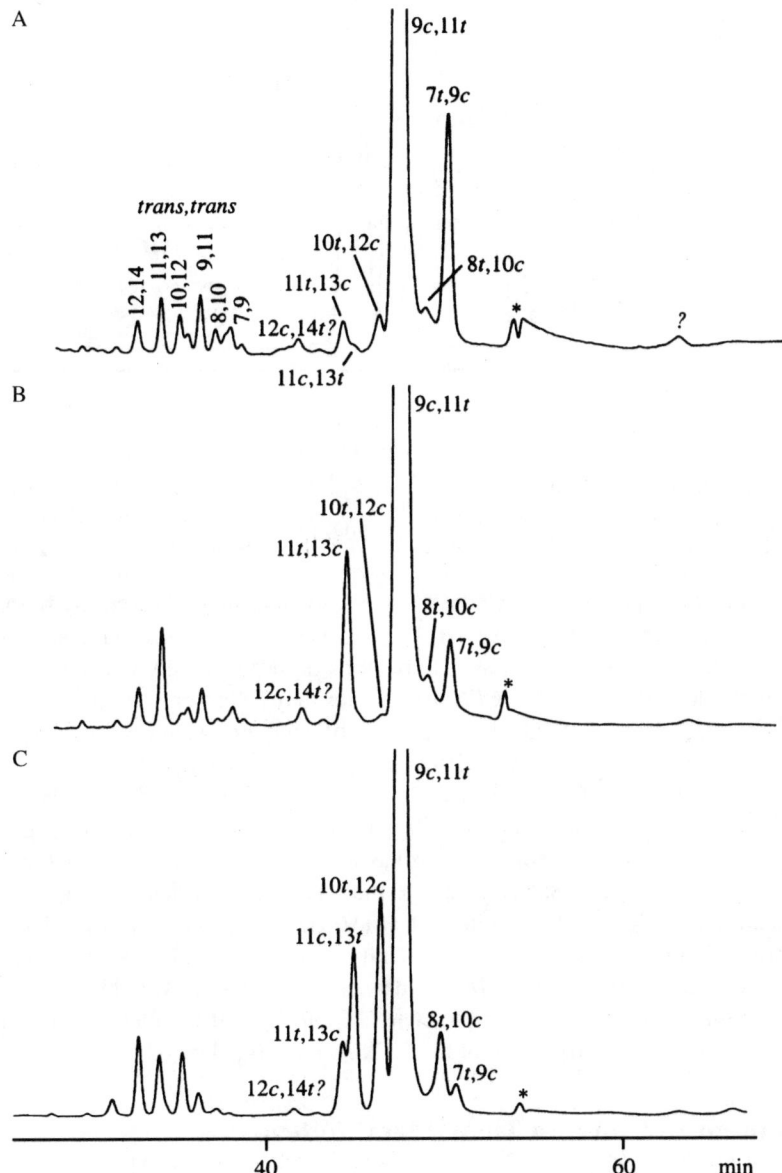

Fig. 7.8. The conjugated fatty acids from (A) total cheese and (B) milk lipids separated by silver-ion high-performance liquid chromatography (Ag$^+$-HPLC) using three Ag$^+$-HPLC columns in series and detection at 233 nm. Panel (C) is a separation of total milk lipids co-injected with a commercial conjugated fatty acid (CFA) methyl ester mixture (71). The main 11,13-18:2 CFA isomer in dairy products is 11*trans*,13*cis*-18:2, whereas that in the commercial CFA mixture is 11*cis*,13*trans*-18:2; *absorption due to methyl oleate. [Reproduced by permission of *Lipids* and redrawn from the original publication.]

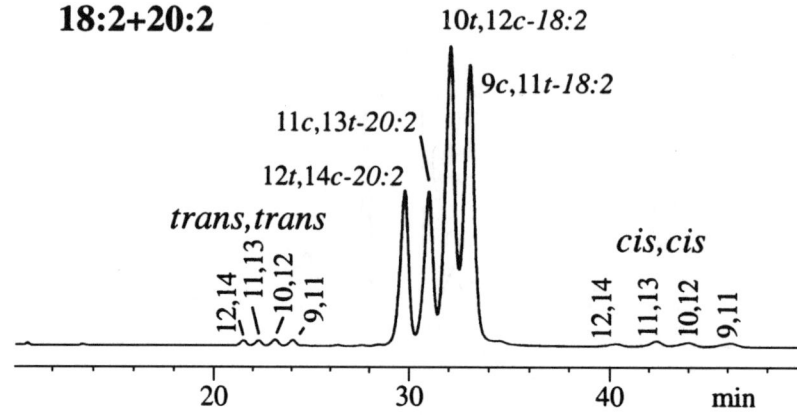

Fig. 7.9. A separation of the alkali-isomerized products of linoleic (9*cis*,12*cis*-18:2) and 11*cis*,14*cis*-eicosadienoic acid by silver-ion high-performance liquid chromatography (Ag$^+$- HPLC) using two Ag$^+$-HPLC columns in series (71). [Reproduced by permission of *Lipids* and redrawn from the original publication.]

for the molecular ion of C-18 CFA of cheese fatty acid methyl esters by high resolution GC-MS (see Chapter 9). Ag$^+$-HPLC with UV detection at 233 nm selectively detects and resolves most CFA. Other fatty acids do not generally interfere because they do not absorb at 233 nm, but CFA metabolites may be present, and they will elute in the same region (Fig. 7.9). GC resolves pairs of geometric isomers (*cis,trans* from *trans,cis*; Fig. 7.5B), whereas Ag$^+$-HPLC separates only those with odd (9,11- and 11,13-18:2) but not even (8,10- and 10,12-18:2) Δ values (Fig. 7.6B). Ag$^+$-HPLC resolves all of the *trans,trans* and *cis,cis* CFA isomers without any overlapping interference from *cis/trans* CFA isomers, whereas in GC, the *trans,trans* isomers with Δ values from 7,9- to 10,12-18:2 coelute, and there is partial overlap of some *cis/trans* and *cis,cis* isomers. In GC, the dominant 9*cis*,11*trans*-18:2 peak in dairy products and meats from ruminants severely overlaps with of number of CFA isomers, such as 7*trans*,9*cis*-18:2 and 8*trans*,10*cis*-18:2, whereas three Ag$^+$-HPLC columns in series clearly resolve these isomers (Fig. 7.8). The GC and Ag$^+$-HPLC methods have been used successfully to identify many of the CFA isomers (4,53,60,62,71).

Enrichment of CFA and Their Metabolites

The concentration of CFA and their possible metabolites in biological matrices is usually very small. During routine fatty acid analysis of dairy products or biological tissues, the presence of the major CFA isomer (rumenic acid, 9*cis*,11*trans*-18:2) often occurs as a minor peak at 0.1–0.5%. The other CFA isomers and their metabolites are at least one order of magnitude less and can be found only by amplifying the response of the GC detector. But the problem is where to look. Sébédio *et al.* (46) optimized the formation of CFA metabolites by feeding C-18 CFA to rats maintained on a fat-free diet. They were able to see the CFA metabolites in the GC chromatogram. We have used the GC-DD-FTIR method to screen FAME mixtures and their DMOX derivatives from biological matrices looking for the characteristic IR doublet of *cis/trans* CFA at 988 and 949 cm^{-1} (unpublished data). We did find certain regions of the GC chromatogram that showed promise but subsequently, we were unable to identify these fatty acid by GC-EIMS as their DMOX derivatives. The peaks were simply too weak without prior concentration.

HPLC using reversed-phase columns is an excellent approach to concentrate CFA and their metabolites on the basis of their partition number by monitoring for UV absorption at 233 nm. These fractions can be analyzed subsequently by GC and Ag$^+$-HPLC as outlined above. This method was used to separate the active anticarcinogenic component from beef fat (20) and cheese (21), and subsequently for the analysis of minor CFA in cheese (54) and CFA metabolites in rat tissue lipids (46). Banni *et al.* (74,75) used reversed-phase HPLC to quantitate CFA and their metabolites in a number of dairy products and meats from ruminants.

What Was the Isomeric Composition of CFA Reported in Published Studies?

The CFA isomer composition of all chemically synthesized CFA mixtures were at first believed to consist of two major *cis/trans* CFA isomers (9*cis*,11*trans*-18:2 and 10*trans*,12*cis*-18:2) with minor amounts of other CFA isomers. This was based on GC analysis first reported in 1989 (21) using a 60-m Supelcowax-10 fused capillary column. Unfortunately, this GC column does not separate the 8,10- from 9,11-18:2, nor does it separate 10,12- from 11,13-18:2. This separation was not questioned thereafter and the same GC column conditions and CFA assignments were used extensively and are still being used today (Table 7.2). However, three recent reports provide evidence that there were additional CFA isomers in commercial CFA mixtures. In April 1997, Sébédio *et al.* (46) reported that the two major *cis/trans* CFA isomers of a commercial CFA mixture separated using a 50-m BPX 70 capillary column each consisted of two CFA isomers. The first eluting peak was a mixture of 8,10-18:2 and 9,11-18:2; the second peak was a mixture of 10,12-18:2 and 11,13-18:2. No information on the composition or the double-bond geometry was given. In November 1997, Christie *et al.* (63) reported that a selected ion scan of an unresolved GC-MS peak of the methyl triazoline dione adduct of a commercial CFA mixture was composed of four positional isomers. This report did not give a separation of the CFA isomers nor was there any information as to the geometry of the double bond. In February 1998, Sehat *et al.* (58) reported for the first time the complete separation of 12 CFA isomers in these commercial mixtures. There were four *cis/trans*, *trans,trans* and *cis,cis* positional CFA isomers each in commercial CFA mixtures as determined by one Ag$^+$-HPLC column (Fig. 7.4, right side). Geometric pairs could not be separated with the use of only one Ag$^+$- HPLC column; however, with the use of three Ag$^+$-HPLC column in series, the odd Δ geometric pairs of isomers could be separated (60). Sehat *et al.* (58) also reported that a synthetic CFA mixture, obtained as a gift from Dr. Pariza, showed only two major *cis/trans* CFA isomers. It would be easy to conclude that all CFA preparations performed on a small scale in the laboratory consisted of two major *cis/trans* CFA, whereas commercial preparations, possibly using more drastic alkali-isomerization conditions (76), consisted of four major *cis/trans* CFA isomers. However, an analysis of several commercial CFA mixtures showed that both the two- and the four-isomer mixtures are found in commercial products (77). Therefore, the dilemma.

It is impossible to determine with certainty how many major *cis/trans* CFA isomers were present in the CFA mixtures used in all of the reported studies. Without doubt, this question is absolutely critical in the interpretation of which isomer is responsible for any given response. However, no matter what the composition of CFA mixtures, if the mixture gave a response and it was reproducible, further studies will have to be undertaken to determine which CFA isomer is responsible for the observed response. Individual CFA isomers were not commercially available to perform such tests until recently, unless they were isolated from natural products. Now, with the availability of pure CFA isomers and methods to analyze them effectively, many of the previous studies will need to be repeated to determine the mechanism of action of each CFA isomer.

Synthesis of CFA from Alkali-Isomerization of Linoleic Acid

The method of alkali-isomerization with the resultant increase in UV absorption at 233 nm dates back to the 1930s. On the basis of theoretical considerations, Nichols *et al.* (8) postulated the formation of two CFA isomers, 9*cis*,11*trans*-18:2, 10*trans*,12*cis*-18:2, and by means of an elaborate crystallization procedure indeed obtained these two products. However, Mounts *et al.* (76) showed that this reaction is also temperature dependent. At 60 and 90°C, the reaction products were mainly *cis/trans* 9,11-18:2 and 10,12-18:2, but at 140°C, there was significant scattering of the double bonds to give four CFA isomers at a relative concentration of 9.8, 39.3, 40.2, and 7.0% for the 8,10-, 9,11-, 10,12-, and 11,13-18:2 isomers, respectively (76). The structure of these CFA isomers was established by oxidative fission of the double bonds and analysis of the aldehyde esters produced.

A comparison of the relative proportion of the CFA isomers products described by Mounts *et al.* (76) with that observed in typical commercial preparations today (58) suggests that the latter concentrations of 8,10- and 11,13-18:2 are significantly higher. To investigate this possibility, one of the commercial suppliers provided us with aliquots from a typical reaction mixture with time and portions taken from subsequent purification steps (Kramer, unpublished data). The GC results indicate that linoleic acid (18:2n-6) is rapidly depleted, whereas 9*cis*,11*trans*-18:2 and 10*trans*,12*cis*-18:2 increased at approximately equal rates (Fig. 7.10) . The relative concentration of the two other CFA [8*trans*,10*cis*-18:2 (from Ag$^+$-HPLC results) and 11*cis*,13*trans*-18:2] also increased equally but they did not exceed ~5% each of the total CFA after 6 h when the reaction was terminated. The Ag$^+$-HPLC separations for the reaction products at 0.5, 2.5, and 6 h are presented in Figure 7.11. The reac-

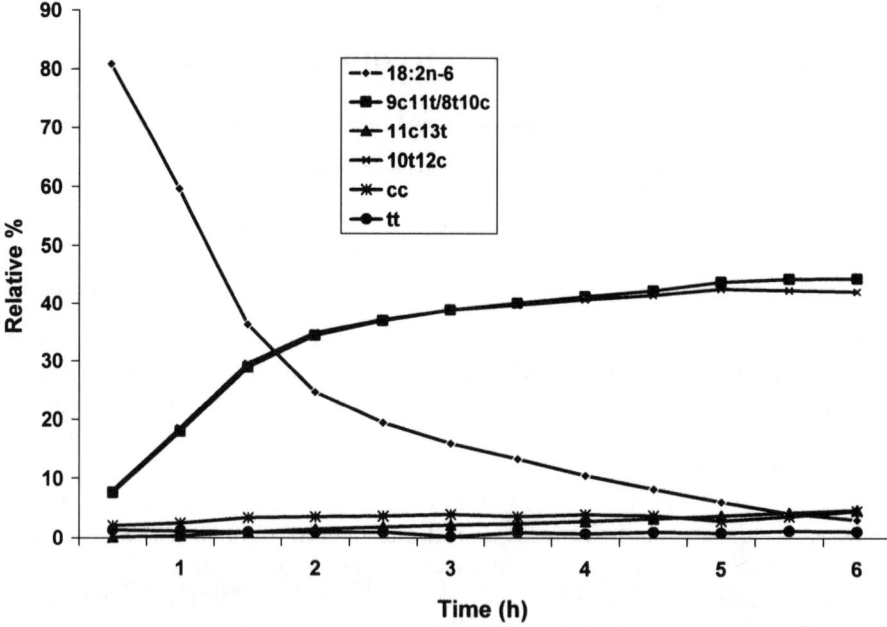

Fig. 7.10. The products of a commercial alkali-isomerization reaction of linoleic acid were determined by gas chromatography (GC) using a 100-m CP Sil 88 fused silica capillary column and flame ionization detection (FID). Aliquots were sampled every 30 min over a 6-h period. The conjugated fatty acid (CFA) isomers 9*cis*,11*trans*-18:2 and 8*trans*,10*cis*-18:2 were combined because they were not resolved by GC. All of the *cis,cis* and *trans,trans* CFA isomers were combined (unpublished results). [The authors wish to thank Nu-Chek-Prep, Elysian, MN for supplying samples from one of their commercial preparations.]

tion product was then purified by crystallization; the final precipitate and the corresponding filtrate are also shown for comparison in Figure 7.11. The results of the precipitate (Fig. 7.11) showed the characteristic pattern of several commercial CFA isomeric mixtures available from unknown sources (77), as well as from three major commercial suppliers of CFA, Biooriginal (Saskatoon, SK), Natural Lipids (Hovdebygda, Norway), and Nu-Chek-Prep (Elysian, MN) (unpublished data). The results indicate that the concentration of the two minor CFA isomers increased from ~5% in the reaction product after 6 h to ~20% each in the purified product. Differences in physical and solubility properties of CFA isomers appear to discriminate against certain CFA isomers during crystallization.

Conclusion

The separation of all the CFA isomers has been a challenge, but is now possible with the use of the complementary GC and Ag$^+$-HPLC techniques. The next challenge

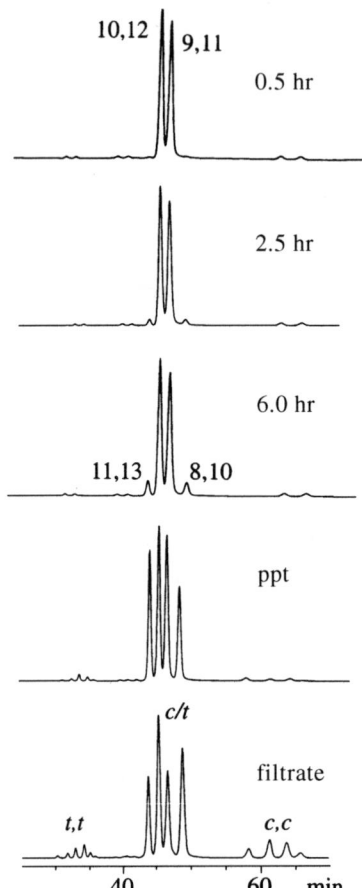

Fig. 7.11. The products of selected samples at 0.5, 2.5 and 6 h of a commercial alkali-isomerization reaction of linoleic acid are presented determined by silver-ion high-performance liquid chromatography (Ag+-HPLC) using three Ag+-HPLC columns in series and detection at 233 nm. The reaction mixture was subsequently purified by crystallization and the final precipitate (ppt) and filtrate are shown. [The authors wish to thank Nu-Chek-Prep, Elysian, MN for supplying samples from one of their commercial preparations.]

will be to analyze for metabolites by combining methods to enrich and separate groups of CFA structures before the analysis of individual CFA isomers. There is a need to reinvestigate many of the biological effects using pure CFA isomers; this will require appropriate separation methods and combinations thereof and subsequent confirmation by GC-MS, GC-FTIR, and NMR.

References

1. Parodi, P.W. (1977) Conjugated Octadecadienoic Acids of Milk Fat, J. *Dairy Sci. 60*, 1550–1553.
2. Kramer, J.K.G., Parodi, P.W., Jensen, R.G., Mossoba, M.M., Yurawecz, M.P., and Adlof, R.O. (1998) Rumenic Acid: A Proposed Name for the Major Conjugated Linoleic Acid Isomer Found in Natural Products, *Lipids 33*, 835.
3. Hopkins, C.Y., and Chisholm, M.J. (1968) A Survey of the Conjugated Fatty Acids of Seed Oils, *J. Am. Oil Chem. Soc. 45*, 176–182.

4. Sehat, N., Kramer, J.K.G., Mossoba, M.M., Yurawecz, M.P., Roach, J.A.G., Eulitz, K., Morehouse, K.M., and Ku, Y. (1998) Identification of Conjugated Linoleic Acid Isomers in Cheese by Gas Chromatography, Silver-Ion High-Performance Liquid Chromatography and Mass Spectral Reconstructed Ion Profiles. Comparison of Chromatographic Elution Sequences, *Lipids 33,* 963–971.
5. Jensen, R.G., Ferris, A.M., and Lammi-Keefe, C.J. (1991) The Composition of Milk Fat, *J. Dairy Sci. 74,* 3228–3243.
6. Lee, K.N., Pariza, M.W., and Ntambi, J.M. (1998) Conjugated Linoleic Acid Decreases Hepatic Stearoyl-CoA Desaturase mRNA Expression, *Biochem. Biophys. Res. Commun. 248,* 817–821.
7. Gillman, A.E., Heilbron, I.M., Hilditch, T.P., and Morton, R.A. (1931) V. Spectroscopic Data of Natural Fats and Their Fatty Acids in Relation to Vitamin A, *Biochem. J. 25,* 30–38.
8. Nichols, P.L., Herb, S.F., and Riemenschneider, R.W. (1951) Isomers of Conjugated Fatty Acids. I. Alkali-Isomerized Linoleic Acid, *J. Am. Chem. Soc. 73,* 247–252.
9. Jackson, J.E., Paschke, R.F., Tolberg, W., Boyd, H.M., and Wheeler, D.H. (1952) Isomers of Linoleic Acid. Infrared and Ultraviolet Properties of Methyl Esters, *J. Am. Oil Chem. Soc. 29,* 229–234.
10. Riel, R.R. (1963) Physico-Chemical Characteristics of Canadian Milk Fat. Unsaturated Fatty Acids, *J. Dairy Sci. 46,* 102–106.
11. Kepler, C.R., Hirons, K.P., McNeill, J.J., and Tove, S.B. (1966) Intermediates and Products of the Biohydrogenation of Linoleic Acid by *Butyrivibrio fibrisolvens, J. Biol. Chem. 241,* 1350–1354.
12. Christie, W.W. (1973) The Structures of Bile Phosphatidylcholines, *Biochim. Biophys. Acta 316,* 204–211.
13. Emken, E.A., Scholfield, C.R., Davison, V.L., and Frankel, E.N. (1967) Separation of Conjugated Methyl Octadecadienoate and Trienoate Geometric Isomers by Silver-Resin Column and Preparative Gas-Liquid Chromatography, *J. Am. Oil Chem. Soc. 44,* 373–375.
14. Ackman, R.G., Eaton, C.A., Sipos, J.C., and Crewe, N.F. (1981) Origin of *cis*-9, *trans*-11- and *trans*-9, *trans*-11-Octadecadienoic Acids in the Depot Fat of Primates Fed a Diet Rich in Lard and Corn Oil and Implications for the Human Diet, *Can. Inst. Food Sci. Technol. J. 14,* 103–107.
15. Scholfield, C.R., and Dutton, H.J. (1971) Equivalent Chain Lengths of Methyl Octadecadienoates and Octadecatrienoates, *J. Am. Oil Chem. Soc. 48,* 228–231.
16. Scholfield, C.R. (1981) Gas Chromatographic Equivalent Chain Lengths of Fatty Acid Methyl Esters on a Silar 10C Glass Capillary Column, *J. Am. Oil Chem. Soc. 58,* 662–663.
17. Adlof, R.O. (1997) Preparation of Methyl *cis*-9, *trans*-11- and *trans*-9, *trans*-11-Octadecadienoate-17,17,18,18-d_4, Two of the Isomers of Conjugated Linoleic Acid, *Chem. Phys. Lipids 88,* 107–112.
18. Adlof, R.O., and Lamm, T. (1998) Fractionation of *cis*- and *trans*-Oleic, Linoleic, and Conjugated Linoleic Fatty Acid Methyl Esters by Silver-Ion High-Performance Liquid Chromatography, *J. Chromatogr. A, 799,* 329–332.
19. Smith, G.N., Taj, M., and Braganza, J.M. (1991) On the Identification of a Conjugated Diene Component of Duodenal Bile as 9*Z*,11*E*-Octadecadienoic Acid, *Free Radic. Biol. Med. 10,* 13–21.
20. Ha, Y.L., Grimm, N.K., and Pariza, M.W. (1987) Anticarcinogens from Fried Ground Beef: Heat-Altered Derivatives of Linoleic Acid, *Carcinogenesis 8,* 1881–1887.

21. Ha, Y.L., Grimm, N.K., and Pariza, M.W. (1989) Newly Recognized Anticarcinogenic Fatty Acids: Identification and Quantification in Natural and Processed Cheeses, *J. Agric. Food Chem. 37,* 75–81.
22. Werner, S.A., Luedecke, L.O., and Shultz, T.D. (1992) Determination of Conjugated Linoleic Acid Content and Isomer Distribution in Three Cheddar-Type Cheeses: Effects of Cheese Cultures, Processing, and Aging, *J. Agric. Food Chem. 40,* 1817–1821.
23. Park, Y., and Pariza, M.W. (1998) Evidence That Commercial Calf and Horse Sera Can Contain Substantial Amounts of *trans*-10,*cis*-12 Conjugated Linoleic Acid, *Lipids 33,* 817–819.
24. Shantha, N.C., Decker, E.A., and Hennig, B. (1993) Comparison of Methylation Methods for the Quantitation of Conjugated Linoleic Acid Isomers, *J. Am. Off. Anal. Chem. Int. 76,* 644–649.
25. Kramer, J.K.G., Fellner, V., Dugan, M.E.R., Sauer, F.D., Mossoba, M.M., and Yurawecz, M.P. (1997) Evaluating Acid and Base Catalysts in the Methylation of Milk and Rumen Fatty Acids with Special Emphasis on Conjugated Dienes and Total *trans* Fatty Acids, *Lipids 32,* 1219–1228.
26. Fogerty, A.C., Ford, G.L., and Svoronos, D. (1988) Octadeca-9,11-dienoic Acid in Foodstuffs and in the Lipids of Human Blood and Breast Milk, *Nutr. Rep. Int. 38,* 937–944.
27. Bligh, E.G., and Dyer, W.J. (1959) A Rapid Method of Total Lipid Extraction and Purification, *Can. J. Biochem. Physiol. 37,* 911–917.
28. Ha, Y.L., Storkson, J., and Pariza, M.W. (1990) Inhibition of Benzo(a)pyrene-Induced Mouse Forestomach Neoplasia by Conjugated Dienoic Derivatives of Linoleic Acid, *Can. Res. 50,* 1097–1101.
29. Ip, C., Chin, S.F., Scimeca, J.A., and Pariza, M.W. (1991) Mammary Cancer Prevention by Conjugated Dienoic Derivative of Linoleic Acid, *Can. Res. 51,* 6118–6124.
30. Chin, S.F., Liu, W., Storkson, J.M., Ha, Y.L., and Pariza, M.W. (1992) Dietary Sources of Conjugated Dienoic Isomers of Linoleic Acid, a Newly Recognized Class of Anticarcinogens, *J. Food Comp. Anal. 5,* 185–197.
31. Shantha, N.C., Decker, E.A., and Ustunol, Z. (1992) Conjugated Linoleic Acid Concentration in Processed Cheese, *J. Am. Oil Chem. Soc. 69,* 425–428.
32. Folch, J., Lees, M., and Sloane-Stanley, G.H. (1957) A Simple Method for the Isolation and Purification of Total Lipids from Animal Tissues, *J. Biol. Chem. 226,* 497–509.
33. Chin, S.F., Storkson, J.M., Liu, W., Albright, K.J., and Pariza, M.W. (1994) Conjugated Linoleic Acid (9,11- and 10,12-Octadecadienoic Acid) Is Produced in Conventional but Not Germ-Free Rats Fed Linoleic Acid, *J. Nutr. 124,* 694–701.
34. Shantha, N.C., Crum, A.D., and Decker, E.A. (1994) Evaluation of Conjugated Linoleic Acid Concentrations in Cooked Beef, *J. Agric. Food Chem. 42,* 1757–1760.
35. Lin, H., Boylston, T.D., Chang, M.J., Luedecke, L.O., and Shultz, T.D. (1995) Survey of the Conjugated Linoleic Acid Contents of Dairy Products, *J. Dairy Sci. 78,* 2358–2365.
36. Shantha, N.C., Ram, L.N., O'Leary, J., Hicks, C.L., and Decker, E.A. (1995) Conjugated Linoleic Acid Concentrations in Dairy Products as Affected by Processing and Storage, *J. Food Sci. 60,* 695–697,720.
37. Ip, C., Briggs, S.P., Haegele, A.D., Thompson, H.J., Storkson, J., and Scimeca, J.A. (1996) The Efficacy of Conjugated Linoleic Acid in Mammary Cancer Prevention Is Independent of the Level or Type of Fat in the Diet, *Carcinogenesis 17,* 1045–1050.

38. Jiang, J., Bjoerck, L., Fondén, R., and Emanuelson, M. (1996) Occurrence of Conjugated *cis*-9, *trans*-11-Octadecadienoic Acid in Bovine Milk: Effects of Feed and Dietary Regimen, *J. Dairy Sci. 79,* 438–445.
39. Belury, M.A., and Kempa-Steczko, A. (1997) Conjugated Linoleic Acid Modulates Hepatic Lipid Composition in Mice, *Lipids 32,* 199–204.
40. Chen, Z.Y., Chan, P.T., Kwan, K.Y., and Zhang, A. (1997) Reassessment of the Antioxidant Activity of Conjugated Linoleic Acids, *J. Am. Oil Chem. Soc. 74,* 749–753.
41. Ip, C., Jiang, C., Thompson, H.J., and Scimeca, J.A. (1997) Retention of Conjugated Linoleic Acid in the Mammary Gland is Associated with Tumor Inhibition During the Post-Initiation Phase of Carcinogenesis, *Carcinogenesis 18,* 755–759.
42. Jahreis, G., Fritsche, J., and Steinhart, H. (1997) Conjugated Linoleic Acid in Milk Fat: High Variation Depending on Production System, *Nutr. Res. 17,* 1479–1484.
43. Jiang, J., Björck, L., and Fondén, R. (1997) Conjugated Linoleic Acid in Swedish Dairy Products with Special Reference to the Manufacture of Hard Cheeses, *Int. Dairy J. 7,* 863–867.
44. McGuire, M.K., Park, Y., Behre, R.A., Harrison, L.Y., Shultz, T.D., and McGuire, M.A. (1997) Conjugated Linoleic Acid Concentrations of Human Milk and Infant Formula, *Nutr. Res. 17,* 1277–1283.
45. Precht, D., and Molkentin, J. (1997) Effect of Feeding on Conjugated *cis* Δ9, *trans* Δ11-Octadecadienoic Acid and Other Isomers of Linoleic Acid in Bovine Milk Fats, *Nahrung 41,* 330–335.
46. Sébédio, J.L., Juanéda, P., Dobson, G., Ramilison, I., Martin, J.C., Chardigny, J.M., and Christie, W.W. (1997) Metabolites of Conjugated Isomers of Linoleic Acid (CLA) in the Rat, *Biochim. Biophys. Acta 1345,* 5–10.
47. Shantha, N.C., Moody, W.G., and Tabeidi, Z. (1997) Conjugated Linoleic Acid Concentration in Semimembranosus Muscle of Grass- and Grain-Fed and Zeranol-Implanted Beef Cattle, *J. Muscle Foods 8,* 105–110.
48. Stanton, C., Lawless, F., Kjellmer, G., Harrington, D., Devery, R., Connolly, J.F., and Murphy, J. (1997) Dietary Influences on Bovine Milk *cis*-9,*trans*-11-Conjugated Linoleic Acid Content, *J. Food Sci. 62,* 1083–1086.
49. Sugano, M., Tsujita, A., Yamasaki, M., Yamada, K., Ikeda, I., and Kritchevsky, D. (1997) Lymphatic Recovery, Tissue Distribution, and Metabolic Effects of Conjugated Linoleic Acid in Rats, *J. Nutr. Biochem. 8,* 38–43.
50. Griinari, J.M., Dwyer, D.A., McGuire, M.A., Bauman, D.E., Palmquist, D.L., and Nurmela, K.V.V. (1998) *Trans*-Octadecenoic Acids and Milk Fat Depression in Lactating Dairy Cows, *J. Dairy Sci. 81,* 1251–1261.
51. Jensen, R.G., Lammi-Keefe, C.J., Hill, D.W., Kind, A.J., and Henderson, R. (1998) The Anticarcinogenic Conjugated Fatty Acid, 9*c*,11*t*-18:2, in Human Milk: Confirmation of Its Presence, *J. Hum. Lact. 14,* 23–27.
52. Kelly, M.L., Berry, J.R., Dwyer, D.A., Griinari, J.M., Chouinard, P.Y., Van Amburgh, M.E., and Bauman, D.E. (1998) Dietary Fatty Acid Sources Affect Conjugated Linoleic Acid Concentrations in Milk from Lactating Dairy Cows, *J. Nutr. 128,* 881–885.
53. Kramer, J.K.G., Sehat, N., Dugan, M.E.R., Mossoba, M.M., Yurawecz, M.P., Roach, J.A.G., Eulitz, K., Aalhus, J.L., Schaefer, A.L., and Ku, Y. (1998) Distributions of Conjugated Linoleic Acid (CLA) Isomers in Tissue Lipid Classes of Pigs Fed a Commercial CLA Mixture Determined by Gas Chromatography and Silver-Ion High-Performance Liquid Chromatography, *Lipids 33,* 549–558.

54. Lavillonnière, F., Martin, J.C., Bougnoux, P., and Sébédio, J.-L. (1998) Analysis of Conjugated Linoleic Acid Isomers and Content in French Cheeses, *J. Am. Oil Chem. Soc. 75,* 343–352.
55. Li, Y., and Watkins, B.A. (1998) Conjugated Linoleic Acids Alter Bone Fatty Acid Composition and Reduce *ex vivo* Bone Prostaglandin E_2 Biosynthesis in Rats Fed n-6 or n-3 Fatty Acids, *Lipids 33,* 417–425.
56. Lin, H., Boylston, T.D., Luedecke, L.O., and Shultz, T.D. (1998) Factors Affecting the Conjugated Linoleic Acid Content of Cheddar Cheese, *J. Agric. Food Chem. 46,* 801–807.
57. Salminen, I., Mutanen, M., Jauhiainen, M., and Aro, A. (1998) Dietary *trans* Fatty Acids Increase Conjugated Linoleic Acid Levels in Human Serum, *J. Nutr. Biochem. 9,* 93–98.
58. Sehat, N., Yurawecz, M.P., Roach, J.A.G., Mossoba, M.M., Kramer, J.K.G., and Ku, Y. (1998) Silver-Ion High-Performance Liquid Chromatographic Separation and Identification of Conjugated Linoleic Acid Isomers, *Lipids 33,* 217–221.
59. Bascetta, E., Gunstone, F.D., and Scimgeour, C.M. (1984) Synthesis, Characterization, and Transformations of Lipid Cyclic Peroxide, *J. Chem. Soc. Perkin Trans. I,* 2199–2205.
60. Eulitz, K., Yurawecz, M.P., Sehat, N., Fritsche, J., Roach, J.A.G., Mossoba, M.M., Kramer, J.K.G., Adlof, R.O., and Ku, Y. (1999) Preparation, Separation and Confirmation of the Eight Geometrical *cis/trans* Conjugated Linoleic Acid Isomers 8,10- Through 11,13-18:2, *Lipids 34,* (in press).
61. Snyder, J.M., and Scholfield, C.R. (1982) *Cis-trans* Isomerization of Unsaturated Fatty Acids with *p*-Toluenesulfinic Acid, *J. Am. Oil Chem. Soc. 59,* 469–470.
62. Yurawecz, M.P., Roach, J.A.G., Sehat, N., Mossoba, M.M., Kramer, J.K.G., Fritsche, J., Steinhart, H., and Ku, Y. (1998) A New Conjugated Linoleic Acid Isomer, 7 *trans*,9 *cis*-Octadecadienoic Acid, in Cow Milk, Cheese, Beef and Human Milk and Adipose Tissue, *Lipids 33,* 803–809.
63. Christie, W.W., Dobson, G., and Gunstone, F.D. (1997) Isomers in Commercial Samples of Conjugated Linoleic Acid, *Lipids 32,* 1231.
64. Dobson, G., Christie, W.W., and Nikolova-Damyanova, B. (1995) Siver Ion Chromatography of Lipids and Fatty Acids, *J. Chromatogr. B, 671,* 197–222.
65. Adlof, R.O. (1997) Normal-Phase Separation Effects with Lipids on a Silver Ion High Performance Liquid Chromatography Column, *J. Chromatogr. A, 764,* 337– 340.
66. Adlof, R.O., Copes, L.C., and Emken, E.A. (1995) Analysis of Monoenoic Fatty Acid Distribution in Hydrogenated Vegetable Oils by Silver-Ion High-Performance Liquid Chromatography, *J. Am. Oil Chem. Soc. 72,* 571–574.
67. Nikolova-Damyanova, B., Herslöf, B.G., and Christie, W.W. (1992) Silver Ion-High Performance Liquid Chromatography of Derivatives of Isomeric Fatty Acids, *J. Chromatogr. 609,* 133–140.
68. Nikolova-Damyanova, B., Christie, W.W., and Herslöf, B. (1996) Mechanistic Aspects of Fatty Acid Retention in Silver Ion Chromatography, *J. Chromatogr. A, 749,* 47–54.
69. Juanéda, P., Sébédio, J.L., and Christie, W.W. (1994) Complete Separation of the Geometrical Isomers of Linolenic Acid by High Performance Liquid Chromatography with a Silver Ion Column, *J. High Res. Chromatogr. 17,* 321–324.
70. Adlof, R.O., and Emken, E.A. (1994) Silver Ion High-Performance Liquid Chromatographic Separation of the Fatty Acid Methyl Esters Labelled with Deuterium Atoms on the Double Bonds, *J. Chromatogr. A, 685,* 178–181.

71. Sehat, N, Rickert, R., Mossoba, M.M., Kramer, J.K.G., Yurawecz, M.P., Roach, J.A.G., Adlof, R.O., Morehouse, K.M., Fritsche, J., Eulitz, K.D., Steinhart, H., and Ku, Y. (1999) Improved Separation of Conjugated Fatty Acid Methyl Esters by Silver-Ion High-Performance Liquid Chromatography, *Lipids 34*, 407–413.
72. Gunstone, F.D. (1967) *An Introduction to the Chemistry and Biochemistry of Fatty Acids and Their Glycerides,* 2nd edn., pp. 67–69, Richard Clay (The Chauser Press) Ltd., Bangay, Suffolk, UK.
73. Pariza, M.W., Park, Y., Albright, K.-J., Liu, W., Storkson, J.M., and Cook, M.E. (1998) Synthesis and Biological Activity of Conjugated Eicosadienoic Acid, Abstracts, 89th Am. Oil Chem. Soc. Annual Meeting, May 10–13, Chicago, IL.
74. Banni, S., Day, B.W., Evans, R.W., Corongiu, F.P., and Lombardi, B. (1995) Detection of Conjugated Diene Isomers of Linoleic Acid in Liver Lipids of Rats Fed a Choline-Devoid Diet Indicates That the Diet Does Not Cause Lipoperoxidation, *J. Nutr. Biochem. 6*, 281–289.
75. Banni, S., Carta, G., Contini, M.S., Angioni, E., Deiana, M., Dessì, M.A., Melis, M.P., and Corongiu, F.P. (1996) Characterization of Conjugated Diene Fatty Acids in Milk, Dairy Products, and Lamb Tissues, *J. Nutr. Biochem. 7*, 150–155.
76. Mounts, T.L., Dutton, H.J., and Glover, D. (1970) Conjugation of Polyunsaturated Acids, *Lipids 5*, 997–1005.
77. Yurawecz, M.P., Sehat, N., Mossoba, M.M., Roach, J.A.G., Kramer, J.K.G., and Ku, Y. (1999) Variations in Isomer Distribution in Commercially Available Conjugated Linoleic Acid, *Fett/Lipid 101* (in press).

Chapter 8

Gas Chromatography/(Electron Impact) Mass Spectrometry Analysis of Conjugated Linoleic Acid (CLA) Using Different Derivatization Techniques

Volker Spitzer

Institut für Tierhaltung und Tierzüchtung–Fachgebiet Tierhaltung und Leistungsphysiologie, Universität Hohenheim, 70599 Stuttgart, Germany

Introduction

Conjugated linoleic acid (CLA) is a term for a mixture of positional and geometric isomers of linoleic acid with a conjugated double bond system. For elucidation of the complete structure of CLA and their metabolites, information is required concerning the chain length, the double bond equivalents, and the position and geometry of the unsaturated double bonds. Gas chromatography/(electron impact) mass spectrometry (GC/MS) is one of the most successful methods for this purpose because it combines a high-resolution separation technique with sensitive and mass-selective detection. Both the GC retention characteristics and the mass spectral data of the individual peaks provide a number of valuable hints for structure determination.

The most widely used derivative for GC/MS analysis of fatty acids is formation of fatty acid methyl esters (FAME). Unfortunately, the structural information obtained from the mass spectra of unsaturated FAME are of limited value. The reason for this restriction is the migration of the double bonds during the ionization process in the ion source of the mass spectrometer (1). This prevents the formation of characteristic fragment ions for location of the double bonds. Due to this bond migration, the mass spectra of the positional and geometric isomers of conjugated fatty acids and related metabolites are similar to those of the analogous nonconjugated compounds and do not enable the location of the conjugated double bond system (2–4). In most cases, the only valuable information from such FAME mass spectra is the molecular weight, which provides information on the chain length and the number of double bond equivalents.

Several derivatization techniques have been developed to manage the problem of excessive bond migration during ionization. In this short survey, these techniques are summarized with special emphasis on the analysis of CLA derivatives.

Derivatization Reactions for the GC/MS Analysis of CLA Isomers and Their Metabolites

The objective of GC/MS derivatization reactions is to obtain volatile GC compounds that produce mass spectra with structure-specific fragmentation patterns and recognizable molecular ions. It is also desirable that the reaction be simple, fast, and reasonably priced. There are two distinct forms of fatty acid derivatizations used for the analysis of CLA isomers and metabolites:

1. The double bonds are converted into new functional groups ("on-site derivatization"). In this way, the unsaturated system is "fixed," and the mass spectra of the modified compounds point to the original location of the double bonds (for reviews, see refs. 5–9).
2. The carboxylic group is modified by introducing a nitrogenous function ("remote-site derivatization"). In these derivatives, the nitrogenous function is preferentially ionized, thus minimizing the ionization and double bond migration of the unsaturated system in the alkyl chain. The mass spectra can provide information on the location of the double bonds and other functional groups, depending on the type of derivative and the fatty acid structure.

Both types of derivatization procedure have their merits for the determination of the double bond position of CLA and their metabolites. The combined interpretation of the GC retention data and the MS data may also give useful hints for the configuration of the double bonds, especially when compared with authentic standards. However, to obtain more unambiguous information on the stereochemistry, it is advisable to complement the GC/MS data with a GC/Fourier transform infrared spectroscopy (FTIR) study (4,10–12). This is because GC/MS methods provide information useful only for the elucidation of double bond position, whereas GC/FTIR data provide information about the absolute double bond configuration.

Preparation of Remote-Site Derivatizations of CLA. The pyrrolidide, picolinyl ester, and 4,4-dimethyloxazoline derivatives have been used for the GC/MS analysis of CLA isomers. Their structures are summarized in Table 8.1. Preparation methods of the different remote-site derivatives of fatty acids have been reviewed recently (13–18). All of the described methods are applicable to the derivatization of microgram to milligram quantities of material.

The pyrrolidides are easily prepared in quantitative yield by heating the FAME mixture with an excess of freshly distilled pyrrolidine in the presence of acetic acid catalyst in a sealed tube for 30 min at 100°C (19). Such derivatives were made from a CLA isomer found in human blood lipids (20) and a series of synthetic CLA isomers (21). A method starting with the free fatty acids has also been published (22).

The carboxylic group of the fatty acids is condensed with 2-amino-2-methyl-1-propanol (AMP) to form DMOX derivatives. Various methods have been described for the formation of DMOX derivatives. The method proposed by Zhang *et al.* (23) in Scheme 8.1 consists of heating one part of free fatty acids with five parts of AMP at 170–180°C for 0.5–1 h in screw-capped micro-vials.

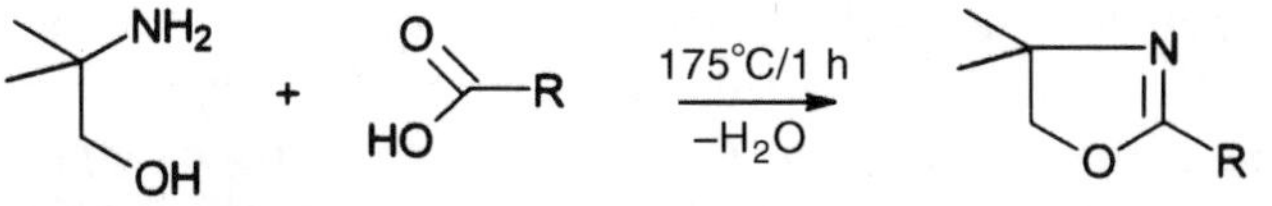

SCHEME 8.1

To avoid oxidation, it is advisable to flush the vials with nitrogen or argon before heating. The excess of AMP is necessary to avoid the formation of *N-O*-diacylated

TABLE 8.1 Derivatives for the Gas Chromatography/Mass Spectrometry (GC/MS) Analysis of Fatty Acid with Conjugated Double Bonds

Derivative	Structures [drawn from all-*trans* 18:2(9,11)]
I. Remote-site derivatives Pyrrolidide	
4,4-Dimethyloxazoline	
Picolinyl ester	
II. On-site derivatives 4-Methyl-1,2,4-triazoline-3-5-dione	
Trimethylsilyl derivatives of the hydroxylation products	
Deuterated picolinyl ester[a]	

[a]This compound can be classified as both a remote- and on-site derivative.

products. Reactions at lower temperatures may cause the formation of less pure derivatives due to incomplete ring cyclization of the *N*-acyl intermediate. This method has been applied successfully to the GC/MS analysis of CLA isomers and other conjugated fatty acids (10,12,24–32).

Fay and Richli (33) proposed a direct derivatization of the FAME mixture and applied it to the analysis of conjugated trienoic fatty acids. In brief, an excess of AMP is added to the FAME mixture in a reaction vial and heated overnight at 180°C. This method was also used for the GC/MS analysis of conjugated dienes and metabolites (4,11,34–38). No discrimination effects or decomposition products were reported. In most reports, shorter heating times (6–8 h) were used. Christie (39) observed the formation of *N*-acyl intermediates that elute much later than the DMOX derivatives from the GC column but produce a mass spectrum almost identical to that of the required derivative.

The fatty acids can be transformed to the more reactive acid anhydrides before derivatization to avoid relatively high reaction temperatures. Briefly, a reaction of equal parts of the fatty acid mixture with dicyclohexylcarbodiimide (DCC) and AMP is carried out in dichlormethane at room temperature for 1-4 h, followed by treatment with thionylchloride for 0.5-1 h (40). The AMP can also be added to the fatty acid/dicyclohexylcarbodiimide solution after 10 min (41). The yields of these reaction methods are comparable to derivatization at 170°C. The reaction may not be safer than the other methods because of the drastic acid conditions (18). To the best of our knowledge, a method using DCC and thionylchloride has never been used for making derivatives from CLA isomers.

Because small residues of unreacted fatty acids and other nonpolar contaminants can interfere in the chromatogram or in the mass spectra, it is advisable to include a clean-up step by using small columns containing silica (40) or Florisil (17,42). Contamination can be a problem especially when small samples are to be derivatized (18). No interferences in the mass spectrum are to be expected from the AMP reagent because it shows only one peak at m/z 58.

The preparation of picolinyl esters can also be done by different methods. In contrast to the pyrrolidine and DMOX derivatives, the picolinyl esters can be made only with free acids (14). Thus, hydrolysis of intact lipids or FAME is compulsory before the conversion to the picolinyl ester is carried out. The derivatization method proposed for polyunsaturated fatty acids (43) consists of conversion of the free fatty acids into the corresponding acid chlorides by reaction with an excess of thionyl chloride. The mixture is then transformed into the picolinyl esters by reaction with 3-hydroxymethylpyridine in acetonitrile. An even milder method involves a conversion of the fatty acids to the corresponding imidazoles by reaction with 1,1'-carbonyldiimidazole (44). The imidazoles are then reacted with 3-hydroxymethyl-pyridine under alkaline conditions (triethylamine) in the presence of 4-pyrrolidonopyridine as a catalyst. Final clean-up procedures with silica or Florisil columns are also recommended (17).

Another method is a combination of an on-site (deuteration) and a remote-site (picolinyl ester) derivatization. In brief, a dioxane solution of the FAME mixture and Wilkinson's catalyst [$(Ph_3P)RhCl(I)$] is degassed with helium and then purged with deuterium gas. The reaction in the deuterium atmosphere is carried out at 60°C for 2 h (42). The deuterated fatty acid mixture is then converted to the picolinyl esters by using the method of Balazy and Nies (44). This method has been applied successfully for the analysis of a 20:3 and 20:4 CLA metabolite found in liver lipids of rats after they were fed CLA isomers (36). It must be mentioned that it is not possible to deuterate the picolinyl esters directly (42).

Preparation of On-Site Derivatizations of CLA. Two different on-site derivatives for the MS analysis of CLA isomers have been published; their structures are summarized in Table 8.1. A classical method is the complete hydroxylation of the double bonds with osmium tetroxide followed by trimethylsilylation of the resulting hydroxyl groups. The method has been applied to methyl esters and trimethylsilyl esters of synthetic all-*trans*-9,11 and 10,12-18:2 isomers and a naturally occurring 9,11-18:2 extracted from a *Eubacterium* sp. (45).

A recently published method (37) for the on-site derivatization of CLA isomers is a Diels-Alder cycloaddition (Scheme 8.2). The FAME mixture is reacted with 4-methyl-1,2,4-triazoline-3,5-dione (MTAD) in dichloromethane solution at 0°C for <10 s. The reaction is then stopped by addition of 1,3-hexadiene. MTDA forms Diels-Alder product with conjugated double bond systems, but when an excess of the reagent is present, it may also react with other nonconjugated fatty acids to form unidentified polar compounds (37) . In contrast to the classical Diels-Alder reaction of conjugated double bonds with maleic acid anhydride that is selective for conjugated *trans, trans* double bonds, the MTAD adducts can be formed readily from all geometric isomers. By choosing certain conditions, MTDA reacts better with the *cis*-9,*trans*-11- isomer than with the *cis*-9,*cis*-11 isomer. This might be useful for differentiation among different isomers. MTDA adducts have also been made from conjugated trienoic acids (37).

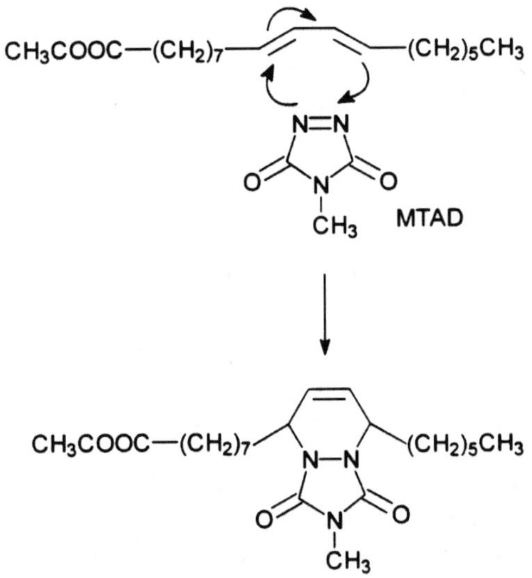

SCHEME 8.2

Another class of on-site derivatives, the dimethyl disulfide adducts, were used successfully for the identification of monoenoic and dienoic acids, but cannot be applied to conjugated fatty acids (18).

GC Properties of the Remote-Site and On-Site Derivatives from Conjugated Fatty Acids

The GC retention properties are an important point for the evaluation of fatty acid derivatives. The FAME are the most volatile fatty acid derivatives, and in many cases, their separation characteristics are also superior to other derivatives. Consequently, the

GC properties of the FAME are the "GC quality reference" for a comparison between the different fatty acid derivatives. It must be noted that the separation of CLA positional and geometric isomers as methyl esters from natural sources and synthetic mixtures is difficult, particularly for minor compounds. Peak overlapping can not be avoided, even when 100-m capillary columns with polar liquid phases are used (4,12,46). Therefore, some groups use additional separation procedures, such as reversed-phase high-performance liquid chromatography (HPLC) and silver ion–HPLC, before GC analysis to further simplify the fatty acid fractions (4,12,31,36,38,47). Separation techniques are discussed in more detail elsewhere in this book.

A remote-site derivatization procedure derivatizes all fatty acids in the test portion. This means that the GC peak pattern of the derivatized test portion should be similar to that of the FAME when the same GC conditions are applied. A major advantage of the DMOX derivatives in comparison to other remote-site derivatives is their high volatility. Typically, the elution temperatures are only ~10° higher than those from the corresponding FAME (23). Thus, almost the same GC conditions as used for a FAME analysis can be applied, and the use of relatively temperature-labile high-resolution GC capillary columns with polar phases, such as CP-SIL 88 or BPX-70, is possible. These polar GC columns permit the separation of a great number of chain length and unsaturated bond number homologues, including a number of monoene and polyunsaturated isomers (39). There is no need to use high temperatures (>250°C) or long analysis times. Excessive column bleeding that can interfere with the GC/MS analysis is thus also avoided. In comparison, the GC temperatures for the analysis of pyrrolidide or picolinyl ester derivatives are ~50°C higher than those for the corresponding methyl esters (48). Some years ago, this was a major problem because only the more temperature-stable GC capillary columns with nonpolar phases could be used. On this type of column, the fatty acid derivatives are separated mainly by their boiling points, and the unsaturated compounds elute in front of the corresponding saturated acids with the same chain length. In most cases, a partial overlapping of unsaturated fatty acids cannot be avoided. However, a high-resolution separation of all remote-site derivatives can be achieved using temperature-stable, low-bleed, cross-linked polar GC columns, when no very long-chain fatty acids are present (18).

The GC analysis of on-site derivatives differs from the GC analysis of remote-site derivatives. The boiling points of on-site derivatives are much higher than those of the original FAME. Temperature-stable nonpolar GC columns must be used in most cases because of the relatively high oven temperatures required. Moreover, only the fatty acids containing the specific structural features required for the reaction are derivatized. Thus, the GC elution pattern differs from that of the original FAME mixture.

The GC properties of the methyl or TMS tetrakis-TMSO-octadecanoates from CLA isomers have not been described in detail in the literature (45). For GC analysis of the MTDA adducts of CLA and metabolites, a nonpolar DB-5 capillary column was employed and the temperature was programmed from 160 to 350° (5°C/min) (36,37). At a temperature of ~270°C, the isomers of CLA eluted as overlapping peaks, and a complete separation was not possible. However, careful interpretation of the mass spectra within the unresolved peak allowed the assignment of double bond positions of the individual compounds. Because the formation of the MTDA adducts is almost selective for conjugated double bonds, the detection of unknown conjugated fatty acids is possible using only GC (37).

Mass Spectrometry of Remote-Site and On-Site Derivatives of CLA and Metabolites

The Fragmentation Mechanisms of Pyrrolidines, Dimethyloxazolines and Picolinyl Esters of CLA and Metabolites. The mass spectra of the pyrrolidines, dimethyloxazolines, and picolinyl esters of 18:2(9,11) are compared in Figures 8.1–8.3. All of the remote-site derivatives show prominent molecular ions. The fragmentation mechanism of these remote-site derivatives is essentially similar.

In the low-mass range, all of the compounds show prominent ions corresponding to fragments containing the ring structure. Some of these fragments are summarized in Table 8.2. The fragmentation of the hydrocarbon chain can be explained as follows (14): Due to preferred ionization at the nitrogenous function, the ionization of the hydrocarbon chain is suppressed, resulting in reduced bond migration along the alkyl chain. However, some of the hydrogen atoms can migrate to the charged nitrogen site, in spite of their energetically unfavorable position. In this way, a radical site at any position in the hydrocarbon chain is produced with a more or less equal probability. Through this mechanism, radical-induced cleavages occur at every carbon-carbon linkage, and in the case of methylene-methylene connections, the respective fragment ions arise in a mass interval of 14 amu within the mass spectrum (see Figs. 8.1–8.3). This radical-induced cleavage mechanism is illustrated in Scheme 8.3.

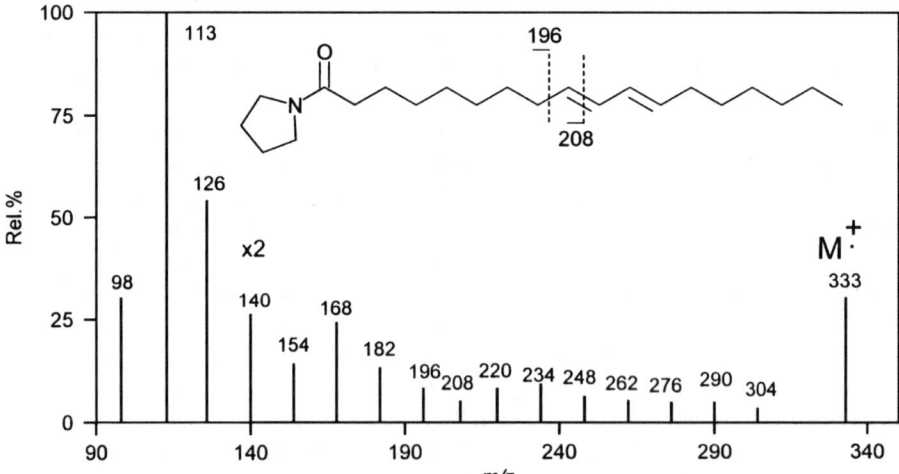

Fig. 8.1. The schematic mass spectrum of the pyrollidine derivative of 18:2(9,11). *Source:* Ref. 20.

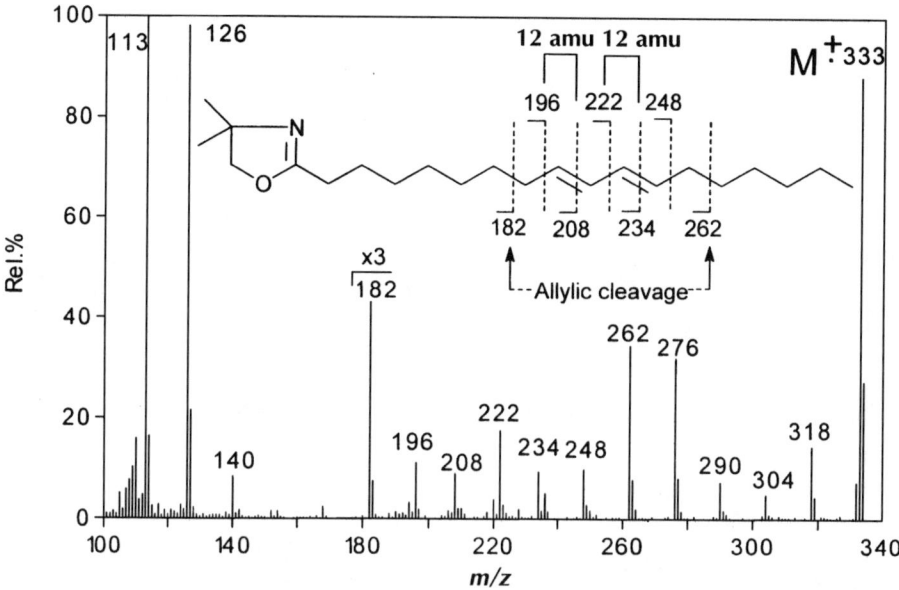

Fig. 8.2. The electron impact mass spectrum of the 4,4-dimethyloxazoline derivative of 18:2(9,11).

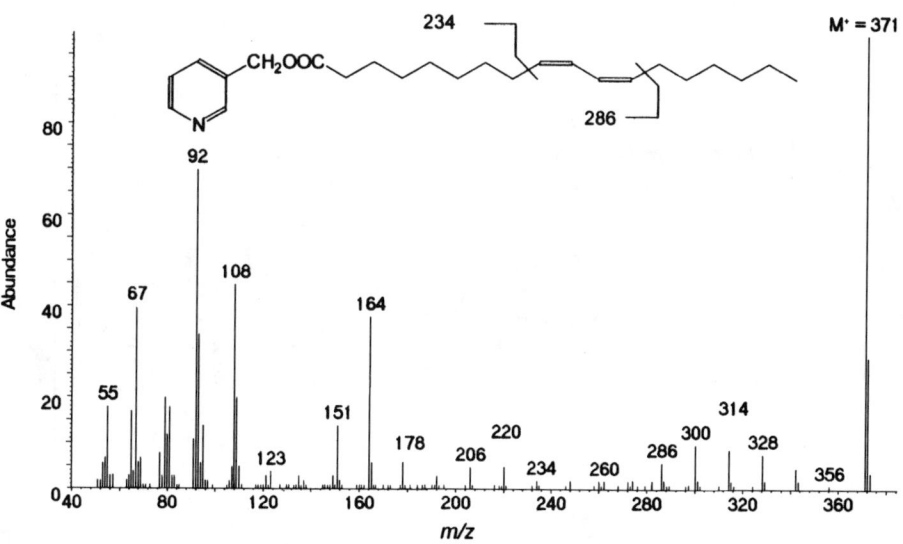

Fig. 8.3. The electron impact mass spectrum of the picolinyl ester of 18:2(9,11) (reproduced with kind permission of Dr. W.W. Christie, Dundee, Scotland).

TABLE 8.2 Main Fragments in the Low-Mass Range of Remote-Site Derivatives

Derivative	Main fragment in the low-mass range
Pyrrolidine[a]	
1,4-Dimethyloxazoline[b]	
Picolinyl ester[c]	

[a]Source: Ref. 19.
[b]Source: Ref. 23.
[c]Source: Ref. 14.

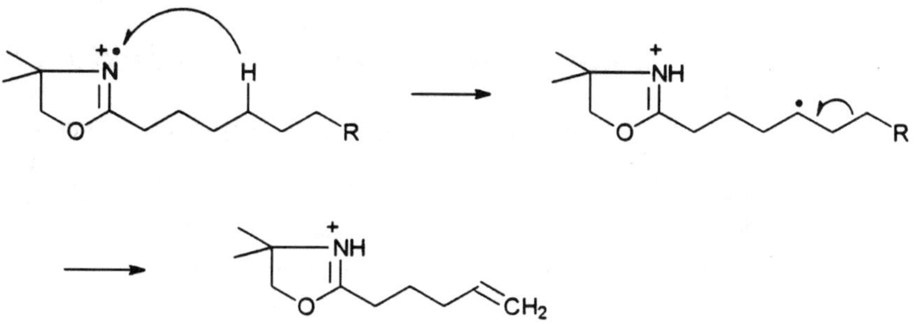

SCHEME 8.3

For pyrrolidines, an analysis of metastable ions indicated that the individual fragments arise directly from the molecular ion. Alternative stepwise cleavages were not observed (19).

The regular 14 amu series is interrupted at the olefinic systems (and other functional groups) in all remote-site derivatives. Formal cleavage between a methylene and a methine group leads to a 12-amu gap in the pyrrolidine and DMOX derivatives and enables the location of the double bond(s) in the hydrocarbon chain of many unsaturated fatty acids (Figs. 8.1,8.2). In the picolinyl ester derivatives (Fig. 8.3), the position of the double bond is more distinctly indicated by a 26-amu gap (two methine groups) between either side of the double bond (14). In all of these derivatives, the fragments that contain double bond(s) show a reduction by 2 amu for each double bond relative to the ion series observed for a corresponding saturated fatty acid derivative.

Anderson (19) explained the fragmentation pattern in the double bond range of pyrrolidine derivatives by a movement of the olefinic system towards the ring by one or more steps and subsequent cleavages of the newly formed isomers. In Figure 8.4, the proposed mechanism is illustrated for 18:1(9). A similar behavior can be postulated for the DMOX derivatives.

Unfortunately, the theoretically expected fragmentation pattern is not always identical to the actual mass spectra recorded for some remote-site derivatives of CLA

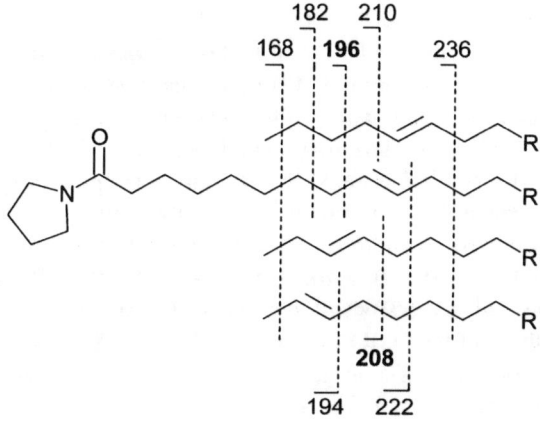

Fig. 8.4. Postulated double bond migration in the pyrollidine derivative of 18:1(9) during electron impact and the resulting fragmentation pattern. *Source:* Ref. 19.

and metabolites. The mass spectra of the pyrrolidine derivatives of conjugated C_{18} diene acids (Fig. 8.1) show the expected 12-amu interval for the cleavage between the methylene and methine group closer to the carbonyl group, indicating the first double bond position. However, the fragmentation at the remote double bond does not obey the theoretic rules (20,21). Instead, the second double bond is represented by another 12-amu interval immediately following the first 12-amu gap. This unexpected behavior rules out these derivatives for the identification of unknown compounds.

In contrast, the mass spectra of the DMOX derivatives of CLA (Fig. 8.2) clearly reveal the location of both double bonds in the conjugated diene system (28,34). The even mass homologous series m/z 126+14 amu is interrupted in the region of the double bonds, and for 9,11-unsaturated derivatives, a mass difference of 12 amu between m/z 196/208 and m/z 222/234 is observed (Fig. 8.2). Furthermore, two prominent peaks deriving from allylic cleavages at m/z 182/262 support the assignment of the double bond location (28). The mass spectra of other positional CLA isomers [18:2(8,10)/10,12/11,13] as DMOX derivatives show an analogous fragmentation pattern and allow the easy recognition of the double bond positions (12,28). This assignment was additionally confirmed for some CLA isomers by analyzing the DMOX derivatives of the monoenes, formed after partial hydrazine reduction of isolated CLA fractions (4). Structure determination with DMOX derivatives is simple and reliable; thus, the method has been applied successfully in a number of works that deal with CLA analysis (e.g., 12,28,31,34,35,37).

The position of the double bonds could also be determined in C:20 metabolites of CLA by using DMOX derivatives (36). The respective mass spectra for 20:3(8,12,14) and 20:4(5,8,12,14) indicate the position of the isolated and the conjugated double bonds by 12-amu gaps as shown in Scheme 8.4 for 20:4(5,8,12,14).

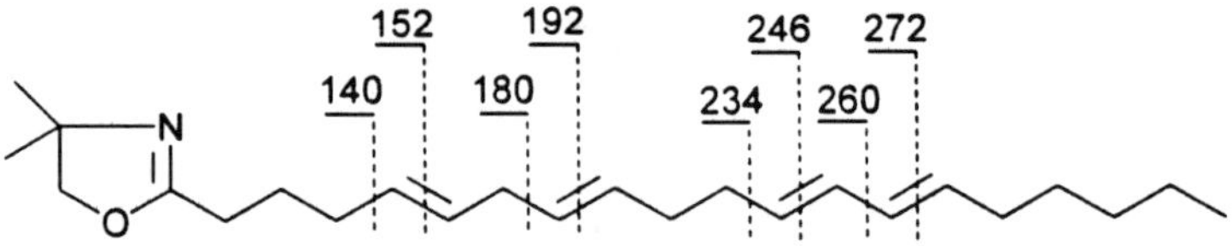

SCHEME 8.4

The structure of the latter compound was additionally confirmed by the mass spectra of the picolinyl esters of the deuterated acid and by the MTDA adduct. Interestingly, the mass spectrum of the DMOX derivative did not allow the complete identification of the double bonds for 20:4(5,8,11,13) (36). The fragmentation pattern revealed only the double bonds at C-5 and C-13, and it was necessary to analyze the picolinyl ester of the fully deuterated 20:4 derivative and the respective MTDA adduct to obtain the desired structural information (36). In contrast, the elucidation of the double bond positions in conjugated trienes is possible using this method (10,26–28,33,38). The picolinyl derivatives of CLA can also be used for the analysis of CLA but the mass spectra are not as unambiguous as those of the DMOX derivatives. For the 18:2(9,11) derivative, the conjugated system produce a 52-amu gap between m/z 234 and 286 (Fig. 8.3). The ions below m/z 234 can be assigned to a saturated fatty acid chain, and the ions above m/z 272 appear 4 amu lower than in saturated compounds due to the conjugated double bonds. In the range of the double bonds (between m/z 234 and 286), the ion peak intensities are low and a clear attribution of the fragmentation at each of the conjugated double bonds is difficult. In contrast, the picolinyl esters of deuterated CLA compounds allow an unmistakable assignment of the original double bond position; the method has been applied for the identification of the CLA metabolite 20:4(5,8,11,13) (36). The mass spectrum of this derivative shows a molecular ion at m/z 411 corresponding to a 20:4 ester with eight deuterium atoms; the 15-amu intervals indicate the original position of the double bonds as illustrated in Scheme 8.5.

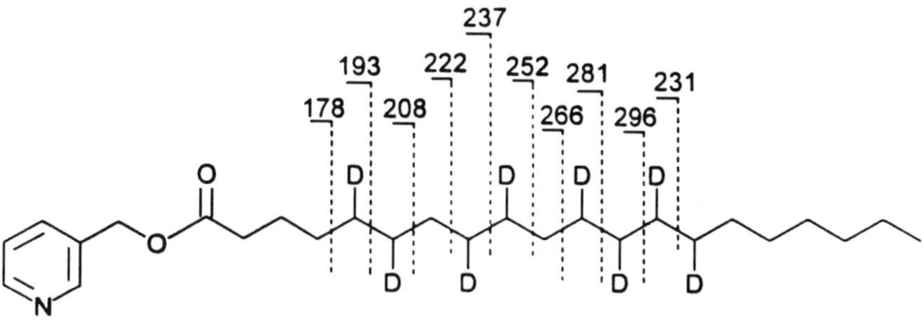

SCHEME 8.5

The Fragmentation Mechanism of the On-Site Derivatives of CLA and Metabolites. The mass spectra of the methyl or TMS tetrakis-TMSO-octadecanoates from CLA isomers are characterized mainly by α-cleavage between the vicinal TMSO-substituted carbon atoms (45). Only the fragments containing one TMSO-substituted carbon are prominent; the other α-cleavage ions, bearing three substituted carbon atoms, are of low relative abundance. The molecular ion is not detectable (FAME) or very weak (TMS esters). Because this method has not been used in recent papers on CLA, no further details are given in this survey.

In contrast, the MTAD adducts of CLA and metabolites have been employed successfully in some recent works (36,37,49). The mass spectra of these compounds show strong molecular ions and give unambiguous information on the position of the original conjugated double bonds, with prominent ions formed through α-cleavages initiated by placement of the radical site on either of the nitrogen atoms in the ring (Fig.8. 5). For the MTAD derivative of 18:2(10,12), the position of the nitrogen-containing ring between C-10 and C-13 of the hydrocarbon chain is indicated by the α-cleavage fragments at m/z 236 and 336. The adduct formation reaction makes it possible to deduce that the conjugated diene system was located between C-10/C-11 and C-12/C-13. The other prominent ions at m/z 304 and 376 can be assigned to the loss of a methoxy radical and methanol from the molecular ion and the α-fragmentation product at m/z 336, respectively (37). Although the GC separation of positional isomers of the MTDA adducts is difficult and peak overlapping cannot be avoided, the assignment of the positions of the conjugated double bonds is nevertheless possible using a selected ion scan technique (49). The characteristic α-fragmentation base peaks for the 18:2(8,10),

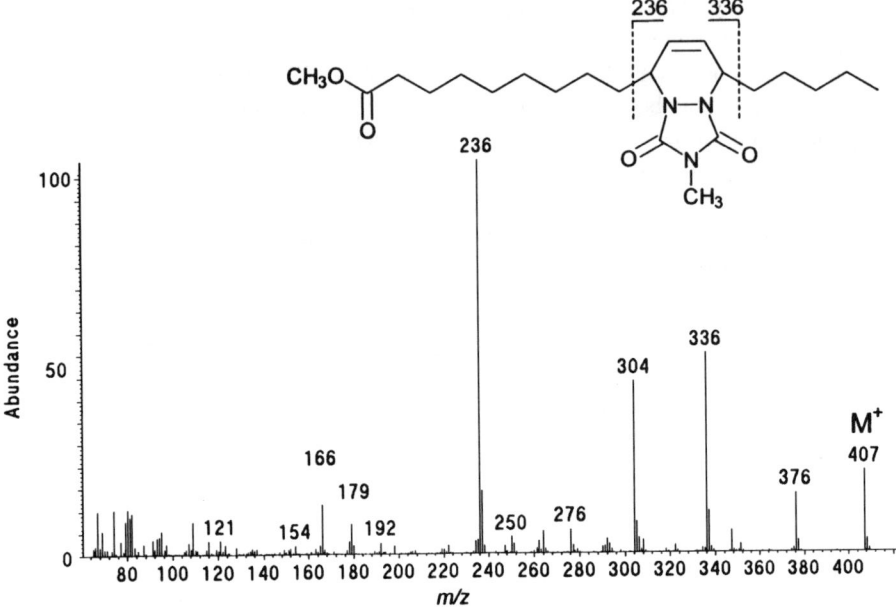

Fig. 8.5. The electron impact mass spectrum of the 4-methyl-1,2,4-triazoline-3,5-dione (MTDA) adduct of 18:2(10,12) (reproduced with kind permission of Dr. G. Dobson, Dundee, Scotland).

18:2(9,11), 18:2(10,12), and 18:2(11,13) can be found at m/z 264, 250, 236, and 222, respectively. The MTDA adducts have also been successfully applied to the analysis of conjugated 20:4(5,8,11,13) and 20:4(5,8,12,14) metabolites (36). By its base peak at m/z 250, the mass spectrum of the 20:4(5,8,11,13) isomer confirms unambiguously the original location of the conjugated double bonds at C-11/C-3 (36).

Conclusion

GC/MS analysis of remote-site and on-site derivatives is an essential tool for the unmistakable location of the double bonds in CLA and metabolites. DMOX derivatives are favored by many researchers because these derivatives have good GC properties and produce readily interpretable mass spectra. If uncertain or new compounds have to be characterized, it is advisable to also record the mass spectra of the deuterated picolinyl ester derivatives and the MTDA adducts of the compounds in question. The latter method seems to be very suitable for trace analysis because the base peak in a MTDA adduct mass spectrum clearly indicates the position of the conjugated double bond system, permitting the use of selected ion monitoring to detect trace amounts of these derivatives.

Acknowledgments

The author is grateful to Dr. W.W. Christie and G. Dobson, Scottish Crop Research Institute, Invergrowrie, Dundee, Scotland, UK, for providing the mass spectra of the picolinyl ester of

18:2(9,11) and of the 4-methyl-1,2,4-triazoline-3,5-dione (MTDA) adduct of the methyl ester of 18:2(10,12), respectively.

References

1. Biemann, K. (1962) *Mass Spectrometry Organic Chemical Applications*, McGraw Hill, New York.
2. Vetter, W., Walther, W., and Vecchi, M. (1971) Pyrrolidide als Derivate fur die Strukturaufklarung aliphatischer und alicyclischer Carbonsauren miters Massenspektrometrie, *Helv. Chim. Acta 54*, 1599–1605.
3. Spitzer, V., Marx, F., Maia, J.G.S., and Pfeilsticker, K. (1991) Identification of Conjugated Fatty Acids in the Seed Oil of *Acioa edulis* (Prance) syn. *Couepia edulis* (*Chrysobalanaceae*), *J. Am. Oil Chem. Soc. 68*, 183–187.
4. Lavillonniere, F., Martin, J.C., Bougnoux, P., and Sébédio, J.-L. (1998) Analysis of Conjugated Linoleic Acid Isomers and Content in French Cheese, *J. Am. Oil Chem. Soc 75*, 343–352.
5. Minnikin, D.E. (1978) Location of Double Bonds and Cyclopropane Rings in Fatty Acids by Mass Spectrometry, *Chem. Phys. Lipids 21*, 313–347.
6. Budzikiewicz, H. (1985) in *Analytiker-Taschenbuch*, vol. 5, pp. 135–158, Springer, Heidelberg.
7. Schmitz, B., and Klein, R.A. (1986) Mass Spectrometric Localization of Carbon-Carbon Double Bonds: A Critical Review of Recent Methods, *Chem. Phys. Lipids 39*, 285–311.
8. Jensen, N. J., and Gross, M.L. (1987) Mass Spectrometry Methods for Structural Determination and Analysis of Fatty Acids, *Mass Spectrom. Rev. 6*, 497–537.
9. Christie, W. W. (1989) in *Gas Chromatography and Lipids*, pp. 161–184, The Oily Press, Ayr, Scotland.
10. Yurawecz, M.P., Molina, A.A., Mossoba, M., and Ku, Y. (1993) Estimation of Conjugated Octadecatrienes in Edible Fats and Oils, *J. Am. Oil Chem. Soc. 70*, 1093–1099.
11. Fritsche, J., Mossoba, M.M., Yurawecz, M.P., Roach, J.A.G., Sehat, N., Ku, Y., and Steinhart, H. (1997) Conjugated Linoleic Acid (CLA) Isomers in Human Adipose Tissue, *Z. Lebensm.-Unters.-Forsch. A 205*, 415–418.
12. Sehat, N., Yurawecz, M.P., Roach, J.A.G., Mossoba, M.M., Kramer, J.K.G., and Ku, Y. (1998) Silver Ion–High-Performance Liquid Chromatographic Separation and Identification of Conjugated Linoleicic Acid Isomers, *Lipids 33*, 217–221.
13. Andersson, B. A. (1978) Mass Spectrometry of Fatty Acid Pyrrolidides, *Prog. Chem. Fats Other Lipids 16*, 279–308.
14. Harvey, D.J. (1992) in *Advances in Lipid Methodology-One*, (Christie, W.W., ed.) pp. 19–80, The Oily Press, Dundee, Scotland.
15. Christie, W.W. (1993) Determination of Fatty Acid Structures, *INFORM 4*, 85–91.
16. Spitzer, V. (1996) Structure Analysis of Fatty Acids by Gas Chromatography-Low Resolution Electron Impact Mass Spectrometry of Their 4,4-Dimethyloxazoline Derivatives—A Review, *Prog. Lipid Res. 35*, 387–408.
17. Dobson, G., and Christie, W.W. (1996) Structural Analysis of Fatty Acids by Mass Spectrometry of Picolinyl Esters and Dimethyloxazoline Derivatives, *Trends Anal. Chem. 15*, 130–136.
18. Christie, W.W. (1998) Gas Chromatography-Mass Spectrometry Methods for Structural Analysis of Fatty Acids, *Lipids 33*, 343–353.

19. Andersson, B.A., and Holman, R.T. (1974) Pyrrolidides for Mass Spectrometric Determination of the Position of the Double Bond in Monounsaturated Fatty Acids, *Lipids 9*, 185–190.
20. Iversen, S.A., Cawood, P., Madigan, M.J., Lawson, A.M., and Dormandy, T.L. (1984) Identification of a Diene Conjugated Component of Human Lipids as Octadeca-9,11-dienoic Acid, *FEBS Lett. 171*, 320–324.
21. Anderson, B.A., Christie, W.W., and Holman, R. T. (1975) Mass Spectrometric Determination of Positions as Double Bonds in Polyunsaturated Fatty Acid Pyrrolidides, *Lipids 10*, 215–219.
22. Vetter, W., and Walthur, W. (1990) Preparation of Pyrrolidides for Fatty Acids via Trimethylsilyl Esters for Gas Chromatography-Mass Spectrometric Analysis, *J. Chromatogr. 513*, 405–407.
23. Zhang, J.Y., Yu, Q T., Liu, B.N., and Huang, Z.H. (1988) Chemical Modification in Mass Spectrometry IV: 2-Alkenyl-4,4-dimethyloxazolines as Derivatives for the Double Bond Location of Long-Chain Olefinic Acids, *Biomed. Environ. Mass Spectrom. 15*, 33–44.
24. Zhang, J.Y., Yu. X., Wang, H.Y., Liu, B.N., Yu, Q.T., and Huang, Z.H. (1989) Location of Triple Bonds in the Fatty Acids from the Kernel Oil of *Pyrularia edulis* by GC-MS of Their 4,4-Dimethyloxazoline Derivatives, *J. Am. Oil Chem. Soc. 66*, 256–259.
25. Zhang, J.Y., Yu, Q.T., Yang, Y.M., and Huang, Z.H. (1988) Chemical Modification in Mass Spectrometry 11. A Study on the Mass Spectra of 4,4-Dimethyloxazoline Derivatives of Hydroxy Fatty Acids, *Chem. Scr. 28*, 357–363.
26. Spitzer, V., Marx, F., Maia, J.G.S., and Pfeilsticker, K. (1991) Occurrence of Conjugated Fatty Acids in the Seed Oil of *Couepia longipendula* (*Chrysobalanaceae*), *J. Am. Oil Chem. Soc. 68*, 440–442.
27. Spitzer, V., Marx, F., Maia, J.G.S., and Pfeilsticker, K. (1992) Occurrence of α-Eleostearic Acid in the Seed Oil of *Parinarium montana* (*Chrysobalanaceae*), *Fat Sci. Technol. 94*, 58–60.
28. Spitzer, V., Marx, F., and Pfeilsticker, K. (1994) The Electron Impact Mass Spectra of Theoxazoline Derivatives of Some Conjugated Diene and Triene C-18 Fatty Acids, *J.Am. Oil Chem. Soc. 71*, 873–876.
29. Spitzer, V. (1996) The Mass Spectra of the 4,4-Dimethyloxazoline Derivatives of Some Conjugated Hydroxy Eneyne C-17 and C-18 Fatty Acids, *J. Am. Oil Chem. Soc. 73*, 489–492.
30. Spitzer, V., Tomberg, W., Hartmann, R., and Aichholz, R. (1997) Analysis of the Seed Oil of *Heistena silvanii* (*Olacaceae*)—A Rich Source of a Novel C_{18} Acetylenic Fatty Acid, Lipids 32, 1189–1200.
31. Kramer, J.K.G., Sehat, N., Dugan, M.E.R., Mossoba, M.M., Yurawecz, M.P., Roach, J.A.G., Eulitz, K., Aalhus, J.L., Schaefer, A L., and Ku, Y. (1998) Distributions of Conjugated Linoleic Acid (CLA) Isomers in Tissue Lipid Classes of Pigs Fed a Commercial CLA Mixture Determined by Gas Chromatography and Silver Ion–High-Performance Liquid Chromatography, *Lipids 33*, 549–558.
32. Spitzer, V., Bordignon. S.A., Schenkel, E.P., and Marx, F. (1994) Identification of Nine Acetylenic Fatty Acids, 9-Hydroxystearic Acid and 9,10-Epoxystearic Acid in the Seed Oil of *Jodina rhombifolia* Hook et Arn. (*Santalaceae*), *J. Am. Oil Chem. Soc. 71*, 1343–1348.
33. Fay, L., and Richli, U. (1991) Location of Double Bonds in Polyunsaturated Fatty Acids by Gas Chromatography-Mass Spectrometry After 4,4-Dimethyloxazoline Derivatization, *J. Chromatogr. 541*, 89–98.

34. Yurawecz, M P., Hood, J.K., Roach, J.A., Mossoba, M.M., Daniels, D.H., Ku, Y., Pariza, M.W., and Chin, S.F. (1994) Conversion of Allylic Hydroxy Oleate to Conjugated Linoleic Acid and Methoxy Oleate by Acid-Catalyzed Methylation Procedures, *J. Am. Oil Chem. Soc. 71,* 1149–1155.
35. Berdeaux, O., Christie, W.W., Gunstone, F.D., Sébédio, J.-L. (1997) Large-Scale Synthesis of Methyl *cis*-9, *trans*-11-Octadecadienote from Methyl Ricinoleate, *J. Am. Oil Chem. Soc. 74,* 1011–1015.
36. Sébédio, J.-L., Juaneda, P., Dobson, G., Ramilison, I., Martin, J.C., and Chardigny, J.M. (1997) Metabolites of Conjugated Isomers of Linoleic Acid (CLA) in the Rat, *Biochim. Biophys. Acta 1345,* 5–10.
37. Dobson, G. (1998) Identification of Conjugated Fatty Acids by Gas Chromatography-Mass Spectrometry of 4-Methyl-1,2,4-triazoline-3,5-dione Adducts, *J. Am. Oil Chem. Soc. 75,* 137–142.
38. Yurawecz, M. P., Roach, J.A.G., Sehat, N., Mossoba, M.M., Kramer, J.K.G., Fritsche, J., Steinhart, H., and Kua, Y. (1998) A New Conjugated Linoleic Acid Isomer, 7-*trans*,9-*cis*-Octadecadienoic Acid, in Cow Milk, Cheese, Beef, and Human Milk and Adipose Tissue, *Lipids 33,* 803–809.
39. Christie, W.W. (1997) in *Advances in Lipid Methodology-Four,* (Christie, W.W., ed.) pp. 119–169, The Oily Press, Dundee, Scotland.
40. Yu, Q.T., Liu, B.N., Zhang, J.Y., and Huang, Z.H. (1989) Location of Double Bonds in Fatty Acids of Fish Oil and Rat Testis Lipids. Gas Chromatography-Mass Spectrometry of the Oxazoline Derivatives, *Lipids 24,* 79–83.
41. Rezanka, T., Zlatkin, I.V., Viden, I., Slabova, O.I., and Nikitin, D.I. (1991) Capillary Gas Chromatography-Mass Spectrometry of Unusual and Very Long-Chain Fatty Acids from Soil Oligotrophic Bacteria, *J. Chromatogr. 558,* 215–221.
42. Dobson, G., Christie, W.W., Brechany, E.Y., Sébédio, J.-L., and Quere, J.-L.L. (1995) Silver Ion Chromatography and Gas Chromatography-Mass Spectrometry in the Structural Analysis of Cyclic Dienoic Acids Formed in Frying Oils, *Chem. Phys. Lipids 75,* 171–182.
43. Harvey, D.J. (1984) Picolinyl Derivatives for the Structural Determination of Fatty Acids by Mass Spectrometry: Application to Polyenoic Acids, Hydroxy Acids, Di-Acids and Related Compounds, *Biol. Mass Spectrom. 11,* 340–347.
44. Balazy, M., and Nies, A.S. (1989) Characterization of Epoxides of Polyunsaturated Fatty Acids by Mass Spectrometry via 3-Pyridinylmethyl Esters, *Biomed. Environ. Mass Spectrom. 18,* 328–336.
45. Janssen, G., and Parmentier, G. (1978) Determination of Double Bond Positions in Fatty Acids with Conjugated Double Bonds, *Biomed. Mass Spectrom. 5,* 439–443.
46. Christie, W.W. (1997) The Analysis of Conjugated Fatty Acids, *Lipid Technol. 9,* 73–75.
47. Kramer, J.K.G., Fellner, V., Dugan, M.E.R., Sauer, F.D., Mossoba, M.M., and Yurawecz, M.P. (1997) Evaluating Acid and Base Catalysts in the Methylation of Milk and Rumen Fatty Acids with Special Emphasis on Conjugated Dienes and Total *trans* Fatty Acids, *Lipids 32,* 1219–1228.
48. Christie, W.W., Brechany, E.Y., Johnson, S.B., and Holman, R.T. (1986) A Comparison of Pyrrolidide and Picolinyl Ester Derivates for the Identification of Fatty Acids in Natural Samples by Gas Chromatography-Mass Spectrometry, *Lipids 21,* 657–661.
49. Christie, W.W., Dobson. G., and Gunstone, F.D. (1997) Isomers in Commercial Samples of Conjugated Linoleic Acid, *J. Am. Oil Chem. Soc. 74,* 1231.

Chapter 9

Identification of CLA Isomers in Food and Biological Extracts by Mass Spectrometry

John A.G. Roach

U.S. Food and Drug Administration, Washington, DC 20204

Introduction

The general analytical approach that we use to analyze conjugated linoleic acid (CLA) by mass spectrometry (MS) is described elsewhere (1). The recent advances in the identification of isomers of CLA are in large part due to the application of silver ion–high-performance liquid chromatography (HPLC) to fractionate the CLA mixture (2). The fractions are then derivatized and analyzed by gas chromatography (GC), GC/MS and GC/Fourier transform infrared spectroscopy (FTIR). HPLC, followed by derivatization and GC of the derivatized components in each fraction, resolves the individual components in the CLA mixture much more effectively than GC alone (3,4).

Gas Chromatography Conditions for CLA

The capillary columns for which we have the most data are 0.2-µm film CP-Sil 88 columns with dimensions of 50 m × 0.2 mm and 100 m × 0.2 mm. Typical GC conditions are splitless injection in isooctane at a column oven temperature of 75°C, injector and interface temperatures 220°C, hold 2 min with split flow off, heat oven at 5°C/min to 180°C, hold 30 min, heat oven at 5°C/min to 220°C, and hold 20 min. Septum purge flow is 3 mL/min, split flow is 100 mL/min.

Helium carrier gas at a head pressure of 12 lb/in^2 is used with the 50-m column. Hydrogen carrier gas at a head pressure of 25 lb/in^2 is used with the 100-m column. The derivatized CLA retention times differ by <3 min for the two columns because of the use of hydrogen rather than helium in the 100-m column. Hydrogen and the 100-m column provide better derivatized CLA isomer resolution.

Elution of the dimethyloxazoline (DMOX) derivatives of CLA extracted from cheese is preceded by the DMOX derivatives of linoleic acid, 9,12-18:2, and a 9,12,15-18:3 acid (Fig. 9.1). The typically abundant 9*cis*,11*trans*-18:2 isomer is a good marker for the leading edge of the CLA window. The 9*trans*,11*trans*-18:2 isomer is found at the trailing edge of the window. The general GC elution order of geometric isomers of CLA is *cis/trans*, *trans/cis*, *cis/cis*, and *trans/trans*. Methoxylated CLA artifacts and furans elute shortly after CLA (5). Furan oxidation products elute long after the CLA (6).

The GC conditions were adapted from those used for GC analysis of the methyl esters of CLA with flame ionization detection (FID). However, the vacuum within a mass spectrometer tends to pull analyte through the column so that methyl esters of CLA elute ~10 min faster in a GC/MS system than in a GC-FID system. Lowering

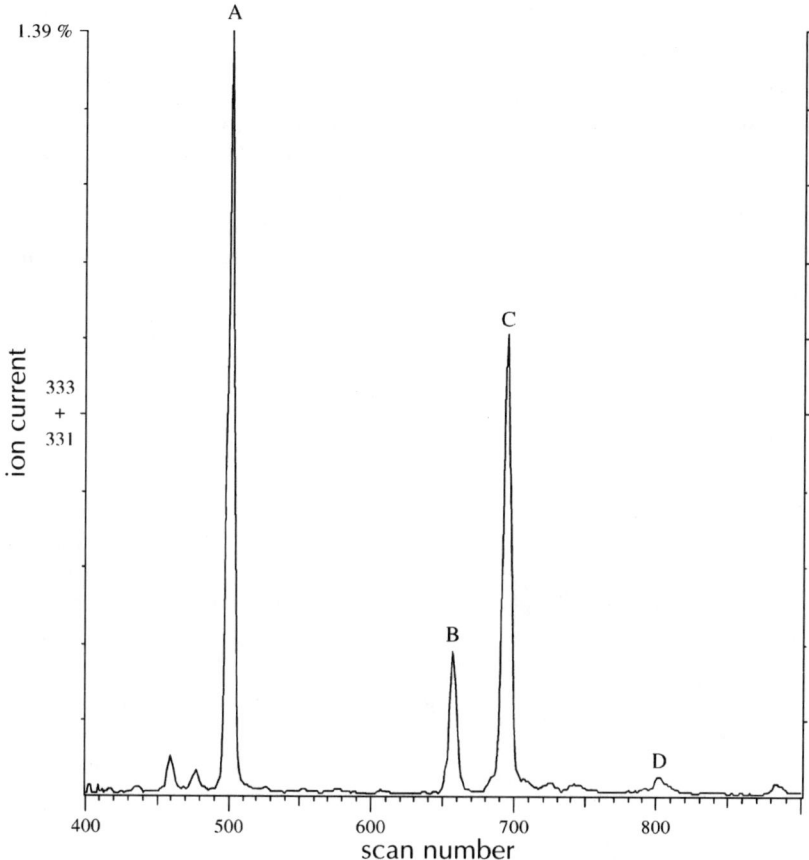

Fig. 9.1. Sum of reconstructed ion profiles of m/z 333 and 331 for dimethyloxazoline (DMOX)-derivatized cheese extract. Identified responses are DMOX derivatives of (A) linoleic acid, (B) 9,12,15-18:3 acid, (C) 9*cis*,11*trans*-18:2 acid, and (D) 9*trans*, 11*trans*-18:2 acid. Profiles are ion trap (Thermoquest GCQ) data.

the temperature 5°C or reducing the carrier flow will bring the CLA GC-FID and GC/MS retention times into reasonable agreement, but the costs include longer analysis times and broader chromatographic peak widths.

Defining the GC/MS Detection Limit

A recurrent challenge in the analyses of biological matrices for specific CLA isomers is the detection and identification of minor components. Unlike an analytical standard, the limit of detection for an analyte in a food or biological extract is the baseline chemical noise arising from coextractive materials. Signals for minor components of interest have to be distinguished from this background continuum with various MS techniques that selectively enhance diagnostic analyte signals.

Appearance of CLA Mass Spectra

The electron impact (EI) mass spectra of the methyl esters of CLA are alike and limit distinctions that can be inferred from GC/MS data to differences in retention times. DMOX derivatives of CLA (DMOX CLA) yield distinctive EI mass spectra (Fig. 9.2) that may be used to identify the positions of the double bonds in the CLA isomers (7).

Methylene losses from the saturated parts of the molecule occur at 14-Da intervals in the mass spectrum. Losses of 12 Da, separated from one another by 14 Da, mark the locations of the conjugated double bonds in the molecule. Adjacent abundant ions bracket the losses of 12 Da and enhance the certainty of the identification (8).

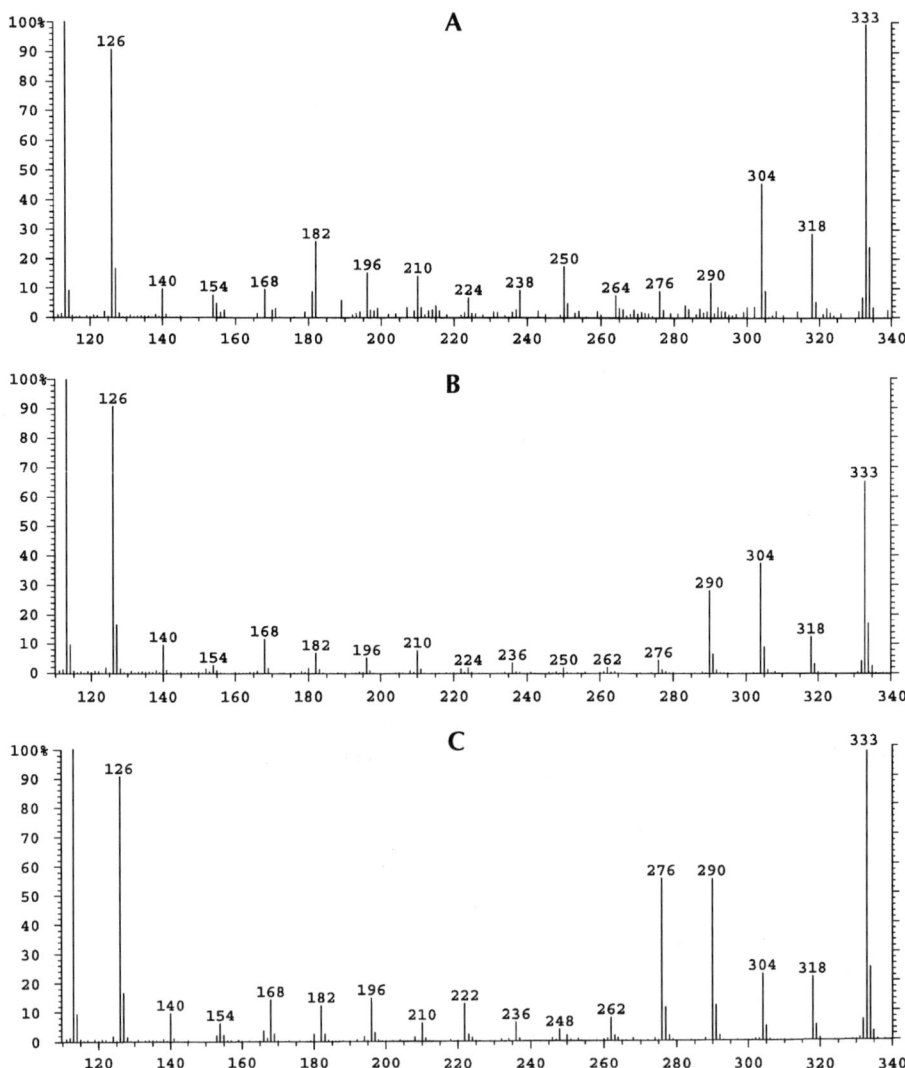

If we assign the letters "n" and "m" to the locations of the double bonds, then n-1, n-2 and m+1, m+2, and m+3 identify nearby carbons in the chain relative to the conjugated double bonds. Table 9.1 lists the m/z of these ions in the EI spectrum (Fig. 9.2, spectrum D) of a DMOX derivative of 9,11-18:2, with designations of each ion relative to "n" and "m" and their designations by carbon number in the chain.

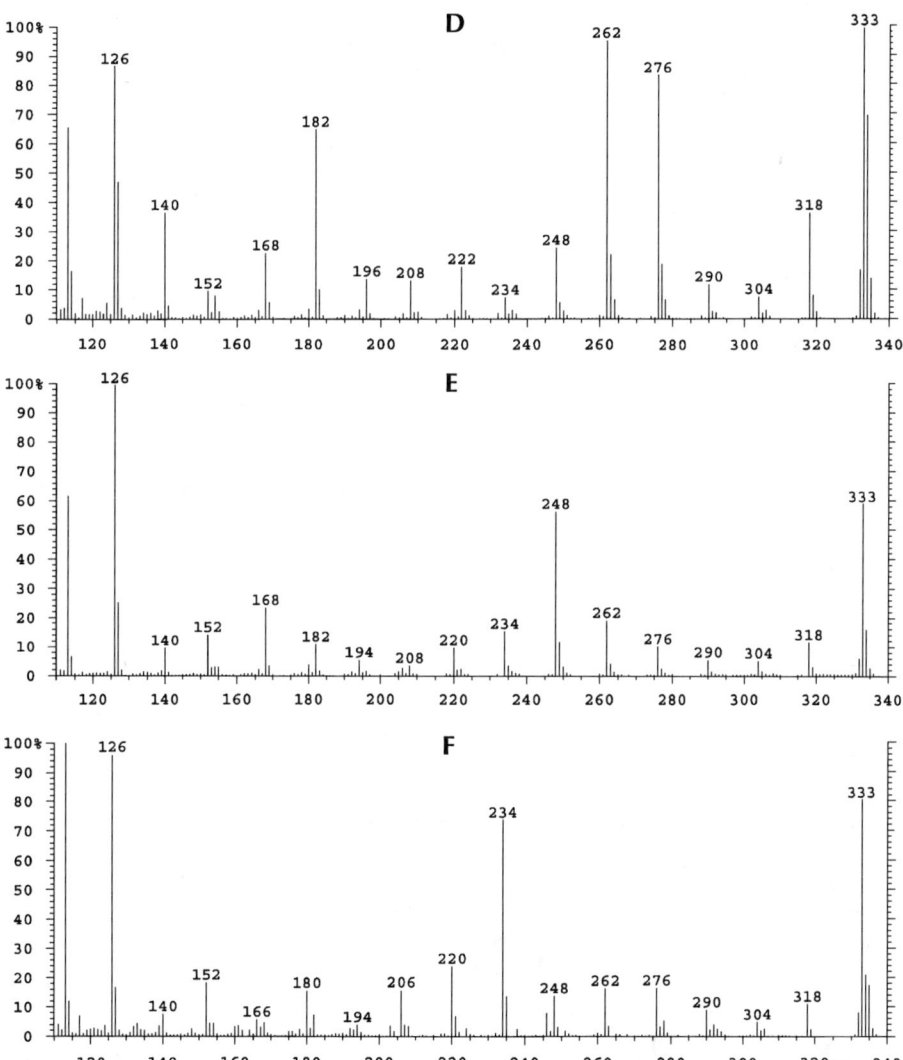

Fig. 9.2. Electron impact (EI) spectra of dimethyloxazoline (DMOX) derivatives of conjugated linoleic acid (CLA) isomers identified in cheese: (A) 12,14-18:2, (B) 11,13-18:2, (C) 10,12-18:2, (D) 9,11-18:2, (E) 8,10-18:2, (F) 7,9-18:2. Gas chromatography/mass spectrometry (GC/MS) data were recorded with a magnetic sector (Autospec Q) instrument.

The losses of 12 Da from carbons "n" and "m" to form m/z 196 and m/z 222 identify the locations of the double bonds in the 9,11-18:2 isomer as positions 9 and 11. The ions corresponding to n-2 (m/z 182), m+2 (m/z 262), and m+3 (m/z 276) are

TABLE 9.1 Partial Ion Series Identifications in EI Mass Spectrum of DMOX Derivative of 9,11-18:2 CLA Relative to Locations of Double Bonds at Carbons "n" and "m" and by Fatty Acid Carbon Number[a]

n-2	n-1	n	m-1	m	m+1	m+2	m+3			
182	196	208	222	234	248	262	276	290	304	318	333
7	8	9	10	11	12	13	14	15	16	17	18

Annotated Structures of DMOX 9,11-18:2 CLA Derivative and 9,11-18:2 CLA[b]

[a]Abbreviations used: EI, electron impact; DMOX, dimethyloxazoline; CLA, conjugated linoleic acid.
[b]The carbons are labeled in both structures according to the carbon numbers of the 9,11-18:2 fatty acid.

A

thought to occur in increased relative abundance because carbons allyl to the double bonds are favorable locations for a radical site in the carbon chain. The general appearance of the mass spectrum may be explained as remote-site fragmentation with retention of the positive charge on the nitrogen and radical-induced cleavage of the carbon chain (1). Suggested remote-site fragmentations facilitated by these allylic radical sites are shown in Figure 9.3.

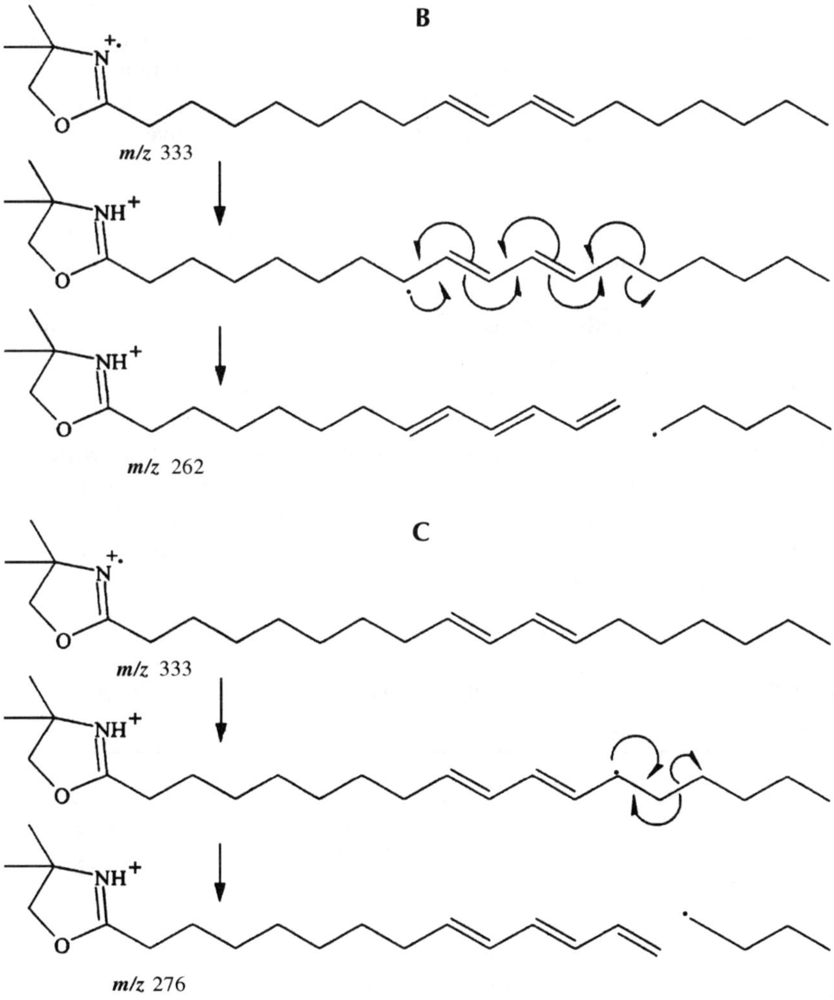

Fig. 9.3. Suggested fragmentation schemes leading to (A) *m/z* 182, (B) *m/z* 262, and (C) *m/z* 276 in mass spectrum of the 9,11 isomer of DMOX CLA. Formation of ions (A) n-2, (B) m+2, and (C) m+3 may be explained in each case by participation of a radical site allyl to a double bond in the carbon chain.

In Figure 9.3, scheme A, the neutral loss in the formation of *m/z* 182 is an allylic radical. In Figure 9.3, schemes B and C, allyl positions for the radical site favor formation of *m/z* 262 and *m/z* 276. Thus, the complete set of characteristic ions for double bond positions in isomers of DMOX CLA are the molecular ion, 333, and ions corresponding to positions n-2, n-1, n, m-1, m, m+2 and m+3 (9). A table of 7 diagnostic ions for each CLA isomer reported in cheese (10,11) contains 42 ions (Table 9.2). The ions occur in the DMOX mass spectra of more than one CLA isomer; thus the list of 42 ions that distinguish these spectra actually consists of only 24 ions.

Most of the ions in the diagnostic list of ions for each DMOX CLA isomer are also found in the mass spectra of the DMOX derivatives of many fatty acids. Thus, confirmation of the identity of a specific CLA isomer requires sufficient chromatographic separation from other congeners to obtain a representative mass spectrum that unambiguously indicates the positional identity of the isomer. In other words, there must be more proof for the presence of a specific isomer in a test portion than the presence of signals for each of its diagnostic ions. It must be possible to produce a mass spectrum from the data that is consistent with a known mass spectrum of that isomer.

Recording data for only a limited set of ions that includes the diagnostic ions for the CLA isomers of interest is an effective way to lower the GC/MS detection limit for CLA. The specific approaches available depend on the type of GC/MS system used for the analysis.

Low-resolution quadrupole (Finnigan TSQ-46, San Jose, CA), magnetic sector (Micromass Autospec Q, Manchester, UK), and ion trap (Thermoquest GCQ, San Jose, CA) MS instrumentation have been used in this laboratory to analyze for CLA. Each instrument has a unique capability that may be used to detect low levels of CLA in challenging matrices.

TABLE 9.2 Diagnostic Ions for DMOX Derivatives of CLA Isomers Reported in Cheese[a,b]

Isomer	n-2	Losses of 12 Da				m+2	m+3
7,9[c]	154	168	180	194	206	234	248
8,10	168	182	194	208	220	248	262
9,11	182	196	208	222	234	262	276
10,12	196	210	222	236	248	276	290
11,13	210	224	236	250	262	290	304
12,14	224	238	250	264	276	304	318

[a]Diagnostic DMOX CLA ions should include the molecular ion, 333, and ions characteristic of losses of 12 Da as well as abundant ions attributed to remote-site fragmentations facilitated by allylic carbon radical sites.
[b]See Table 9.1 for abbreviations.
[c]154 is weak, 234 is more abundant than 248.

Quadrupole MS Detection of CLA

A quadrupole mass spectrometer can rapidly step between ions across its entire mass range without degradation of performance. Thus, it is a simple task to devise a scan algorithm for a quadrupole instrument that will record data only for the diagnostic ions of CLA. Eliminating replicate ions in a mass list devised with Table 9.2 and including 333, the molecular ion of DMOX CLA, results in a list of the following 24 ions: 154, 168, 180, 182, 194, 196, 206, 208, 210, 220, 222, 224, 234, 236, 238, 248, 250, 262, 264, 276, 290, 304, 318, 333. If each ion is monitored for the same amount of time with a total scan time equal to that required for a mass scan of 100–340 Da, the result is a selective multiple ion DMOX CLA assay.

The expected improvement in signal-to-noise ratio for monitoring a limited set of ions as long as it would normally take to record a full mass scan is estimated by taking the square root of the result of dividing the number of ions in a full mass scan by the number of ions in the limited mass set. For example, recording data for only four ions in each 1-s scan rather than recording data for 400 ions in each 1-s scan, improves the signal-to-noise ratio for the four monitored ions ~10-fold. Thus, only a modest threefold improvement is expected for monitoring 24 diagnostic CLA ions rather than scanning from 100 to 340 Da.

In those laboratories that are equipped with a quadrupole mass spectrometer data system that will permit monitoring 24 ions simultaneously, the approach is subject to an additional constraint. The data record will not contain information about other components in the extract. Post-analysis questions about possible metabolites or interferences cannot be answered by reexamining a limited data record. For those questions, it will be necessary to repeat the analysis with different MS acquisition parameters.

Magnetic Sector MS Detection of CLA

High-resolution selected ion recording (SIR) is a useful check on the validity of tentative GC peak identifications based on FID data. This technique is widely used for the detection of femtogram quantities of chlorinated dioxins (12). The resolution of the instrument is adjusted to 10,000 or more. At this resolution, ions of defined elemental composition are accurately selected for detection with minimal matrix interference. During a GC/MS analysis, the magnetic field is held constant and the accelerating voltage is quickly stepped to bring the selected ions sequentially into focus at the detector.

Operating a double-focusing magnetic sector instrument at high resolution discards 90% of the ion current before it reaches the detector. This signal loss is offset by a corresponding improvement in the signal-to-noise ratio. The detected signal for a selected ion increases >80-fold compared with low-resolution limited mass scan data for the same ion at the same detector sensitivity. A mass spectrometer operating at high resolution can detect the molecular ions of C_{18} and C_{20}

fatty acid methyl esters or their corresponding DMOX derivatives with better sensitivity than a flame ionization detector.

High-resolution SIR is a useful tool for sorting CLA and non-CLA signals in the CLA elution window. Figure 9.4A is the methyl ester CLA region of an esterified cheese extract GC trace with FID. Figure 9.4B is the methyl ester CLA region of the same cheese extract recorded by GC/MS with high resolution SIR detection of m/z 294.2559. The dissimilarities between Figure 9.4A and Figure 9.4B demonstrate that not all FID responses in the region are due to methyl esters of CLA.

The high-resolution SIR data record is intentionally limited to molecular ion information to maximize sensitivity; thus the potential for misinterpretation of the SIR data exists. The molecular ions of 18:0, 18:1, 18:2, 18:3, and 18:4 congeners contribute to the signals of each other to a limited extent with isotope cluster contributions and losses of H_2. The contributions of 18:1 and 18:3 to the 18:2 signal are recognized by comparing the absolute abundances of the signals for m/z 296.2715 and 292.2402 to the m/z 294.2559 signal. A more interesting response is provided by artifacts of methylation with acid catalyst (5). The artifacts elute just beyond the CLA region and fragment in several ways, including expulsion of the elements of methanol to form m/z 294.

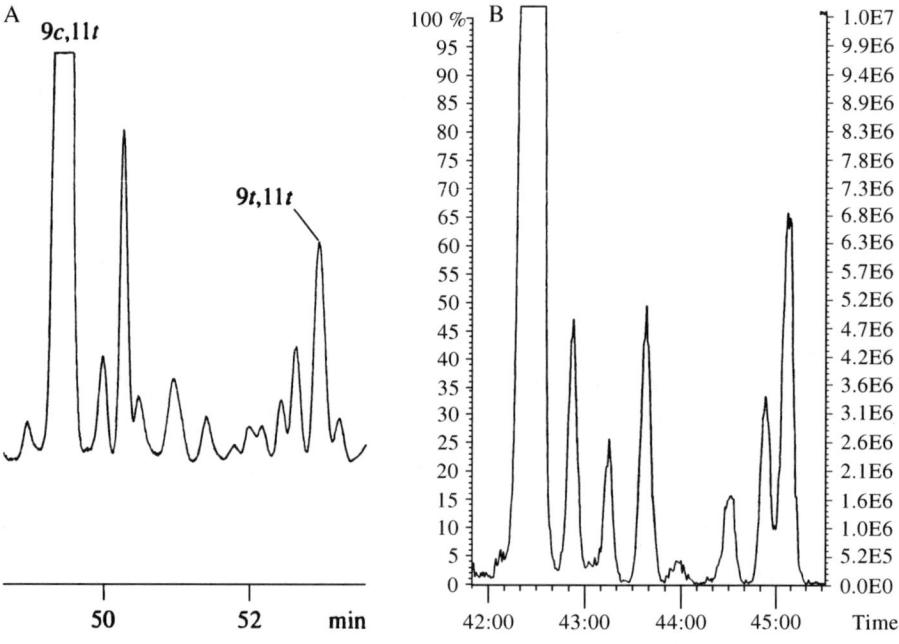

Fig. 9.4. The methyl ester conjugated linoleic acid (CLA) region of a gas chromatography (GC) chromatogram for cheese with (A) flame ionization detection, and (B) mass spectrometry (MS) detection in high-resolution selected ion recording mode, ion profile for m/z 294.2559.

Concentration of the sample ~40-fold after FID or high-resolution SIR yields a test concentration that is suitable for the identification of CLA by low-resolution GC/MS. The low-resolution acquisition algorithm for the analysis of CLA and DMOX CLA is a magnet scan from 440 to 40 Da at 1 s/decade at a resolution of 1600. The data contain information about CLA and coextractives in the test portion. These data may be used to examine questions about most of the responses observed in the chromatogram.

For example, a composite platelet extract obtained from a patient contained a large unidentified GC FID peak eluting beyond arachidonic acid that was not observed in extracts obtained from the patient's sibling or parent. Examination of the GC/MS data recorded for the extract showed that the large peak in the GC FID data was squalene.

If the intent of the analyses is to obtain data about the DMOX derivatives of positional isomers of CLA, a scan from 440 to 40 Da contains a great deal of unnecessary information. Limiting the mass scan to one of 340–110 Da results in a modest improvement in sensitivity without a loss of information about DMOX CLA. But in this case, there is a better reason to limit the mass scan than a small improvement in sensitivity. A limited mass scan can be completed in less time; thus more scans can be recorded during the elution of each GC peak. The increased number of scans provides chromatographic data that more closely resemble FID chromatograms. The GC/MS data are examined using abundant m+2 and m+3 diagnostic ions to establish the locations of individual DMOX CLA isomers in the data record (Fig. 9.5).

The coelution of the selected abundant diagnostic ions in the data guide the analyst in the selection of adjacent background spectra to subtract from the summed spectra of the isomer of interest. The idea is to remove signal contributions of adjacent peaks to obtain a representative signal for the GC peak. The task is not as easy as it sounds. The required data for Figure 9.5 can be improved slightly by limiting the scan to 340–190 Da to obtain more than two scans per second, but that is only because the identified isomers, 11,13-18:2, 12,14-18:2, and 9,11-18:2 in Figure 9.5 are sufficiently separated from one another in the chromatogram to obtain representative spectra of each isomer.

The predictable spectra of the DMOX derivatives of CLA permit examination of GC/MS data for previously unreported isomers. For example, in Table 9.2, the ions at m/z 180 and 206 of the 7,9 isomer are not subject to interference by the other five DMOX CLA isomers in Figure 9.2. The coelution of these two ions in the data led to the detection of the 7,9 isomer in cheese, milk, and other biological extracts (11). The 7,9 isomer occurs at levels as high as 16% of total CLA but went undetected because of its incomplete separation from other components until the distinctive features of its predicted spectrum were used to search for it in the recorded data.

The 7*trans*,9*cis* and 9*trans*,11*cis* 18:2 isomers comprise the front and back of the largest chromatographic peak in the ion profile of m/z 333 in Figure 9.6A. The locations

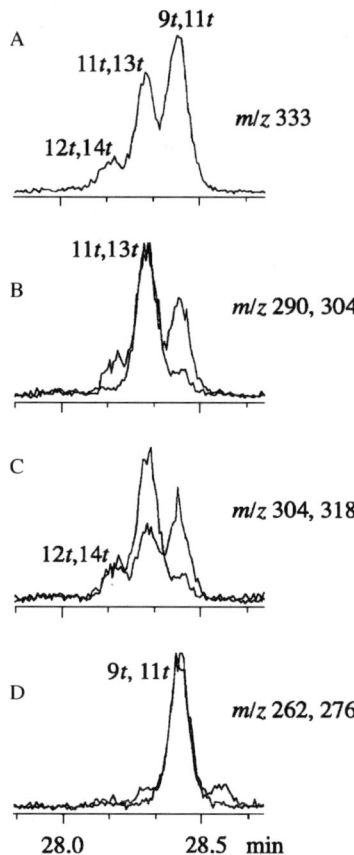

Fig. 9.5. Reconstructed ion chromatograms for dimethyloxazoline (DMOX) derivatives of *trans,trans* conjugated linoleic acid (CLA) isomers isolated from cheese using silver ion–liquid chromatography. Profiles are as follows: (A) molecular ion, m/z 333, (B) allylic ions (m+2) and (m+3) for 11,13-18:2, corresponding ions for (C) 12,14-18:2, and (D) 9,11-18:2. In each case, the ion pairs were normalized for the CLA isomer indicated.

of the two isomers in the chromatogram are revealed with ion profiles of their relatively abundant m+2 diagnostic ions, m/z 234 and 262. The responses of the 7*trans*,9*cis* and 9*trans*,11*cis* 18:2 and isomers elute separately as seen in Figure 9.6B and provide satisfactory spectra for each isomer after background subtraction. It is difficult to extract representative spectra of two isomers from the front and back of "one" apparent chromatographic peak, such as the close elution of the 7*trans*,9*trans* and 9*trans*,11*trans* 18:2 isomers in Figure 9.6, because their spectra are very similar. Subtracting one side of the peak from the other rarely produces a good spectrum for both isomers.

Ion Trap MS Detection of CLA

The quadrupole ion trap is a remarkably sensitive mass spectrometer. In direct comparisons in this laboratory, the trap is not as sensitive for CLA analysis as a good magnetic sector instrument, but it is considerably cheaper.

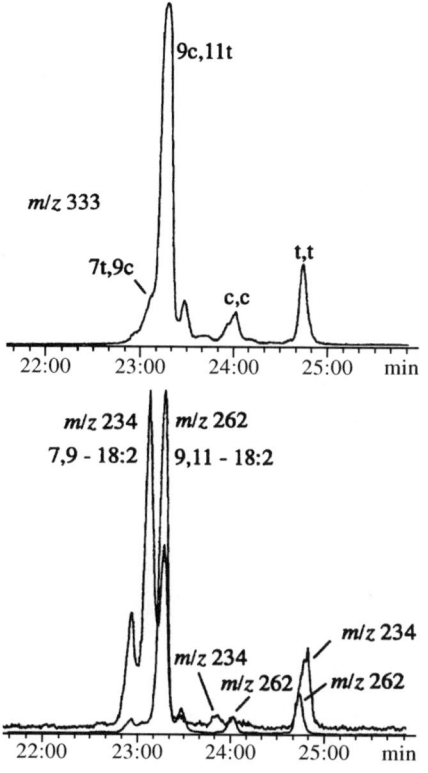

Fig. 9.6. Reconstructed ion chromatograms for dimethyloxazoline derivatives of conjugated linoleic acid (CLA) isomers isolated from human adipose tissue. Ion profiles of (A) m/z 333 and the superimposed ion profiles of (B) m/z 234 and 262 reveal the presence of the 7 trans,9 cis isomer in the leading edge and the 9 trans,11 cis isomer in the trailing edge of the largest peak of the m/z 333 ion profile, and the 9 trans,11 trans isomer in the front half and the 7 trans,9 trans isomer in the last half of the last peak of the m/z 333 ion profile.

Ions within the trap are cooled by successive collisions with helium buffer gas. Hydrogen can be used as the carrier gas in place of helium if the trap is provided with a separate source of helium buffer gas.

The process by which an ion trap is scanned is such that full scan data are superior to multiple ion data if it is necessary to detect more than two ions (12). Representative DMOX CLA spectra recorded with a GCQ ion trap are shown in Figure 9.7.

An ion trap records good data over a mass range of 110–340 Da at two scans per second. This is very good performance, but the credibility of ion trap data is still subject to the limitations of the chromatographic separation. If the components in a mixture cannot be resolved chromatographically, the mixed spectra obtained cannot be relied upon to identify the components in the mixture with accuracy.

Conclusions

Significant strides have been made in the separation and subsequent identification of individual isomers associated with CLA. A set of 24 diagnostic ions may be used to identify individual isomers of CLA reported in cheese as their DMOX derivatives. Recording GC/MS data for a limited set of ions that includes the diagnostic ions for

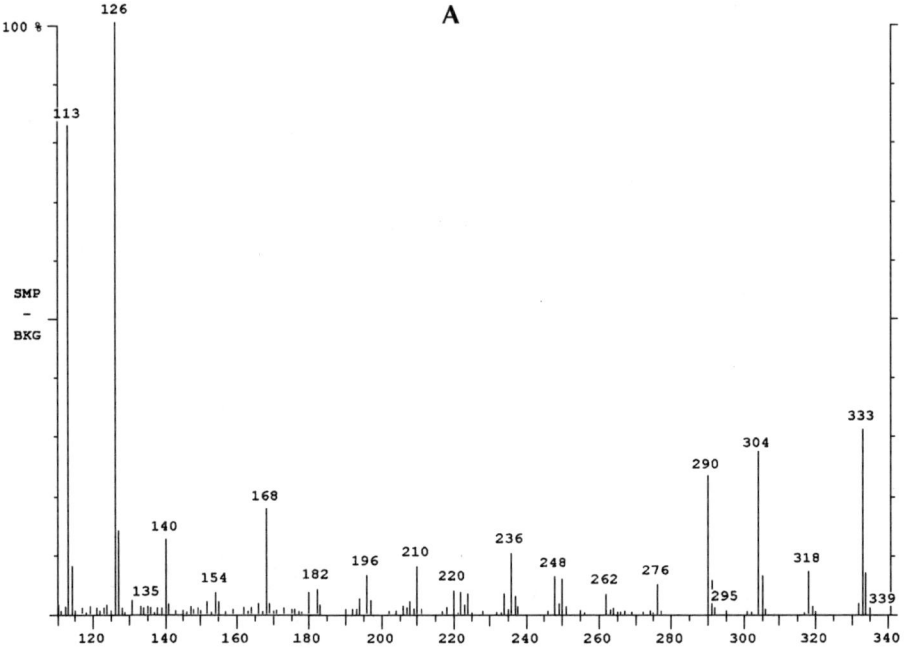
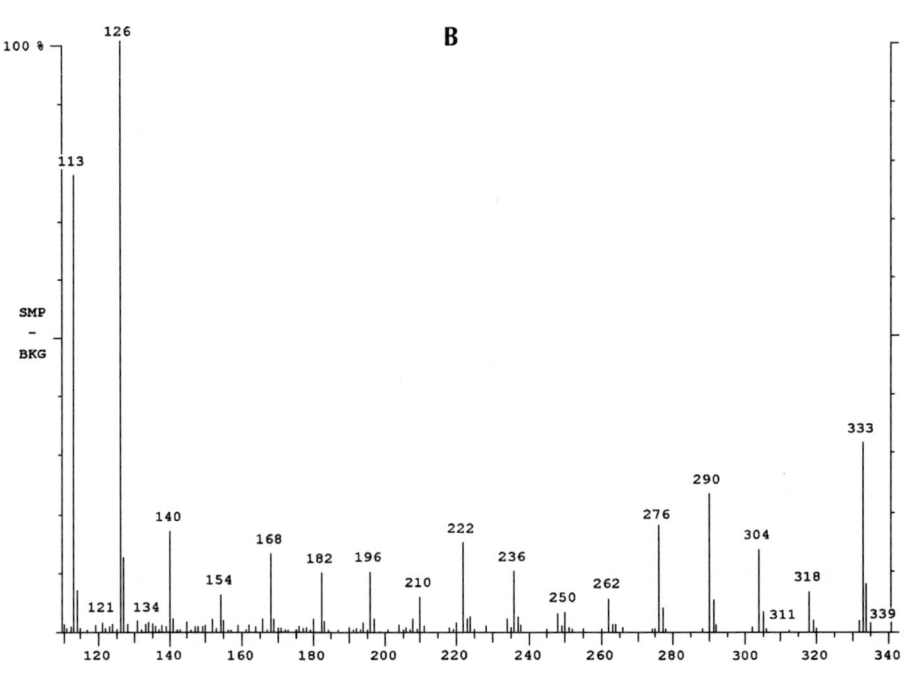

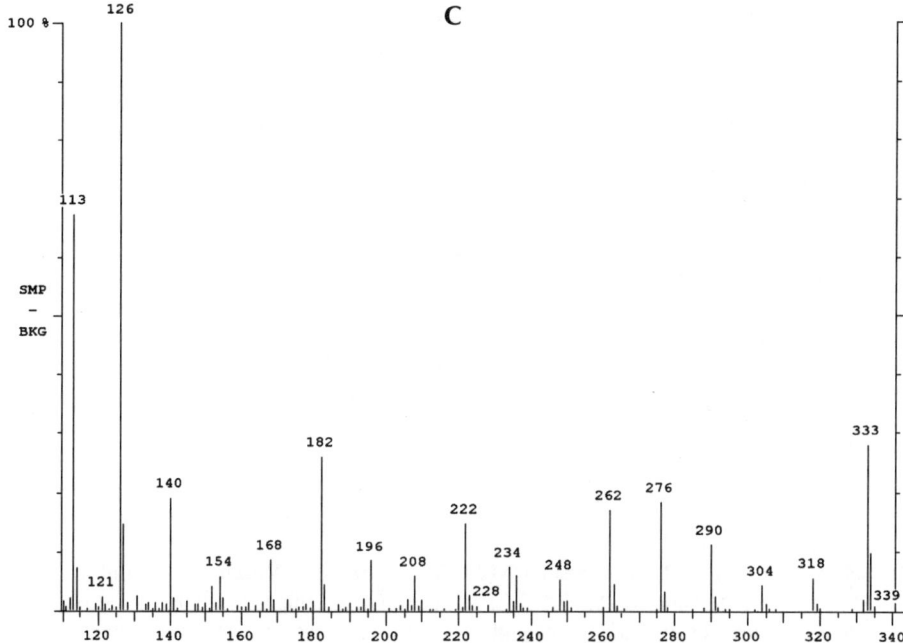

Fig. 9.7. Electron impact (EI) spectra of dimethyloxazoline (DMOX) derivatives of conjugated linoleic acid (CLA) isomers: (A) 11,13-18:2, (B) 10,12-18:2, (C) 9,11-18:2. Gas chromatography/mass spectrometry (GC/MS) data recorded with an ion trap (Thermoquest GCQ) instrument. Displayed spectra were auto-subtracted.

the CLA isomers of interest is an effective way to lower the GC/MS detection limit for CLA. Identification of a specific CLA isomer by GC/MS requires sufficient chromatographic separation from other congeners and coextractives to obtain a representative mass spectrum of the isomer.

Acknowledgements

The author thanks Martha L. Gay for her review and comments. Peter M. Yurawecz, Magdi M. Mossoba, and John K.G. Kramer are thanked for their persistent interest in CLA and support of CLA research.

References

1. Roach, J.A.G., Yurawecz, M.P., Mossoba, M.M., and Eulitz, K. (1999) in *Spectral Properties of Lipids* (Hamilton, R.J., and Casteds, J.), CRC Press, Boca Raton, FL, pp. 191–234.
2. Kramer, J.K.G., Sehat, N., Dugan, M.E.R., Mossoba, M.M., Yurawecz, M.P., Roach, J.A.G., Eulitz, K., Aalhus, J.L., Schaefer, A.L., and Ku, Y (1998) Distributions of Conjugated Linoleic Acid (CLA) Isomers in Tissue Lipid Classes of Pigs Fed a Commercial CLA Mixture Determined by Gas Chromatography and Silver Ion–High-Performance Liquid Chromatography, *Lipids 33,* 549–558.

3. Sehat, N., Yurawecz, M.P., Roach, J.A.G., Mossoba, M., Kramer, J.K.G., Ku, Y. (1998) Silver Ion–High Performance Liquid Chromatographic Separation and Identification of Conjugated Linoleic Acid Isomers, *Lipids 33,* 217–221.
4. Sehat, N., Rickert, R., Mossoba, M.M., Kramer, J.K.G., Yurawecz, M.P., Roach, J.A.G., Adlof, R.O., Morehouse, K.M., Fritsche, J., Steinhart, H., and Ku, Y. (1999) Improved Separation of Conjugated Fatty Acid Methyl Esters by Silver Ion–High-Performance Liquid Chromatography, *Lipids 34,* 407–413.
5. Yurawecz, M.P., Sehat, N., Mossoba, M.M., Roach, J.A.G., and Ku, Y. (1997) in *New Techniques and Applications in Lipid Analysis,* (McDonald, R.E., and Mossoba, M.M., eds.) pp. 183–215, AOCS Press, Champaign, IL.
6. Sehat, N., Yurawecz, M.P., Roach, J.A.G., Mossoba, M.M., Eultiz, K., Mazzola, E.P., and Ku, Y. (1998) Autoxidation of the Furan Fatty Acid Ester, Methyl 9,12-Epoxyoctadeca-9,11-Dienoate, *J. Am. Oil Chem. Soc. 75,* 1313-1319.
7. Spitzer, V. (1997) Structure Analysis of Fatty Acids by Gas Chromatography-Low Resolution Electron Impact Mass Spectrometry of Their 4,4-Dimethyloxazoline Derivatives—A Review, *Prog. Lipid Res. 35,* 387–408.
8. Fay, L., and Richli, U. (1991) Location of Double Bonds in Polyunsaturated Fatty Acids by Gas Chromatography-Mass Spectrometry after 4,4-Dimethyloxazoline Derivatization, *J. Chromatogr. 541,* 89–98.
9. Sehat, N., Kramer, J.K.G., Mossoba, M.M., Yurawecz, M.P., Roach, J.A.G., Eulitz, K., Morehouse, K.M., and Ku, Y. (1998) Identification of Conjugated Linoleic Acid Isomers in Cheese by Gas Chromatography and Mass Spectral Reconstructed Ion Profiles. Comparison of Chromatographic Elution Sequences, *Lipids 33,* 963–971.
10. Lavillonniere, F., Martin, J.C., Bougnoux, P., and Sébédio, J.-L. (1998) Analysis of Conjugated Linoleic Acid Isomers and Content in French Cheeses, *J. Am. Oil Chem. Soc. 75,* 343–352.
11. Yurawecz, M.P., Roach, J.A.G., Sehat, N., Mossoba, M.M., Kramer, J.K.G., Fritsche, J., Steinhart, H., and Ku, Y. (1998) A New Conjugated Linoleic Acid Isomer, 7*trans*, 9*cis*-Octadecadienoic Acid, in Cow Milk, Cheese, Beef and Human Milk and Adipose Tissue, *Lipids 33,* 803–809.
12. Roach, J.A.G. (1998) in *Spectral Methods in Food Analysis: Instrumentation and Applications,* (Mossoba, M.M., ed.) pp. 159-250, Marcel Dekker, New York.

Chapter 10

Confirmation of Conjugated Linoleic Acid Geometric Isomers by Capillary Gas Chromatography-Fourier Transform Infrared Spectroscopy

Magdi M. Mossoba, Martin P. Yurawecz, John K.G. Kramer, Klaus D. Eulitz, Jan Fritsche, Najib Sehat, John A.G. Roach, and Yuoh Ku

> Food and Drug Administration, Center for Food Safety and Applied Nutrition, Washington, DC 20204

Introduction

Infrared spectroscopy is also known as the fingerprint method because it can distinguish between closely related chemical compounds including conjugated linoleic acid (CLA) geometric isomers. Gas chromatography/Fourier transform infrared (GC-FTIR) spectroscopy and GC-mass spectrometry (MS) are complementary analytical techniques. GC-MS can confirm the molecular weight and the position of double bonds (Fig. 10.1A), whereas GC-FTIR can confirm the double bond configuration, *cis/trans* (Fig. 10.1B), *cis,cis*, or *trans,trans*, for complex mixtures of CLA geometric isomers. In addition to a chromatographic profile, GC-FTIR can also provide structural information on mixtures of unknown reaction products, such as methoxy fatty acids and furan fatty acids. Methoxy fatty acids are artifacts that may be formed during acid-catalyzed methylation procedures. Furan fatty acids are CLA oxidation products and markers of oxidative damage. Two GC-FTIR interfaces that operate at cryogenic temperatures, the matrix isolation (MI) and direct deposition (DD), have been used in this research. Unique IR bands were observed and used to discriminate between CLA geometric isomers and identify related reaction products. (For detailed discussions of CLA derivatization, chromatographic separation, artifact formation, GC-MS analysis, and related analytical topics, see the relevant chapters in this book.)

Dispersive Infrared and Fourier Transform Infrared Spectroscopy

Dispersive infrared spectroscopy has long been used to distinguish between fatty acid geometric isomers, namely, between the *cis* and *trans* double bond configurations in fats and oils. Well-characterized transmission infrared absorption spectra for conjugated diene fatty acid isomers were reported about half a century ago (1); these spectra were also used for the quantitative determination of conjugated *cis/trans* and *trans,trans* geometric isomers (2,3). In the last decade, a number of instrumental advances have had a positive impact on the creation of hyphenated techniques, such as

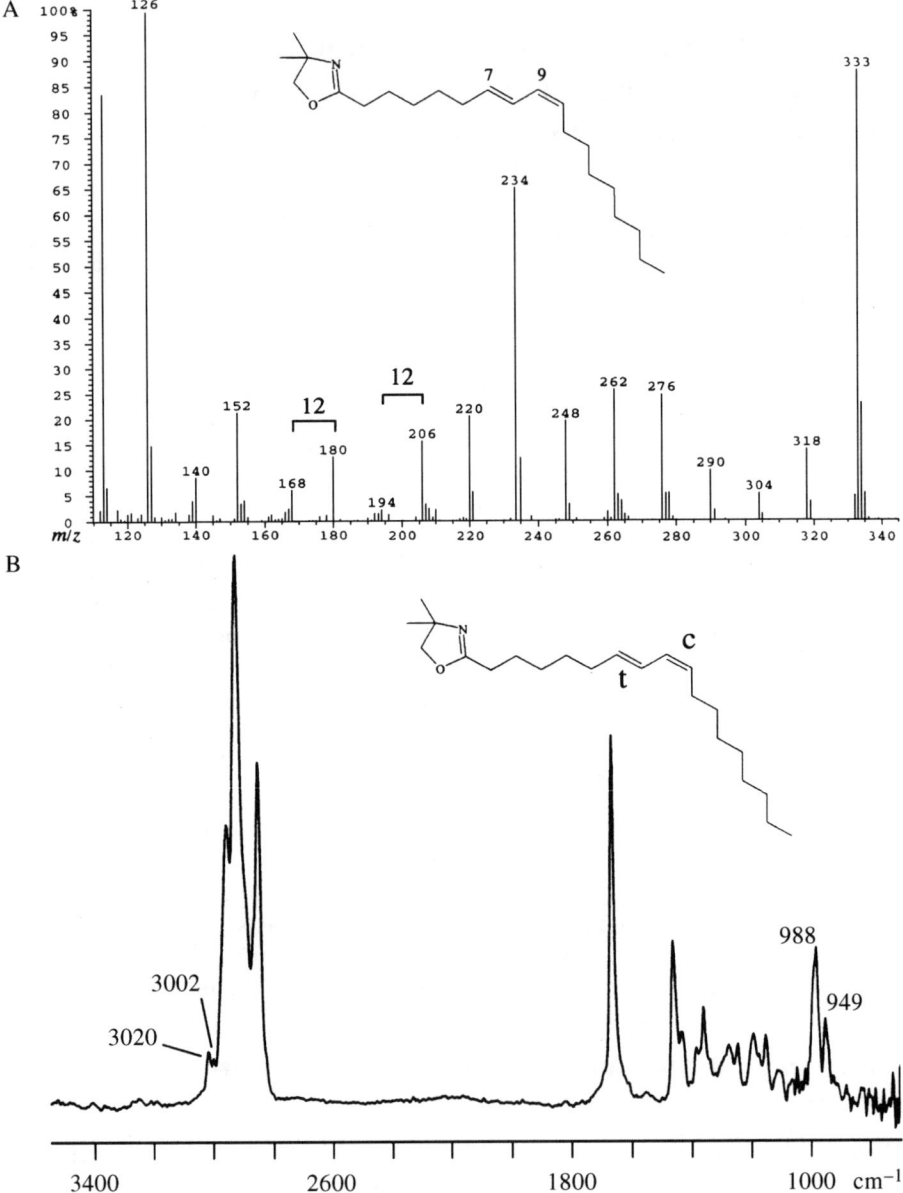

Fig. 10.1. The combination of (A) mass and (B) infrared spectral data can indicate both the position (C-7 and C-9) and configuration (*t*, *trans* and *c*, *cis*) of conjugated double bonds as shown for the 4,4-dimethyloxazoline (DMOX) derivative of a conjugated linoleic acid (CLA) isomer.

GC-FTIR (4), that allowed the on-line measurement of vibrational spectra for individual analytes eluting from a gas chromatograph. Hence, fatty acids, including CLA isomers, from complex fat and oil mixtures could be separated as fatty acid methyl esters (FAME) or other derivatives and measured by FTIR (5–12).

Capillary GC-FTIR Interfaces

Capillary GC-FTIR is a technique that has provided useful structural information about double bond configuration and functional groups for fats and oils. Its usefulness in lipid research was first demonstrated in 1987 on complex mixtures of cyclic fatty acid monomers isolated from heated linseed and sunflower oils (13). Using a light-pipe (LP) GC-FTIR interface, the size of monounsaturated cyclic fatty acid rings could be determined by differentiating between a *cis* double bond in a 5-membered ring (712 cm^{-1}) and one in a 6-membered ring (660 cm^{-1}).

The LP interface (4) consists of a heated cylindrical tube with alkali halide IR-transparent windows. This glass tube is coated internally with a highly reflective gold layer. As analytes exit the GC column and pass through the LP, transmittance IR spectra are measured continuously on-the-fly. Even under optimum conditions, some mixing of the separated components takes place in the flow-through cell and leads to degradation of chromatographic resolution. The long path length of the LP (10–20 cm) reduces the transmission of IR light, and its narrow aperture (1 mm i.d.) increases spectral noise, which adversely influences the IR signal-to-noise ratio (SNR). Although the LP interface is commonly found in GC-FTIR systems, it suffers from its low sensitivity for fatty acids. Its minimum identifiable quantity (MIQ) is in the range of 5–25 ng per analyte. The GC-FTIR data presented in this chapter were acquired with two other interfaces that provided improvements in MIQ by about an order of magnitude, down to the subnanogram levels. They are the matrix isolation (MI) and the direct deposition (DD) interfaces (4), which were found to be more adequate for CLA analysis (7).

The GC effluent is collected with MI for subsequent off-line measurement by FTIR. This is achieved by spraying the GC effluent during a chromatographic run onto the outer rim of a slowly rotating gold-plated disk held at ~12K under vacuum. The separated analytes are individually frozen on the moving collection disk as they exit the capillary column. The mobile phase consists of a mixture of 98.5% helium and 1.5% argon. At 12K, helium is still a gas and it is pumped away by the vacuum, whereas the argon atoms are condensed along with the separated analytes on the cryogenic disk. This leads to the formation of a solid matrix in which analyte molecules are isolated from each other by a large excess of IR-transparent argon atoms. Each analyte peak is deposited on an area with a diameter as small as 0.5 mm; the result is greater IR absorbance because absorbance is inversely proportional to the cross-sectional area on which an analyte is trapped. Post-GC-run FTIR data acquisition is then carried out for several minutes (instead of ~2 s with the LP) for each analyte of interest. This extensive signal averaging improves the SNR dramatically.

The DD is also a sensitive mobile-phase elimination interface, except that the GC effluent is deposited directly without an argon matrix on a rectangular zinc selenide IR window cooled to ~77K. During a chromatographic run, the window is moved in small increments. Each frozen analyte passes through the focused IR beam ~15 s after deposition on the window. The transmitted IR beam is collected and focused by microscope objectives onto a remote detector. A major advantage of the DD interface is that post-GC-run signal averaging is usually not required (although it is still available) because the on-the-fly measurement in this case is sufficiently sensitive for most fatty acid applications. Although GC-FTIR instrumentation is complex, it is a useful analytical tool in CLA research (7).

CLA Geometric Isomers in Chemical and Biological Matrices

Like other C_{18} diene fatty acids, CLA isomers share structural similarities, such as chain length and degree of unsaturation, yet their double bonds have different configurations, giving rise to unique FTIR spectral features that make their geometric isomers readily distinguishable (5,7). Initial GC-MI-FTIR data on CLA (7) were published in 1991 as part of a study on *trans* fatty acids in partially hydrogenated vegetable oils in which the (then newly reported) anticarcinogenic (14) CLA isomers were also identified (7).

In general, the characteristic GC-FTIR bands that allow the differentiation of FAME are usually an order of magnitude weaker than the maximum absorbances observed in these spectra (5). Most FAME bands are usually similar, except for a few subtle, yet characteristic, differences due to the number, position, and configuration of double bonds. For long hydrocarbon chain FAME, the common bands obtained by GC-MI-FTIR at 4 cm^{-1} resolution are due to the CH_3 asymmetric (2961 cm^{-1}) and symmetric (2880 cm^{-1}) and CH_2 asymmetric (2935 cm^{-1}) and symmetric (2863 cm^{-1}) stretching vibrations, the CH_2 in-plane bend (1463 cm^{-1}), the CH_3 symmetric scissors (1381 cm^{-1}), the CH_3 in-plane rock (1123 cm^{-1}) and CH_2 rock (727 cm^{-1}) deformation vibrations, the ester symmetric C-O stretch (1176 cm^{-1}), and the strong ester carbonyl stretch at 1754 cm^{-1}. The intensity of the strong CH_2 asymmetric stretch band usually decreases relative to that of the ester carbonyl stretch as the degree of unsaturation increases. With 4,4-dimethyloxazoline (DMOX) derivatives, the oxazoline ring gives rise to several features that are common to all fatty acid DMOX spectra (12,15); see Figure 10.2. These include the ring C=N stretching vibration at 1678 cm^{-1}, the C-O cyclic ether band, which is split with a maximum at 1002 cm^{-1}, and three weaker components near 1018, 979, and 954 cm^{-1}. Several weak bands in the fingerprint region that are attributed to the ring skeletal vibrations are also found.

More importantly, the position of the weaker, yet discriminating, unsaturated group =C-H stretch bands is found by GC-MI-FTIR for the minor CLA FAME geometric isomers at 3028 and 3009 cm^{-1} (conjugated *cis* and *trans* double bonds), and 3023 and 3005 cm^{-1} (two conjugated *trans* double bonds) (5). The corresponding

GC-DD-FTIR data (*cis/trans*: 3020 and 3002 cm^{-1}; *cis,cis*: 3037 and 3005 cm^{-1}; and *trans,trans*: 3017 cm^{-1}) for CLA DMOX derivatives (15) are given in Figure 10.3. For comparison purposes, bands for other nonconjugated FAME (5) were observed at 3035 and 3005 cm^{-1} (one or two *trans* double bonds), 3010 cm^{-1} (*cis* double bond), 3018 cm^{-1} (two or three *cis* double bonds), 3035, 1010, and 3005 cm^{-1} (*cis* and *trans* double bonds separated by more than one methylene group), 3035, 3018, and 3005 cm^{-1} (*cis* and *trans* double bonds separated by a single methylene group).

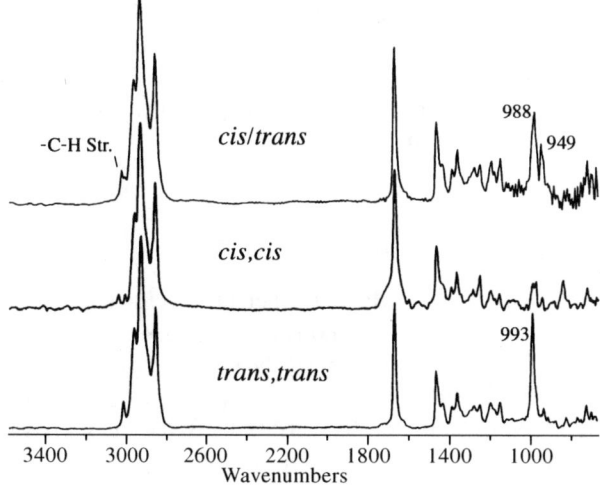

Fig. 10.2. Gas chromatography-direct deposition-Fourier transform infrared spectroscopy (GC-DD-FTIR) spectra observed at 8 cm^{-1} resolution can readily discriminate between conjugated linoleic acid (CLA) geometric isomers as demonstrated for 4,4-dimethyloxazoline (DMOX) derivatives.

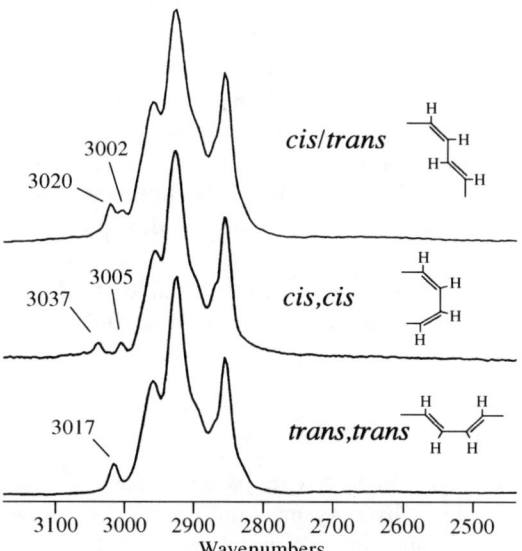

Fig. 10.3. Expanded gas chromatography-direct deposition-Fourier transform infrared spectroscopy (GC-DD-FTIR) hydrocarbon stretch spectral region, observed at 8 cm^{-1} resolution for conjugated linoleic acid (CLA) 4,4-dimethyloxazoline (DMOX) geometric isomers, exhibited unique differences in band number, position, and relative intensity.

Similarly, with the use of GC-MI-FTIR, the C-H out-of-plane deformation vibrations in $R_1HC=CHR_2$ groups exhibited highly characteristic band positions for CLA FAME isomers (7): 986 and 950 cm^{-1} (conjugated *cis* and *trans* double bonds) and 990 cm^{-1} (two conjugated *trans* double bonds). The corresponding GC-DD-FTIR spectra for CLA DMOX derivatives (15) are given in Figure 10.2. For nonconjugated FAME (5), GC-MI-FTIR bands were found at 971 cm^{-1} (one or two *trans* double bonds), 730 cm^{-1} (one or two *cis* double bonds), 721 cm^{-1} (three *cis* double bonds), 971 and 730 cm^{-1} (*cis* and *trans* double bonds interrupted by one or more methylene groups).

Recently, GC-DD-FTIR spectroscopy was applied to the identification of conjugated geometric CLA isomers obtained from several commercial sources (16), as well as those found in a number of foods and biological matrices (17–19). These included commercial reference mixtures from chemical suppliers, dietary supplements, cow milk, cheese, beef, human milk and adipose tissue, as well as organs of pig that were fed CLA diets.

Conjugated Octadecatrienes in Edible Fats and Oils

An investigation was conducted to determine whether other fatty acids similar to CLA in structure were also present in edible fats and oils (20). It was reported that conjugated octadecatriene (COT) fatty acids are formed during the processing of vegetable oils as a result of the dehydration of secondary oxidation products of linoleic acid. Their levels were found to be up to ~0.2% (20). COT reportedly have adverse physiologic effects at low levels (21) and are not generally considered acceptable food components (22). When hydroxy conjugated dienoic fatty acids such as coriolic acid (13 hydroxy, 9 *cis*, 11 *trans*-octadecadienoic acid), were exposed to acidic conditions, they converted to COT. To avoid the formation of conjugated triene fatty acid artifacts, transmethylation with sodium methoxide/methanol has been recommended (20). COT fatty acids from heated cottonseed oil, including α-eleostearic acid (9 *cis*, 11 *trans*, 13 *trans*-octadecatrienoic acid) and β-eleostearic acid (all-*trans* 9,11,13-octadecatrienoic acid), were separated and characterized as FAME by GC-MI-FTIR spectroscopy (20). Characteristic =C-H stretching vibration bands were found at 3026 and 3005 cm^{-1} for α-eleostearic acid, and at 3020 and 3002 cm^{-1} for β-eleostearic acid. Bands at 995, 968, and 730 cm^{-1} for α-eleostearic acid, and at 998 cm^{-1} for β-eleostearic acid were attributed to =C-H out-of-plane deformation. Ethyl esters of COT fatty acids derived from sunflower and tung oils were also characterized (20).

CLA Artifacts and Methoxy Fatty Acid Artifacts

As observed with COT, acid-catalyzed methylation procedures have been found to lead to the formation of CLA artifacts from allylic hydroxy monounsaturated fatty acids (23); this reaction also resulted in the isomerization of *cis/trans* to *trans,trans* CLA iso-

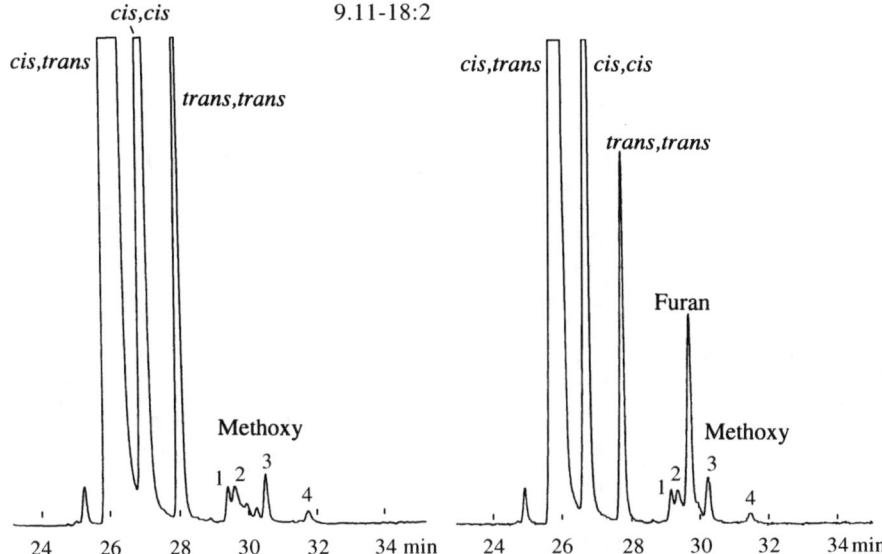

Fig. 10.4. Flame ionization detector profile for the gas chromatography (GC) separation of the geometric isomers of the 9 *cis*, 11 *trans*-18:2 fatty acid methyl ester (FAME) positional isomer. Methoxy FAME products (peaks labeled 1–4) were formed as a result of BF_3 methylation. Autoxidation of the same conjugated linoleic acid (CLA) mixture gave rise to the corresponding furan FAME 9,12-epoxy-9,11-octadecadienoic ($F_{9,12}$). $F_{9,12}$ eluted in the same retention time range as methoxy FAME products, but could be distinguished by FTIR (see Fig. 10.6).

mers and the formation of methoxy fatty acid artifacts (Fig. 10.4). Confirmatory evidence was provided by infrared spectroscopy. Characteristic GC-DD-FTIR bands were observed for methoxy fatty acids at 2829 and 1094 cm^{-1} due to the CH_3 symmetric stretch in -O-CH_3 and the C-O-C asymmetric stretch, respectively (Fig. 10.5).

CLA Oxidation Products

It has been known that singlet oxygen will react with conjugated dienes to form furan fatty acid (FFA) (24). The occurrence and biological functions of FFA are active areas of research (25–27). For example, the FFA 9,12-epoxy-9,11-octadecadienoic ($F_{9,12}$) is a plant lipid found in *Exocarpus cupressiformis* (25) and has been the subject of oxidation studies (26–27). Surprisingly, similar FFA oxidation products were reportedly obtained *via* the autoxidation of CLA (28). The oxidation of a commercial reference mixture (Nu-Chek-Prep, Elysian, MN) that consisted of four CLA positional isomers yielded four FFA that were readily base-line resolved as FAME by GC on a 50-m highly polar CP Sil 88 (Chrompack, Raritan, NJ) capillary column (28). They were identified as 8,11-epoxy-8,10-octadecadienoic

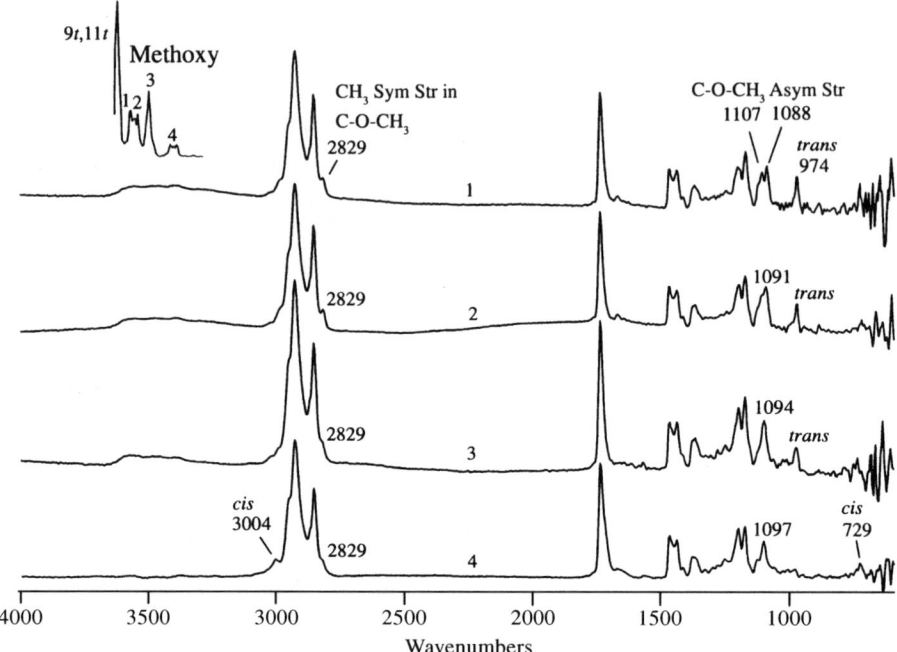

Fig. 10.5. Gas chromatography-direct deposition-Fourier transform infrared spectroscopy (GC-DD-FTIR) spectra observed at 8 cm^{-1} resolution confirmed the identity of the methoxy fatty acid methyl ester (FAME) products (peaks labeled 1–4 in Fig. 10.4). The inset (upper left corner) shows the corresponding GC trace measured by FTIR. The double bond configuration was determined to be *trans* (974 cm^{-1}) for compounds 1–3, and *cis* (3004 and 729 cm^{-1}) for compound 4.

($F_{8,11}$); $F_{9,12}$; 10,13-epoxy-10,12- octadecadienoic ($F_{10,13}$); and 11,14-epoxy-11,13-octadecadienoic ($F_{11,14}$) acids. The hyphenated GC-MI-FTIR technique was also used to confirm the identity of these FFA products (28,29). Unique furan ring bands were found for FAME products at 3111, 3031, and 3000 cm^{-1} (=C-H stretch); 1574 cm^{-1} (C=C stretch); 1013 cm^{-1} (C-O stretch); and 780 cm^{-1} (=C-H out-of-plane bend). Typical GC-DD-FTIR data for furan FAME are given in Figure 10.6.

Conclusions

GC-FTIR complements GC-MS in confirming the identity of unknown CLA and other fatty acid mixture components. GC-FTIR can confirm the double bond configuration (*cis/trans*; *cis,cis*; *trans,trans*) of CLA geometric isomers. GC-FTIR can also provide structural information on fatty acid reaction products such as methoxy and furan fatty acids.

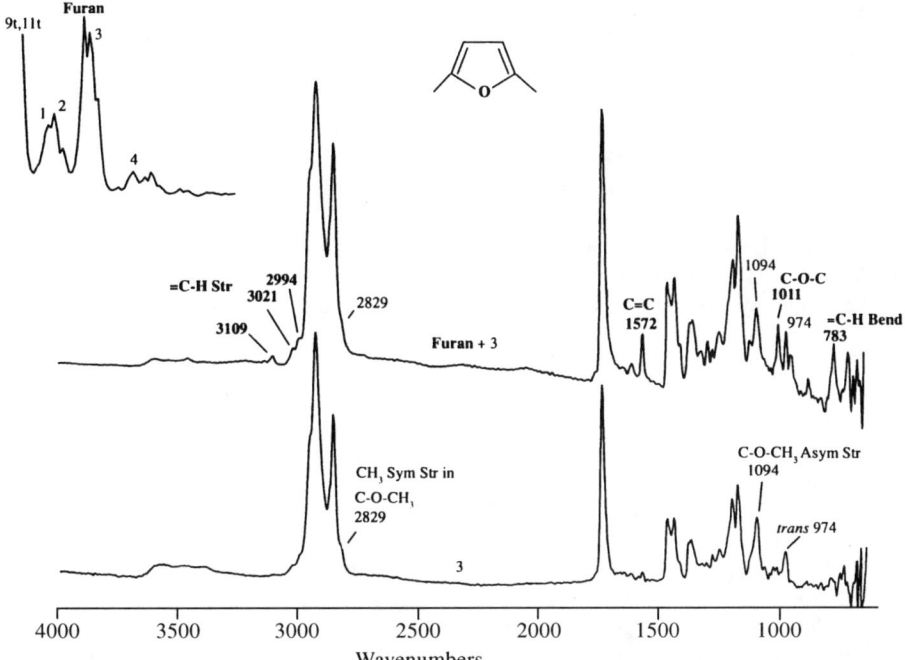

Fig. 10.6. Gas chromatography-direct deposition-Fourier transform infrared spectroscopy (GC-DD-FTIR) spectra observed at 8 cm^{-1} resolution near the GC retention time of one of the methoxy fatty acid methyl ester (FAME) products (peak labeled 3 in Fig. 10.4) before (bottom spectrum) and after (upper spectrum) autoxidation of the 9,11-18:2 positional isomer to the corresponding furan FAME 9,12-epoxy-9,11-octadecadienoic ($F_{9,12}$). The inset (upper left corner) shows the corresponding GC trace measured by FTIR.

References

1. Jackson, J.E., Paschke, R.F., Tolberg, W., Boyd, H.M., and Wheeler, D.H. (1952) Isomers of Linoleic Acid. Infrared and Ultraviolet Properties of Methyl Esters, *J. Am. Oil Chem. Soc. 29*, 229–234.
2. Chipault, J.R, and Hawkins, J.M. (1959) The Determination of Conjugated *cis-trans* and *trans-trans* Methyl Octadecadienoates by Infrared Spectroscopy, *J. Am. Oil Chem. Soc. 36*, 535–539.
3. Scholfield, C.R. (1974) Infrared Absorption of Methyl *cis*-9,*trans*-11-, and *trans*-10,*cis*-12-Octadecadienoates, *J. Am. Oil Chem. Soc. 51*, 33–34.
4. White, R. (1990) *Chromatography/Fourier Transform Infrared Spectroscopy and Its Applications,* Practical Spectroscopic Series, vol. 10, Marcel Dekker, New York.
5. Mossoba, M.M., McDonald, R.E., Chen, J.-Y.T., Armstrong, D.J., and Page, S.W. (1990) Identification and Quantitation of *trans*-9, *trans*-12-Octadecadienoic Acid Methyl Ester and Related Compounds in Hydrogenated Soybean Oil and Margarines by GC-MI-FTIR Spectroscopy, *J. Agric. Food Chem. 38*, 86–92.

6. Mossoba, M.M., McDonald, R.E., Armstrong, D.J., and Page, S.W. (1991) Hydrogenation of Soybean Oil: A Thin-Layer Chromatography and GC-MI-FTIR Study, *J. Agric. Food Chem. 39,* 695–699.
7. Mossoba, M.M., McDonald, R.E., Armstrong, D.J., and Page, S.W. (1991) Identification of Minor C18 Triene and Conjugated Diene Isomers in Hydrogenated Soybean Oil and Margarine by GC-MI-FTIR Spectroscopy, *J. Chromatogr. Sci. 29,* 324–330.
8. Mossoba, M.M., McDonald, R.E., and Prosser, A.R. (1993) GC-MI-FTIR Spectroscopic Determination of *trans*-Monounsaturated and Saturated Fatty Acid Methyl Esters in Partially Hydrogenated Menhaden Oil, *J. Agric. Food Chem. 41,* 1998–2002.
9. Mossoba, M.M., and McDonald, R.E., (1993) Applications of Capillary GC-FTIR Spectroscopy to Fatty Acid Analysis, *INFORM 4,* 854–859.
10. Mossoba, M.M., Yurawecz, M.P., Roach, J.A.G., Lin, H.S., McDonald, R.E., Flickinger, B.D., and Perkins, E.G. (1994) Rapid Determination of Double Bond Configuration and Position Along the Hydrocarbon Chain in Cyclic Fatty Acid Monomers, *Lipids 29,* 893–896.
11. Mossoba, M.M., Yurawecz, M.P., Flickinger, B.D., Lin, H.S., McDonald, R.E., and Perkins, E.G. (1995) Application of GC-MI-FTIR to the Structural Elucidation of Cyclic Fatty Acid Monomers, *Am. Lab.,* September, 16 K–O.
12. Mossoba, M.M., Yurawecz, M.P., McDonald, R.E., Flickinger, B.D., and Perkins, E.G. (1996) Analysis of Cyclic Fatty Acid Monomer 2-Alkenyl-4,4-dimethyl-oxazoline Derivatives by GC-MI-FTIR Spectroscopy, *J. Agric. Food Chem. 44,* 3193–3196.
13. Sébédio, J.L., J.-L. Le Quere, Semon, E., Morin, O., Prevost, J., and Grandgirard, A. (1987) Heat Treatment of Vegetable Oils. II. GC-MS and GC-FTIR Spectra of Some Isolated Cyclic Fatty Acid Monomers, *J. Am. Oil Chem. Soc. 64,* 1324–1333.
14. Ha, Y.L., Grimm, N.K., and Pariza, M.W. (1987) Newly Recognized Anticarcinogenic Fatty Acids: Identification and Quantification in Natural and Processed Cheese, *J. Agric. Food Chem. 37,* 75–81.
15. Fritsche, J., Mossoba, M.M., Yurawecz, M.P., Roach, J.A.G., Sehat, N., Ku, Y., and Steinhart, H. (1997) Conjugated Linoleic Acid (CLA) Isomers in Human Adipose Tissue, *Z. Lebensm.-Unters.-Forsch.-A 205,* 415–518.
16. Sehat, N., Yurawecz, M.P., Roach, J.A.G., Mossoba, M.M., Kramer, J.K.G., and Ku, Y. (1998) Silver Ion–High-Performance Liquid Chromatographic Separation and Identification of Conjugated Linoleic Acid Isomers, *Lipids 33,* 217–221.
17. Yurawecz, M.P., Roach, J.A.G., Sehat, N., Mossoba, M.M., Kramer, J.K.G., Fritsche, J., Steinhart, H., and Ku, Y. (1998) A New Conjugated Linoleic Acid (CLA) Isomer, 7 *trans*,9 *cis*-Octadecadienoic Acid, in Cow Milk, Cheese, Beef and Human Milk and Adipose Tissue, *Lipids 33,* 803–809.
18. Sehat, N., Kramer, J.K.G., Mossoba, M.M.,Yurawecz, M.P., Roach, J.A.G., Eulitz, K. Morehouse, K.M., and Ku, Y. (1998) Identification of Conjugated Linoleic Acid Isomers in Cheese by Gas Chromatography, Silver Ion–High Performance Liquid Chromatography and Mass Spectral Reconstructed Ion Profiles. Comparison of Chromatographic Elution Sequences, *Lipids 33,* 963–971.
19. Kramer, J.K.G., Sehat, N., Dugan, M.E.R., Mossoba, M.M., Yurawecz, M.P., Roach, J.A.G., Eulitz, K., Aalhus, J.L., Schaefer, A.L., and Ku, Y. (1998) Distribution of Conjugated Linoleic Acid (CLA) Isomers in Tissue Lipid Classes of Pigs Fed a Commercial CLA Mixture Analyzed by Gas Chromatography and Silver Ion–High Performance Liquid Chromatography, *Lipids 33,* 549–558.

20. Yurawecz, M.P., Molina, A.A., Mossoba, M.M., and Ku, Y. (1993) Estimation of Conjugated Octadecatrienes in Edible Fats and Oils, *J. Agric. Food Chem. 70,* 1093–1099.
21. Nutgeren, D.H., and Christ-Hazelhof, E. (1987) Naturally Occuring Conjugated Octadecatrienoic Acids are Strong Inhibitors of Prostaglandin Biosynthesis, *Prostaglandins, 33,* 403–417.
22. Spitzer, V., Mara, J.G.S., and Pfeilstiker, K. (1991) Identification of Conjugated Fatty Acids in the Seed Oil of *Acioa edulis* (France) syn. *Couepia edulis* (Chryssobalanaceae), *J. Amer. Oil Chem. Soc., 68,* 183–189.
23. Yurawecz, M.P., Hood, J.K., Roach, J.A.G., Mossoba, M.M., Daniels, D.H., and Ku, Y. (1994) Conversion of Allylic Hydroxy Oleate to Conjugated Linoleic Acid and Methoxy Oleate by Acid-Catalyzed Methylation Procedures, *J. Am. Oil Chem. Soc. 71,* 1149–1155.
24. Ideses, R., Shani, A., and Klug, J.T. (1982) Cyclic Peroxide: An Isolable Intermediate in Singlet Oxygen Oxidation of Pheromones to the Furan System, *Chem. Ind. 19,* 409–410.
25. Morris, L.J., Marshall, M.O., and Kelly, V. (1966) A Unique Furanoid Fatty Acid from Exocarpus Seed Oil, *Tetrahedron Lett. 36,* 4249–4253.
26. Rosenblat, G., Tabak, M., Lie Ken Jie, M.S.F., and Neeman, I. (1993) Inhibition of Bacterial Urease by Autoxidation of Furan C-18 Fatty Acid Methyl Ester Products, *J. Am. Oil Chem. Soc. 70,* 501-505.
27. Boyer, R.F., Litts, D., Kostishak, J., Wijesundera, R.C., and Gunstone, F.D. (1979) The Action of Lipoxygenase-1 on Furan Derivatives, *Chem. Phys. Lipids 25,* 237–346.
28. Yurawecz, M.P., Hood, J.K., Mossoba, M.M., Roach, J.A.G., and Ku, Y. (1995) Furan Fatty Acids Determined as Oxidation Products of Conjugated Octadecadienoic Acid, *Lipids 30,* 595–598.
29. Yurawecz, M.P., Sehat, N., Mossoba, M.M., Roach, J.A.G., and Ku, Y. (1997) Oxidation Products of Conjugated Linoleic Acid and Furan Fatty Acids, in *New Techniques and Applications in Lipid Analysis,* (McDonald, R.E., and Mossoba, M.M., eds.) AOCS Press, Champaign, IL.

Chapter 11

Nuclear Magnetic Resonance Spectroscopic Analysis of Conjugated Linoleic Acid Esters

Marcel S.F. Lie Ken Jie, M.K. Pasha, and M.S. Alam

Department of Chemistry, The University of Hong Kong, Pokfulam Road, Hong Kong

Introduction

Nuclear magnetic resonance spectroscopy is one of the most powerful analytical techniques available to the organic chemist. High-resolution proton nuclear magnetic resonance (^{1}H NMR) spectroscopy of organic molecules permitted structural, kinetic, and equilibrium studies of individual compounds in pure state or in mixtures to be carried out (1–7). ^{1}H NMR spectroscopy was used to determine the degree of unsaturation in triacylglycerols and the average molecular weight of natural fats (8). Analysis of the positional distribution of fatty acids in triacylglycerols by ^{1}H NMR spectroscopy was successfully performed with the help of shift reagents (9). The identification of 2-isovaleroyl and 1,3-diisovaleroyl triacylglycerol structures in dolphin, porpoise, and toothed whale fats (10), the differentiation between unsaturated and isomeric triacylglycerols (11,12), and the determination of positional distribution of linoleate and the linolenate chain in triacylglycerols (13) are a few examples of the application of shift reagents to the analysis of triacylglycerols by ^{1}H NMR spectroscopy. However, this application became limited by the broadening effect of signals when larger quantities of shift reagents were used in attempts to induce greater shift changes (14). The technique was further handicapped by the narrow proton magnetic resonance range of the spectrum.

Taking advantage of the development of digital computers, pulse Fourier transform techniques, and the availability of superconducting magnets, ^{13}C NMR techniques complemented ^{1}H NMR techniques in many ways. The low natural abundance (1.1%) of the ^{13}C isotope in nature was compensated by the low possibility of ^{13}C-^{13}C spin-spin coupling. Broadband heteronuclear decoupling (i.e., elimination of all ^{1}H-^{13}C couplings) resulted in sharp singlets for all ^{13}C absorptions, which allowed carbon nuclei with small carbon chemical shift differences to be measured. With the introduction of advanced electronic techniques, it became possible to run a ^{13}C NMR analysis under conditions that allowed for a quantitative integration of the intensity of carbon signals, a technique similar to ^{1}H NMR spectroscopy for the quantitative measurements of protons. The quantitative application of ^{13}C NMR spectroscopy to lipid molecules was reviewed by Shoolery (15). ^{13}C NMR spectroscopy also offered a means to determine the composition of mixtures of fatty acids and lipid molecules in much greater detail than with ^{1}H NMR spectroscopy. Gunstone has published several valuable reviews that deal specifically with the use of high-resolution ^{13}C NMR techniques in the analysis of lipid mixtures (16–19). Lie Ken Jie and Mustafa (20) have recently reviewed the applications of ^{13}C NMR to the analyses of unsaturated fatty acids and triacylglycerols.

High-resolution ^{1}H NMR spectroscopy seemed at first to be of limited use to fatty acid analysis because of the small range of chemical shifts covered by protons. However, the splitting patterns observed in ^{1}H NMR spectra offer unique spectral features, which show fine structural details of lipid molecules under investigation (21). ^{13}C NMR spectral analysis of fatty acids provides a large number of signals spread over a wide range of chemical shifts, which make the spectrum appear complicated but are very informative. Techniques of correlating signals between ^{1}H and ^{13}C NMR spectra provide two-dimensional correlation spectra (^{13}C-^{1}H COSY), which permit confirmation of signals. Other techniques, such as INADEQUATE (incredible natural abundance double quantum transfer experiment), HMQC (heteronuclear multiple quantum correlation), and HMBC (heteronuclear multiple bond correlation), are some of the latest techniques in NMR spectroscopy from which structural details can be determined with a high degree of certainty. Such NMR experiments can be carried out readily by using the latest models of high-field NMR instruments.

Nuclear Magnetic Resonance Spectroscopic Analysis of Conjugated Linoleic Acid Esters

Conjugated linoleic acids (CLA) consist of a mixture of geometric isomers of 18:2(9Z,11E), 18:2(9E,11Z), 18:2(9Z,11Z), and 18:2(9E,11E), among which 18:2(9Z,11E) is the dominant component. Some other positional isomers of conjugated octadecadienoic acids have also been reported in mixtures of CLA (22). Gunstone and Said (23) reported the first synthesis of 18:2(9Z,11E) starting from methyl ricinoleate. Berdeaux *et al.* (24) extended the procedure to a large-scale synthesis of methyl 18:2(9Z,11E) (83% purity after urea fractionation). Lie Ken Jie *et al.* (25) have prepared the four geometric isomers of CLA in the pure state from either methyl ricinoleate or methyl santalbate as the starting material. Adlof (26) has synthesized a mixture of 18:2(9Z,11E) and 18:2(9E,11E) (in a ratio of 46:54) with deuterium atoms on the 17- and 18-carbon atoms.

Bus *et al.* (27) were the first group of researchers to report on the carbon shifts of conjugated octadecadienoate isomers. The carbon shifts of the olefinic carbon nuclei and the adjacent methylene carbon atoms of three conjugated 18:2 isomers are reproduced in Table 11.1. The carbon chemical shifts of these three conjugated 18:2 isomers [*viz.,* 18:2(9Z,11E), 18:2(10Z,12E), and 18:2(10E,12Z)] were obtained from one-dimensional spectra run on a Bruker WH90 Fourier transform NMR spectrometer, operating at 22.63 MHz with proton noise decoupling. The accuracy of the δ_C values was claimed to be ±0.1 ppm. The assignments of the various chemical shift values were based on results obtained from studying many series of positional isomers of unsaturated Z- and E-alkenoates. From such studies, it was possible to determine the shift effects of the carboxylic acid (in fatty acids) or carbomethoxy (in methyl esters) on the shifts of the various methylene and olefinic carbon nuclei. Shift effects of the ω-methyl

TABLE 11.1 Selected Carbon Shifts of Conjugated Octadecadienoate Isomers[a]

Carbon nucleus	Isomers		
	18:2(9Z,11E)	18:2(10Z,12E)	18:2(10E,12Z)
C-8	27.70		
C-9	130.00	27.75	32.90
C-10	125.75	130.00	134.55
C-11	128.90	125.75	128.80
C-12	134.80	128.80	125.90
C-13	32.95	134.70	130.10
C-14		32.90	27.75

[a]Source: Ref. 27.

group at the end of the alkyl chain and the effects of olefinic systems on the shifts of the carbon nuclei of neighboring methylene groups or methine carbons of double bonds in polyunsaturated fatty ester were also determined. The small, but significant differences observed in the shifts of the carbon nuclei of methylene groups and olefinic carbon nuclei were carefully noted. These induced effects were referred to as shift parameters, which can be negative or positive in value. A positive value indicated a deshielding effect, whereas a negative value indicated a shielding effect. Application of the additivity rule allowed shift values to be predicted. Thus, by adding or subtracting the respective parameter from a basic value (29.75 and 130.0 ppm for unperturbed methylene and olefinic carbon atoms, respectively), the resultant would give a rather accurate estimation of the carbon shifts of many of the methylene and olefinic carbon atoms of polyunsaturated fatty acids or ester. It was through such studies that the carbon shifts of the methylene carbons adjacent to conjugated 18:2 isomers and those of the olefinic carbon atoms were determined as shown in Table 11.1.

Lie Ken Jie *et al.* (25) analyzed each isomer of CLA [*viz.,* 18:2(9Z,11E), 18:2(9E,11Z), 18:2(9Z,11Z), and 18:2(9E,11E)] in its pure form by using a combination of various techniques in NMR spectroscopy. The shift values of the various carbon shifts are reproduced in Table 11.2. Note: In this table, we have now made definitive assignments for C-8 to C-13 of the 18:2(9Z,11E) and 18:2(9E,11Z) isomers, which were originally reported to be "interchangeable." The definitive assignments of the shifts of C-8/C-13, C-9/C-12, C-10/C-11 for these geometric isomers were based on data from Davis *et al.* (see Chapter 12) using HMQC-TOCSY (a technique that combines heteronuclear multiple quantum correlation and total correlation spectroscopy).

The assignments of the signals of the ^{1}H and ^{13}C NMR spectra of each geometric isomer are discussed in detail below. Note: Bus *et al.* (27) have erroneously assigned some of the shifts of the olefinic carbon atoms of 18:2(9Z,11E), 18:2(10Z,12E), and 18:2(10E,12Z).The shifts of the "inner" carbon atoms were assigned in reversed order (see Table 11.1).

TABLE 11.2 Carbon Shift Values of Conjugated Octadecadienoate Isomers[a,b]

Carbon nucleus	Isomers			
	18:2(9Z,11E)	18:2(9E,11Z)	18:2(9E,11E)	18:2(9Z,11Z)
C-1	174.32	174.34	174.22	174.27
C-2	34.10	34.10	34.09	34.10
C-3	24.95	24.95	24.98	24.97
C-4	29.06	28.97	29.04	29.14
C-5/C-6/C-7	29.12–29.67	29.13–29.45	29.14–29.77	29.11–29.60
C-8	27.66	32.86	32.61[a]	27.46[d]
C-9	129.89	134.51	132.16[b]	131.87[e]
C-10	128.71	125.72	130.37[c]	123.58[f]
C-11	125.58	128.57	130.51[c]	123.72[f]
C-12	134.76	130.17	132.43[b]	132.14[e]
C-13	32.92	27.72	32.68[a]	27.54[d]
C-14	29.41	29.73	29.40	29.68
C-15	28.95	28.97	28.97	29.04
C-16	31.77	31.77	31.82	31.81
C-17	22.65	22.65	22.68	22.69
C-18	14.12	14.12	14.13	14.13
COOCH$_3$	51.44	51.45	51.39	51.42

[a]Source: Ref. 25. [b]a-a, b-b, c-c, d-d, e-e, f-f are interchangeable (definitive assignments have been made on the basis of data from Chapter 12).

Analysis of Methyl (9Z,11E)-Octadecadienoate

The assignments of the various signals were accomplished by using a combination of two-dimensional correlation spectral techniques, viz., ^{13}C-^{1}H COSY (Fig. 11.1), INADEQUATE (Fig. 11.2), and HMBC (Fig. 11.3) correlation experiments. The ^{1}H NMR spectrum of 18:2(9Z,11E) shows four distinct signals for the four olefinic protons. These signals appear as two multiplets at δ_H 5.32 and 5.65, which correspond to the shifts of the protons of the "outer" (9-H, 12-H) positioned olefinic protons of the conjugated diene system. An apparent triplet at δ_H 5.82 (J = 10.9 Hz) and a double doublet at δ_H 6.24 (J = 9.9 Hz) correspond to the shifts of the protons of the inner (10-H, 11-H) positioned olefinic protons of the conjugated diene system. The ^{13}C NMR spectrum shows the following four signals in the olefinic region: δ_C 125.58, 128.71, 129.89, and 134.76. From the ^{13}C-^{1}H NMR COSY correlation spectrum (Fig. 11.1), it is clear that the double doublet at δ_H 6.24 is connected to the carbon at δ_C 125.58, and the triplet at δ_H 5.82 to the carbon at δ_C 128.71. These connections support the assignments of the shifts of the inner carbon atoms (C-10, C-11) at δ_C 125.58 and 128.71. The multiplets at δ_H 5.32 and 5.65 are connected to the carbon atoms at δ_C 129.89 and 134.76, respectively, which can be assigned to the shifts of the outer positioned carbon atoms (C-9, C-12).

The same spectrum also shows connections of the allylic carbon atoms (signals at δ_C 27.66 and 32.92) with the proton signals at δ_H 2.1 and 2.0, respectively, which appear as multiplets. From the INADEQUATE spectrum (Fig. 11.2), which is used

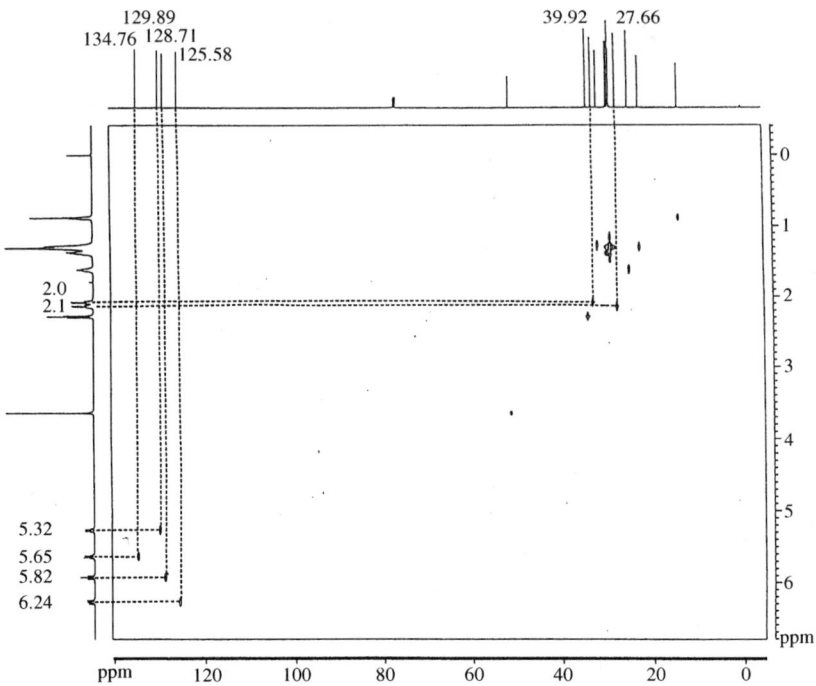

Fig. 11.1. ^{13}C-^{1}H nuclear magnetic resonance (NMR) correlation (COSY) spectrum of 18:2(9Z,11E).

to show connections between one carbon nucleus and its adjacent carbon nucleus, the signal at δ_C 134.76 has connections with two carbon atoms, viz., δ_C 125.58 and 32.92. The latter signal is characteristic of the shift of an allylic methylene carbon atom adjacent to an E-double bond, which means that the signal at δ_C 134.76 arises from the shift of the outer carbon of the E-double bond (i.e., C-12). Because the signal at δ_C 134.76 is also connected to δ_C 125.58, this correlation implies that the latter signal is due to the shift of C-11 (one of the inner carbon atoms of the conjugated diene system). The signal at δ_C 129.89 is connected to that of δ_C 27.66, which originates from the shift of the allylic methylene carbon adjacent to the Z-double bond. Hence the signal at δ_C 129.89 can be assigned to the remaining outer carbon signal of the diene system (i.e., C-9). The spectrum shows further a connection between the signal at δ_C 125.58 (C-11) and that of δ_C 128.71. The latter signal, therefore, can be due only to the shift of the C-10 carbon nucleus. From the INADEQUATE spectrum (Fig. 11.2), it is therefore possible to assign unambiguously the four carbon atoms of the Z,E-diene system as follows: δ_C 129.89 (C-9), 128.71 (C-10), 125.58 (C-11), and 134.76 (C-12). Referring back to the ^{13}C-^{1}H COSY spectrum (Fig. 11.1), it is now possible to correlate unambiguously the signals of the various olefinic carbon atoms with the corresponding proton signals. The shifts of the olefinic protons are confirmed, therefore, as the following: δ_H 5.32 (9-H), 5.82 (10-H), 6.24 (11-H), and

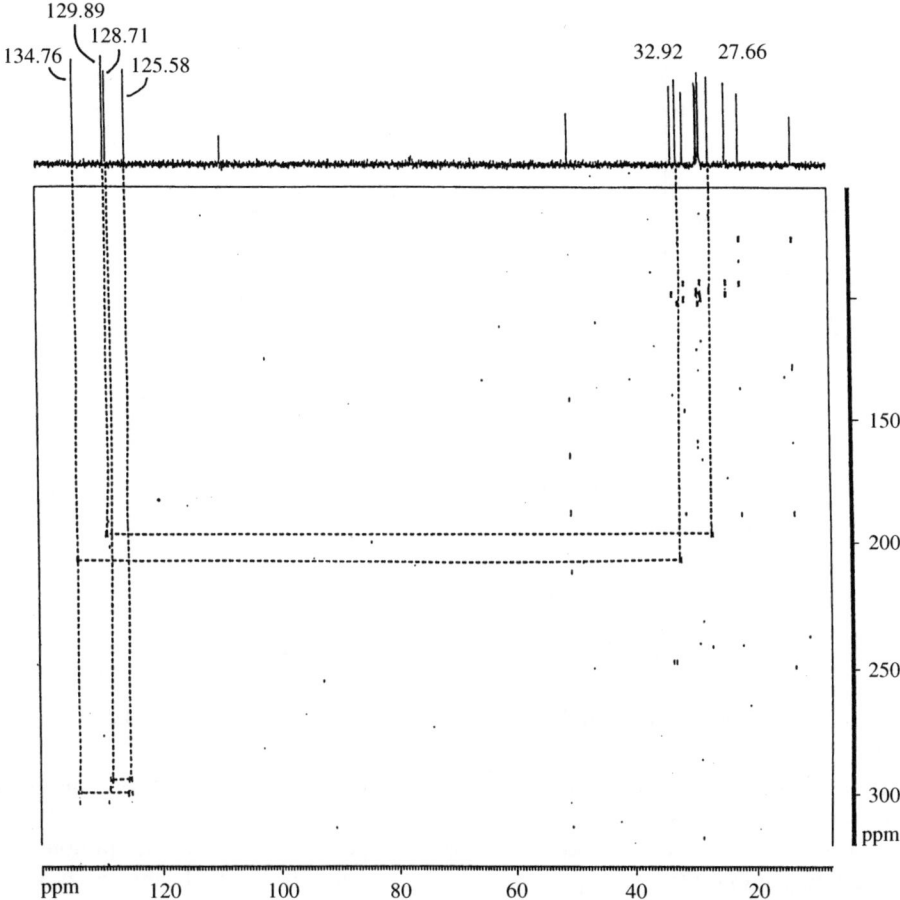

Fig. 11.2. INADEQUATE spectrum of 18:2(9Z,11E). Abbreviation: INADEQUATE, incredible natural abundance double quantum transfer experiment.

5.65 (12-H). In addition, it is also possible to assign the signals for the 8-H and 13-H protons as appearing at δ_H 2.1 and 2.0, respectively.

To confirm these assignments, the HMBC spectrum (Fig. 11.3) was recorded. This technique is well known for its ability to show correlation of nuclei (carbon to proton) that are two or three bonds apart. The signal at δ_C 125.58 (C-11) is connected to the proton signals at δ_H 5.82 (t, 10-H), 5.32 (m, 9-H) and also to the signal at 2.0 (m, 13-H). The signal at δ_C 128.71 (C-10) is connected to the proton at δ_H 5.65 (m, 12-H) and to the signal at δ_H 2.1 (m, 8-H); the signal at δ_C 129.89 (C-9) is connected to the proton signal at δ_H 6.24 (dd, 11-H); and the signal at δ_C 134.76 (C-12) is connected to the proton signals at δ_H 5.82 (t, 10-H) and also to the signal at δ_H 2.0 (m, 13-H). Furthermore, the carbon signal at δ_C 32.92 (C-13) is connected to the proton signals at δ_H 6.24 (dd, 11-H) and 5.65 (m, 12-H), whereas the signal at δ_C 27.66 (C-8) is connected to δ_H 5.82

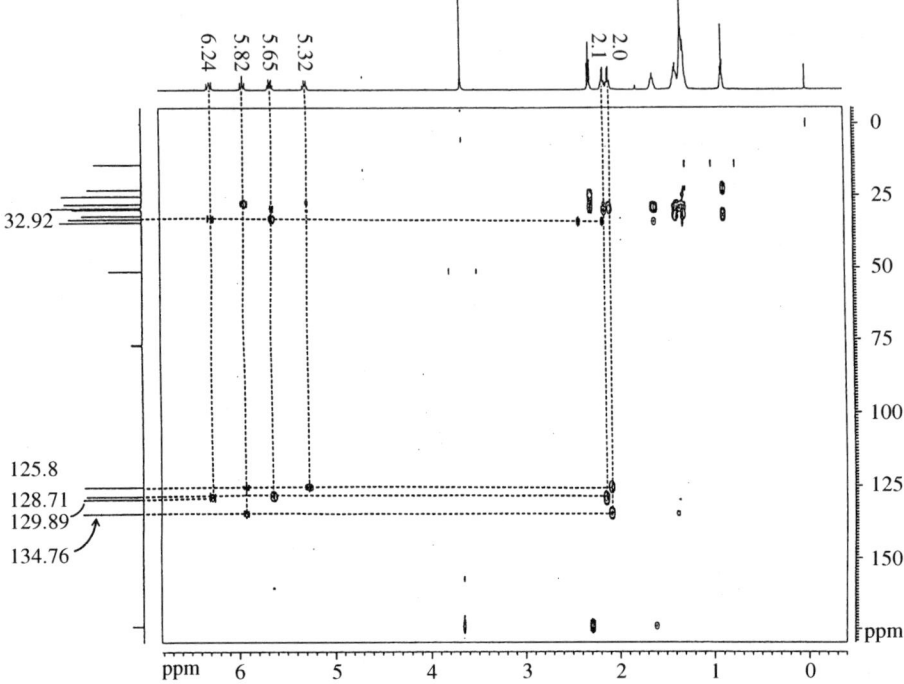

Fig. 11.3. Heteronuclear multiple bond correlation (HMBC) spectrum of 18:2 (9Z,11E).

(t, 10-*H*). These two- or three-bond connections reconfirm the assignments of the carbon atoms of the 9Z,11E-diene system.

It should be borne in mind that the splitting patterns for both inner protons of the conjugated system are expected to produce double doublets. In this case, only one such double doublet (11-*H*) is observed. The apparent triplet (10-*H*) is due to the coalescing of signals as a result of extended coupling. By examining the coupling constants of these signals more closely by running the sample in a 500 MHz instrument, the J values for inner protons in this case agreed with the expected coupling constants associated with *E*- and *Z*-double bond protons. We are convinced from the various NMR techniques used in this analysis that our assignments of the four olefinic protons and carbon atoms are correct.

Analysis of Methyl (9E,11Z)-Octadecadienoate

This geometric isomer of CLA displays an almost identical spectral pattern to that of 18:2(9Z,11E), which has been described above in detail. It is not possible to use NMR techniques alone to determine the positions of the *Z*- and *E*-double bonds in the alkyl chain of such conjugated diene isomers. By using similar NMR techniques (^{13}C-^{1}H COSY, INADEQUATE, and HMBC) to study the NMR properties of this isomer, the following carbon shift values were confirmed: δ_C 32.86 (C-8), 134.51

(C-9), 125.72 (C-10), 128.57 (C-11), 130.17 (C-12), and 27.72 (C-13). The ^{1}H NMR spectral analysis showed signals for the olefinic protons at δ_H 5.66 (m, 9-*H*), 6.24 (dd, *J* = 10 Hz, 10-*H*), 5.93 (t, *J* = 10.8 Hz, 11-*H*), and 5.30 (m, 12-*H*).

Analysis of Methyl (9Z,11Z)-Octadecadienoate

The ^{13}C-^{1}H COSY spectrum of 18:2(9Z,11Z) is shown in Figure 11.4. A distorted triplet appears at δ_H 0.88 (*J* = 6 Hz, 3H) for the shift of the protons of the terminal methyl group, which is connected to the signal at δ_C 14.13. The shift of the methyl protons of the ester group is a singlet at δ_H 3.65, which is confirmed by its connection to the signal at δ_C 51.42 of the carbon spectrum. The shift of the methylene protons (2-*H*) adjacent to the ester group results in a triplet at δ_H 2.30 (*J* = 7 Hz, 2H), which is connected to the signal at δ_C 34.10. The multiplet at δ_H 2.15 (4H) is due to the overlap of signals arising from the 8-*H* and 13-*H* methylene protons (methylene groups adjacent to the conjugated diene system). The assignments are supported by the connections of these signals to the carbon signals at δ_C 27.46 and 27.54, which are characteristic carbon shifts of methylene carbon atoms adjacent to a *Z*-olefinic system. The remaining methylene protons give rise to a multiplet at δ_H 1.18–1.51 (18H), which is

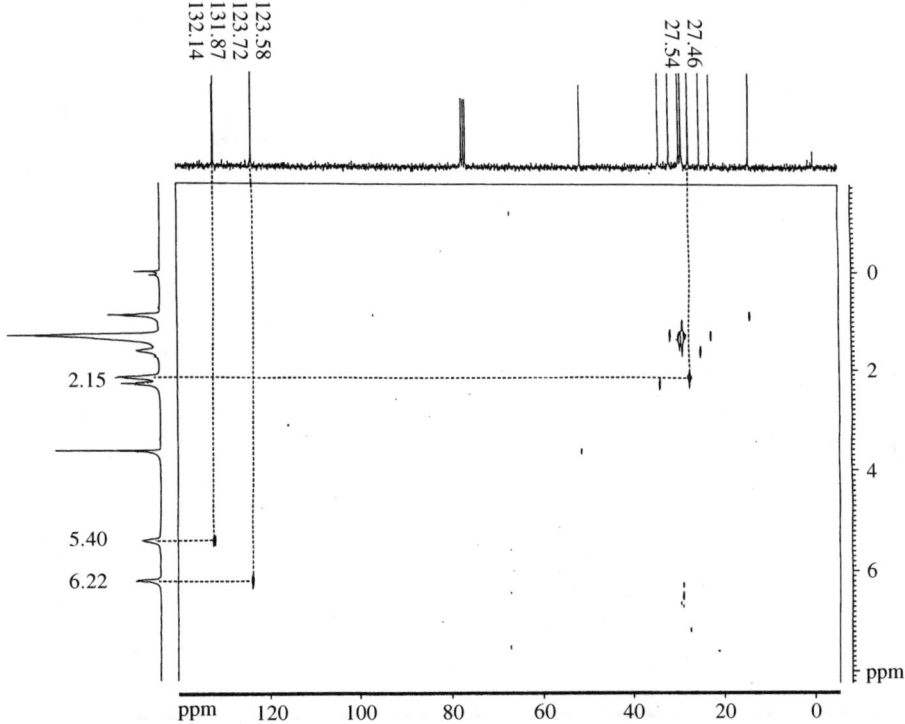

Fig. 11.4. ^{13}C-^{1}H nuclear magnetic resonance (NMR) correlation (COSY) spectrum of 18:2(9Z,11Z).

connected to carbon signals stacked in the region of δ_C 29.11–29.60. Further differentiation of the methylene protons or carbon atoms is not possible.

The shifts of the olefinic protons give a multiplet at δ_H 5.40 (2H) and a double doublet at δ_H 6.22 (2H). The latter signal is most likely due to the coupling of the two inner positioned olefinic protons of the diene system (i.e., 10-*H* and 11-*H*). This assumption is supported by the fact that this proton signal (double doublet) is connected to the pair of carbon signals at δ_C 123.58 and 123.72. The signal of the other set of olefinic protons (multiplet at δ_H 5.40) is connected to the pair of carbon signals that appear at δ_C 131.87 (C-9) and 132.14 (C-12) for the outer carbon atoms of the olefinic system. The outer carbon shifts at δ_C 131.87 and 132.14 are confirmed from the INADEQUATE spectrum (Fig. 11.5), which shows connections of these signals with the shifts of C-8 and C-13 at δ_C 27.46 and 27.54, respectively.

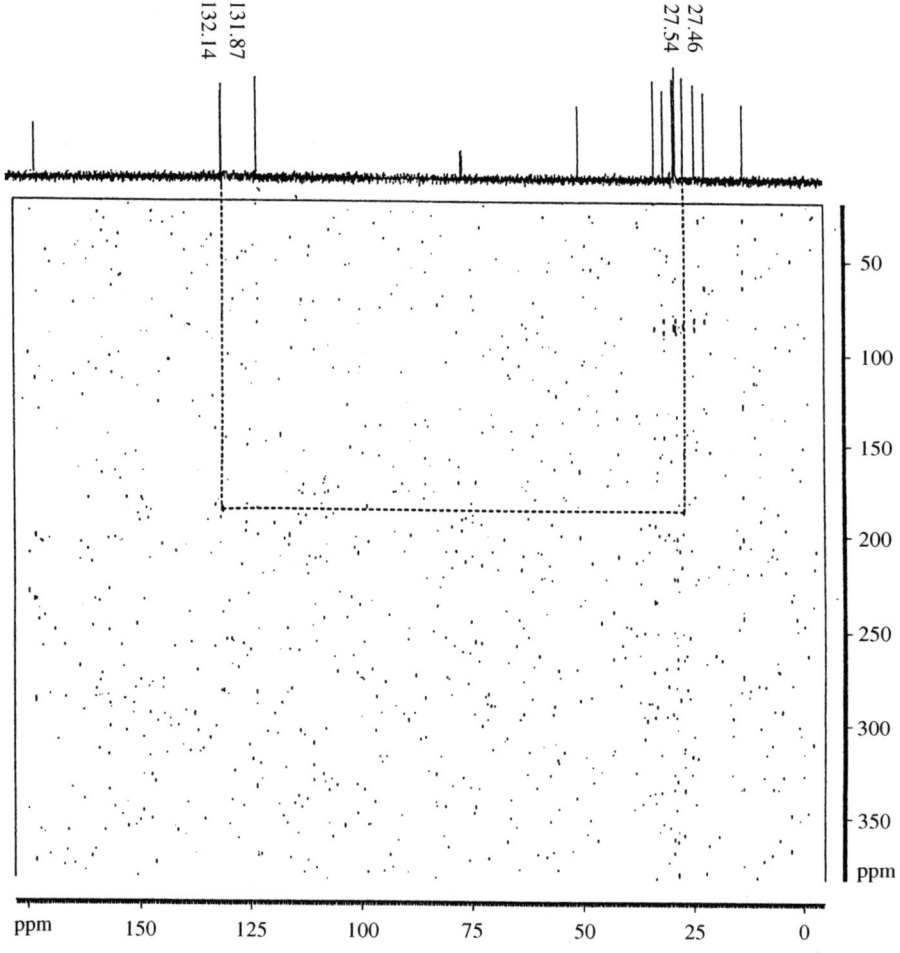

Fig. 11.5. INADEQUATE spectrum of 18:2(9*Z*,11*Z*). See Figure 11.2 for abbreviation.

Analysis of Methyl (9E,11E)-Octadecadienoate

Similar NMR techniques (^{13}C-^{1}H COSY, INADEQUATE, and HMBC) have been used to study the NMR properties of this geometric isomer. The results show that the signals of the inner positioned olefinic protons (10-*H*, 11-*H*) are slightly more upfield (δ_H 5.96) than those observed for the similarly positioned protons of the 18:2(9Z,11Z) isomer. The shifts of the outer positioned olefinic protons are found at δ_H 5.56 (m). The *E,E*-diene nature of the conjugated system is evident from the appearance of two signals at δ_C 32.61 and 32.68, which are the carbon shifts due to C-8 and C-13, respectively. Similar to the characteristics displayed by the Z,Z-diene isomer, the NMR spectroscopic analysis of the *E,E*-isomer distinguishes the inner carbon atoms (C-10, C-11) from those of the outer carbon atoms (C-9, C-12). These shifts appear at δ_C 130.37 (C-10) and 130.51 (C-11) for the inner olefinic carbons and at δ_C 132.16 (C-9) and 132.43 (C-12) for the outer olefinic carbon atoms, respectively. The ^{13}C-^{1}H NMR correlation (COSY) spectrum of 18:2(9*E*,11*E*) is shown in Figure 11.6.

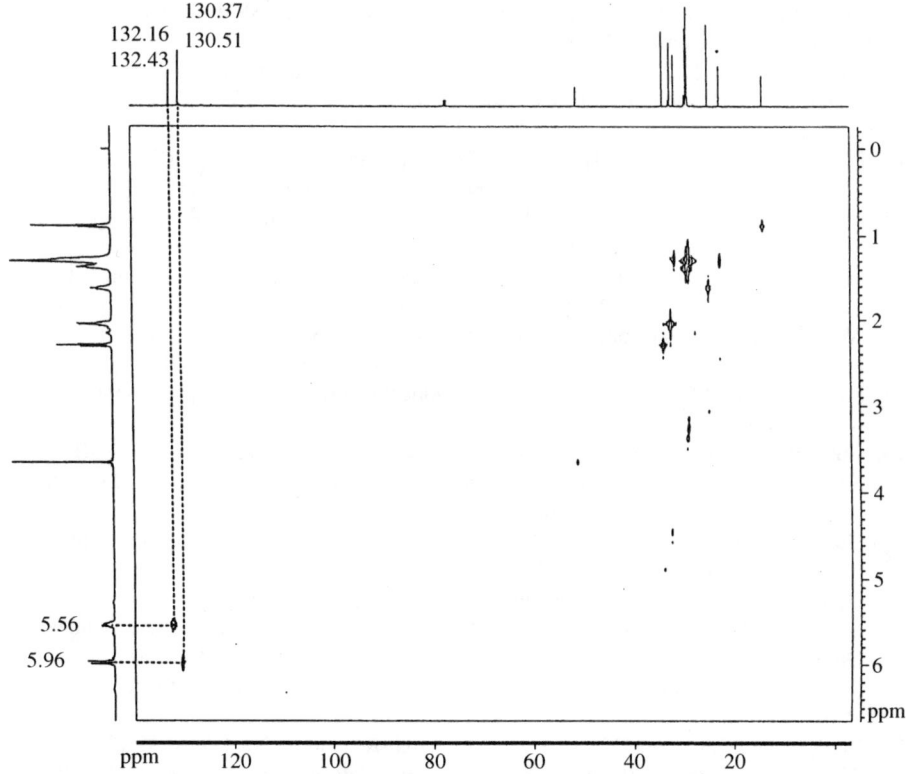

Fig. 11.6. ^{13}C-^{1}H nuclear magnetic resonance (NMR) correlation (COSY) spectrum of 18:2(9*E*,11*E*).

References

1. Williams, D.H., and Fleming, I. (1995) *Spectroscopic Methods in Organic Chemistry,* 5th edn., McGraw Hill, New York.
2. Kemp, W. (1987) *Organic Spectroscopy,* MacMillan, Basingstoke.
3. Derome, A.E. (1990) *Modern NMR Techniques for Chemistry Research,* Pergamon Press, Oxford.
4. Canet, D. (1996) *Nuclear Magnetic Resonance: Concepts and Methods,* John Wiley & Sons, New York.
5. Paudler, W.W. (1987) *Nuclear Magnetic Resonance: General Concepts and Applications,* John Wiley & Sons, New York.
6. Sanders, J.K.M., and Hunter, B.K. (1987) *Modern NMR Spectroscopy: A Guide for Chemists,* Oxford University Press, Oxford.
7. Croasmun, W.R., and Carlson, R.M.K. (1994) *Two-Dimensional NMR Spectroscopy,* 2nd edn., VCH, New York.
8. Johnson, L.F., and Shoolery, J.N. (1962) Determination of Unsaturation and Average Molecular Weight of Natural Fats by Nuclear Magnetic Resonance, *Anal. Chem. 34,* 1136–1139.
9. Wedmid, Y., and Litchfield, C. (1975) Positional Analysis of Isovaleroyl Triglycerides Using Proton Magnetic Resonance with Eu(fod)$_3$ and Pr(fod)$_3$ Shift Reagents: I. Model Compounds, *Lipids 10,* 145–151.
10. Wedmid, Y., and Litchfield, C. (1976) Positional Analysis of Isovaleroyl Triglycerides Using Proton Magnetic Resonance with Eu(fod)$_3$ and Pr(fod)$_3$ Shift Reagents: II. Cetacean Triglycerides, *Lipids 11,* 189–193.
11. Pfeffer, P.E., and Rothbart, H.L. (1972) PMR Spectra of Triglycerides: Discrimination of Isomers with the Aid of a Chemical Shift Reagent, *Tetrahedron Lett.,* 2533–2536.
12. Pfeffer, P.E., and Rothbart, H.L. (1973) Effects of a Europium-Shift Reagent upon the PMR Spectrum of Some Triglycerides, *Acta Chem. Scand. 27,* 3131–3132.
13. Frost, D.J., Bus, J., Keuning, R., and Sies, I. (1975) PMR Analysis of Unsaturated Triglycerides Using Shift Reagents, *Chem. Phys. Lipids 14,* 189–192.
14. Almqvist, S.O., Andersson, R., Shahab, Y., and Olsson, K. (1972) Lanthanide-Induced PMR Chemical Shifts in Triglycerides, *Acta Chem. Scand. 26,* 3378–3380.
15. Shoolery, J.N. (1977) Some Quantitative Applications of ^{13}C NMR Spectroscopy, *Prog. NMR Spectroscopy 11,* 79–93.
16. Gunstone, F.D. (1992) Structural Analysis of Lipids by High Resolution ^{13}C NMR Spectroscopy, in *Contemporary Lipid Analysis,* (Olsson, N.U., and Herslof, B.G., eds.) pp. 7–22, LipidTeknik, Stockholm.
17. Gunstone, F.D. (1992) High Resolution ^{1}H and ^{13}C NMR, in *Lipid Analysis,* (Hamilton, R.J., and Hamilton, S., eds.) pp. 243–262, Oxford University Press, Oxford.
18. Gunstone, F.D. (1993) High Resolution ^{13}C NMR Spectroscopy of Lipids, in *Advances in Lipid Methodology - Two,* (Christie, W.W., ed.) pp. 1–68, The Oily Press, Dundee.
19. Gunstone, F.D. (1994) ^{13}C NMR of Lipids, in *Developments in the Analysis of Lipids* (Tyman, J.H.P., and Gordon, M.H., eds.) pp. 109–122, Royal Society of Chemistry, Cambridge.
20. Lie Ken Jie, M.S.F., and Mustafa, J. (1997) High-Resolution Nuclear Magnetic Resonance Spectroscopy—Applications to Fatty Acids and Triacylglycerols, *Lipids 32,* 1019–1034.

21. Frost, D.J., and Gunstone, F.D. (1975) The PMR Analysis of Non-Conjugated Alkenoic and Alkynoic Acids and Esters, *Chem. Phys. Lipids 15*, 53–85.
22. Sehat, N., Yurawecz, M.P., Roach, J.A.G., Mossoba, M.M., Kramer, J.K.G., and Ku, Y. (1998) Silver-Ion High-Performance Liquid Chromatographic Separation and Identification of Conjugated Linoleic Acid Isomers, *Lipids 33*, 217–221.
23. Gunstone, F.D., and Said, A.I. (1971) Methyl 12-Mesyloxyoleate as a Source of Cyclopropane Esters and of Conjugated Octadecadienoates, *Chem. Phys. Lipids 7*, 121–134.
24. Berdeaux, O., Christie, W.W., Gunstone, F.D., and Sébédio, J.L. (1997) Large-Scale Synthesis of Methyl *cis*-9, *trans*-11-Octadecadienoate from Methyl Ricinoleate, *J. Am. Oil Chem. Soc. 74*, 1011–1015.
25. Lie Ken Jie, M.S.F., Pasha, M.K., and Alam, M.S. (1997) Synthesis and Nuclear Magnetic Resonance Properties of All Geometric Isomers of Conjugated Linoleic Acids, *Lipids 32*, 1041–1044.
26. Adlof, R. (1997) Preparation of Methyl *cis*-9, *trans*-11 and *trans*-9, *trans*-11-Octadecadienoate-17,17,18,18-d_4, Two of the Isomers of Conjugated Linoleic Acid, *Chem. Phys. Lipids 88*, 107–112.
27. Bus, J., Sies, I., and Lie Ken Jie, M.S.F. (1976) ^{13}C-NMR of Methyl, Methylene and Carbonyl Carbon Atoms of Methyl Alkenoates and Alkynoates, *Chem. Phys. Lipids 17*, 501–518.

Chapter 12

Identification and Quantification of Conjugated Linoleic Acid Isomers in Fatty Acid Mixtures by ^{13}C NMR Spectroscopy

Adrienne L. Davis[a], Gerald P. McNeill[b], and David C. Caswell[a]

[a]Unilever Research Colworth, Colworth House, Sharnbrook, Bedford MK44 1LQ, UK
[b]Loders Croklaan, Channahon, IL 60410

Introduction

Conjugated linoleic acid (CLA) is a collective term referring to a mixture of naturally occurring positional and geometric isomers of linoleic acid containing a conjugated double bond system. The most common naturally occurring CLA is the 9Z,11E isomer, found in milk and beef fat (1). Other CLA isomers have been found in food products (e.g., the 10E,12Z isomer), possibly as a result of the effects of processing techniques on linoleic acid (2). Certain synthetically prepared CLA samples have been shown by gas chromatography-mass spectrometry (GC-MS) to consist of mixtures of positional isomers (3). However, information about the full range and identity of CLA isomers (i.e., positional and geometric) present in complicated CLA mixtures has not been available by any analytical method (including GC-MS) until recently.

Nuclear magnetic resonance (NMR) spectroscopy is the obvious technique to apply to the problem of identifying isomers in complicated CLA mixtures. The sensitivity and versatility of NMR spectroscopy as a probe of molecular structure is extremely well established (e.g., see Refs. 4–7), and the utility of NMR spectroscopy in the analysis of fatty acids and their derivatives is also widely recognized (see Chapter 11 and references therein). Furthermore, NMR is a quantitative technique (provided certain experimental constraints are observed) (8) and therefore has the potential to be able to quantify CLA isomers in addition to identifying them. This chapter describes the development of an NMR method for CLA analysis.

To develop an NMR method for analysis of any type of mixture, it is necessary to identify in the NMR spectrum of that mixture at least one quantifiable resonance from each of the different compounds of interest. Clearly, in a mixture of structurally similar compounds such as CLA isomers, many of the resonances from the different isomers will be coincident (or nearly so) and therefore unidentifiable or unquantifiable. Consequently, it is necessary to seek out a group of resolvable *and* identifiable peaks: the so-called "reporter" resonances.

Because the structural variation among CLA isomers occurs in the olefinic part of the molecules, the olefinic NMR signals are the most likely candidates for reporter resonances. However, the strong similarity among isomers means that the chemical shift differences even between reporter resonances may, in some circumstances, be very small indeed (e.g., in a comparison of the 8Z,10Z and 9Z,11Z isomers, the environment of a given olefinic proton or carbon differs between isomers only in its proximity to the rather remote carboxyl and methyl end groups). Therefore, ^{13}C NMR spectroscopy, with its large chemical shift dispersion (~20 times greater than that of ^{1}H NMR), is the NMR method of choice for this problem.

It is clear then that the main requirement for development of an NMR method for analysis of CLA mixtures is assignment of the NMR resonances of the olefinic carbons of these compounds. To reliably assign these signals, it is necessary to first study some pure materials and simple mixtures. This allows unequivocal signal assignments to be obtained for at least some of the components of the mixture, thereby enabling chemical shift trends to be related to structural differences, which is a first step toward deducing a complete spectral assignment.

Unequivocal NMR signal assignments are obtainable by two-dimensional (2D) NMR spectroscopy (5) in which connectivity information can be used to eliminate ambiguity. Several 2D experiments were found to be particularly useful for this application, including the following: TOCSY (*total correlation spectroscopy*), a homonuclear method that allows proton resonances from scalar coupling networks to be correlated together regardless of whether or not they are directly coupled (9) [*cf.,* COSY, which allows correlation between directly coupled proton partners (10)]; HMQC (*heteronuclear multiple quantum correlation*), a heteronuclear method for correlating protons with their directly attached carbons (11); HMQC-TOCSY, a combination of TOCSY and HMQC, which enables a given carbon to be correlated with its directly attached proton *and* other protons along the scalar coupling network of the attached proton (11); and, finally, HMBC (*heteronuclear multiple bond correlation*), which allows correlation of protons with carbons two or three bonds distant (11,12). The use of these techniques to facilitate assignment of ^{13}C NMR spectra of standard samples of individual isomers and simple mixtures is described in the next section.

NMR Studies of Standard Materials and Simple Mixtures

^{13}C NMR Assignment of the 9,11 Isomers of Octadecadienoic Acid

The ^{13}C NMR spectra of the 9,11 CLA (*ZZ, EE* and *ZE*) were assigned using samples of the individual isomers and the methods described below for the 10*E*,12*Z* isomer (13); results are given in Table 12.1. These assignments are in agreement with data published by Lie Ken Jie *et al.* (14) during the course of this work and are not discussed further except to note that uncertainties in the olefin assignments of the *ZZ* and *EE* isomers are resolved here.

Identification and ^{13}C NMR Assignment of Octadeca-10E,12Z-dienoic Acid

This was achieved using a combination of ^{1}H, ^{13}C, and 2D NMR spectroscopy on a sample containing predominantly this isomer (13). ^{13}C NMR data for octadeca-10*E*,12*Z*-dienoic acid are included in Table 12.1.

TABLE 12.1 ^{13}C NMR Data for Unequivocally Identified CLA Isomers (including Me esters)[a]

Atom	9Z,11Z	9E,11E	9Z,11E[b]	9Z,11E
1	179.98	179.86	174.32	180.09
2	34.06	34.00	34.10	34.05
3	24.68	24.67	24.95	24.67
4	28.99–29.11	28.92–29.10	29.06	29.03
5	28.99–29.11	28.92–29.10	29.12–29.67	29.12
6	28.99–29.11	28.92–29.10	29.12–29.67	29.03
7	29.57	29.36[d]	29.12–29.67	29.66
8	27.52	32.56	27.66	27.65
9	131.87	132.21	129.89	129.89
10	123.72	130.46	128.71	128.73
11	123.55	130.30	125.58	125.58
12	132.19	132.53	134.76	134.80
13	27.43	32.63	32.92	32.90
14	29.64	29.42[d]	29.41	29.40
15	28.99–29.11	28.92–29.10	28.95	28.93
16	31.76	31.77	31.77	31.76
17	22.64	22.63	22.65	22.63
18	14.09	14.10	14.12	14.10
OMe	—	—	51.44	—

[a]Unless otherwise stated, data are for conjugated linoleic acids at 2.5% w/w in CDCl3 solvent at 300K, measured in units of ppm and referenced against internal TMS (tetramethysilane). Note that slight variations in the C-1, C-2, C-3 shifts from sample to sample have been noted.
[b]Data for the Me ester from Ref. 14.
[c]Identified as described in the text.
[d]Assignment interchangeable.
[e]Assigned by analogy to data from 9E,11Z; 9Z,11E; and 10E,12Z.

The ^{1}H NMR spectrum of (the mixture containing) 10E,12Z CLA (Fig. 12.1) is almost identical to that of the 9Z,11E isomer, and the following groups of signals are readily recognizable from their chemical shifts and multiplicities: the olefinics (H-10 to H-13), the allylics (H-9 and H-14), H-2, H-3, and the Me signal, H-18 (see Chapter 11 and references therein). It is a simple task to establish which olefinic resonances are *cis* (Z) and which are *trans* (E) by the size of the proton-proton coupling constants across the double bond (~11 Hz for *cis* and 15 Hz for *trans*); further examination of the coupling constants also identifies the inner (H-11, H-12) and outer (H-10, H-13) olefin resonances.

To assign the olefinic proton signals unequivocally, it is necessary to establish which of the outer olefin resonances is from the proton closest to the Me group (H-13) and which is from the proton closest to the COOH group (H-10). This is readily deduced from a 2D TOCSY experiment performed with a long mixing time (~90–120 ms). Under these conditions, magnetization transfer from the Me signal, H-18, occurs more strongly to H-13 than to the more remote H-10. Consequently, the outer olefinic resonances can be distinguished by the relative intensities of their crosspeaks to the Me signal (Fig. 12.2). Because the signal correlating most

11Z,13E[c]	8E,10Z[c]	9E,11Z[b]	10E,12Z[c]
180.09	180.05	174.34	180.09
34.06	34.03	34.10	34.06
24.69	24.63	24.95	24.67
29.07–29.45	28.81–29.25	28.97	29.06
29.07–29.45	28.81–29.25	29.13–29.45	29.16–29.29
29.07–29.45	28.81–29.25	29.13–29.45	29.16–29.29
29.07–29.45	32.78	29.13–29.45	29.16–29.29
29.07–29.45	134.32	32.86	29.39
29.73[e]	125.83	134.51	32.87
27.69[e]	128.54	125.72	134.58
130.04	130.27	128.57	125.70
128.66	27.71[e]	130.17	128.60
125.65	29.76[e]	27.72	130.16
134.66	28.81–29.25	29.73	27.68
32.57	28.81–29.25	28.97	29.44
31.59	31.87	31.77	31.50
22.29	22.68	22.65	22.57
13.95	14.11	14.12	14.06
—	—	51.45	—

strongly with the Me signal is known to be a *cis* (Z) proton (from its coupling constants), confirmation that the compound is an *EZ* isomer is obtained by this method. Other correlations in the TOCSY spectrum (from the H-2 and H-3 signals) similarly allow the allylic protons resonances of H-9 and H-14 to be distinguished.

The identified proton signals can be correlated with their directly attached carbons in a 2D HMQC spectrum allowing identification of C-2, C-3, C-9, C-10, C-11, C-12, C-13, C-14 and C-18. Further ^{13}C assignments are made by observing two- and three-bond correlations in the 2D HMBC spectrum (listed in Table 12.2). Correlations from the olefinic protons H-10 and H-13 allow the carbons adjacent to the allylic carbons (i.e., C-8 and C-14, respectively) to be identified, whereas correlations from H-2 and H-3 allow C-4 to be identified.

The most important correlations given in Table 12.2 are from both H-18 *and* an allylic proton (H-15) to the same carbon resonance. Clearly, because long-range correlations are transmitted over two or three bonds and this carbon correlates strongly to H-18, it must be either C-16 or C-17. The latter possibility is easily ruled out on chemical shift grounds; therefore, the carbon in question must be C-16. A strong long-range correlation from an allylic proton to C-16 is proof that this *EZ* isomer can only be a 10,12 or 11,13 CLA.

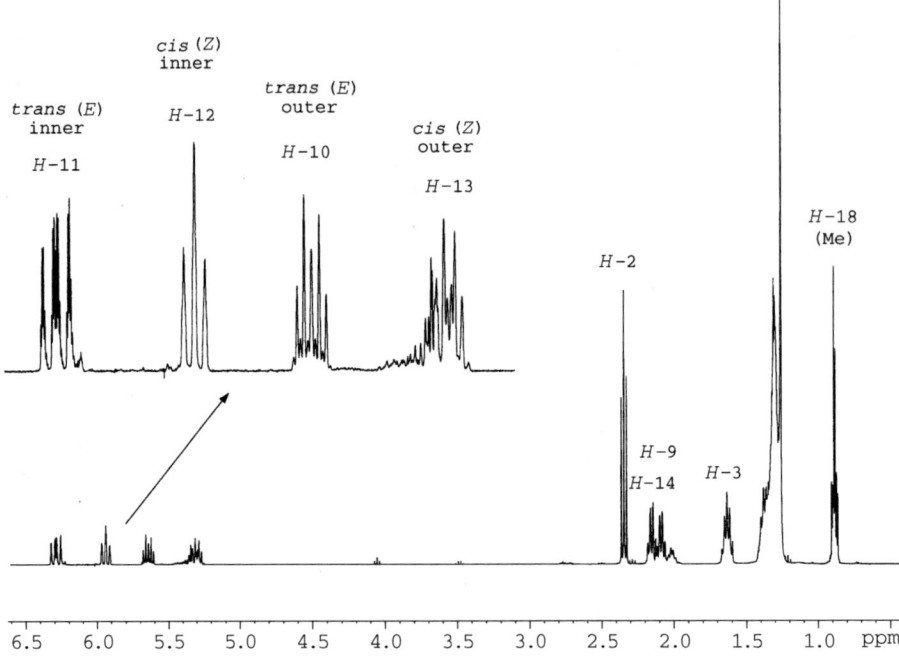

Fig. 12.1. 400 MHz ¹H nuclear magnetic resonance (NMR) spectrum of a mixture of fatty acids containing octadeca-10*E*,12*Z*-dienoic acid as the main component.

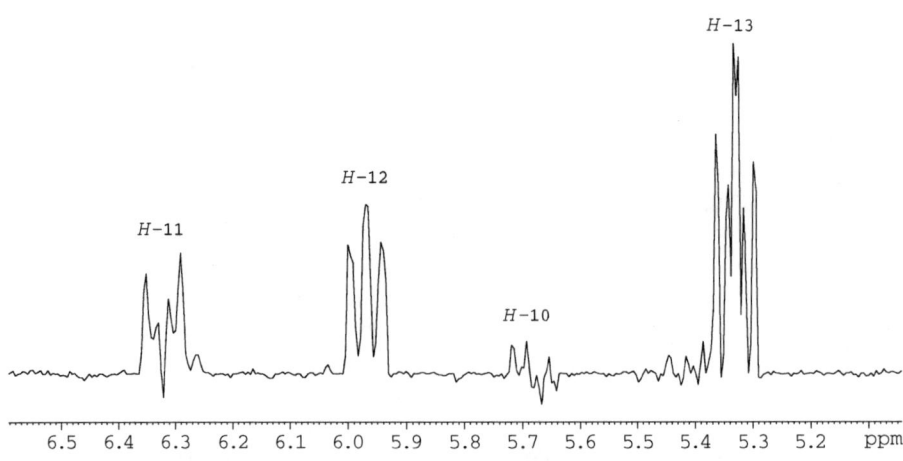

Fig. 12.2. Slice through the Me peak of the two-dimensional total correlation (2D TOCSY) spectrum of octadeca-10*E*,12*Z*-dienoic acid showing the different extent of magnetization transfer to the olefinic resonances. *Source:* Reprinted from Ref. 13 with permission of Elsevier Science.

TABLE 12.2 Selected Long-Range ^{1}H-^{13}C Correlations Observed for Octadeca-10E,12Z-dienoic Acid[a,b]

Proton	Carbon
2	1, 3, 4
3	1, 2, (4,5)
9	8, 10, 11
10	8, 9, 12
11	9, 12, 13
12	10, 11, 14
13	11, 14, 15
14	12, 13, 15, 16
18	16, 17

[a]Measured by the heteronuclear multiple bond correlation (HMBC) experiment at 400 MHz in CDCl$_3$ solvent. Correlations in parentheses could not be resolved.
[b]*Source:* reprinted from Ref. 13 with permission of Elsevier Science.

The distinction can be made by reference to chemical shift data for octadecenoic acid isomers shown in Table 12.3 (15). These data show that the chemical shifts of C-17 and C-16 alter by –0.04 ppm and –0.23 ppm, respectively, when a (*cis*) double bond moves from position 11 to position 12 in the chain, whereas movement of the double bond from position 11 to position 13 causes C-17 to change by a much greater amount, –0.32 ppm, whilst C-16 is increased by +0.19 ppm. Clearly, from the point of view of the C-17 and C-16 chemical shifts, the above changes in double bond position are equivalent to going from a 9,11 to a 10,12 isomer and a 9,11 to 11,13 isomer, respectively. In particular, it is worth noting from the data in Table 12.3 that the shift of C-17 changes significantly when the double bond arrives at the 13 position, and therefore this carbon resonance is a reporter for 11,13 CLA.

The changes observed for δC-17 and δC-16 for the CLA in question relative to the 9,11 isomers are clearly of the correct magnitude and sign to demonstrate that the unknown is a 10,12 and not an 11,13 CLA (–0.08 ppm and –0.27 ppm relative to 9E,11Z; –0.07 ppm and –0.26 ppm relative to 9Z,11Z). Therefore, the identity of this compound as octadeca-10E,12Z-dienoic acid is confirmed. All assignments given for 10E,12Z CLA in Table 12.1 are consistent with this conclusion combined with the long-range correlations listed in Table 12.2.

Identification and NMR Assignment of Octadeca-11Z,13E- and -8E,10Z-dienoic Acids

These compounds were present in a simple mixture containing four major CLA isomers. The two main components were readily identified by ^{13}C NMR as the 9Z,11E and 10E,12Z isomers. Identification and assignment of the NMR spectra of the 11Z,13E and 8E,10Z acids was more problematic than in previous examples due to signal overlap (see below). ^{13}C NMR data for these compounds are given in Table 12.1.

TABLE 12.3 Selected ^{13}C NMR Data from Octadecenoic Acids[a]

Double bond position (n)	E isomer				Z isomer			
	C-17	C-16	C-n	C-(n+1)	C-17	C-16	C-n	C-(n+1)
5	22.78	32.04	128.72	131.98	22.79	32.06	128.21	131.41
6	22.76	32.04	129.49	131.13	22.78	32.04	129.01	130.62
7	22.78	32.05	129.89	130.81	22.82	32.10	129.45	130.31
8	22.76	32.01	130.13	130.67	22.80	32.06	129.65	130.17
9	22.76	32.01	130.23	130.54	22.80	32.06	129.78	130.09
10	22.76	32.01	130.31	130.47	22.76	32.01	129.83	130.00
11	22.73	31.88	130.35	130.43	22.72	31.87	129.89	129.96
12	22.62	31.49	130.41	130.41	22.68	31.64	129.94	129.94
13	22.27	31.96	130.39	130.39	22.40	32.06	129.90	129.90
14	22.82	34.79	130.13[b]	130.63[b]	22.96	29.69	129.66[c]	130.16[c]

[a]Data (in ppm) obtained from Ref. 15, measured in CDCl$_3$ solvent.
[b,c]Assignment interchangeable.

Signals from the 11Z,13E isomer were distinguished from those of the other components of the mixture by the relative signal intensities. The C-17 shift of the 11Z,13E CLA is 22.29 ppm (–0.34 ppm relative to the 9Z,11E isomer), which is diagnostic of an 11,13 isomer as described earlier; therefore, once the double bond geometry is established, the identity of the isomer is known. Observation of the ^{1}H NMR spectrum showed the 11,13 isomer to be either ZE or EZ. The distinction was made with a 2D HMQC-TOCSY experiment. The reporter carbon at 22.29 ppm clearly correlated with a resolvable peak in the group of Me triplets in the ^{1}H NMR spectrum, identifying the Me triplet from the 11,13 isomer as the most downfield in the overlapping group (Fig. 12.3). The resolvable line of the 11,13 Me triplet correlates most strongly with the pair of *trans* (E) protons in the 2D TOCSY spectrum (Fig. 12.4) and also with the *trans* carbons (identified by HMQC) in the HMQC-TOCSY. Therefore, the unknown component in question must be the 11Z,13E CLA (and the olefinic carbon signals can be assigned by the relative intensities of the correlations in the 2D spectra). Other correlations in the HMQC-TOCSY are consistent with this conclusion, e.g., the ^{1}H Me peak correlates with the allylic carbon at 32.57 ppm (i.e., adjacent to the *trans* bond, see data for 9Z,11E and 10E,12Z in Table 12.1) and not to the one at 27.69 ppm (adjacent to the *cis* bond).

GC-MS data (Christie, W.W., personal communication), obtained while this work was in progress and measured by the recently published method (3) indicated that the remaining major component of the mixture was an 8,10 CLA. Observation of the C-16 and C-17 chemical shifts is consistent with this conclusion, i.e., δC-16 and δC-17 are both increased slightly from the values obtained from the 9,11 isomers, by ~0.10 ppm and 0.03 ppm, respectively (*cf.*, data in Table 12.3 where the analogous values for migration of the double bond from position 11 to 10 are 0.14 and 0.04 ppm). It was clear from the olefinic proton and carbon signals that this isomer is also either EZ or ZE; however, because no ^{1}H resonance was resolvable for the 8,10 CLA,

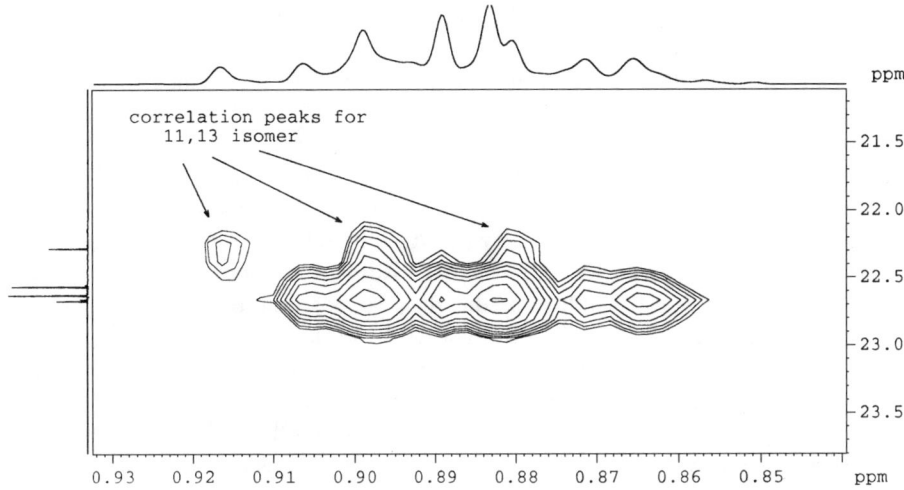

Fig. 12.3. Portion of the two-dimensional heteronuclear multiple quantum correlation and total correlation (2D HMQC-TOCSY) spectrum of a conjugated linoleic acid (CLA) mixture showing that the diagnostic 11,13 carbon peak at 22.3 ppm correlates with the most downfield Me triplet in the proton spectrum. *Source:* Reprinted from Ref. 13 with permission of Elsevier Science.

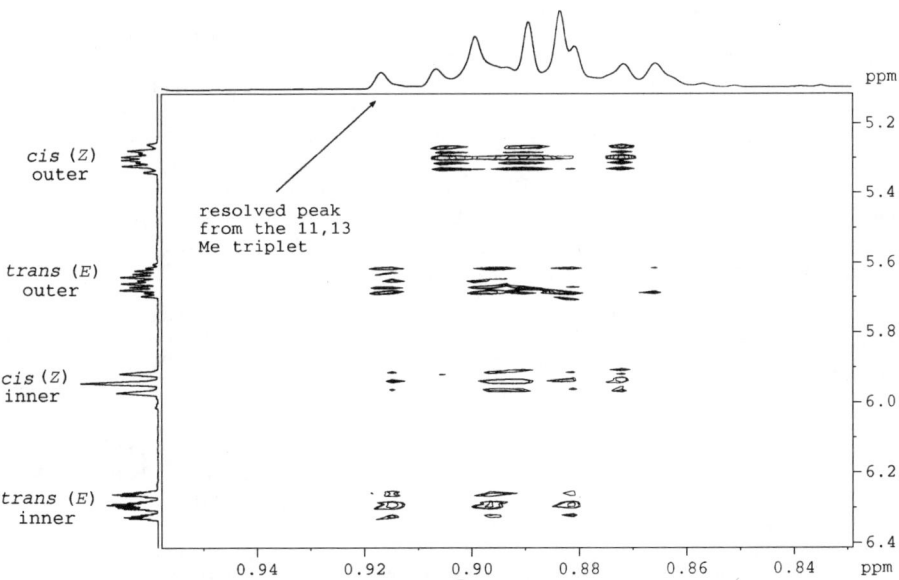

Fig. 12.4. Portion of the two-dimensional total correlation (2D TOCSY) spectrum of a conjugated linoleic acid (CLA) mixture showing that the most downfield Me triplet correlates most strongly with the *trans* olefinic protons. *Source:* Reprinted from Ref. 13 with permission of Elsevier Science.

it was impossible to establish unequivocally the stereochemistry of the molecule by the 2D methods described above.

Examination of the trends in the chemical shift data established by ourselves and other workers made it possible to confidently identify this compound as the 8E,10Z isomer. Comparison of the literature values for the chemical shifts of the olefinic carbons of the 9Z,11E and 9E,11Z methyl esters (14) suggests the following shift increments for ZE to EZ conversion (starting with the carbon closest to the CO group): +4.62, –2.99, +2.99 and –4.59 ppm. Assuming that these increments are approximately the same regardless of the actual position of the conjugated double bond system in the chain, it is possible to predict the peak positions of other CLA isomers, as shown in Table 12.4 for the 11E,13Z and 10Z,13E isomers.

Data for the EZ/ZE 8,10 CLA can be added to Table 12.4 in one of two ways, as shown in parts (a) and (b): The unknown can be identified as 8E,10Z, its olefinic carbon chemical shifts assigned by analogy to those of other EZ isomers, and then the shifts for 8Z,10E predicted as described above; alternatively, the unknown can be identified as 8Z,10E and the shifts of 8E,10Z can be predicted using the shift incre-

TABLE 12.4 Identification of the 8E,10Z Isomer by Examination of ^{13}C NMR Chemical Shift Trends for Selected Conjugated Linoleic Acid (CLA) Isomers[a]

(a) Unknown CLA correctly identified as 8E,10Z.

Olefinic carbon	EZ isomers				ZE isomers			
	8,10	9,11[b]	10,12	11,13[c]	8,10[c]	9,11	10,12[c]	11,13
OL-1	134.32	134.51	134.58	134.66	129.70	129.89	129.96	130.04
OL-2	125.83	125.72	125.70	125.67	128.82	128.73	128.69	128.66
OL-3	128.54	128.57	128.60	128.64	125.55	125.58	125.61	125.65
OL-4	130.27	130.17	130.16	130.07	134.86	134.80	134.75	134.66

(b) Unknown CLA incorrectly identified as 8Z,10E.

Olefinic carbon	EZ isomers				ZE isomers			
	8,10[c]	9,11[b]	10,12	11,13[c]	8,10	9,11	10,12[c]	11,13
OL-1	134.89	134.51	134.58	134.66	130.27	129.89	129.96	130.04
OL-2	125.55	125.72	125.70	125.67	128.54	128.73	128.69	128.66
OL-3	128.82	128.57	128.60	128.64	125.83	125.58	125.61	125.65
OL-4	129.73	130.17	130.16	130.07	134.32	134.80	134.75	134.66

[a]The olefinic carbon closest to the carboxyl group is labeled OL-1; the one closest to the Me group is labeled OL-4, and so on. All data listed were obtained from samples at 2.5% w/w in CDCl$_3$ at 300K, measured in units of ppm and referenced against internal TMS.
[b]Data for the Me ester taken from Ref. 14.
[c]Predicted from the value observed for the alternative isomer (EZ or ZE as appropriate), as described in the text (data from samples examined subsequently were found to match the predictions for 8Z,10E, 10Z,12E, and 11E,13Z extremely well).

ments. A comparison of the shifts for each olefinic carbon along the rows of Table 12.4 indicates that identification of the unknown as 8E,10Z [Table 12.4(a)] fits the trends in the data for other positional isomers far better than assigning the unknown as 8Z,10E [Table 12.4(b)]. Therefore, the unknown is identified as octadeca-8E,10Z-dienoic acid, and the olefinic reporter resonances of this compound are assigned as in Table 12.4(a).

Identification of CLA Isomers in Complex Mixtures

The ^{13}C chemical shift information given in Tables 12.1 and 12.4, obtained by a combination of 2D NMR techniques and reference to related data in the literature, is sufficient to allow the spectra of more complex CLA isomer mixtures to be analyzed.

Selected reporter resonances for *EZ* and *ZE* isomers from the ^{13}C NMR spectra of two commercially available synthetic CLA samples are shown in Figure 12.5.

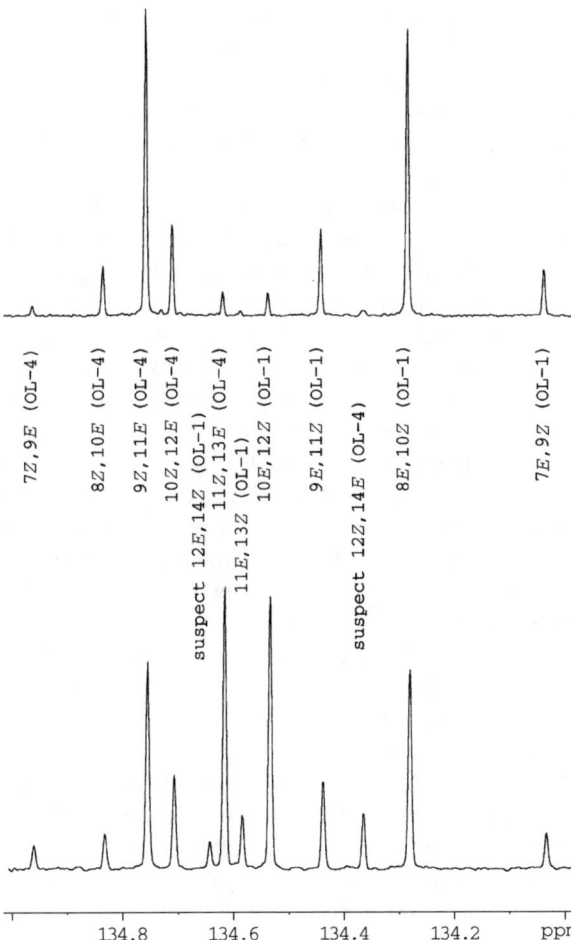

Fig. 12.5. Selected region of the ^{13}C-{^{1}H} nuclear magnetic resonance (NMR) spectra of two commercially available samples of synthetically prepared conjugated linoleic acid (CLA) showing reporter resonances for *EZ* and *ZE* isomers. Spectra were measured at 100.61 MHz (^{13}C frequency) on samples containing 10% w/w CLA in CDCl$_3$. OL-1 refers to the olefinic carbon closest to the carboxyl group, whereas OL-4 is the most remote olefinic carbon.

Clearly, both samples contain quite a complex mixture of these isomers. Identification of known CLA (8E,10Z; 9Z,11E; 9E,11Z, 10E,12Z; 11Z,13E) is trivial in such a spectrum. Previously unidentified CLA such as 8Z,10E, 10Z,12E, and 11E,13Z can be assigned by using the tabulated predicted peak positions, which agree almost precisely with observable peaks (*N.B.*, the data in Table 12.4 and the spectra in Figure 12.5 were obtained at different concentrations; see comments below). Furthermore, it is possible to confidently identify 7,9 isomers from observation of the chemical shift trends.

Peaks from the 6,8 isomers are more difficult to identify because they are generally present only at very low levels and can occur in regions of the spectrum where peaks from non-CLA fatty acids fall. Reporter resonances for the 12E,14Z and 12Z,14E isomers are also difficult to identify because the end group effect of the Me is expected to have a significant effect on the chemical shifts of the olefin carbons as they approach it. Chemical shift data for C-3 in 2-heptene and 2-hexene can be used to make some tentative assignments for these isomers because C-3 is olefinic and separated from the Me group by 3 and 2 methylene units, respectively, which is analogous to the environment of the fourth olefinic carbon (closest to the Me group; OL-4) in 11,13 and 12,14 CLA isomers, respectively. The C-3 resonance is shielded by 0.2 ppm on going from (E)-2-heptene to (E)-2-hexene (0.1 ppm for the Z isomers) (4), and therefore a shift increment of similar magnitude is expected for OL-4 on going from an 11,13 to 12,14 isomer. This allows the peak from the olefinic carbon OL-4 of 12Z,14E to be tentatively assigned as shown in Figure 12.5. The other 12,14 peak in Figure 12.5 (OL-1 from 12E,14Z) is assigned by a process of elimination (it cannot be from 6,8 or other isomers with the olefins closer to the carboxyl group).

A set of olefinic carbon chemical shift increments diagnostic of migration of the conjugated system up and down the fatty acid chain can be calculated by using the unequivocally identified reporter resonances for EZ and ZE isomers. Making the assumption that these increments are the same regardless of the double bond geometries, it is possible to predict the peak positions of the 7,9 to 11,13 ZZ and EE isomers because the olefinic ^{13}C resonances of the 9Z,11Z and 9E,11E isomers are known from the standard samples (Table 12.1). Once again, observed peaks coincide almost precisely with the predicted positions. Tentative assignments for reporter resonances of the 12E,14E and 12Z,14Z isomers were obtained in the same way. Selected reporter resonances for the EE and ZZ isomers from the ^{13}C NMR spectra of two commercially available synthetic CLA are shown in Figure 12.6.

A complete set of assignments for the olefinic carbons of all the isomers of 7,9 to 11,13 octadecadienoic acids is given in Table 12.5. These data were obtained on samples measured at 10% w/w concentration (data in Tables 12.1 and 12.4 were obtained at 2.5% w/w). Although peak positions were found to show variations at the two different concentrations, the chemical shift increments were almost invariant, and therefore assignments at the higher concentration could be deduced easily by running one sample at both concentrations and comparing the two spectra.

As stated above, many of the assignments in Table 12.5 were made by assuming that the chemical shift increments calculated for migration of the conjugated double

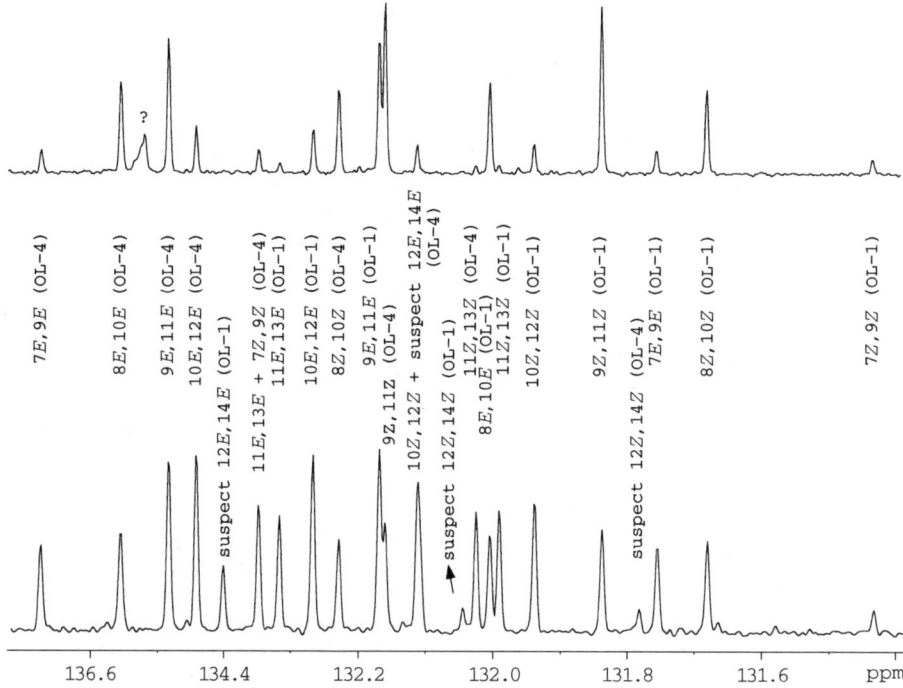

Fig. 12.6. Selected region of the ^{13}C-{^{1}H} nuclear magnetic resonance (NMR) spectra of two commercially available samples of synthetically prepared conjugated linoleic acid (CLA) showing reporter resonances for *EE* and *ZZ* isomers. Spectra were measured at 100.61 MHz (^{13}C frequency) on samples containing 10% w/w CLA in CDCl$_3$. OL-1 refers to the olefinic carbon closest to the carboxyl group, whereas OL-4 is the olefinic carbon closest to the Me group.

bonds within the chain are the same regardless of the geometry of the conjugated system. Figure 12.7 illustrates that this is true. Further confirmation that the assignments in Table 12.5 are correct comes from the opposite case, i.e., consistent increments are calculated for *EZ→ZE* and *EE→ZZ* isomerization regardless of the position of the conjugated system in the chain (at least for the 7,9 to 11,13 compounds); see Table 12.6. Therefore, the data in Table 12.5 are completely self-consistent and the peak assignments are secure.

Quantification of CLA Isomers

In the previous section, we described how NMR can be used to confidently identify a large range of CLA isomers, both positional and geometric, in complex fatty acid mixtures. Under ideal conditions, the integral of the NMR signal is proportional to the concentration of the species producing it; therefore, to quantify the different CLA isomers in a sample, reporter peaks from the ^{13}C-{^{1}H} NMR spectrum are integrated. In

TABLE 12.5 Chemical Shifts for the Olefinic Carbon Resonances of Elected Conjugated Linoleic Acid (CLA) Isomers[a,b]

(a) ZE isomers					
Olefinic Carbon	7Z,9E	8Z,10E	9Z,11E	10Z,12E	11Z,13E
OL-1	129.439	129.690	129.849	129.946	129.994
OL-2	128.988	128.865	128.780	128.728	128.697
OL-3	125.532	125.587	125.619	125.649	125.683
OL-4	134.960	134.833	134.756	134.707	134.616
(b) EZ isomers	7E,9Z	8E,10Z	9E,11Z	10E,12Z	11E,13Z
OL-1	134.034	134.281	134.439	134.534	134.585
OL-2	125.979	125.861	125.784	125.728	125.698
OL-3	128.528	128.583	128.618	128.642	128.673
OL-4	—[c]	130.225	130.155	130.115	130.024
(c) ZZ isomers	7Z,9Z	8Z,10Z	9Z,11Z	10Z,12Z	11Z,13Z
OL-1	131.433	131.680	131.838	131.939	131.991
OL-2	123.955	123.837	123.752	123.693	123.661
OL-3	123.490	123.546	123.580	123.609	123.641
OL-4	132.348[d]	132.229	132.161	132.112	132.025
(d) EE isomers	7E,9E	8E,10E	9E,11E	10E,12E	11E,13E
OL-1	131.755	132.005	132.169	132.267	132.316
OL-2	130.720	130.594	130.505	130.452	130.421
OL-3	130.257	130.307	130.341	130.367	130.403
OL-4	132.674	132.555	132.483	132.442	132.348

[a] All data listed were obtained from samples at 10% w/w in $CDCl_3$ at 300 K, measured in units of ppm and referenced against internal TMS. The olefinic carbon closest to the carboxyl group is labeled OL-1 and the one closest to the Me group is labeled OL-4 and so on.
[b] Source: reprinted from Ref. 13 with permission of Elsevier Science.
[c] Peak not assignable due to interfering signals from other fatty acids present in the sample.
[d] Coincident with OL-4 from 11E,13E.

proton-decoupled ^{13}C NMR experiments, it is necessary to employ inverse gated decoupling and relaxation delays of >5T_1 to obtain strictly quantitative data (8); unfortunately, these conditions lead to a reduction in the signal-to-noise ratio of the resulting spectrum, thus lowering detection limits; they are therefore not practical for analysis of CLA mixtures. However, the use of ^{13}C-{^1H} conditions that are not strictly quantitative does not lead to significant error in the measurements because the CLA are structurally very similar and only olefin signals are quantified. Therefore, factors potentially leading to imperfect quantification [relaxation times, nuclear Overhauser enhancements (NOE)] are expected to be almost identical for each peak, giving the correct *relative* amount for each CLA regardless of the precise experimental conditions.

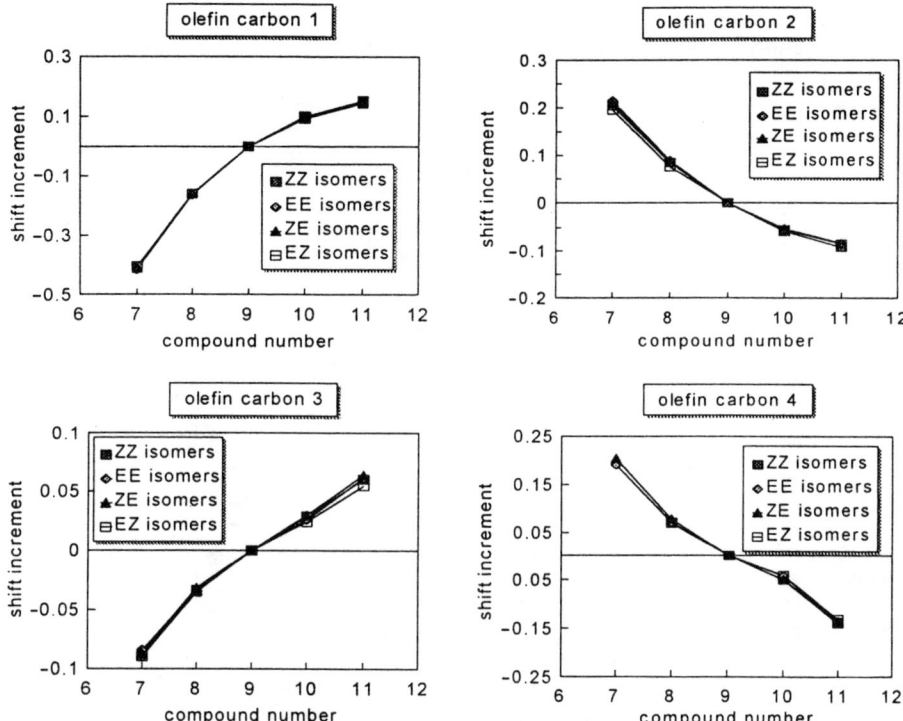

Fig. 12.7. Olefinic carbon chemical shift increments observed for migration of the conjugated double bond system along the fatty acid chain for *EZ, ZE, EE* and *ZZ* isomers of octadecadienoic acids. Compound number n refers to the n, n+2 isomer. All increments are given relative to the ^{13}C shifts of the 9,11 isomer. Olefinic carbon 1 is the closest to the carboxyl group. *Source:* Reprinted from Ref. 13 with permission of Elsevier Science.

It should be noted, however, that this may not apply to other resonances (e.g., from non-olefinic peaks, peaks from other species and so on).

Quantitative NMR data for a complex CLA mixture and the corresponding results obtained by GC-MS (Christie, W.W., personal communication) are given in Table 12.7. A comparison of these data shows that the two techniques agree very well with regard to the relative proportions of 7,9 to 11,13 isomers present. The GC-MS technique has the advantage over ^{13}C NMR in that it allows measurement of positional isomers outside of the range in which confident NMR assignments are currently available (e.g., 6,8). However, the NMR method has clear advantages over GC-MS, i.e., it gives information about the quantities of the different geometrical isomers that are present in the mixture, as well as requiring no sample derivatization.

^{13}C NMR spectroscopy can give unequivocal information about the identity and quantity of isomers (double bond positions and geometries) in complex CLA mixtures.

TABLE 12.6 Comparison of the Consistent Differences Observed in Olefinic ^{13}C Chemical Shifts Among Selected Conjugated Linoleic Acid (CLA) Configurational Isomers[a,b]

Positional isomer	^{13}C Chemical shift increment				
	7,9	8,10	9,11	10,12	11,13
Olefinic carbon		ZE to EZ interconversion			
OL-1	4.60	4.59	4.59	4.59	4.59
OL-2	–3.01	–3.00	–3.00	–3.00	–3.00
OL-3	3.00	3.00	3.00	2.99	2.99
OL-4	—[c]	–4.61	–4.60	–4.59	–4.59
		ZZ to EE interconversion			
OL-1	0.32	0.33	0.33	0.33	0.33
OL-2	6.77	6.76	6.75	6.76	6.76
OL-3	6.77	6.76	6.76	6.76	6.76
OL-4	0.33	0.33	0.32	0.33	0.32

[a]All data are in units of ppm and calculated from the data listed in Table 12.5. The olefinic carbon closest to the carboxyl group is labeled OL-1 and the one closest to the Me group is labeled OL-4, and so on.
[b]*Source:* reprinted from Ref. 13 with permission of Elsevier Science.
[c]Not measurable.

TABLE 12.7 Comparison of Analytical Data Obtained by GC-MS and NMR for a CLA Mixture (Expressed as Percentage of Total CLA Content)[a,b]

Positional isomer	GC-MS	NMR				
			Geometric isomer			
	Total	EZ	ZE	EE	ZZ	Total
7, 9	5.2	1.9	1.3	2.5	0.6	6.3
8, 10	18.1	10.4	1.9	2.7	2.3	17.3
9, 11	20.8	4.6	10.2	4.2	2.7	21.7
10, 12	27.6	13.1	4.6	4.4	3.4	25.5
11, 13	20.2	2.5	13.5	2.8	2.8	21.6
Other	8.1					7.8

[a]NMR data measured at a concentration of 10% w/w in CDCl$_3$ at 300 K. Spectra were acquired without gated decoupling using a 55° pulse and a 4.5-s recycle time (2.5-s acquisition time and 2-s relaxation delay). These conditions are not optimized for signal-to-noise ratio. A digital resolution of 0.2 Hz is necessary to resolve the NMR peaks adequately. GC-MS data (Christie, W.W., personal communication) measured by the published method (3).
[b]Abbreviations: GC-MS, gas chromatography-mass spectrometry; NMR, nuclear magnetic resonance; CLA, conjugated linoleic acid.

The experiment can be run overnight on a moderate field NMR system (400 MHz ^{1}H frequency) and information is acquired in "one-shot" with no sample derivatization required. Clearly, NMR spectroscopy has the potential to be an extremely powerful analytical tool in the field of CLA research.

References

1. Kepler, C.R., Hirons, K.P., McNeill, J.J., and Tove, S.B. (1966) Intermediates and Products of the Biohydrogenation of Linoleic Acid by *Butyrivibrio fibrisolvens, J. Biol. Chem. 241,* 1350–1354.
2. Ha, Y.L., Grimm, N.K., and Pariza, M.W. (1987) Anticarcinogens from Fried Ground Beef: Heat-Altered Derivatives of Linoleic Acid, *Carcinogenesis 8,* 1881–1887.
3. Christie, W.W., Dobson, G., and Gunstone, F.D. (1997) Isomers in Commercial Samples of Conjugated Linoleic Acid, *Lipids 32,* 1231.
4. Breitmaier, E., and Voelter, W. (1989) *Carbon-13 NMR Spectroscopy, High Resolution Methods and Applications in Organic Chemistry and Biochemistry,* 3rd ed., VCH Verlagsgesellschaft mbH, Weinheim.
5. Croasmun, R., and Carlson M.K., (eds.) (1994) *Two Dimensional NMR Spectroscopy, Applications for Chemists and Biochemists,* 2nd edn., VCH Publishers Inc., New York.
6. Sanders, J.K.M., and Hunter, B.K. (1987) *Modern NMR Spectroscopy A Guide for Chemists,* Oxford University Press, Oxford.
7. Neuhaus, D., and Williamson M.P. (1989) *The Nuclear Overhauser Effect in Structural and Conformational Analysis,* VCH Publishers Inc., New York.
8. Harris, R.K. (1983) *Nuclear Magnetic Resonance Spectroscopy A Physicochemical View,* pp. 78 and 111–113, Pitman, London.
9. Bax, A., and Davis, D.G. (1985) MLEV-17-Based Two-Dimensional Homonuclear Magnetization Transfer Spectroscopy, *J. Magn. Reson. 65,* 355–360.
10. Marion, D., and Wüthrich, K. (1983) Application of Phase Sensitive Two-Dimensional Correlated Spectroscopy (COSY) for Measurements of Proton-Proton Spin-Spin Coupling Constants in Proteins, *Biochem. Biophys. Res. Commun. 113,* 967–974.
11. Lerner, L., and Bax, A. (1987) Application of New, High-Sensitivity Proton and Carbon-13 NMR Spectral Techniques to the Study of Oligosaccharides, *Carbohydr. Res. 166,* 35–46.
12. Bax, A., and Marion, D. (1988) Improved Resolution and Sensitivity in ^{1}H-Detected Heteronuclear Multiple-Bond Correlation Spectroscopy, *J. Magn. Reson. 78,* 186–191.
13. Davis, A.L., McNeill, G.P., and Caswell, D.C. (1999)Analysis of Conjugated Linoleic Acid Isomers by ^{13}C NMR Spectroscopy, *Chem. Phys. Lipids, 97,* 155–165.
14. Lie Ken Jie, Marcel S.F., Pasha, M.K., and Alam, M.S. (1997) Synthesis and Nuclear Magnetic Resonance Properties of All Geometrical Isomers of Conjugated Linoleic Acids, *Lipids 32,* 1041–1044.
15. Gunstone, F.D., Pollard, M.R., Scrimgeour, C.M., and Vedanayagam, H.S. (1977) Fatty Acids Part 50. Carbon-13 Nuclear Magnetic Resonance Studies of Olefinic Fatty Acids and Esters, *Chem. Phys. Lipids 18,* 115–129.

Chapter 13

Biosynthesis of Conjugated Linoleic Acid and Its Incorporation into Meat and Milk in Ruminants

J. Mikko Griinari[a] and Dale E. Bauman[b]

[a]Valio Ltd., FIN-00039, Helsinki, Finland
[b]Department of Animal Science, Cornell University, Ithaca, NY 14853-4801

Introduction

Meat and milk from ruminant animals represent the major dietary sources of conjugated linoleic acid (CLA). Because potential health benefits have been associated with dietary consumption of CLA, enhancement of CLA concentrations in meat and milk has become an important objective in animal nutrition research. Dietary intake of CLA has decreased over recent years in many Western societies as ruminant fats have been replaced by plant lipids. In addition, decreases in total energy intake from dietary fat have occurred mainly in the portion of saturated animal fats as reviewed by Pietinen *et al.* (1).

Parodi (2) first established *cis*-9, *trans*-11 octadecadienoic acid as the major CLA component in bovine milk fat. This same CLA isomer had been identified previously as the first intermediate of linoleic acid biohydrogenation in the rumen (3). It is generally accepted that CLA in ruminant meat and milk originates from incomplete biohydrogenation of linoleic acid in the rumen (4,5). However, in accepting this simple explanation for the presence of CLA in ruminant lipids, we have failed to recognize several important features in the ruminal biohydrogenation of unsaturated fatty acids. These suggest that the amount of CLA escaping rumen biohydrogenation and being absorbed may be inadequate to account for CLA levels in milk and body fat.

Milk fat concentrations of CLA are known to vary among herds and individuals, and between seasons (2,4,6,88). Underlying factors resulting in this variation are related predominantly to diet as reviewed by Jahreis *et al.* (Chapter 16). The relative contribution of ruminal production vs. tissue synthesis of CLA is not known. Understanding the mechanisms of CLA biosynthesis will allow us to design diets and improve feeding strategies that will enhance CLA concentrations in milk fat.

CLA Synthesis in the Rumen

Ruminal Biohydrogenation

Hydrogenation of dietary unsaturated fatty acids and formation of *trans* fatty acids in rumen contents were first demonstrated by Reiser (8) and Shorland *et al.* (9). Subsequent work using radiolabeled linoleic and linolenic acids (10–12) confirmed the formation of "a bewildering number of radioactive C_{18} acids with various degrees of unsaturation and positional isomerization" (13). Shorland *et al.* (14) observed that conjugated dienoic acids accumulated when linoleic acid was incubated with rumen contents, but none were formed from linolenic acid. Bartlet and Chapman (15) found a constant relationship between *trans*-octadecenoic acids and conjugated unsaturation in a large number of butter samples as determined by differential infrared spectroscopy. On the basis of this constant ratio, they suggested a sequence of reactions that explained the biohydrogenation of linoleic acid.

Experiments with radiolabeled or pure fatty acids *in vivo* and *in vitro* allowed identification of the major pathways of ruminal biohydrogenation. The predominant pathways of linoleic and α- and γ-linolenic acids are presented in Figure 13.1. The sequence of linoleic acid biohydrogenation to stearic acid involves at least two steps and at least two distinct populations of ruminal bacteria (reviewed in Ref. 16). The sequence begins with isomerization of linoleic acid to *cis*-9, *trans*-11 octadecadienoic acid and is followed by hydrogenation of the *cis*-double bond of the conjugated diene to a *trans*-monoenoic acid (3). Similarly, the conjugated *cis*-double bonds of α- and γ-linolenic acids are hydrogenated after initial isomerization of the *cis*-12 double bond.

Isomerization of the *cis*-12 double bond is considered a prerequisite step in the biohydrogenation of fatty acids containing a *cis*-9, *cis*-12 double bond system (17). *trans*-11 Octadecenoic acid is a common intermediate in the biohydrogenation of α- and γ-linolenic and linoleic acids. Its reduction appears to be rate limiting in the complete biohydrogenation sequence of unsaturated C_{18} fatty acids. Therefore, this penultimate biohydrogenation intermediate accumulates in the rumen (18).

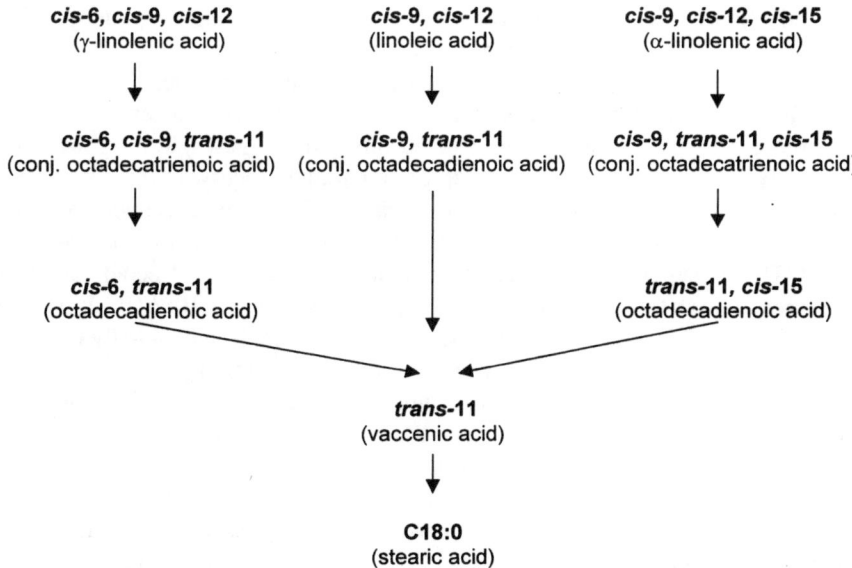

Fig. 13.1. Predominant pathways of biohydrogenation of unsaturated C_{18} fatty acids. *Source:* Ref. 16.

Cis/trans Isomerization

A diverse range of rumen bacterial species have been isolated; they demonstrate a capacity to isomerize cis-double bonds of unsaturated fatty acids to form conjugated cis/trans double bond systems and further hydrogenate these conjugated acids (16). However, the enzyme, linoleate isomerase (EC 5.2.1.5), responsible for the conjugation of the cis-9, cis-12 double bond structure of linoleic as well as α- and γ-linolenic acids, has been studied in some detail only in a limited number of bacterial species (19,20). Linoleate isomerase is a particulate enzyme bound to the bacterial cell membrane. Partial purification and description of its kinetic properties were performed by Kepler and Tove (19). The enzyme demonstrates an absolute substrate requirement for a cis-9, cis-12 diene system and a free carboxyl group (21).

The simple pathways of ruminal biohydrogenation presented in Figure 13.1 do not account for the array of trans-octadecenoic acids that is found in rumen digesta and tissue lipids. Analyses of rumen digesta and milk fat have demonstrated the presence of trans-octadecenoic acid isomers in which the double bond is located in every position from $\Delta 4$ to $\Delta 16$ (22–25). The formation of these positional trans-isomers of octadecenoic acid in enzymatic biohydrogenation by anaerobic ruminal bacteria has been suggested to result from double bond migration (7,13,14). When radioactive linolenic acid was incubated with rumen contents in an artificial rumen, a large number of positional trans-isomers of dienoic and monoenoic fatty acids were detected (10). The authors suggested that the hydrogenation of dienoic acid to a monoenoic acid was apparently accompanied by considerable double bond migration. This suggestion was based on parallels with the metal-catalyzed process in the manufacture of partially hydrogenated oils in which extensive double bond migration occurs. However, to our knowledge, the extent to which double bonds migrate in enzymatic biohydrogenation in the rumen has not been investigated in any detail. The consistent profiles and distinct differences in trans-octadecenoic acid isomers in ruminant lipids compared with partially hydrogenated oils (26) suggest that enzymatic and metal-catalyzed hydrogenation do not represent parallel processes.

An alternative explanation for the wide range of trans-octadecenoic acid isomers in the rumen is that ruminal bacteria possess several specific cis,trans isomerases. This is supported by recent identification of conjugated octadecadienoic acid isomers in milk fat with double bonds in positions $\Delta 7,9$; $\Delta 8,10$; $\Delta 10,12$; $\Delta 11,13$; and $\Delta 12,14$ (27–30). These conjugated octadecadienoic acids undoubtedly originate from the rumen, and their further reduction in the rumen would produce a range of trans-monoenes. In addition, the rates of hydrogenation of trans-octadecenoic acids to stearic acid by ruminal bacteria also vary. Rates were greatest for trans-octadecenoic acids with double bonds in positions $\Delta 8$ through $\Delta 10$, and lowest for positions $\Delta 5$ through $\Delta 7$ and $\Delta 11$ through $\Delta 13$ (17). Differences in the rates of hydrogenation of specific trans-octadecenoic acid isomers would further enhance concentrations of some of the more unusual isomerase products found in rumen digesta and milk fat.

Evidence regarding specific isomerase activity has also been obtained in studies with pure cultures of rumen bacteria. These bacteria isomerize cis-double bonds with or without associated double bond migration. *Trans*-isomerization of the *cis*-9 double bond in the conjugated 9,11 double bond system was suggested by Fujimoto *et al.* (31). Incubation of linoleic acid with *Butyrivibrio fibrisolvens* and *Selenomonas ruminantium* isolated from sheep produced both *trans*-9, *trans*-11 octadecadienoic acid and *trans*-9 octadecenoic acid. A *Propionibacterium acnes* strain isolated from the cecum of a mouse was shown to isomerize the *cis*-9 double bond of linoleic acid and γ-linolenic acid to a *trans*-10 position (32). The first evidence that an isolated *cis*-double bond in octadecenoic acid could be isomerized by a bacterial enzyme was provided by Mortimer and Niehaus (33). In their study, a soluble enzyme preparation from a *Pseudomonad* was shown to convert oleic acid to *trans*-10 octadecenoic acid. When bacteria isolated from a defaunated sheep were grown in culture with linolenic acid, octadecenoic acids accumulated (34). The predominant isomer was *cis*-15 (83%) and additional isomers were *trans*-15 (12%), *trans*-14 (4%), and *trans*-13 (1%). The mere existence of these isomers suggests that the rumen is populated by bacteria possessing several different specific isomerases.

The relatively constant profiles of *trans*-octadecenoic acids found in ruminant meat and milk fat in which *trans*-11 octadecenoic acid is the predominant isomer (35) suggest a relatively stable rumen bacteria population and associated enzyme activities involving fatty acid biohydrogenation. However, recent studies have demonstrated that an altered rumen environment induced by feeding low-fiber diets is associated with a change in the *trans*-octadecenoic acid profile of milk fat (22). In this situation, *trans*-10 octadecenoic acid became the predominant *trans*-octadecenoic acid isomer in milk fat. Production of *trans*-10 octadecenoic acid would presumably involve a specific *cis*-9, *trans*-10 isomerase in rumen bacteria with the formation of a *trans*-10, *cis*-12 conjugated double bond structure as the first reaction. Putative pathways in the ruminal biohydrogenation involving *cis*-9, *trans*-10 isomerization are presented in Figure 13.2. Although *B. fibrisolvens* is known to express only *cis*-12, *trans*-11 isomerase, it can hydrogenate *trans*-10, *cis*-12 octadecadienoic acid and thus produce *trans*-10 octadecenoic acid in pure culture (3).

Further evidence in support of a specific bacterial *cis*-9, *trans*-10 isomerase is provided by observations that low-fiber diets increase the proportion of *trans*-10, *cis*-12 CLA isomer in the rumen digesta (Griinari, J.M., and Bauman, D.E., unpublished data) and in milk fat from cows fed low-fiber diets (22).The isomer, *trans*-10, *cis*-12 CLA, has also been observed as one of three major CLA isomers in digesta obtained from continuous flow-through ruminal fermenters (36). Changes in ruminal biohydrogenation, characterized by increased *cis*-9, *trans*-10 isomerization, were associated with a dramatic reduction in the rate of milk fat synthesis, and a role for *trans*-10 octadecenoic acid and/or *trans*-10, *cis*-12 CLA as specific inhibitors of milk fat synthesis was proposed (22). Consistent with this, milk fat synthesis was reduced when mixtures of CLA isomers were infused postruminally into dairy cows (37,38). We have recently demonstrated that the *trans*-10, *cis*-12 CLA isomer causes milk fat depression, whereas the *cis*-9, *trans*-11 CLA isomer does not (Baumgard, L.H., and Bauman, D.E., unpublished data).

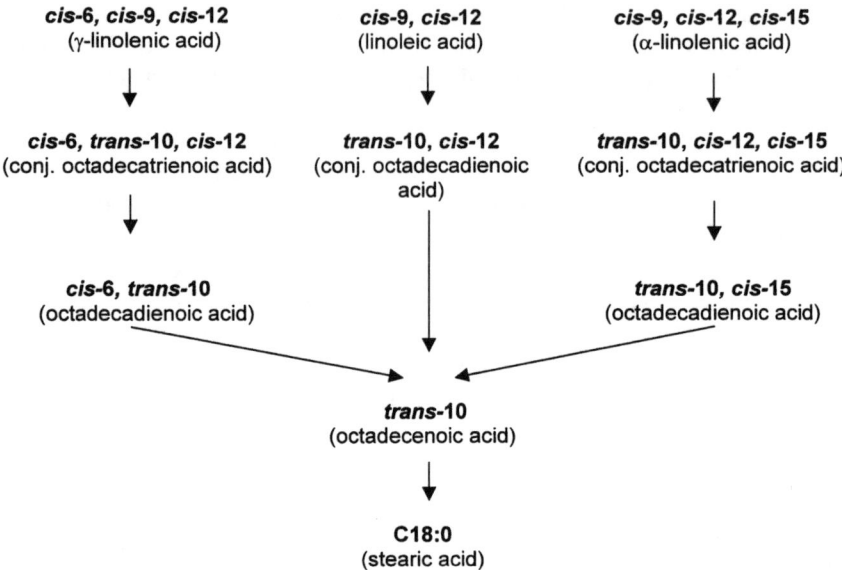

Fig. 13.2. Proposed pathways of ruminal biohydrogenation of unsaturated C_{18} fatty acids involving *cis*-9, *trans*-10 isomerase.

When low-fiber diets are fed to ruminants, the rumen environment is characterized by a decrease in pH. This results in a shift in the bacterial population and related changes in the pattern of fermentation end products such as an increase in the rate of propionate production (39, 40). Leat *et al.* (41) provided compelling evidence that the change in rumen bacteria populations that occurs when low-fiber diets are fed is associated with modifications in ruminal biohydrogenation consistent with the altered profile of *trans*-octadecenoic acids in ruminal digesta and tissue lipids. Gnotobiotic lambs, maintained on low-fiber diets, were inoculated with seven species of rumen bacteria including *B. fibrisolvens* or fresh rumen contents obtained from a conventional sheep containing a spectrum of rumen bacteria. Lambs inoculated with the specific mixture of bacteria species produced significant amounts of *trans*-11 octadecenoic acid in the rumen, but no other *trans*-octadecenoic acid isomers. In contrast, *trans*-10 became the predominant octadecenoic acid isomer in lambs inoculated with fresh rumen contents and only small amounts of *trans*-9 and *trans*-11 isomers were produced (41). The effect of a low-fiber diet on the rumen environment and the selective pressure it creates were clearly manifested among the complex mixture of bacteria contained in the inoculant of rumen contents.

Ruminal Origin of Tissue and Milk Fat CLA

Viviani (5) suggested that the presence of conjugated dienoic acids in milk indicated their formation in the rumen. The first premise for the ruminal origin of tissue and milk fat CLA is that CLA is an intermediate of linoleic acid biohydrogenation (Fig. 13.1). However, this does not explain the increased CLA concentrations observed when cows are fed diets that are low in linoleic acid, e.g., pasture feeding or fish oil supplements (6,42). The second premise is that a portion of CLA formed in the rumen escapes further biohydrogenation and is absorbed from the digestive tract. The extent to which CLA escapes ruminal biohydrogenation has not been quantified, but earlier studies suggest it may be minimal. Investigations of biohydrogenation with rumen bacterial cultures revealed that *cis*-9, *trans*-11 CLA was a transient intermediate in the process, whereas *trans*-11 octadecenoic acid accumulated (see review in Ref. 16). *In vitro* studies using labeled linoleic acid with rumen contents demonstrated isomerization of linoleic acid and a rapid conversion of CLA to *trans*-11 octadecenoic acid. Hydrogenation of *trans*-11 octadecenoic acid occurred less rapidly and therefore the acid accumulated (43,44). Similar results were obtained in time course studies of linoleic and linolenic acid biohydrogenation (45,46).

When linolenic acid is hydrogenated in the rumen, the predominant initial intermediate is conjugated octadecatrienoic acid, *cis*-9, *trans*-11, *cis*-15 (Fig. 13.1). The pattern of conjugated triene formation *in vitro* is similar to the formation of the predominant conjugated diene, *cis*-9, *trans*-11 octadecadienoic acid from linoleic acid hydrogenation (46). Again, if the primary source of milk fat conjugated fatty acids is from ruminal escape of these biohydrogenation intermediates, higher levels of conjugated trienes should occur in milk fat when diets rich in linolenic acid are fed. However, this is not observed. Banni *et al.* (47) found that lactating sheep fed only pasture, which typically contains linolenic acid as the predominant fatty acid, produced a high level of *cis*-9, *trans*-11 octadecadienoic acid in milk, whereas conjugated octadecatrienoic acids were detected only at trace levels. Furthermore, the ratio of *cis*-9, *trans*-11 CLA to *trans*-11 was very low (1:40) in rumen digesta from pasture-fed cows compared with a relatively high ratio (1:3) in milk fat (Griinari, M., Nurmela, K., Korpela, R., and Sairanen, A., unpublished data). The ratio found in milk fat was within the range previously reported (48–50). Overall, these data suggest that a relatively small portion of CLA formed in the rumen escapes further ruminal biohydrogenation and is subsequently absorbed from the digestive tract. Thus, a major portion of the CLA in milk and tissue lipids must originate from another source.

Tissue Synthesis of CLA

The levels of CLA in meat and milk from ruminants compared with nonruminants (51) suggest a close association between rumen function and CLA levels in tissue and milk fat. However, as discussed earlier, the availability of CLA originating from the rumen appears inadequate to account for tissue and milk fat levels. Therefore, CLA must also be synthesized in tissues from a precursor of rumen origin.

Desaturase Hypothesis

The relatively constant ratio of *trans*-11 octadecenoic acid and CLA found in milk fat across a wide range of diets (15,48–50) led us to speculate that *trans*-11 octadecenoic acid is the rumen-derived precursor of milk fat CLA. We proposed that CLA could be produced endogenously from *trans*-11 octadecenoic acid in tissues by Δ^9-desaturase (52,79). This hypothesis is consistent with *in vitro* studies, demonstrating that *trans*-11 octadecenoic acid is converted to CLA by microsomal preparations of rat liver (53,54) and the presence of Δ^9-desaturase activity in bovine mammary tissue (55). According to our hypothesis, *trans*-11 octadecenoic acid accumulates in the rumen and a portion escapes further biohydrogenation. *Trans*-11 octadecenoic acid is efficiently absorbed from the digestive tract (56) and utilized by different tissues, including the mammary gland (57). In tissues, a portion of the *trans*-11 octadecenoic acid is desaturated to CLA and incorporated into tissue and milk lipids (Fig. 13.3).

Our hypothesis of *trans*-11 octadecenoic acid conversion to CLA was tested in lactating dairy cows using a postruminal infusion protocol. The study consisted of a 3-d baseline period during which only the vehicle (skim milk) was infused, a 3-d test period with infusion of 25 g/d of a mixture of *trans*-11 and *trans*-12 octadecenoic acids (1:1) emulsified in skim milk, and a 5-d washout period during which only vehicle was infused (Griinari, J.M., Nurmela, K.V.V., Corl, B.A, and Bauman, D.E., unpublished data). Results demonstrated that tissue conversion of *trans*-11 octadecenoic acid to CLA is an active pathway in lactating ruminants. Both *trans*-11 and *trans*-12 octadecenoic acids increased in milk fat, and concentrations of related desaturase products, *cis*-9, *trans*-11 and *cis*-9, *trans*-12 octadecadienoic acids, also increased (Fig. 13.4). Desaturation of octadecenoic acid with a *trans*-11 double bond was greater than desaturation of the *trans*-12 double bond, suggesting some substrate specificity. This finding is consistent with the observed *in vitro* desaturation and incorporation into triacylglycerols by liver microsomal preparations expressing Δ^9-desaturase activity (54,58),

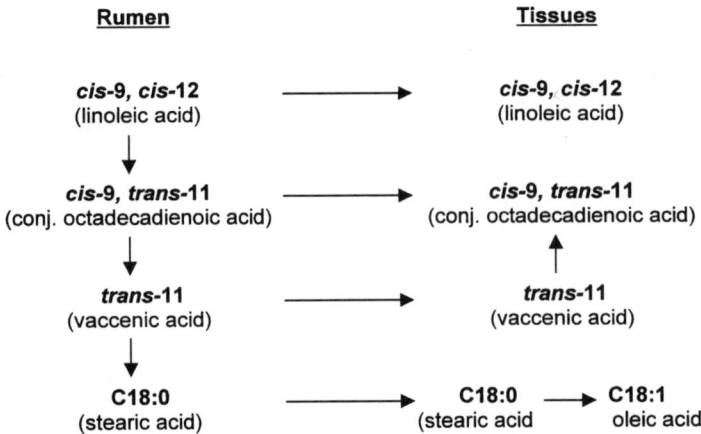

Fig. 13.3. Pathways of conjugated linoleic acid (CLA) biosynthesis.

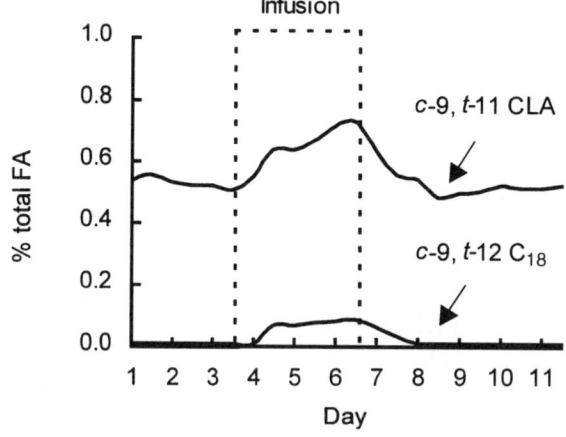

Fig. 13.4. Enrichment of Δ^9-desaturase products in milk fat during a 3-d postruminal infusion of a mixture of trans-11 + trans-12 octadecenoic acid isomers (25 g/d). Abbreviations: FA, fatty acids; CLA, conjugated linoleic acid. Source: Griinari, J.M., Nurmela, K.V.V., Corl, B.A., and Bauman, D.E., unpublished data.

but in contrast with the rate of desaturation in another similar system (53). The presence of other cis-9, trans-n octadecadienoic acids in milk fat also supports the role of an active Δ^9-desaturase. Recently, Yurawecz et al. (30) identified a new CLA isomer, trans-7, cis-9 octadecadienoic acid, and Ulberth and Henninger (59) have identified cis-9, trans-13 octadecadienoic acid in milk fat. By analogy, the presence of several other cis-9, trans-n octadecadienoic isomers in milk fat can be predicted. Overall, it is clear that CLA synthesis via Δ^9-desaturase occurs, but the quantitative significance of endogenous CLA synthesis relative to incorporation of preformed CLA has yet to be determined.

Δ^9-Desaturase and trans-11 Substrate

The desaturase system is a multienzyme system that includes NADH-cytochrome b_5 reductase, cytochrome b_5, acyl-CoA synthase, and the terminal Δ^9-desaturase (60). The Δ^9- desaturase reaction introduces a cis-double bond between carbons 9 and 10, and Δ^9-desaturase is the key enzyme in the desaturase system that varies in amount. Stearoyl-CoA and palmitoyl-CoA are the major substrates for Δ^9-desaturase, resulting in the production of oleoyl-CoA and palmitoleoyl-CoA, respectively. However, a wide range of saturated and unsaturated acyl CoA can serve as substrates (61), including trans-11 octadecenoic acid (53,54). Overall, Δ^9-desaturase affects the fatty acid composition of phospholipids and triglycerides. Effects on phospholipid composition are important in the maintenance of membrane fluidity, and alterations have been implicated in a variety of disease states (60,62). Similarly, the pattern of fatty acids in milk fat are thought to be of special importance in maintaining fluidity characteristics that allow for normal synthesis and secretion of milk fat (55,63).

Differences exist among species and tissues in the distribution of Δ^9-desaturase. In rodents, mRNA levels and activity are highest in the liver (60,62). In contrast, growing sheep and cattle have substantially greater Δ^9-desaturase in adipose tissue as indicated by mRNA abundance and enzyme activity (64–68). This suggests that adipose tissue is the primary site of trans-11 octadecenoic acid conversion to CLA in growing ruminants. In contrast, the mammary gland is apparently the primary site of endogenous synthesis of CLA during lactation. In lactating ruminants, the mammary tissue has a high activity of Δ^9-desaturase (55,69). Ward et al. (70) determined tissue-specific changes in mRNA abundance of Δ^9-desaturase in sheep for different physiologic states. They observed that there was a decrease in mRNA abundance for Δ^9- desaturase in adipose tissue and an increase in mammary tissue with the onset of lactation. In vivo results are also consistent with the lactating mammary gland being of greatest importance in endogenous synthesis of CLA during lactation. Bickerstaffe and Johnson (71) demonstrated that intravenous infusion of sterculic acid, a specific inhibitor of Δ^9-desaturase, resulted in a marked decrease in the oleic acid:stearic acid ratio in milk fat, but only minimal differences in plasma fatty acid composition in lactating goats. Because intravenously infused sterculic acid could pass through all organs, the authors concluded that the mammary gland must be the major site of desaturation for fatty acids found in milk fat.

The vast majority of research on the regulation of Δ^9-desaturase has utilized rats and focused on hepatic tissue. These investigations established that abundance of mRNA and enzyme activity for Δ^9-desaturase are responsive to changes in diet, hormone balance, physiologic state, and other activating and inhibiting factors (see reviews in Refs. 60,62). However, other than the aforementioned studies of Ward et al. (70), investigations of the regulation of Δ^9-desaturase have not been extended to ruminants. Differences in tissue levels of Δ^9-desaturase may help explain the substantial individual variation in milk fat content of CLA observed even when cows are presented with the same diet (4,6). For example, one cow consuming a diet containing sunflower oil produced milk fat in which CLA represented an impressive 5.5% of total milk fatty acids, a concentration that is >10-fold greater than typically observed (4). Dairy breed variations in milk fat CLA content may also be related to differences in Δ^9-desaturase activity (72–74). However, breed differences appear to be minor compared with dietary influences. Therefore, particular attention is required to control for diet effects in studies examining breed differences.

Trans-11 is the predominant trans-octadecenoic acid isomer formed as a result of ruminal biohydrogenation of C_{18} polyunsaturated fatty acids (PUFA) (23,56). It is also the only intermediate common to all C_{18} PUFA in the predominant pathways of ruminal biohydrogenation (Fig. 13.1). Among intermediates, trans-octadecenoic acids are also the most likely to accumulate in the rumen and escape biohydrogenation as discussed earlier. Dietary fish oil supplementation will increase ruminal formation of trans-octadecenoic acids (75,76), specifically trans-11 octadecenoic acids (77). Fish oils typically contain low levels of C_{18} PUFA and high levels of C_{20} and longer PUFA (e.g., eicosapentaenoic acid and docosahexaenoic acid); biohydrogenation of these long-chain PUFA is extensive (78). It is not known whether long-chain PUFA could serve as direct precursors of trans-11 octadecenoic acids in the rumen. However, trans-11 octadecenoic acid is produced in the rumen when cows are fed a wide range of diets. When this occurs in conjunction with elevated CLA levels in milk, it becomes clear that ruminal formation of trans-11 octadecenoic acid could in large part explain dietary responses in milk fat CLA.

The trans-11 to cis-9, trans-11 Ratio in Milk Fat

A close linear relationship between milk fat *trans*-octadecenoic acids and conjugated dienoic fatty acids was first observed in Canadian butter samples on the basis of differential infrared spectroscopy (15). The authors attributed the constant ratio of these fatty acids to their common source as intermediates of ruminal biohydrogenation. Butter samples (n = 300) examined in this study demonstrated a remarkably constant relationship between milk fat *trans*-octadecenoic acids and conjugated fatty acids over a wide range of concentrations. We have proposed that this close relationship suggests a precursor-product association in which *trans*-octadecenoic acid is the precursor and CLA is the product (79).

Examination of the relationship between *trans*-11 octadecenoic acid and CLA in milk fat and tissue lipids for different dietary situations could provide information on the role and function of Δ^9-desaturase in the biosynthesis of CLA. Typical results are presented in Figure 13.5. The correlation coefficient between *trans*-11 octadecenoic acid and CLA concentrations in milk fat is high, and the slope is within the range of published values (48–50). The linear relationship between milk fat *trans*-11 octadecenoic acid and CLA concentrations across a wide range of concentrations (Fig. 13.5) suggests a high capacity for Δ^9-desaturase. If it is assumed that incorporation of rumen-derived CLA in milk fat is minimal, then the ratio between *trans*-11 octadecenoic acid and CLA in milk fat reflects the extent to which *trans*-11 is desaturated to CLA. The slope of 0.5, similar to that in Figure 13.5, would suggest that roughly one third of the circulating *trans*-11 octadecenoic acid taken up by the mammary gland is desaturated by Δ^9- desaturase. This estimate is consistent with the extent to which preformed stearic acid is desaturated to oleic acid in the mammary gland. Estimates on stearic acid desaturation vary from 31% with *in vitro* studies (55) to 52% on the basis of *in vivo* measurements of mammary arterial-venous difference (80).

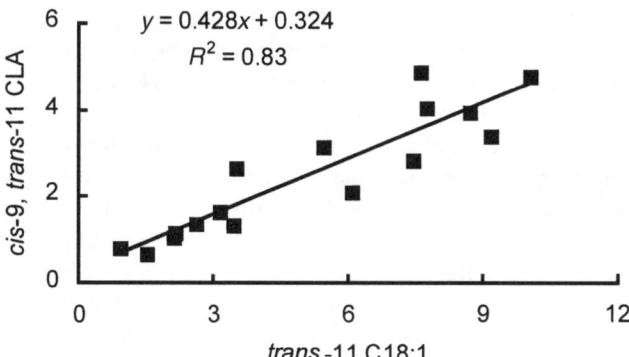

Fig. 13.5. Relationship between *trans*-11 octadecenoic acid and *cis*-9, *trans*-11 conjugated linoleic acid (CLA) (g/100 g of total fatty acids) in milk fat. *Source:* Griinari, J.M., and Bauman, D.E., unpublished data.

The linear relationship between *trans*-11 octadecenoic acid and CLA in milk fat is remarkably close over a wide range of dietary situations; however, the slope term varies from 0.25 to 0.57 (48–50). Although the quantitative significance of the endogenous conversion of *trans*-11 octadecenoic acid to CLA has not been assessed, variation in the slope term may reflect differences in Δ^9-desaturase activity. Overall, data demonstrate that *trans*-11 octadecenoic acid is the major predictor of milk fat CLA concentration and suggest that a substantial portion of CLA may be of endogenous origin. Recently, evidence suggests a role for Δ^9-desaturase in the biosynthesis of CLA in humans (81) and mice (see Chapter 14).

Factors That Affect CLA Content in Meat and Milk

The Effect of Dietary Factors

Many dietary conditions are known to affect CLA content in milk fat (see Chapter 16). In this section, we will focus on the relationship of observed variation in meat and milk fat concentrations of CLA to possible mechanisms described in the previous sections.

In contrast to studies on milk products, there are few data on factors affecting CLA content in ruminant meat. Limited reports suggest that CLA concentration in beef fat varies among countries (82–85). Australian beef had the highest values, with CLA representing 1% of total fatty acids; German values were intermediate, with CLA concentrations at an average of 0.65% of total fat; and U.S. beef had the lowest values, with CLA representing 0.3–0.5% of total fat. The CLA content of milk fat in these countries can be placed in the same relative order (50,51,82). Thus, there appear to be no major differences in enrichment of CLA in meat fat and milk fat.

Differences in CLA content of body fat in growing animals most likely reflect differences in type of diet. Lower concentrations of CLA in U.S. beef probably relate to the current practice of feeding low-fiber diets during the finishing period. Finishing diets, consisting mostly of grain and concentrates with minimal fiber, are fed over the period in which animals gain body condition and intramuscular fat. On the basis of studies with lactating cows (22), this type of diet would lead to a change in the rumen environment that would result in decreased formation of *trans*-11 octadecenoic in the rumen and, as a consequence, decreased CLA concentration in tissues. Consistent with this, Wood (86) found that U.S. beef fat contained *trans*-10 as the major octadecenoic acid isomer rather than *trans*-11 as in fat from European beef (35). In addition, the effect of dietary fiber on CLA content in meat of growing cattle was demonstrated in a study in which pasture feeding increased CLA content of body fat compared with that observed with traditional grain diets (85).

The *cis*-9, *trans*-11 CLA isomer typically represents ≥90% of the CLA isomers in milk fat, but its proportion in beef fat is only 60–85% (51,84). Again, the lower relative proportion of the *cis*-9, *trans*-11 isomer is probably related to a change in ruminal biohydrogenation induced by traditional high-concentrate/low-fiber finishing diets. An altered profile of CLA isomers, characterized by an elevated level of *trans*-10, *cis*-12 isomer, was found in the milk fat of lactating dairy cows fed a low-fiber diet (22). Recently, subcutaneous and intermuscular fat samples from German Simmental cattle were analyzed; the *cis*-9, *trans*-11 isomer represented >90% of total CLA (87). In this case, cattle were fed a corn silage-based diet with a moderate level of grain supplement. This type of diet should provide sufficient fiber to maintain normal rumen fermentation and typical biohydrogenation. Fritsche and Steinhart (83) observed a close linear relationship between the concentrations of *trans*-octadecenoic acid and CLA in beef fat, consistent with the relatively constant ratio found in milk fat. If we apply the same reasoning used in examining the *trans*-11 vs. *cis*-9, *trans*-11 ratio in milk fat in the previous section, these data suggest that endogenous desaturation of *trans*-11 octadecenoic acid could also be a major precursor for CLA in body fat.

Milk fat concentration of CLA can vary widely (2,4,6,88). Specific factors that cause this variation have not been investigated extensively, although it is clear that dietary factors are of major importance. Data from 13 studies reporting effects of specific dietary factors on CLA concentration in milk fat are listed in Table 13.1. In this summary, dietary factors are grouped into categories relative to the mechanism by which they may exert an effect. The first category includes dietary conditions that increase milk fat CLA by providing lipid substrate for formation of CLA or *trans*-11

TABLE 13.1 Summary of Dietary Factors That Affect CLA Concentrations in Milk Fat

Dietary factor	Effect on CLA content of milk fat	Reference
Lipid substrate		
Unsaturated vs. saturated fat	Increased by addition of unsaturated fat	22
Type of plant oil	Greatest with oils high in 18:2	4, 42
Level of plant oil	Dose-dependent increase	89
Ca salts of plant oils	Increased as with free oils	42
Fat in animal by-products	Minimal effect	42
High-oil plant feeds		
High-oil corn	Minimal effect	42
Soybeans	Heat processing will increase	42
Rapeseed vs. soybean	Similar effect	90
Modifiers of biohydrogenation		
Forage:concentrate ratio	Increased with high ratio	22
Nonstructural carbohydrate (NSC) level	Minor effect (possible oil × NSC interaction)	92
Restricted feeding	Increased with restricted feeding	49, 91
Fish oils	Greater increase than with plant oils	42
Monensin-ionophore	Variable effect	92, 93
Dietary buffers	Little effect	92
Combination		
Pasture vs. conserved forages	Higher on pasture	6, 48, 50, 91, 94
Growth stage of forage	Increased with less mature forage	92

octadecenoic acid in the rumen. The second category consists of dietary factors that have an effect on the rumen population of bacteria involved in the biohydrogenation of unsaturated fatty acids. The third category includes the effects of pasture feeding. These are not well understood, but may involve both lipid substrate and modification of bacterial populations associated with biohydrogenation.

Comparisons between different types of plant oils suggest that those high in linoleic acid increase CLA concentrations most effectively (Table 13.1). This is consistent with the major pathways of ruminal biohydrogenation of PUFA (Fig. 13.1). However, as discussed earlier, ruminal supply of *trans*-11 octadecenoic acid may be more important in modulating milk fat CLA levels than the ruminal contribution of CLA. Therefore, it is important to recognize that in both studies (4,42) the supply of *trans*-11 octadecenoic acid from the rumen may have been altered because oil feeding resulted in a depression in milk fat synthesis. Milk fat depression is associated with a relative increase in *trans*-10 octadecenoic acid content of milk fat and a decline in *trans*-11 content (22). This characteristic shift in the *trans*-octadecenoic acid isomer profile was also observed in samples of milk fat from the study by Kelly *et al.* (4) (Griinari, J.M., unpublished data). However, effects of different types of plant oils in accumulation of *trans*-11 octadecenoic acid in the rumen have not been described in conjunction with milk fat depression, and this complicates interpretation.

Adding graded levels of rapeseed oil to the diet of dairy cows results in a clear dose dependent increase in milk fat CLA (89). This response could be explained simply as an effect of substrate supply. However, rumen culture studies demonstrate that at high levels of linoleic acid, an unusual pattern of biohydrogenation occurs where *trans*-octadecenoic acid rather than stearic acid accumulates (95,96). This phenomenon was attributed to linoleic acid acting as a competitive inhibitor for the biohydrogenation of the monoenoic acid (96). Consistent with this, Harfoot *et al.* (95) found that high levels of linoleic acid irreversibly inhibited hydrogenation of *trans*-11 octadecenoic acid. Interestingly, when high levels of α-linolenic acid were added to rumen cultures, *cis*-15 and *trans*-15 octadecenoic acids rather than *trans*-11 octadecenoic acid accumulated (34). The inhibitory effect of high levels of unsaturated fatty acids on biohydrogenation may be a partial explanation for the exceptionally high milk fat CLA among individual cows when diets high in linoleic acid are fed (4). Sunflower oil diets (high linoleic acid) would result primarily in the accumulation of *trans*-11 octadecenoic acid, whereas linseed oil diets would also result in the accumulation of *cis*- and *trans*-15 octadecenoic acids, which cannot be converted to CLA by Δ^9-desaturase.

Provision of polyunsaturated lipid substrate in the diet of a ruminant animal will also result in increased formation of biohydrogenation intermediates, predominantly formation of *trans*-octadecenoic acids as discussed earlier. However, several dietary factors are known to alter formation of biohydrogenation intermediates without any apparent change in lipid substrate (Table 13.1). Dietary fish oils produce a greater increase in milk fat concentration of CLA compared with an equal amount of plant oils (42). The mechanisms are not known because biohydrogenation of the long-chain

PUFA in fish oils has not been shown to produce CLA or *trans*-11 octadecenoic acid. However, increased ruminal production of *trans*-11 octadecenoic acid has been reported (77). By analogy with the inhibitory effect of high levels of linoleic acid on *trans*-octadecenoic acid hydrogenation (95,96), we propose that long-chain PUFA in fish oil inhibit ruminal biohydrogenation of unsaturated octadecadienoic acids through inhibition of the growth of bacteria that reduce octadecenoate or a specific inhibition of the reductases of these bacteria.

The effect of ionophores (e.g., the antibiotic monensin) on CLA concentration in milk fat has been variable, with minimal effect in one study (92) and a significant increase in another (93). Mechanisms of ionophore action on rumen biohydrogenation have been described in more detail elsewhere (see Chapter 15). Ionophores inhibit the growth of bacteria that produce hydrogen (97), and this decreased rate of hydrogen production may limit biohydrogenation in the rumen. However, the quantitative need for reducing equivalents in biohydrogenation is minimal compared with the reducing capacity in the rumen (16); therefore, reducing equivalent disposal alone cannot explain monensin effects on biohydrogenation. *B. fibrisolvens* and *ruminococci* species are active biohydrogenating bacteria that are particularly sensitive to ionophores *in vitro* (98). Therefore, the observed ionophore effect on ruminal biohydrogenation may be related in part to a direct inhibition of biohydrogenating bacteria. Minimal effect of ionophore feeding on biohydrogenation in one study (92) could be explained by ruminal adaptations in which ionophore-resistant species replaced ionophore-sensitive bacteria in ruminal biohydrogenation. This type of adaptation has previously been described relative to ruminal fiber digestion (99). In addition, variable levels of dietary PUFA may explain differences; however, these data are not available. In this case, bacterial populations involved in biohydrogenation may be altered, but substrate supply may not be adequate to allow the change in biohydrogenation to be expressed.

High-concentrate, low-fiber diets increase the production of *trans*-octadecenoic acids in the rumen. This effect has been attributed to alterations associated with a low ruminal pH (100). Consistent with this hypothesis, addition of buffer to a low-fiber diet increased rumen pH and decreased production of *trans*-octadecenoic acids (101). However, it is not always possible to distinguish between lipid substrate and pH-related effects on ruminal biohydrogenation. Decreasing the dietary forage to concentrate ratio from 50:50 to 20:80 resulted in an increase in milk fat CLA in one study (92) and in a decrease in another (22). High-concentrate diets are typically higher in fat content in comparison to high-fiber diets. To avoid this complication, Griinari *et al.* (22) fed a degermed corn product (containing only 0.7% crude fat) at two levels (50 and 80% of total dry matter) and supplemented both diets with an equal amount of corn oil. Rumen pH was decreased on the high-corn diet, but this did not affect the milk fat content of *trans*-octadecenoic acids. Nevertheless, the profile of *trans*-octadecenoic acids in milk fat was changed, i.e., *trans*-10 was the major isomer and there was a corresponding increase in the milk fat content of *trans*-10, *cis*-12 CLA. These data suggest that low-fiber diets may not change total *trans* fatty acids produced in the rumen, but the proportion of specific isomers is altered.

The effect of pasture feeding on milk fat concentrations of CLA has been described in several studies (Table 13.1). Generally, pasture feeding increases milk fat content of CLA compared with feeding a total mixed ration with a similar lipid content (6). Available lipids in pasture forages consist mainly of glycolipids and phospholipids, which provide only 2% of absorbable fatty acids of the diet dry matter (40). Glycolipids are hydrolyzed and hydrogenated in a similar manner to triglycerides by rumen cultures (43,102,103). In addition, the stage of forage maturity appears to be an important factor affecting milk fat content of CLA. Diets with early growth stage result in a higher level of milk fat CLA than late growth or second cutting (92). However, lipid content and composition of the forage explain differences only partially. In addition, effects of lipid substrate may be synergistic with effects of other components in pasture that alter biohydrogenation in the rumen. Increasing the proportion of non-fiber carbohydrate in the diet of dairy cows numerically decreases milk fat CLA, whereas addition of Ca salts of canola oil fatty acids increases milk fat CLA significantly (42). Interestingly, a trend toward interaction ($P < 0.07$) between the fiber content of the diet and fatty acid addition was observed.

Summary

The CLA in meat and milk fat has been thought to be of rumen origin. However, this view is far too simplistic and fails to consider adequately the complexity of the ruminal biohydrogenation process. Several features in ruminal biohydrogenation of unsaturated fatty acids suggest that the amount of CLA escaping ruminal biohydrogenation is inadequate to account for CLA levels in meat and milk fat. Rather, a major portion of the CLA in tissue and milk lipids appears to originate from endogenous synthesis. A role for Δ^9-desaturase has been proposed in which this enzyme converts rumen-derived *trans*-11 octadecenoic acid to CLA. In growing ruminants, adipose tissue has the greatest activity of Δ^9-desaturase, but in lactating ruminants, the mammary gland appears to be the major site of endogenous synthesis of CLA by Δ^9-desaturase. Differences in desaturase activity may also explain large differences in concentrations of CLA in milk fat observed among animals in response to dietary treatments.

Specific factors that cause the wide variation in meat and milk fat concentrations of CLA have not been delineated, but dietary factors are likely to be of major importance. The close linear relationship between the concentrations of *trans*-11 octadecenoic acid and CLA in meat and milk fat suggest that endogenous desaturation of *trans*-11 octadecenoic acid is an important source of CLA in both growing and lactating ruminants. Thus, novel feeding strategies to enhance concentrations of CLA in meat and milk fat should be designed to increase ruminal formation of *trans*-11 octadecenoic acid as well as CLA.

References

1. Pietinen, P., Vartiainen, E., Seppänen, R., Aro, A., and Puska, P. (1996) Changes in Diet in Finland from 1972 to 1992: Impact on Coronary Heart Disease Risk, *Prev. Med. 25,* 243–250.
2. Parodi, P.W. (1977) Conjugated Octadecadienoic Acids of Milk Fat, *J. Dairy Sci. 60,* 1550–1553.
3. Kepler, C.R., Tucker, W.P., and Tove, S.B. (1966) Intermediates and Products of the Biohydrogenation of Linoleic Acid by *Butyrivibrio fibrisolvens, J. Biol. Chem. 241,* 1350–1354.
4. Kelly, M.L., Berry, J.R., Dwyer, D.A., Griinari, J.M., Chouinard, P.Y., Van Amburgh, M.E., and Bauman, D.E. (1998) Dietary Fatty Acid Sources Affect Conjugated Linoleic Acid Concentrations in Milk From Lactating Dairy Cows, *J. Nutr. 128,* 881–885.
5. Viviani, R. (1970) Metabolism of Long Chain Fatty Acids in the Rumen, *Adv. Lipid Res. 8,* 267–346.
6. Kelly, M.L., Kolver, E.S., Bauman, D.E., Van Amburgh, M.E., and Muller, L.D. (1998) Effect of Intake of Pasture on Concentrations of Conjugated Linoleic Acid in Milk of Lactating Dairy Cows, *J. Dairy Sci. 81,* 1630–1636.
7. Seltzer, S. (1972) *Cis-trans* Isomerization, in *The Enzymes,* (Boyer, P.D., ed.) pp. 381–406, Academic Press, New York.
8. Reiser, R. (1951) Hydrogenation of Polyunsaturated Fatty Acids by the Ruminant, *Fed. Proc. 10,* 236.
9. Shorland, F.B., Weenink, R.O., and Johns, A.T. (1955) Effect of the Rumen on Dietary Fat, *Nature 175,* 1129–1130.
10. Ward, P.F.V., Scott, T.W., and Dawson, R.M.C. (1964) The Hydrogenation of Unsaturated Fatty Acids in the Ovine Digestive Tract, *Biochem. J. 92,* 60–68.
11. Wilde, P.F., and Dawson, R.M.C. (1966) The Biohydrogenation of α-Linolenic Acid and Oleic Acid by Rumen Micro-organisms, *Biochem. J. 98,* 469–475.
12. Wood, R.D., Bell, M.C., Grainger, R.B., and Teekell, R.A. (1963) Metabolism of Labeled [1-^{14}C] Linoleic Acid in the Sheep Rumen, *J. Nutr. 79,* 62–68.
13. Dawson, R.M.C., and Kemp, P. (1970) Biohydrogenation of Dietary Fats in Ruminants, in *Physiology of Digestion and Metabolism in the Ruminant,* (Phillipson, A.T., ed.) pp. 504–518, Oriel Press, Newcastle-upon-Tyne.
14. Shorland, F.B., Weenink, R.O., Johns, A.T., and McDonald, I.R.C. (1957) The Effect of Sheep-Rumen Contents on Unsaturated Fatty Acids, *Biochem. J. 67,* 328–333.
15. Bartlett, J.C., and Chapman, D.G. (1961) Detection of Hydrogenated Fats in Butter Fat by Measurement of *cis-trans* Conjugated Unsaturation, *Agric. Food. Chem. 9,* 50–53.
16. Harfoot, C.G., and Hazelwood, G.P. (1988) Lipid Metabolism in the Rumen, in *The Rumen Microbial Ecosystem,* (Hobson, P.N., ed.) pp. 285–322, Elsevier Science Publishers, London.
17. Kemp, P., Lander, D.J., and Gunstone, F.D. (1984) The Hydrogenation of Some *cis-* and *trans-*Octadecenoic Acids to Stearic Acid by a Rumen *Fusocillus* Sp., *Br. J. Nutr. 52,* 165–170.
18. Keeney, M. (1970) Lipid Metabolism in the Rumen, in *Physiology of Digestion and Metabolism in the Ruminant,* (Phillipson, A.T., ed.) pp. 489–503, Oriel Press, Newcastle-upon-Tyne.

19. Kepler, C.R., and Tove, S.B. (1967) Biohydrogenation of Unsaturated Fatty Acids. III. Purification and Properties of a Linoleate Δ^{12}-cis, Δ^{11}-trans-Isomerase from *Butyrivibrio fibrisolvens, J. Biol. Chem. 242,* 5685–5692.
20. Yokoyama, M.T., and Davis, C.L. (1971) Hydrogenation of Unsaturated Fatty Acids by *Treponema (Borrelia)* Strain $B_2 5$, a Rumen Spirochete, *J. Bacteriol. 107,* 519–527.
21. Kepler, C.R., Tucker, W.P., and Tove, S.B. (1970) Biohydrogenation of Unsaturated Fatty Acids. IV. Substrate Specificity and Inhibition of Linoleate Δ^{12}-cis, Δ^{11}-trans Isomerase from *Butyrivibrio fibrisolvens, J.Biol.Chem. 245,* 3612–3620.
22. Griinari, J.M., Dwyer, D.A., McGuire, M.A., Bauman, D.E., Palmquist, D.L., and Nurmela, K.V.V. (1998) *Trans*-Octadecenoic Acids and Milk Fat Depression in Lactating Dairy Cows, *J. Dairy Sci. 81,* 1251–1261.
23. Katz, I., and Keeney, M. (1966) Characterization of the Octadecenoic Acids in Rumen Digesta and Rumen Bacteria, *J. Dairy Sci. 49,* 962–966.
24. Molkentin, J.A., and Precht, D. (1995) Optimized Analysis of *trans*-Octadecenoic Acids in Edible Fats, *Chromatographia 41,* 267–272.
25. Parodi, P.W. (1976) Distribution of Isomeric Octadecenoic Fatty Acids in Milk Fat, *J. Dairy Sci. 59,* 1870–1873.
26. Precht, D., and Molkentin, J. (1995) *Trans* Fatty Acids: Implications For Health, Analytical Methods, Incidence in Edible Fats and Intake, *Die Nahrung 39,* 343–374.
27. Ha, Y.L., Grimm, N.K., and Pariza, M.W. (1989) Newly Recognized Anticarcinogenic Fatty Acids: Identification and Quantification in Natural and Processed Cheeses, *J. Agric. Food Chem. 37,* 75–81.
28. Lavillonnière, F., Martin, J.C., Bougnoux, P., and Sébédio, J.L. (1998) Analysis of Conjugated Linoleic Acid Isomers and Content in French Cheeses, *J. Am. Oil Chem. Soc. 75,* 343–352.
29. Sehat, N., Kramer, J.K.G., Mossoba, M.M., Yurawecz, M.P., Roach, J.A.G., Eulitz, K., Morehouse, K.M., and Ku, Y. (1998) Identification of Conjugated Linoleic Acid Isomers in Cheese by Gas Chromatography, Silver Ion–High-Performance Liquid Chromatography and Mass Spectral Reconstructed Ion Profiles. Comparison of Chromatographic Elution Sequences, *Lipids 33,* 963–971.
30. Yurawecz, M.P., Roach, J.A.G., Sehat, N., Mossoba, M.M., Kramer, J.K.G., Fritsche, J., Steinhart, H., and Ku, Y. (1998) A New Conjugated Linoleic Acid Isomer, 7 *trans,* 9 *cis*-Octadecadienoic Acid, in Cow Milk, Cheese, Beef and Human Milk and Adipose Tissue, *Lipids 33,* 803–809.
31. Fujimoto, K., Kimoto, H., Shishikura, M., Endo, Y., and Ogimoto, K. (1993) Biohydrogenation of Linoleic Acid by Anaerobic Bacteria Isolated from Rumen, *Biosci. Biotechnol. Biochem. 57,* 1026–1027.
32. Vehulst, A., Janssen, G., Parmentier, G., and Eyssen, H. (1987) Isomerization of Polyunsaturated Long Chain Fatty Acids by *Propionibacteria, Syst. Appl. Microbiol. 9,* 12–15.
33. Mortimer, C.E., and Niehaus, W.G., Jr. (1972) Enzymatic Isomerization of Oleic Acid to *trans*-10 Octadecenoic Acid, *Biochem. Biophys. Res. Commun. 49,* 1650–1656.
34. White, R.W., Kemp P., and Dawson, R.M.C. (1970) Isolation of a Rumen Bacterium That Hydrogenates Oleic Acid as Well as Linoleic Acid and Linolenic Acid, *Biochem. J. 116,* 767–768.
35. Wolff, R.L. (1995) Content and Distribution of *trans*-18:1 Acids in Ruminant Milk and Meat Fats. Their Importance in European Diets and Their Effect on Human Milk, *J. Am. Oil Chem. Soc. 72,* 259–272.

36. Fellner, V., Sauer, F.D., and Kramer, J.K.G. (1997) Effect of Nigericin, Monensin, and Tetronasin on Biohydrogenation in Continuous Flow-Through Ruminal Fermenters, *J. Dairy Sci. 80*, 921–928.
37. Chouinard, P.Y., Corneau, L., Barbano, D.M., Metzger, L.E., and Bauman, D.E., (1999) Conjugated Linoleic Acids Alter Milk Fatty Acid Composition and Inhibit Milk Fat Secretion in Dairy Cows, *J. Nutr.,* in press.
38. Loor, J.J., and Herbein, J.H. (1998) Exogenous Conjugated Linoleic Acid Isomers Reduce Bovine Milk Fat Concentration and Yield by Inhibiting de novo Fatty Acid Synthesis, *J. Nutr. 128,* 411–419.
39. Bauman, D.E., Davis, C.L., and Bucholtz, H.F. (1971) Propionate Production in the Rumen of Cows Fed Either a Control or High Grain, Low Fiber Diet, *J. Dairy Sci. 54,* 1282–1287.
40. VanSoest, P.J. (1994) *Nutritional Ecology of the Ruminant,* 2nd edn., Cornell University Press, Ithaca, NY.
41. Leat, W.M.F., Kemp, P., Lysons, R.J., and Alexander, T.J.L. (1977) Fatty Acid Composition of Depot Fats from Gnotobiotic Lambs, *J. Agric. Sci. (Camb.) 88,* 175–179.
42. Chouinard, P.Y., Corneau, L., Bauman, D.E., Butler, W.R., Chilliard, Y., and Drackley, J.K. (1998) Conjugated Linoleic Acid Content of Milk From Cows Fed Different Sources of Dietary Fat, *J. Dairy Sci., 81* (Suppl. 1):*223* (Abstr.).
43. Singh, S., and Hawke, J.C. (1979) The *in vitro* Lipolysis and Biohydrogenation of Monogalactosyldiglycerides by Whole Rumen Contents and Its Fractions, *J. Sci. Food Agric. 30,* 603–612.
44. Tanaka, K., and Shigeno, K. (1976) The Biohydrogenation of Linoleic Acid by Rumen Micro-Organisms, *Jpn. J. Zootech. Sci. 47,* 50–53.
45. Harfoot, C.G., Noble, R.C., and Moore, J.H. (1973) Factors Influencing the Extent of Biohydrogenation of Linoleic Acid by Rumen Micro-Organisms *in vitro, J. Sci. Food Agric. 24,* 961–970.
46. Kellens, M.J., Goderis, H.L., and Tobback, P.P. (1986) Biohydrogenation of Unsaturated Fatty Acids by a Mixed Culture of Rumen Microorganisms, *Biotechnol. Bioeng. 28,* 1268–1276.
47. Banni, S., Carta, C., Contini, M.S., Angioni, E., Deiana, M., Dessi, M.A., Melis, M.P., and Corongiu, F.P. (1996) Characterization of Conjugated Diene Fatty Acids in Milk, Dairy Products, and Lamb Tissues, *J. Nutr. Biochem. 7,* 150–155.
48. Jahreis, G., Fritsche, J., and Steinhart, H. (1997) Conjugated Linoleic Acid in Milk Fat: High Variation Depending on Production System, *Nutr. Res. 17,* 1479–1484.
49. Jiang, J.L., Björck, L., Fondén, R., and Emanuelson, M. (1996) Occurrence of Conjugated *cis*-9, *trans*-11-Octadecadienoic Acid in Bovine Milk: Effects of Feed and Dietary Regimen, *J. Dairy Sci. 79,* 438–445.
50. Precht, D., and Molkentin, J. (1997) Effect of Feeding on Conjugated *cis*-9, *trans*-11 Octadecadienoic Acid and Other Isomers of Linoleic Acid in Bovine Milk Fats, *Die Nahrung 41,* 330–335.
51. Chin, S.F.W., Liu, J.M., Storkson, Y., Ha, L., and Pariza, M.W. (1992) Dietary Sources of Conjugated Dienoic Isomers of Linoleic Acid, a Newly Recognized Class of Anticarcinogens, *J. Food Comp. Anal. 5,* 185–197.
52. Griinari, J.M., Nurmela, K.V.V., and Bauman, D.E. (1997) *Trans*-10 Isomer of Octadecenoic Acid Corresponds with Milk Fat Depression, *J. Dairy Sci. 80 (Suppl. 1),* 204 (Abstr.).

53. Mahfouz, M.M., Valicenti, A.J., and Holman, R.T. (1980) Desaturation of Isomeric *trans*-Octadecenoic Acids by Rat Liver Microsomes, *Biochim. Biophys. Acta 618*, 1–12.
54. Pollard, M.R., Gunstone, F.D., James, A.T., and Morris, L.J. (1980) Desaturation of Positional and Geometric Isomers of Monoenoic Fatty Acids by Microsomal Preparations from Rat Liver, *Lipids 15*, 306–314.
55. Kinsella, J.E. (1972) Stearyl CoA as a Precursor of Oleic Acid and Glycerolipids in Mammary Microsomes from Lactating Bovine: Possible Regulatory Step in Milk Triglyceride Synthesis, *Lipids 7*, 349–355.
56. Bickerstaffe, R., Noakes, D.E., and Annison, E.F. (1972) Quantitative Aspects of Fatty Acid Biohydrogenation, Absorption and Transfer into Milk Fat in the Lactating Goat, with Special Reference to the *cis*- and *trans*-Isomers of Octadecenoate and Linoleate, *Biochem. J. 130*, 607–617.
57. Thompson, G.E., and Christie, W.W. (1991) Extraction of Plasma Triacylglycerols by the Mammary Gland of the Lactating Cow, *J. Dairy Res. 58*, 251–255.
58. Riisom, T., and Holman, R.T. (1981) Incorporation into Lipid Classes of Products from Microsomal Desaturation of Isomeric *trans*-Octadecenoic Acids, *Lipids 16*, 647–654.
59. Ulberth, F., and Henninger, M. (1994) Quantitation of *trans* Fatty Acids in Milk Fat Using Spectroscopic and Chromatographic Methods, *J. Dairy Res. 61*, 517–527.
60. Ntambi, J.M. (1995) The Regulation of Stearoyl-CoA Desaturase (SCD), *Prog. Lipid Res. 34*, 139–150.
61. Enoch, H.G., Catala, A., and Strittmatter, P. (1976) Mechanism of Rat Liver Microsomal Stearyl-CoA Desaturase, *J. Biol. Chem. 251*, 5096–5103.
62. Tocher, D.R., Leaver, M.J., and Hodgson, P.A. (1998) Recent Advances in the Biochemistry and Molecular Biology of Fatty Acyl Desaturases, *Prog. Lipid Res. 37*, 73–117.
63. Parodi, P.W. (1982) Positional Distribution of Fatty Acids in the Triglyceride Classes of Milk Fat, *J. Dairy Res. 49*, 73–80.
64. Cameron, P.J., Rogers, M., Oman, J., May, S.G., Lunt, D.K., and Smith, S.B. (1994) Stearoyl Coenzyme A Desaturase Activity and mRNA Levels Are Not Different in Subcutaneous Adipose Tissue from Angus and American Wagyu Steers, *J. Anim. Sci. 72*, 2624–2628.
65. Chang, J.H.P., Lunt, D.K., and Smith, S.B. (1992) Fatty Acid Composition and Fatty Acid Elongase and Stearoyl-CoA Desaturase Activities in Tissues of Steers Fed High Oleate Sunflower Seed, *J. Nutr. 122*, 2074–2080.
66. Page, A.M., Sturdivant, C.A., Lunt, D.K., and Smith, S.B. (1997) Dietary Whole Cottonseed Depresses Lipogenesis but Has No Effect on Stearoyl Coenzyme Desaturase Activity in Bovine Subcutaneous Adipose Tissue, *Comp. Biochem. Physiol. 118B*, 79–84.
67. St. John, L.C., Lunt, D.K., and Smith, S.B. (1991) Fatty Acid Elongation and Desaturation Enzyme Activities of Bovine Liver and Subcutaneous Adipose Tissue Microsomes, *J. Anim. Sci. 69*, 1064–1073.
68. Wahle, K.W.J. (1974) Desaturation of Long-Chain Fatty Acids by Tissue Preparations of the Sheep, Rat and Chicken, *Comp. Biochem. Physiol. 48B*, 87–105.
69. Bickerstaffe, R., and Annison, E.F. (1970) The Desaturase Activity of Goat and Sow Mammary Tissue, *Comp. Biochem. Physiol. 35*, 653–665.
70. Ward, R.J., Travers, M.T., Richards, S.E., Vernon, R.G., Salter, A.M., Buttery, P.J., and Barber, M.C. (1998) Stearoyl-CoA Desaturase mRNA Is Transcribed from a Single Gene in the Ovine Genome, *Biochim. Biophys. Acta 1391*, 145–156.

71. Bickerstaffe, R., and Johnson, A.R. (1972) The Effect of Intravenous Infusions of Sterculic Acid on Milk Fat Synthesis, *Br. J. Nutr. 27*, 561–570.
72. Beaulieu, A.D., and Palmquist, D.L. (1995) Differential Effects of High Fat Diets on Fatty Acid Composition in Milk of Jersey and Holstein Cows, *J. Dairy Sci. 78*, 1336–1344.
73. DePeters, E.J., Medrano, J.F., and Reed, B.A. (1995) Fatty Acid Composition of Milk Fat from Three Breeds of Dairy Cattle, *Can. J. Anim. Sci. 75*, 267–269.
74. Lawless, F., Stanton, C., L'Escop, P., Devery, R., and Murphy, J.J. (1999) The Influence of Breed on Bovine Milk *cis*-9, *trans*-11 Conjugated Linoleic Acid Content, *Livest. Prod. Sci.* (in press).
75. Pennington, J.A., and Davis, C.L. (1975) Effects of Intraruminal and Intra-Abomasal Additions of Cod Liver Oil on Milk Fat Production in the Cow, *J. Dairy Sci. 58*, 49–55.
76. Wonsil, B.J., Herbein, J.H., and Watkins, B.A. (1994) Dietary and Ruminally Derived *trans*-18:1 Fatty Acids Alter Bovine Milk Lipids, *J. Nutr. 124*, 556–565.
77. Chilliard, Y., Chardigny, J.M., Chabrot, J., Ollier, A., Sébédio, J.L., and Doreau, M. (1999) Effects of Ruminal or Postruminal Fish Oil Supply on Conjugated Linoleic Acid (CLA) Content of Cow Milk Fat, *Proc. Nutr. Soc.,* in press.
78. Doreau, M., and Chilliard, Y. (1997) Effects of Ruminal or Postruminal Fish Oil Supplementation on Intake and Digestion in Dairy Cows, *Reprod. Nutr. Dev. 37*, 113–124.
79. Griinari, J.M., Chouinard, P.Y., and Bauman, D.E. (1997) *Trans* Fatty Acid Hypothesis of Milk Fat Depression Revised. *Proc. Cornell Nutr. Conf. Feed Manuf.* pp. 208–216. Ithaca, NY.
80. Enjalbert, F., Nicot, M.-C., Bayourthe, C., and Moncoulon, R. (1998) Duodenal Infusions of Palmitic, Stearic or Oleic Acids Differently Affect Mammary Gland Metabolism of Fatty Acids in Lactating Dairy Cows, *J. Nutr. 128*, 1525–1532.
81. Salminen, I.M., Mutanen, M., Jauhiainen, M., and Aro, A. (1998) Dietary *trans* Fatty Acids Increase Conjugated Linoleic Acid Levels in Human Serum, *J. Nutr. Biochem. 9*, 93–98.
82. Fogerty, A.C., Ford, G.L., and Svoronos, D. (1988) Octadeca-9,11-dienoic Acid in Foodstuffs and in the Lipids of Human Blood and Breast Milk, *Nutr. Rep. Int. 38*, 937–944.
83. Fritsche, J., and Steinhart, H. (1998) Amounts of Conjugated Linoleic Acid (CLA) in German Foods and Evaluation of Daily Intake, *Z. Lebensm.-Unters.-Forsch. A. 206*, 77–82.
84. Shantha, N.C., Crum, A.D., and Decker, E.A. (1994) Evaluation of Conjugated Linoleic Acid Concentrations in Cooked Beef, *J. Agric. Food Chem. 42*, 1757–1760.
85. Shantha, N.C., Moody, W.G., and Tabeidi, Z. (1997) A Research Note: Conjugated Linoleic Acid Concentration in Semimembranosus Muscle of Grass- and Grain-Fed and Zeranol-Implanted Beef Cattle, *J. Muscle Foods 8*, 105–110.
86. Wood, R. (1983) Geometrical and Positional Monoene Isomers in Beef and Several Processed Meats in *Dietary Fats and Health,* (Perkins, E.G., and Visek, W.J., eds.) pp. 341–358, American Oil Chemists' Society, Champaign, IL.
87. Fritsche, S., and Fritsche, J. (1998) Occurrence of Conjugated Linoleic Acid Isomers in Beef, *J. Am. Oil Chem. Soc. 75*, 1449–1451.
88. Riel, R.R. (1963) Physico-Chemical Characteristics of Canadian Milk Fat. Unsaturated Fatty Acids, *J. Dairy Sci. 46*, 102–106.
89. Tesfa, A.T., Tuori, M., and Syrjälä-Qvist, L. (1991) High Rapeseed Oil Feeding to Lactating Dairy Cows and Its Effect on Milk Yield and Composition in Ruminants, *Finn. J. Dairy Sci. 49*, 65–81.

90. Lawless, F., Murphy, J.J., Harrington, D., Devery, R., and Stanton, C. (1998) Elevation of Conjugated *cis*-9, *trans*-11 Octadecenoic Acid in Bovine Milk Because of Dietary Supplementation, *J. Dairy Sci. 81,* 3259–3267.
91. Timmen, H., and Patton, S. (1988) Milk Fat Globules: Fatty Acid Composition, Size and *in vivo* Regulation of Fatty Liquidity, *Lipids 23,* 685–689.
92. Chouinard, P.Y., Corneau, L., Kelly, M.L., Griinari, J.M., and Bauman, D.E. (1998) Effect of Dietary Manipulation on Milk Conjugated Linoleic Acid Concentrations, *J. Dairy Sci., 81,* (Suppl. 1): *233* (Abstr.).
93. Sauer, F.D., Fellner, V., Kinsman, R., Kramer, J.K.G., Jackson, H.A., Lee, A.J., and Chen, S. (1998) Methane Output and Lactation Response in Holstein Cattle with Monensin or Unsaturated Fat Added to the Diet, *J. Anim. Sci. 76,* 906–914.
94. Zegarska, Z., Paszczyk, B., and Borejszo, Z. (1996) *Trans* Fatty Acids in Milk Fat, *Pol. J. Food Nutr. Sci. 5,* 89–96.
95. Harfoot, C.G., Noble, R.C., and Moore, J.H. (1973) Food Particles as a Site of Biohydrogenation of Unsaturated Fatty Acids in the Rumen, *Biochem. J. 132,* 829–832.
96. Polan, C.E., McNeill, J.J., and Tove, S.B. (1964) Biohydrogenation of Unsaturated Fatty Acids by Rumen Bacteria, *J. Bacteriol. 88,* 1056–1064.
97. Van Nevel, C.J., and Demeyer, D.I. (1996) Control of Rumen Methanogenesis, *Environ. Monit. Assess. 42,* 73–97.
98. Chen, M., and Wolin, M.J. (1979) Effect of Monensin and Lasalocid-Sodium on the Growth of Methaonogenic and Rumen Saccarolytic Bacteria, *Appl. Environ. Microbiol. 38,* 72–77.
99. Russell, J.D., and Storbel, H.J. (1989) Mini-Review: The Effect of Ionophores on Ruminal Fermentation, *Appl. Environ. Microbiol. 55,* 1–6.
100. Romo, G.A. (1995) *Trans* Fatty Acids: Rumen *in vitro* Production and Their Subsequent Metabolic Effects on Energy Metabolism and Endocrine Responses in Lactating Dairy Cows, Ph.D. Thesis, University of Maryland, College Park.
101. Kalscheur, K.F., Teter, B.B., Piperova, L.S., and Erdman, R.A. (1997) Effect of Dietary Forage Concentration and Buffer Addition on Duodenal Flow of *trans*-C18:1 Fatty Acids and Milk Fat Production in Dairy Cows, *J. Dairy Sci. 80,* 2104–2114.
102. Dawson, R.M.C., Hemington, N., Grime, D., Lander, D., and Kemp, D. (1974) Lipolysis and Hydrogenation of Galactolipids and the Accumulation of Phytanic Acid in the Rumen, *Biochem. J. 144,* 169–171.
103. Dawson, R.M.C., Hemington, N., and Hazlewood, G.P. (1977) On the Role of Higher Plant and Microbial Lipases in the Ruminal Hydrolysis of Grass Lipids, *Br. J. Nutr. 38,* 225–232.

Chapter 14

Endogenous Synthesis of Rumenic Acid in Rodents and Humans

Donald L. Palmquist and Jamie E. Santora

Department of Animal Sciences, The Ohio State University/Ohio Agricultural Research and Development Center, Wooster, OH 44691–4096

Introduction

Dietary intake of CLA may be lower than optimal for maximum biological effects (1). It was proposed by Parodi (2) that the *trans*-11 octadecenoic acid found in dietary fats could be desaturated to *c*-9, *t*-11 18:2 (rumenic acid, a specific isomer of CLA) by microsomal Δ-9 desaturase, on the basis of the description of such activity by Pollard *et al.* (3) and Holman and Mahfouz (4) in rat liver. *Trans*-11 octadecenoic (*trans*-vaccenic) acid is the predominant *trans* monoene of ruminant fats (5,6), commonly constituting three- to fivefold the values for CLA. Therefore, we investigated whether *trans*-vaccenic acid is a quantitatively important precursor for rumenic acid. The mouse was the model chosen because of the simplicity of working with the total tissues and in consideration of the high cost of purified *trans*-vaccenic acid as a component of the diet.

Present Status of Knowledge

Numerous authors have published estimates of *trans* fatty acid intake by humans (6–11). All have acknowledged the difficulty in making an accurate estimate of such intake; however, estimates are mainly in the range of 5–15 g/d. Intake of *trans*-11 18:1, however, has been estimated less frequently; Emken (8) estimated 1.26 g/d, and Wolff (6) reported intake from ruminant fats in the European Economic Community [except for Spain and Portugal (0.8 g/d)] of 1.3–1.8 g/d total *trans* 18:1 acids. The predominant *trans* monoene in ruminant fats is *trans*-11 18:1 (36–68% of total) (5,6). Thus, *trans*-11 18:1 intake by humans approximates 1 g/d; this is higher than intake of CLA (1), which ordinarily is only 15–50% of the concentration of *trans*-11 18:1 in ruminant milk fat (12,13); CLA concentration may be lower in the fat of ruminant meats than in milk fat (14).

The molecular biology of the superfamily of acyl desaturases has been reviewed comprehensively (15). The regulation of stearoyl-CoA desaturase (Δ-9 desaturase, EC 1.14.99.5) was reviewed in depth by Ntambi (16). Δ-9 Desaturase activity is expressed differentially by two highly homologous genes, SCD1 and SCD2. Mouse and rat SCD1 and SCD2 peptide sequences are >90% identical with human adipose stearoyl-CoA desaturase (16). SCD1 is expressed constitutively in adipose tissue, but not in liver of rodents. Expression in liver is induced by a fat-free diet (16) and peroxisome-proliferating agents, such as clofibric acid (17,18); it is inhibited by dietary polyunsaturated fatty acids (19). In rodents, SCD2 differs markedly from SCD1 in terms of tissue distribution and response to diet modulation; however, it is not expressed in liver.

Characterization of Δ-9 desaturase in human tissues is limited. However, lipid synthesis from carbohydrate in humans is predominately hepatic, rather than in adipose tissues (20,21). Δ-9 Desaturase activity may not necessarily be limited to tissues with active fatty acid synthesis; however, distribution of Δ-9 desaturase activity in human tissues is presently unknown.

Pollard *et al.* (3) reported that *trans*-monoenoic fatty acids with *trans*-7 and *trans*-11 unsaturation were desaturated by rat liver microsomes to *trans*-7, *cis*- 9, and *cis*-9, *trans*-11 dienes, respectively, whereas monoenes with *trans*-8 or *trans*-10 did not yield 8, 9 or 9, 10- allenes. The corresponding 7- and 11-*cis* monoenes were not desaturated further. They reported also that 41% of the 9-*cis*, 11-*trans* diene was further desaturated by a Δ-6 desaturase. Holman and Mahfouz (4) confirmed and extended the observations of Pollard *et al.* (3). The *trans*-11 18:1 was desaturated at only 20–30% of rates observed for the *trans* isomers of 4-, 6-, and 13- monoenes. *Cis* and *trans* monoenes were desaturated differently by rat liver microsomes, whereas the *trans* monoenes and saturated acids were desaturated similarly. Interestingly, after Δ-9 desaturation, the *cis*, *trans* dienes could be isomerized to the *cis*, *cis* isomers without change of bond positions.

Metabolism of deuterium-labeled *trans* 18:1 in two young adult men was reported by Emken *et al.* (8); *cis*-9 and *trans*-11 monoenes were absorbed equally, but turnover of labeled *trans*-11 18:1 in the plasma triacylglycerol pool was higher than that for the *cis*-9 isomer. Incorporation into plasma phosphatidylcholine was similar for both monoenes; however, *trans*-11 18:1 was incorporated preferentially into the *sn*-1 position. They found essentially no incorporation of the *trans* monoene into plasma cholesteryl esters. Chain-shortening was two to three times greater for *trans*-11 18:1 than for *cis*-9 18:1. They found no evidence for desaturation of the *trans*-11 18:1 in the plasma lipids during the 48-h duration of the study. Emken *et al.* (8) reported discrimination of incorporation of *trans*-11 18:1 into certain lipid classes by comparing ratios of label in *trans*-11 and *cis*-9 isomers in plasma lipids with the ratios of these in the diet. They did not consider the possibility of desaturation of *trans*-11 18:1 by intestinal tissues during absorption. If this occurred, interpretation of discrimination against *trans* monoene incorporation into various lipid classes would have been overestimated.

The effects of a diet high in *trans* monoenoic fatty acids on fatty acid profiles in blood lipids were reported by Salminen *et al.* (22). They studied two groups of 40 healthy adults. Following a 5-wk baseline diet high in dairy products, subjects consumed a diet high in stearic acid or high in hydrogenated vegetable fats (providing 3 g of *trans*-11 18:1 per day) for a period of 5 wk. Intakes of CLA for the dairy (preliminary), stearic, and *trans* fatty acid diets were 310, 90, and 40 mg/d, respectively.

At the end of the study, CLA concentration in the serum lipids (percentage of total fatty acids) was 0.33, 0.17, and 0.43, respectively. The authors concluded that the most likely explanation for the higher plasma CLA concentration with hydrogenated vegetable fats was endogenous Δ-9 desaturation of dietary *trans*-11 18:1. At present, any explanation for differences in outcomes between the studies of Emken *et al.* (8) and Salminen *et al.* (22), other than duration of the studies, is obscure. Nevertheless, to understand the role of dietary CLA in health, it is imperative to determine quantitatively any contribution from endogenous synthesis of CLA and the role of dietary Δ-11-*trans* 18:1 in this synthesis.

Quantitative Studies on Endogenous Synthesis of Rumenic Acid

We initiated studies to quantify conversion of *trans*-11 18:1 to rumenic acid, using mice as the model. In all studies, mice were fed a basal semipurified diet composed of dextrose, casein, vitamins, minerals, and 4% corn oil. Control diets contained 1% of stearic acid and experimental diets contained 1% of *trans* vaccenic acid, except where objectives determined otherwise. Mice were fed experimental or control diets for 2 wk; food intake and body weights were determined daily. In two experiments, feces were collected nonquantitatively. Mice (n = 2–6) were killed at the beginning of each experiment to determine initial body composition and content of CLA. After the mice were killed by CO_2 asphyxiation, gut contents were removed and whole carcasses were freeze-dried, followed by homogenization in liquid nitrogen. Lipids in an aliquot of dried carcass were extracted and transmethylated with sodium methoxide. Fatty acid methyl esters were quantified by gas–liquid chromatography, with triheptadecanoin as internal standard.

Trans-vaccenic was 29–58% of fecal fat, compared with 20% of the dietary fat; thus absorption was variable and lower than for corn oil fatty acids. Lower absorption was not unique to *trans*-vaccenic acid; the proportion of stearic acid in fecal fat was 70%, compared with 20% of dietary fat in control diets.

Desaturation of trans-*Vaccenic Acid*

In three experiments, in which 1% *trans*-vaccenic acid was fed with 4% corn oil (n = 4, 6, and 10 mice per group), the percentage conversion of dietary *trans*-vaccenic acid to rumenic acid in the total carcass was 12.3, 12.0, and 10.0, respectively (overall mean ± SD, 11.4 ± 1.25). However, recovery of *trans*-11 18:1 + *cis*-9, *trans*-11 18:2 fatty acids from the carcass was only ~20% of dietary *trans*-vaccenic acid. Because of apparently low absorption and high oxidation of *trans*-vaccenic acid, we computed conversion to rumenic acid by an alternate procedure, in which we considered conversion (desaturation) as a proportion of the *trans*-vaccenic available in the tissues for conversion, using control and treatment group means: Conversion (stored), % = [carcass rumenic acid (treatment – control)]/[carcass rumenic acid + *trans*-vaccenic acid (treatment − control)] × 100. By this analysis, the percentage conversion of the stored *trans*-vaccenic acid in the three experimental groups was 48.8, 52.6, and 51.0, respectively (overall, 50.8 ± 1.91).

Modifiers of Desaturase Activity

We investigated whether desaturation of *trans*-vaccenic acid to rumenic acid could be influenced by including in the diet known modifiers of stearoyl-CoA desaturase activity. A summary of the data is given in Table 14.1. Net gain of rumenic acid in the carcass of animals fed 0.5% clofibrate to stimulate desaturase activity or 10% corn oil to provide increased polyunsaturated fatty acids (PUFA) to inhibit desaturase was one half to one third that of controls. Dietary conversion was lower also, whereas the proportion of stored *trans*-vaccenic acid that was desaturated to rumenic acid was similar to that of controls in mice fed clofibrate and decreased 30% in mice fed increased PUFA. We had postulated that clofibrate would induce the stearoyl-CoA desaturase and increase the proportion of rumenic acid in the carcass. The amount of carcass *trans*-vaccenic acid, as well as rumenic acid, in the clofibrate group was one half that of controls. Possibly the clofibrate induced peroxisomal oxidation of lipids including *trans*-vaccenic acid, thereby decreasing the amount available for desaturation. Concentration of *trans*-vaccenic acid in fecal fat was 29 and 20% for control and clofibrate mice, respectively, compared with 20% of dietary fat. The apparent higher absorption of *trans*-vaccenic acid by mice fed clofibrate could have been caused by increased cholesterol conversion to bile acids by the peroxisomes (18). As postulated, feeding higher amounts of PUFA decreased desaturation of *trans*-vaccenic acid to rumenic acid.

TABLE 14.1 Net Gain of Rumenic Acid in the Carcass and Conversion of *trans*-Vaccenic Acid in Mice (n = 5/treatment) Fed Diets Containing 1% *trans*-Vaccenic Acid with No Modifier of Stearoyl-CoA Desaturase, 0.5% Clofibrate, or 10% Corn Oil

	Control[a]	Clofibrate[b]	PUFA[c]
t-11 18:1 available for desaturation (mg)[d]	122	57	64
Net gain of rumenic acid (mg)[e]	64.2	31.8	23.5
Conversion (%)			
Dietary[f]	12.0	7.0	5.1
Stored[g]	52.6	55.5	37.0

[a]Four percent corn oil + 1% Δ-11-*trans* 18:1 in the diet.
[b]As control, + 0.5% clofibrate in the diet.
[c]As control, + 6% corn oil in the diet.
[d]*trans*-11 18:1 + *cis*-9, *trans*-11 measured in the whole carcass.
[e]Difference between rumenic acid in the whole carcass of treatment groups and similar groups fed stearic acid rather than *trans*-vaccenic acid. Group means used for comparisons.
[f]Carcass rumenic acid as a percentage of total dietary *trans*-vaccenic acid consumed in 2 wk.
[g]Carcass rumenic acid as a percentage of carcass (rumenic acid + *trans*-vaccenic acid).

Rumenic Acid in Lipid Classes

We pooled carcasses of three mice chosen randomly from each group of mice fed 1% stearic acid, elaidic acid, *trans*-vaccenic acid, or CLA. The total lipids were extracted and the distribution of rumenic acid in the triacylglycerols and phospholipids was determined (Table 14.2). In the triacylglycerols, feeding *trans*-vaccenic acid doubled rumenic acid compared with feeding stearic or elaidic acids. In the group fed CLA, its content in triacylglycerol was threefold that of controls. CLA was found in the phospholipids only when it was fed, and its proportion was less than one half that found in triacylglycerols. No *trans*-vaccenic acid was found in the phospholipids, whereas elaidic acid occurred in measurable amounts. Though not possible to test statistically, *cis*-9 18:1 occurred in the phospholipids in higher amounts when elaidic acid was fed, and arachidonic and docosahexaenoic acids were lower. Feeding CLA tended to decrease linoleic, arachidonic, and docosahexaenoic acids in phospholipids.

The distribution of rumenic acid found in lipid classes causes some consideration of differences in the metabolism of endogenous (by Δ-9 desaturation of *trans*-vaccenic acid) vs. exogenous (dietary) CLA. Occurrence of CLA in phospholipids has received attention because the presumed biologically active isomer, *cis*-9, *trans*-11 18:2, was so identified when it was the only isomer found in phospholipids after a mixture of CLA isomers was fed (23). However, most authors report that CLA is a higher proportion of fatty acids in triacylglycerols than in phospholipids (24), and several CLA isomers have been found in lesser amounts in phospholipids (25). Our data are consistent with incorporation of dietary CLA into both triacylglycerols and phospholipids in the intestinal mucosa, with subsequent transport to various tissues, followed by incorporation into plasma membranes or storage lipid.

TABLE 14.2 Distribution of Fatty Acids (wt%) in Carcass Triacylglycerol and Total Phospholipids from Mice Fed Stearic, *trans*-Vaccenic, Elaidic, or Conjugated Linoleic Acids (CLA)

Treatment group	14:0	16:0	16:1	18:0	t18:1	c18:1	18:2	18:3	c,t18:2	20:4	22:6	24:1
Triacylglycerol												
Stearic	1.78	25.9	8.30	5.43	0.11	36.1	18.5	0.75	0.85	0.38	1.73	0.08
trans-Vaccenic	1.71	26.6	7.88	6.09	1.67[a]	36.0	15.1	0.54	1.74	0.10	2.45	0.14
Elaidic	1.65	23.8	8.09	4.88	4.90[b]	36.8	17.8	0.68	0.95	0.33	0.14	0
CLA	1.67	29.2	5.94	2.91	0	37.6	18.7	0.42	2.72	0.44	0.39	0
Phospholipid												
Stearic	3.45	13.6	3.46	19.4	0	18.6	14.7	0	0	11.6	13.1	2.10
trans-Vaccenic	2.86	13.5	4.23	17.4	0	20.2	15.7	0	0	11.6	12.6	1.91
Elaidic	2.98	15.8	5.39	15.1	3.63[b]	26.9	17.0	0	0	5.59	7.64	0
CLA	2.42	17.0	2.25	20.1	0	24.1	12.3	0	1.41	8.05	10.6	1.95

[a]Δ-11 isomer.
[b]Δ-9 isomer.

Occurrence of endogenous rumenic acid only in triacylglycerols is consistent with the expression of Δ-9 desaturase in adipose tissue, but not in liver, when PUFA were included in the diet of mice. Thus, we conclude that *trans*-vaccenic acid was desaturated in significant amounts (50%) when it was taken up by adipose tissue and then was stored as triacylglycerol. We postulate that redistribution of endogenous rumenic acid to other tissues would occur over some time period longer than the 2-wk feeding periods we used.

Emken *et al.* (8) were unable to detect desaturation products of deuterated *trans*-vaccenic acid in blood lipids of young men during 48 h after consumption of 7–8 g of the labeled material. More recently, Salminen *et al.* (22) reported increased proportions of CLA in blood lipids of subjects consuming 3 g/d of *trans*-vaccenic acid in hydrogenated vegetable oils for a period of 5 wk. Whether methodological, dietary, or other factors were responsible for the different outcomes cannot be determined at present; however, our data, together with those of Salminen *et al.* (22), suggest that *trans*-vaccenic acid in the human diet is an important precursor for endogenous rumenic acid.

Implications

Rodent and human fat metabolism differ in many ways, including major sites of fatty acid synthesis (20,21). Although *trans*-vaccenic acid apparently was desaturated mainly in adipose tissue in these studies, fatty acid synthesis in humans occurs primarily in the liver. However, distribution of Δ-9 desaturase activity in various human tissues is less certain. If conversion of *trans*-vaccenic acid occurs in human tissues at an order of magnitude similar to that found in our studies, the amount of CLA available to the body would be higher than current estimates (assuming average daily intake of CLA at 250 mg and of *trans*-vaccenic acid at 1000 mg). Further, concepts of "healthfulness" of the *trans* fatty acids in the diet also would have to change. Clearly, *trans*-vaccenic acid is metabolized differently and must be viewed separately from the "unnatural" *trans* fatty acids formed during chemical hydrogenation of vegetable oils (7).

Further Studies Required

Although our estimates of the conversion of stored *trans*-vaccenic acid were highly repeatable in three experiments, the data were calculated from single points because the estimates were based on mean contents of *trans*-vaccenic acid and rumenic acid in groups of animals. More detailed information on variability of responses could be gained by making comparisons from pair-fed animals. Dose-response curves with changing intakes of *trans*-vaccenic acid should be established. Further investigation of effects of metabolic modifiers of Δ-9 desaturase activity would be useful. We observed decreased digestibility of the high-melting *trans*-vaccenic acid fed at 10–20% of the dietary fat. We do not believe that absorption would be a problem with conventional human diets. Emken *et al.* (8) concluded that deuterated *trans*-vaccenic acid (7–8 g/d and oleic acid were equally well absorbed, although no fecal data were reported.

To establish the importance of *trans*-vaccenic acid as a precursor of endogenous rumenic acid, it is imperative to conduct studies in humans. Both descriptive data (tissue sites and activities of stearoyl-CoA desaturase) and quantitative studies should be undertaken. Such studies will increase not only understanding of CLA metabolism in humans, but also the appreciation for animal products and animal fats in the diet.

References

1. Huang, Y.-C., Luedecke, L.O., and Shultz, T.D. (1994) Effect of Cheddar Cheese Consumption on Plasma Conjugated Linoleic Acid Concentrations in Men, *Nutr. Res. 14*, 373–386.
2. Parodi, P.W. (1994) Conjugated Linoleic Acid: An Anticarcinogenic Fatty Acid Present in Milk Fat, *Aust. J. Dairy Technol. 49*, 93–97.
3. Pollard, M.R., Gunstone, F.D., James, A.T., and Morris, L.J. (1980) Desaturation of Positional and Geometric Isomers of Monoenoic Fatty Acids by Microsomal Preparations from Rat Liver, *Lipids 15*, 306–314.
4. Holman, R.T., and Mahfouz, M.M. (1981) *Cis* and *trans*-Octadecenoic Acids as Precursors of Polyunsaturated Acids, *Prog. Lipid Res. 20*, 151–156.
5. Parodi, P.W. (1976) Distribution of Isomeric Octadecenoic Fatty Acids in Milk Fat, *J. Dairy Sci. 59*, 1870–1873.
6. Wolff, R.L. (1995) Content and Distribution of *trans*-18:1 Acids in Ruminant Milk and Meat Fats. Their Importance in European Diets and Their Effect on Human Milk, *J. Am. Oil Chem. Soc. 72*, 259–272.
7. Craig-Schmidt, M.C. (1992) Fatty Acid Isomers in Foods, in *Fatty Acids in Foods and Their Health Implications,* (Chow, C.K., ed.) Marcel Dekker, New York.
8. Emken, E.A., Rohwedder, W.K., Adlof, R.O., DeJariais, W.J., and Gulley, R.M. (1986) Absorption and Distribution of Deuterium-Labeled *trans*- and *cis*-11-Octadecenoic Acid in Human Plasma and Lipoprotein Lipids, *Lipids 21*, 589–595.
9. Gurr, M.I. (1996) Dietary Fatty Acids with *trans* Unsaturation, *Nutr. Res. Rev. 9*, 259–279.
10. Hunter, J.E., and Applewhite, T.H. (1986) Isomeric Fatty Acids in the US Diet: Levels and Health Perspectives, *Am. J. Clin. Nutr. 44*, 707–717.
11. Mansour, M.P., and Sinclair, A.J. (1993) The *trans* Fatty Acid and Positional (*sn*-2) Fatty Acid Composition of Some Australian Margarines, Dairy Blends and Animal Fats, *Asia Pac. J. Clin. Nutr. 3*, 155–163.
12. Griinari, J.M., Dwyer, D.A., McGuire, M.A., Bauman, D.E., Palmquist, D.L, and Nurmela, K.V.V. (1998) *Trans*-Octadecenoic Acids and Milk Fat Depression in Lactating Dairy Cows, *J. Dairy Sci. 81*, 1251–1261.
13. Precht, D., and Molkentin, J. (1997) Effect of Feeding on Conjugated *cis* Δ9, *trans* Δ11-Octadecadienoic Acid and Other Isomers of Linoleic Acid in Bovine Milk Fats, *Nahrung 41*, 330–335.
14. Leth, T., Ovesen, L., and Hansen, K. (1998) Fatty Acid Composition of Meat from Ruminants with Special Emphasis on *trans* Fatty Acids, *J. Am. Oil Chem. Soc. 75*, 1001–1005.

15. Tocher, D.R., Leaver, M.J., and Hodgson, P.A. (1998) Recent Advances in the Biochemistry and Molecular Biology of Fatty Acyl Desaturases, *Prog. Lipid Res. 37,* 73–117.
16. Ntambi, J.M. (1995) The Regulation of Stearoyl-CoA Desaturase (SCD), *Prog. Lipid Res. 34,* 139–150.
17. Diczfalusy, U., Eggertsen, G., and Alexson, S.E.H. (1995) Clofibrate Treatment Increases Stearoyl-CoA Desaturase mRNA Level and Enzyme Activity in Mouse Liver, *Biochim. Biophys. Acta 1259,* 313–316.
18. Reddy, J.K., and Mannaerts, G.P. (1994) Peroxisomal Lipid Metabolism, *Annu. Rev. Nutr. 14,* 343–370.
19. Clarke, S.D., and Jump, D.B. (1994) Dietary Polyunsaturated Fatty Acid Regulation of Gene Transcription, *Annu. Rev. Nutr. 14,* 83–98.
20. Patel, M.S., Owen, O.E., Goldman, L.I., and Hanson, R.W. (1975) Fatty Acid Synthesis by Human Adipose Tissue, *Metabolism 24,* 161–173.
21. Shrago, E., and Spennetta, T. (1976) The Carbon Pathway for Lipogenesis in Isolated Adipocytes From Rat, Guinea Pig and Human Adipose Tissue, *Am. J. Clin. Nutr. 29,* 540–545.
22. Salminen, I., Mutanen, M., Jauhiainen, M., and Aro, A. (1998) Dietary *trans* Fatty Acids Increase Conjugated Linoleic Acid Levels in Human Serum, *Nutr. Biochem. 9,* 93–98.
23. Ha, Y.L., Storkson, J., and Pariza, M.W. (1990) Inhibition of Benzo(a)pyrene-Induced Mouse Forestomach Neoplasia by Conjugated Dienoic Derivatives of Linoleic Acid, *Can. Res. 50,* 1097–1101.
24. Ip, C., Briggs, S.P., Haegele, A.D., Thompson, H.J., Storkson, J., and Scimeca, J.A. (1996) The Efficacy of Conjugated Linoleic Acid in Mammary Cancer Prevention Is Independent of the Level or Type of Fat in the Diet, *Carcinogenesis 17,* 1045–1050.
25. Kramer, J.K.G., Sehat, N., Dugan, M.E.R., Mossoba, M.M., Yurawecz, M.P., Roach, J.A.G., Eulitz, K., Aalhus, J.L., Schaefer, A.L, and Ku, Y. (1998) Distributions of Conjugated Linoleic Acid (CLA) Isomers in Tissue Lipid Classes of Pigs Fed a Commercial CLA Mixture Determined by Gas Chromatography and Silver Ion–High-Performance Liquid Chromatography, *Lipids 33,* 549–558.

Chapter 15

Effect of Ionophores on Conjugated Linoleic Acid in Ruminal Cultures and in the Milk of Dairy Cows

V. Fellner[a], F.D. Sauer[b], and J.K.G. Kramer[c]

[a]Department of Animal Science, North Carolina State University, Raleigh, NC 27695–7621
[b]Center for Food and Animal Research, Agriculture and Agri-Food Canada, Ottawa, ON K1A 0C6, Canada
[c]Southern Crop Protection, Food Research Center, Agriculture and Agri-Food Canada, Guelph, ON N1G 2W1, Canada

Introduction

Ionophores are feed additives that have been used widely in beef-lot operations; their beneficial effects in improving production efficiency of ruminants have been well documented (1,2). Improvements in animal performance as a result of feeding ionophores are related in part to their effects on ruminal fermentation (3–6). Ionophores alter bacterial fermentation, which has a major effect on the efficiency of energy transformations in the rumen. The exact mode of action of ionophores is not yet clearly understood, but is related to their ability to modify the movement of ions across bacterial membranes (7). Ionophores are lipid soluble and capable of translocating specific ions across the membrane, resulting in the disruption of the membrane potential. The complex outer membrane of the gram-negative bacteria is relatively impermeable to ionophores. Gram-positive bacteria lack this outer membrane and are more susceptible to the ionophores (8). Gram-negative bacteria are the major propionate producers, whereas the gram-positive bacteria are responsible primarily for acetate and hydrogen. Hence, in the presence of ionophores, the gram-negative populations increase at the expense of the more susceptible gram-positive species. This microbial shift is associated with an increase in the ratio of propionate to acetate, decreased proteolysis, and a reduction in methane production, all of which contribute to the beneficial effects of ionophores on animal performance.

The addition of ionophores to lactating dairy cattle diets has not yet been approved for general use in North America. The successful use of ionophores in the beef-lot industry and the positive effects observed in *in vitro* studies clearly demonstrate the potential benefits of feeding such additives to dairy cows. There are limited data on the response of lactating cows to ionophore feeding.

Even less is known about the effects of ionophores on lipid metabolism in the rumen. Dietary unsaturated lipids undergo extensive biohydrogenation in the rumen (9,10). This process requires hydrogen but may be inhibited when ionophores are included in the diet. Ionophores could interfere with the process of ruminal biohydrogenation either directly by inhibiting the growth of gram-positive bacteria or indirectly as a result of reduced amounts of hydrogen produced.

Ionophores Alter Ruminal Biohydrogenation in Continuous Cultures

The effect of ionophores on fermentation and fatty acid biohydrogenation was evaluated in continuous cultures of ruminal microorganisms (11). Four ionophores were tested: monensin, nigericin, tetronasin, and valinomycin. In addition to differences in binding selectivities for cations, these ionophores differed in antiporter or uniporter activity. Monensin, nigericin, and tetronasin have a high binding affinity for Na, K, and Ca, respectively, and all three function as antiporters (12,13). Valinomycin binds preferentially to K but is a uniporter and does not exchange hydrogen ions (14).

We added the ionophores into the fermenters at a rate of 2 mL/h to provide 4 μg/(g feed·h) of respective ionophore (11). After 48 h of ionophore infusion, supplemental linoleic acid was infused continuously into the fermenters. Samples of ruminal culture were obtained before and after the addition of linoleic acid. In general, ionophores had a major effect on fatty acid metabolism by ruminal bacteria, and results clearly indicated that only ionophores that exhibited antiporter activity elicited a response irrespective of cation binding selectivity (Table 15.1). Valinomycin, a uniporter, had no apparent effect on bacterial fermentation or fatty acid biohydrogenation.

The concentration of saturated fatty acids (FA) in ruminal cultures not receiving any ionophore treatment was high (Table 15.1). More than 70% of the total fatty acids consisted of stearic and palmitic acids. The high concentration of the saturated FA is indicative of the extensive biohydrogenation of unsaturated FA by rumen bacteria and is consistent with results reported earlier (15).

All three ionophores (monensin, nigericin, and tetronasin) exhibiting antiporter activity had a similar influence on fatty acid metabolism by ruminal bacteria. They decreased the proportion of saturated FA and increased that of unsaturated FA in the culture medium (Table 15.1). The extent of linoleic acid biohydrogenation in the ab-

TABLE 15.1 Fatty Acid (FA) Composition of Culture Contents Receiving Linoleic Acid (13.8 mg/h) in the Absence (Control) or Presence of Continuous Addition (4 g/h) of Ionophore Compounds[a]

Fatty acid	Control	Monensin	Nigericin	Tetronasin	SEM
			g/100 g total FA		
18:0	45.6[b]	31.0[c]	26.2[d]	28.0[d]	0.80
18:1 trans	12.6[f]	16.8[g]	16.5[g]	15.8[g]	1.12
18:1 cis	7.1[b]	10.6[c]	13.5[d]	10.2[c]	0.50
18:2n-6	3.7[b]	9.5[c]	10.9[d]	13.3[e]	0.49
18:2 conjugated (CLA)	0.5[b]	1.2[c]	1.6[c]	1.9[c]	0.10

[a]Source: Ref. 11.
[b,c,d,e] $P < 0.01$.
[f,g] $P < 0.05$.

sence of ionophores averaged 80% but was reduced significantly to 58, 56, and 38% in the presence of monensin, nigericin, and tetronasin, respectively (Table 15.2). Consequently, large amounts of linoleic acid remained intact when ionophores were present in the culture medium (Table 15.2). In sheep, feeding salinomycin, an ionophore antibiotic with an activity similar to that of monensin, resulted in an increased proportion of the monounsaturated oleate (18:1) and a decreased proportion of the saturated stearate (18:0) in the duodenal digesta (16).

The ionophore-induced decrease in the extent of 18:2 biohydrogenation was also reflected in a significantly lower rate of 18:0 production. Stearic acid production rates were reduced by >80% in the presence of the ionophores (Table 15.2), whereas the production of the monounsaturated 18:1 fatty acid occurred at a much higher rate; 18:1 formation was 20% higher in the presence of monensin or nigericin. The production of *trans* fatty acids was not affected by ionophores, and the increase in 18:1 was due mainly to an increase in the *cis* isomer (Table 15.2). Tetronasin inhibited the rate of 18:0 formation but had no effect on the rates of 18:1 or the *trans* and *cis* isomers of 18:1, which were similar to rates observed in control cultures receiving no ionophore treatment. The level of infused 18:2 that remained intact was highest in cultures receiving tetronasin (Table 15.1).

Conjugated linoleic FA (CLA) represented a small proportion of the total FA in the ruminal culture (Table 15.1). The major CLA isomers were *cis*-9, *trans*-11-18:2, *trans*-10, *cis*-12-18:2, and a *trans*-9, *trans*-11-18:2. These three main isomers comprised nearly 90% of the total conjugated acids in the culture. The increase in total CLA content in ruminal cultures as a result of ionophore infusion was small but significant. From 0.5% in the control fermenters, CLA content increased by more than two, three, and four times in fermenters receiving monensin, nigericin, and tetronasin, respectively (Table 15.1). The increase in total CLA content as a result of feeding ionophores was due primarily to an increase in the *cis*-9, *trans*-11-18:2 isomer.

TABLE 15.2 Extent of Ruminal Biohydrogenation and Rates of Formation of End Products from Linoleic Acid Infused Continuously into Fermenters in the Absence (Control) or Presence of Respective Ionophores[a]

Fatty acid	Control	Monensin	Nigericin	Tetronasin	SEM
		mg/(L·h)[b]			
18:0	7.5[c]	1.4[d]	1.5[d]	0.9[d]	0.57
18:1 *trans*	4.2	4.1	4.0	3.7	0.94
18:1 *cis*	1.7[e]	3.4[f]	3.4[f]	1.6[e]	0.50
Biohydrogenation, %	80.0	58.0	56.0	38.0	—

[a]*Source:* Ref. 11.
[b]Based on the addition of 13.8 mg of 18:2n-6 per h to 700 mL of ruminal culture.
[c,d]$P < 0.05$.
[e,f]$P < 0.10$.

Effect of Ionophores on Fatty Acids in the Milk of Dairy Cows

In a lactating cow study, a herd of 88–109 Holstein dairy cattle was housed in a tie stall barn and fed monensin to determine its effect on total barn methane and carbon dioxide emissions, milk production, and milk fatty acid profiles (17). Monensin was added to the dairy ration gradually over a period of 7 d and then maintained at 24 ppm for 21 d. Milk samples were taken from 10 randomly selected cows before the ionophore was included in the diet. The same 10 cows were sampled 2 wk after receiving the ionophore treatment. During the entire period, total barn methane output was also monitored with the use of infrared detectors (18).

Dairy cows fed monensin consumed less dry matter and produced more milk, resulting in improved efficiency of milk production (17). Total barn methane production by lactating dairy cows decreased significantly by 21% from an average of 658 L/(cow·d) during the control period to 517 L/(cow·d) when monensin was included in the ration (17). However, during the same period of monensin feeding, total carbon dioxide expired by the cows remained unaffected; as a result, the ratio of carbon dioxide to methane in the barn increased from 11:1 before the addition of monensin to 15:1 after monensin was included in the diet (17). The shift in the ratio of the two gases produced by the cows in the barn clearly suggests that the reduction in methane output was indeed due to inhibited rates of methane production and not due to decreased feed consumption. During the period in which monensin was introduced into the lactating ration and total methane output in the barn declined, there was also a significant change in the milk fat percentages (17). Milk fat averaged 3.6% before the addition of monensin and declined to 2.7% as a result of feeding monensin (Table 15.3). The shift in rumen volatile fatty acids to a low acetate:propionate ratio was consistent with the effects of ionophores in increasing propionate concentration in the rumen.

The fatty acid profile of milk from cows in this study (Table 15.3) was relatively similar to values that have been reported earlier in other experiments (19,20). Monensin had a significant effect on the milk fatty acid profile (Table 15.3). The characteristic decrease in biohydrogenation of unsaturated fatty acids observed in mixed cultures (Tables 15.1 and 15.2) was evident in milk fatty acids. Total saturated fatty acids decreased and total unsaturated fatty acids increased in the milk of cows fed monensin. The concentration of 18:0 was lower, but total 18:1 and *trans* 18:1 levels were increased in the milk after ionophore feeding. As a result, total linoleic acid in the milk increased from 3.4 to 4.3% (Table 15.3).

Conjugated linoleic fatty acids made up only 0.8% of the total milk fat and increased to 1.3% as a result of feeding monensin (Table 15.3). Although the increase was small, the change was significant. A more detailed analysis of the milk fatty acids indicated that the 9*c*, 11*t*-18:2 and the 9*t*, 11*t*-18:2 were the major isomers that made up the total CLA (21). The analysis showed that >60% of the total CLA was comprised of the 9*c*, 11*t*-18:2 isomer and that the increase in CLA induced by including monensin in the diet was due mainly to an increase in this isomer (Table 15.3). Dhiman *et al.* (22) reported no increase in CLA content in milk of cows fed monensin. In that study, monensin was included at 12 ppm of dietary dry matter compared with 24 ppm fed in this study.

TABLE 15.3 Change in Milk Fatty Acids (FA) in Dairy Cows in the Absence (Control) or Presence of Ionophore Supplementation[a]

Fatty acid	Control	Monensin	SEM
Total fat, %	3.6	2.7	—
Individual FA		wt%	
10:0[b]	2.4[b]	1.8[c]	0.12
12:0[b]	3.2[b]	2.6[c]	0.19
16:0[b]	28.2[b]	24.4[c]	0.72
18:0[c]	11.2[d]	7.4[e]	0.59
18:1 trans[c]	4.0[d]	11.9[e]	0.90
18:1 cis	23.3	22.1	1.02
18:2n-6[b]	3.4[b]	4.3[c]	0.22
18:2 conjugated[b]	0.8[b]	1.3[c]	0.15
9c,11t-18:2	0.5	1.1	0.03
9t,11t-18:2	0.1	0.1	0.01

[a]Source: Refs. 17 and 21.
[b]$P < 0.05$.
[c]$P < 0.001$.

The changes in milk fatty acid profiles are consistent with those observed in ruminal cultures and clearly indicate that the shift in ruminal biohydrogenation as a result of feeding ionophores is reflected in altered composition of fat in the milk. There is renewed interest in this area as a result of the increased awareness of risks associated with the consumption of fat high in saturated FA. Consumer demand for an increase in the content of unsaturated fatty acids has resulted in research efforts to manipulate milk fat composition. In addition, the increasing evidence of CLA as potent anticarcinogenic agents has focused greater attention on foods that are consumed in large amounts and may be a potential source of CLA. The beneficial effect of feeding ionophores to lactating cattle is evidenced not only in the increase in total unsaturated FA in the milk but more specifically in an increase in linoleic acid, which is an essential FA, as well as the CLA, all of which are desirable attributes that have positive health benefits.

References

1. Goodrich, R.D., Garrett, J.E., Gast, D.R., Kirick, M.A., Larson, D.A., and Meiske, J.C. (1984) Influence of Monensin on the Performance of Cattle, *J. Anim. Sci. 58,* 1484–1498.
2. Stock, R.A., Landert, S.B., Stroup, W.W., Larson, E.M., Parrott, J.C., and Britton, R.A. (1995) Effect of Monensin and Monensin and Tylosin Combination on Feed Intake Variation of Feedlot Steers, *J. Anim. Sci 73,* 39–44.
3. Newbold, C.J., Wallace, R.J., and Walker, N.D. (1993) The Effect of Tetronasin and Monensin on Fermentation, Microbial Numbers and the Development of Ionophore-Resistant Bacteria in the Rumen, *J. Appl. Bacteriol. 75,* 129–134.
4. Sauer, F.D., and Teather, R.M. (1987) Changes in Oxidation Reduction Potentials and Volatile Fatty Acid Production by Rumen Bacteria When Methane Synthesis Is Inhibited, *J. Dairy Sci. 70,* 1835–1840.

5. Schelling, G.T. (1984) Monensin Mode of Action in the Rumen, *J. Anim. Sci. 58,* 1518–1527.
6. Yang, C.J., and Russell, J.B. (1993) The Effect of Monensin Supplementation on Ruminal Ammonia Accumulation *in vivo* and the Numbers of Amino Acid-Fermenting Bacteria, *J. Anim. Sci. 71,* 3470–3476.
7. Russell, J.B., and Strobel, H.J. (1989) Effect of Ionophores on Ruminal Fermentation, *Appl. Environ. Microbiol. 55,* 1–6.
8. Chen, M., and Wolin, M.J. (1979) Effect of Monensin and Lasalocid-Sodium on the Growth of Methanogenic and Rumen Saccharolytic Bacteria, *Appl. Environ. Microbiol. 38,* 72–77.
9. Ferlay, A., Chabrot, J., Elmeddah, Y., and Doreau, M. (1993) Ruminal Lipid Balance and Intestinal Digestion by Dairy Cows Fed Calcium Salts of Rapeseed Oil Fatty Acids or Rapeseed Oil, *J. Anim. Sci. 71,* 2237–2245.
10. Wu, Q., Ohajuruke, O.A., and Palmquist, D.L. (1991) Ruminal Synthesis, Biohydrogenation, and Digestibility of Fatty Acids by Dairy Cows, *J. Dairy Sci. 74,* 3025–3034.
11. Fellner, V., Sauer, F.D., and Kramer, J.K.G. (1997) Effect of Nigericin, Monensin, and Tetronasin on Biohydrogenation in Continuous Flow-Through Ruminal Fermenters, *J. Dairy Sci. 80,* 921–928.
12. Grandjean, J., and Laszlo, P. (1983) Solution Structure and Cation-Binding Abilities of Two Quasi-Isomorphous Antibiotic Ionophores, M 139603 and Tetronomycin, *Tetrahedron Lett. 24,* 3319–3321.
13. Pressman, B.C. (1976) Biological Applications of Ionophores, *Annu. Rev. Biochem. 45,* 501–530.
14. Ramos, S., and Kaback, H.R. (1977) The Electrochemical Proton Gradient in *Escherichia coli* Membrane Vesicles, *Biochemistry 16,* 848–851.
15. Fellner, V., Sauer, F.D., and Kramer, J.K.G. (1995) Steady-State Rates of Linoleic Acid Biohydrogenation by Ruminal Bacteria in Continuous Culture, *J. Dairy Sci. 78,* 1815–1823.
16. Kobayashi, Y., Wakita, M., and Moshino, S. (1992) Effects of the Ionophore Salinomycin on Nitrogen and Long-Chain Fatty Acid Profiles of Digesta in the Rumen and the Duodenum of Sheep, *Anim. Feed Sci. Technol. 36,* 67.
17. Sauer, F.D., Fellner, V., Kinsman, R., Kramer, J.K.G., Jackson, H.A., Lee, A.J., and Chen, S. (1998) Methane Output and Lactation Response in Holstein Cattle with Monensin or Unsaturated Fat Added to the Diet, *J. Anim. Sci. 76,* 906–914.
18. Kinsman, R., Sauer, F.D., Jackson, H.A., and Wolynetz, M.S. (1995) Methane and Carbon Dioxide Emissions from Dairy Cows in Full Lactation Monitored over a Six Month Period, *J. Dairy Sci. 78,* 2760–2766.
19. Jiang, J., Bjoerck, L., Fonden, R., and Emanuelson, M. (1996) Occurrence of Conjugated *cis*-9, *trans*-11-Octadecadienoic Acid in Bovine Milk: Effect of Feed and Dietary Regimen, *J. Dairy Sci. 79,* 438–445.
20. Wonsil, B.J., Herbein, J.H., and Watkins, B.A. (1994) Dietary and Ruminally Derived *trans*-18:1 Fatty Acids Alter Bovine Milk Lipids, *J. Nutr. 124,* 556–562.
21. Kramer, J.K.G., Fellner, V., Dugan, M.E.R., Sauer, F.D., Mossoba, M.M., and Yurawecz, M.P. (1997) Evaluating Acid and Base Catalysts in the Methylation of Milk and Rumen Fatty Acids with Special Emphasis on Conjugated Dienes and Total *trans* Fatty Acids, *Lipids 32,* 1219–1228.
22. Dhiman, T.R., Anand, G.R., and Satter, L.D. (1996) Conjugated Linoleic Acid Content of Milk from Cows Fed Different Diets, *J. Dairy Sci. 79 (Suppl. 1),* 137.

Chapter 16

Species-Dependent, Seasonal, and Dietary Variation of Conjugated Linoleic Acid in Milk

Gerhard Jahreis[a], Jan Fritsche[b], and Jana Kraft[a]

[a]Institute of Nutrition, Friedrich Schiller University, D-07743 Jena, Germany
[b]Institute of Biochemistry and Food Chemistry, University of Hamburg, D-20145 Hamburg, Germany

Introduction

With regard to the anticarcinogenic and other beneficial effects of conjugated linoleic acid (CLA), there is an interest in increasing the CLA content of milk fat rather than the absolute amount of CLA secreted in the milk. In recent years, numerous publications have addressed several factors influencing the CLA content in milk fat. Typical concentrations of CLA in commercial milk fat are 3–6 mg/g of fat (1). The objectives of this review are to summarize different influences on CLA content in milk fat and to evaluate the possibilities of increasing its concentration.

Influence of Species, Breed, Age, and Individual Conditions

Species

CLA is formed mainly *via* isomerization of linoleic acid by the anaerobe *Butyrivibrio fibrisolvens* in the rumen. Therefore, Kramer *et al.* (2) recommended naming the *cis*-9, *trans*-11-isomer "rumenic acid." However, ruminants are not the only producers; monogastrides also can produce CLA. Pollard *et al.* (3) and Salminen *et al.* (4) showed that CLA may be formed primarily by conversion of dietary *trans* fatty acids, possibly by Δ-9 desaturation of *trans* vaccenic acid.

There are enormous differences in the milk CLA level of different ruminants (cow, goat, sheep) compared with nonruminants (horse, pig, human; Fig. 16.1). The highest CLA concentration among the collected bulk milk samples was found in sheep's milk [1.2% of total fatty acid methyl esters (FAME)]. The milk fat of cows fed comparably contained ~0.7%. Goat's milk had the lowest CLA concentration (0.6%) of the ruminants.

Among nonruminants, mare's milk was nearly CLA-free (0.09%) and sow's milk contained only small amounts of CLA (0.23%), caused by contamination *via* the food chain. In comparison to these nonruminants, human milk was relatively rich in CLA (0.39%). Forgerty *et al.* (5) found 0.58% CLA in breast milk from Australian women (Table 16.1). There are, of course, differences between those who do and do

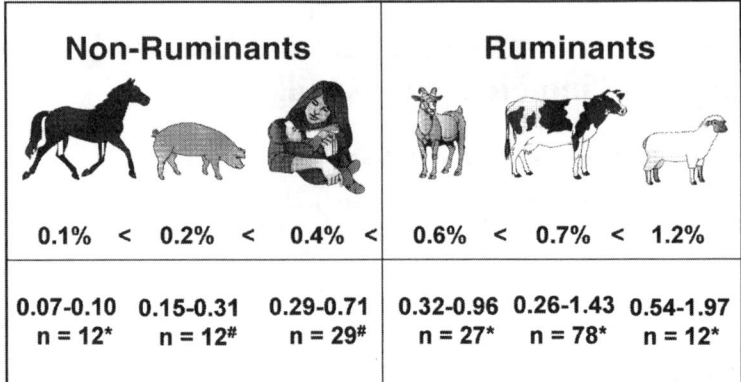

Fig. 16.1. Comparative milk-conjugated linoleic acid (CLA) levels in mammals (% of FAME, mean values and ranges; *bulk milk samples, # individual samples). *Source:* Ref. 25 and unpublished data.

TABLE 16.1 Conjugated Linoleic Acid (CLA) Content in Human Milk Fat

	CLA (% of total fatty acid methyl esters)		
	Mean	Range	Reference
Australian women	0.58	0.31–0.8	5
Hare Krishna	1.12	0.97–1.2	5
U.S. women	0.18	0.14–0.2	26
German women	0.39	0.27–0.53	25
U.S. women	3.64[a]	2.23–5.43[a]	27

[a]Expressed in mg/g fat.

not drink milk. Milk of women who belonged to the Hare Krishna religious sect, whose diet excludes meat, but includes butter and milk, contained twice the amount of CLA.

Age and Genetically Determined Condition of Cows

Stanton *et al.* (6) reported that older cows (>4 lactations) produce more CLA in their milk fat than cows with lactation numbers of 2–4. These results agree with those of Lal and Narayanan (7), whose studies also showed a higher CLA level with increasing age of cows. No effect of stage of lactation on milk fat CLA content could be ascertained (6); however, in this investigation, only lactation periods were compared, which excluded the beginning and the end of a complete lactation. Further results of these authors showed that some individual cows appeared to be more efficient at incorporating CLA into milk (range from 1.5 to 16 mg CLA/g milk fat). A similar degree of variation was reported by Jiang *et al.* (8) among individual cows (between 2.5 and 17.7 mg/g fat). Lawless *et al.* (9) analyzed milk of three different breeds (Holstein-Friesian, Montbeliardes, and Normandes) and found variations in the CLA concentration among individual cows within the investigated herds ranging from 4.8 to 35.0 mg/g fat. The mean CLA level among the breeds (16 animals each) varied only from 14.5 to 18.3 mg/g fat.

Kelly *et al.* (10) also reported a wide variation between different herds. Cows received either a pasture diet or a totally mixed diet of forage and concentrate. The authors observed up to a threefold difference among individuals fed the same diet. The CLA level ranged between 2.4 and 7.0 mg/g of milk fat. A greater variation, between 2.5 and 17.7 mg/g fat, was found by Jiang *et al.* (8). Their analysis determined that the highest CLA level (11.28 mg/g) occurred in cows fed restricted amounts of a diet with a lower forage:concentrate ratio. Decreasing the proportion of fiber and increasing the content of soluble carbohydrates resulted in increased formation of *trans* vaccenic acid and, proportionately, CLA. But the "dilution factor" should also be taken into consideration, i.e., a decrease of milk fat yield and a postulated constant formation level of CLA will lead to higher percentages of CLA in milk fat.

These published results indicate the influence of breed, age of dairy cow, and possibly the stage of lactation. The wide range of individual CLA content in milk fat may be related to factors that are associated with ruminal fermentation, with the formation of CLA from *trans* vaccenic acid in the mammary gland, the total CLA synthesis per day, and the total milk fat produced per day.

Influence of the Feed Regimen

Fatty Acids

In two experiments, Dhiman *et al.* (11) determined the CLA content of milk from cows offered diets rich in polyunsaturated fatty acids. Of course, an increase can only be expected when oil is readily accessible to the rumen microorganisms. In experiment 1, linseed oil at two different concentrations (2.2 and 4.4%) on a dry matter (DM) basis was fed. Milk fat CLA increased from 3.9 mg/g in the control group to 15.9 and 16.3 mg/g fat, respectively (Table 16.2). In experiment 2, rations with four different soybean oil concentrations (0.5, 1, 2, and 4%) on a DM basis were offered. A marked increase of CLA content from 5.0 to 20.8 mg/g of total fatty acids followed.

Chouinard *et al.* (12) examined Ca salts of three different oils, i.e., canola, soybean, and linseed. Feeding Ca salts of fatty acids from soybean oil caused the highest level of CLA (Table 16.2). The application of 300 g sunflower oil per day increased the CLA content from 0.9 to 1.2% of total fatty acids (13).

In the experiment of Offer *et al.* (14), three oil supplements were compared. Cows received 250 g/d linseed oil or marine oil, or 95 g/d of tuna orbital oil. All supplements increased the CLA percentages in milk fat, with marine oil having the greatest effect (Table 16.2).

TABLE 16.2 Changes in Cow's Milk Conjugated Linoleic Acid (CLA) Due to Dietary Oils Compared with the Unsupplemented Control Groups

Oil type	Control	Treatment (mg/g fat)[a]	Δ CLA	Reference
4% Ca salts of FA from canola oil	3.5	13.0	9.5	12
4% Ca salts of FA from soybean oil	3.5	22.0	18.5	12
4% Ca salts of FA from linseed oil	3.5	19.0	15.5	12
200 mL/d fish oil of menhaden type	6.2	17.5	11.3	12
400 mL/d fish oil of menhaden type	6.2	16.5	10.3	12
4.4% linseed oil	3.9	16.3	12.4	28
4.0% soybean oil	5.0	20.8	15.8	28
300 g/d sunflower oil	0.9	1.2	0.3	13
5.3% sunflower oil	[b]	24.4	—	15
5.3% linseed oil	[b]	16.7	—	15
5.3% peanut oil	[b]	13.3	—	15
1650 g/d full-fat rapeseed	4.8	7.9	3.1	6
250 g/d linseed oil	0.16%	0.28%	0.12%	14
250 g/d marine oil	0.16%	1.55%	1.39%	14
95 g/d tuna orbital oil	0.16%	0.52%	0.36%	14

[a]Or percentage of total fatty acid methyl esters.
[b]No control group without oil supplement.

Kelly et al. (15) showed a substantial variation among different dietary oil treatments (peanut oil, sunflower oil, and linseed oil) on milk fat CLA concentration. Sunflower oil contains the most linoleic acid and resulted in the (significantly) highest concentration of CLA (Table 16.2). This concentration exceeded the normal level by 500% for cows consuming a traditional diet. In dairy cows consuming either rapeseed oil or sunflower oil in combination with buffer supplements, we found CLA concentrations between 1.12 and 1.46% of FAME. CLA yield per day was significantly increased in cows fed 400 g sunflower oil plus 200 g of dietary buffer ($NaHCO_3$/MgO/$CaCO_3$, 6:2:1) compared with the unbuffered group (15.7 g vs. 9.8 g).

There were clear differences in CLA percentages in milk from cows fed either 1118 g sunflower oil per day for 2 wk (15) or 400 g/d for 3 mo (Table 16.3). The CLA percentage was inversely correlated with the daily fat yield (2.44% CLA and 730 g fat vs. 1.41% CLA and 1174 g fat). But the CLA yields per day for the sunflower groups were similar in both experiments. It follows that the totally synthesized fat per day determines the "dilution" of CLA.

Among different dietary buffer substances ($NaHCO_3$, $NaHCO_3$ + $KHCO_3$, MgO), only MgO increased the ratio of cis-18:1 to trans-18:1 in milk (17). But the milk CLA content (11.1 mg/g fat) was not affected as a result of supplementation with these buffer substances (18). Cows were fed total mixed ration (TMR) containing 4% Ca salts of canola oil fatty acid in contrast to the above-mentioned experiments with unprotected oils.

TABLE 16.3 Influence of Fat Yield on the Percentage and Yield of Conjugated Linoleic Acid (CLA)

Oil supplement	Milk yield (kg/d)	Fat (%)	Fat yield (g/d)	CLA (%)	CLA yield (g/d)	Reference
Peanut, 1118 g/d	34.2	2.26	770	1.33	10.2	15
Sunflower, 1118 g/d	33.4	2.18	730	2.44	17.8	15
Linseed, 1118 g/d	35.1	2.81	810	1.67	13.5	15
Rapeseed, 400 g/d + 100 g buffer/d	31.8[a]	3.70[a]	1151[a]	1.12[b]	12.0[b]	16
Sunflower, 400 g/d + 200 g buffer/d	31.8[a]	3.73[a]	1174[a]	1.41[b]	15.7[b]	16

[a]Mean values of the 3-mo experimental period.
[b]Results from the second half of the experiment.

CLA content can also be increased by feeding full-fat extruded soybeans and cottonseed (19). Cows of the control group received soybean meal (13.5% of DM). The treatment groups were offered either full-fat extruded soybeans or full-fat extruded cottonseed (12% of DM). CLA concentration of milk increased from 3.4 mg/g fat in the control group to 6.9 and 6.0 mg/g fat (full-fat extruded soybeans or cottonseed, respectively).

A mixture of tallow/yellow grease (7:1) increased the milk-CLA concentration in cows receiving the control diet from 2.2 mg/g fat to 3.8 and 4.7 mg/g fat, respectively (2.2 and 4.4 % of the fat mixture, respectively) (12).

The results of these experiments demonstrated that the most effective supplements for increasing the percentage of CLA in milk were sunflower oil and marine oil, and Ca salts of soybean and linseed oil (Table 16.2). Oil supplements also depress milk fat content while at the same time increasing the percentage of *trans* fatty acids and, in this connection, the CLA level.

Is there a possibility of disrupting the strong correlation between CLA and *trans* fatty acids? The health benefits from increased CLA content in milk fat may be countered by the metabolic effects of *trans* fatty acids, which are generally accepted as unhealthy. In particular, *trans*-10-octadecenoic acid has been shown to have negative effects in human beings as well as in dairy cows, leading to milk fat depression. The relationship between the platelet *trans* fatty acids and the degree of angiographically assessed coronary artery disease (CHD) was examined in 191 patients (20). After adjustment for established CHD risk indicators, *trans*-10-octadecenoic acid and elaidic acid were positively associated with CHD. However, there was no significant correlation between CHD and the *trans* vaccenic acid in platelets. Furthermore, humans and lactating cows can produce cis-9,*trans*-11-octadecadienoic acid endogenously from *trans*-11-octadecenoic acid (4, 21). The *trans* fatty acid fraction of the milk fat consists mainly of *trans* vaccenic acid. Addition of buffer to oil-rich dairy rations decreased the formation of *trans*-10- octadecenoic acid without influencing *trans* vaccenic acid and CLA levels in milk fat (16).

Feed and Dietary Regimen

In most experiments, pasture feeding was shown to be effective in increasing CLA content in milk fat. Dhiman *et al.* (11) found CLA values between 8.4 and 22.7 mg/g in milk fat of cows that consumed one third, two thirds, or all of their daily feed from the pasture (Table 16.4). Cows grazing only from the pasture had the highest CLA content in milk fat. This investigation also measured the influence of high-oil corn silage feeding compared with normal corn silage diet on CLA concentration. The CLA content ranged (insignificantly) between 3.7 (normal corn) and 4.0 mg/g fat (high-oil corn). In another study, the CLA level of the grazing cows was double that of cows consuming a total mixed diet (Table 16.4) (10).

Only a marginal increase of CLA (+10%) was found by Griinari *et al.* (13) when cows were grazed on mixed pasture compared with cows fed grass silage-based diets. However, CLA content in milk tended to increase toward the end of the season, suggesting that change in pasture forage quality may have an effect (or an accumulation effect caused by lower milk-fat yield?). Chouinard *et al.* (18) measured the influence of timothy silage prepared from different grass growth stages (early heading, flowering, and second cutting). The milk CLA content in the respective order was as follows: 11.4, 4.8, and 8.1 mg/g fat with a significantly higher level for the early-heading stage. In another study, an influence of the forage maturity on milk fat CLA content in cows fed mixed grass pasture was not ascertainable (22).

CLA in milk fat can also be increased through changing the ratio of forage to concentrate (Table 16.4). A low-forage diet increased the CLA concentration significantly compared with the high-forage diet (18). Griinari *et al.* (23) also showed that changing the forage to concentrate ratio influences the milk-fat concentration of CLA. Furthermore, a dietary restriction in connection with a lower forage:concentrate

TABLE 16.4 Changes of Conjugated Linoleic Acid (CLA) in Milk Fat Due to Different Feed and Dietary Regimens

Feed and dietary regimen	Control[a] (mg/g fat)	Treatment (mg/g fat)	Reference
Growth stage: early heading	—	11.4	18
Growth stage: flowering	—	4.8	18
Growth stage: second cutting	—	8.1	18
Forage/concentrate ratio, 80:20	—	4.5	18
Forage/concentrate ratio, 20:80	—	7.3	18
One third pasture	—	8.4	11
Two thirds pasture	—	13.7	11
Only pasture	—	22.7	11
Dietary restriction	5.0	11.3	8
Ad libitum consumption	5.0	6.6	8
Grazing on pasture	4.6	10.9	10

[a]—Indicates no control group.

ratio contributes to an increase of CLA content in milk (8). Restricted-fed cows had twice the concentration (11.28 mg/g) of CLA in their milk fat compared with the control group (5.04 mg/g) and cows fed *ad libitum* (6.56 mg/g). This could also be a result of diminished fat synthesis (fewer triacylglycerols increase the proportion of CLA).

Influence of Season and Farm Management

The concentration of milk CLA in ruminant species is season dependent. All ruminant species show a significant decrease of CLA level at the end of the winter season (March) and a considerable increase during the summer period (Fig. 16.2). The exceptions were indoor-fed cows and goats; comparatively, they had the lowest CLA concentration in their milk over the whole year (24,25). The greatest differences were measured in sheep's milk, i.e., 1.28% in summer and 0.54% at the end of the winter period. An impressive season-dependent cycle of CLA content was analyzed (monthly measurement) in a dairy cow herd of an ecological farm (260 animals) in comparison to an indoor herd (1500 animals) fed only fermented feed for the whole year (Fig. 16.3). The highest concentrations of CLA were found especially during the grazing period (May to September). The variation of CLA between bulk milk samples of the different herds was substantial and ranged between 0.26 and 1.14% FAME. The significantly higher content of CLA was measured in the milk fat of the ecological group. This high level likely resulted from the higher proportion of polyunsaturated fatty acids in pasture fodder as well as from the differing composition of dairy rations. In contrast to these results with dairy cows, there were no differences found between indoor-fed goats (small variation during the year, mean 0.7%) and pasture-fed goats (25).

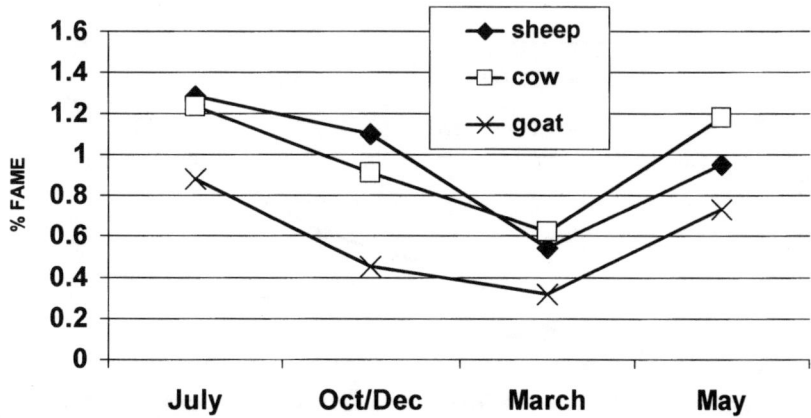

Fig. 16.2. Conjugated linoleic acid (CLA) content of ruminant milk fat in different seasons. *Source:* Ref. 25.

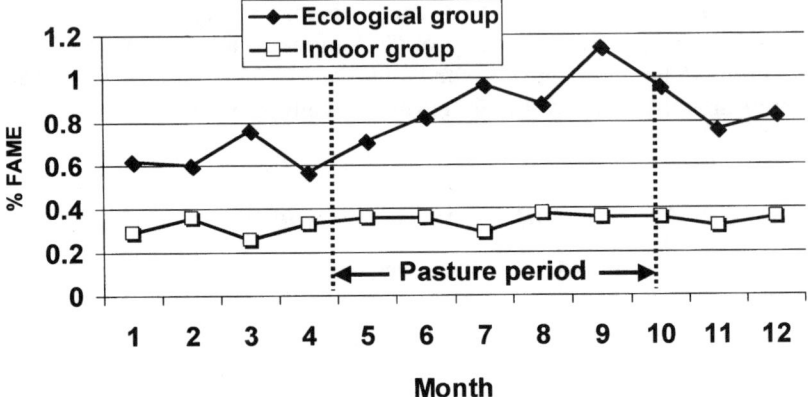

Fig. 16.3. Percentage of conjugated linoleic acid (CLA) in bulk milk fat of two alternatively farmed dairy cow herds over a 1-y period. *Source:* Ref. 24.

Conclusions

The survey results prove that various dietary situations are able to modify the ruminal environment and affect milk content of CLA. The experiments cited clearly indicate that CLA content in milk fat can be increased through dietary manipulation. In particular, supplementation with plant oils high in PUFA significantly enhances the CLA level. These unsaturated fatty acids appear only partly in the milk fat because they also become either biohydrogenated to saturated isomers or partially biohydrogenated to various unsaturated isomers. Two of the most important intermediates are *trans* vaccenic acid and CLA. There are close correlations between these two derivatives of the fed PUFA (8,24,25; Fig. 16.4). Furthermore, feeding cows low-fiber diets plus dietary unsaturated fat increases the content of *trans*-10-octadecenoic acid (21).

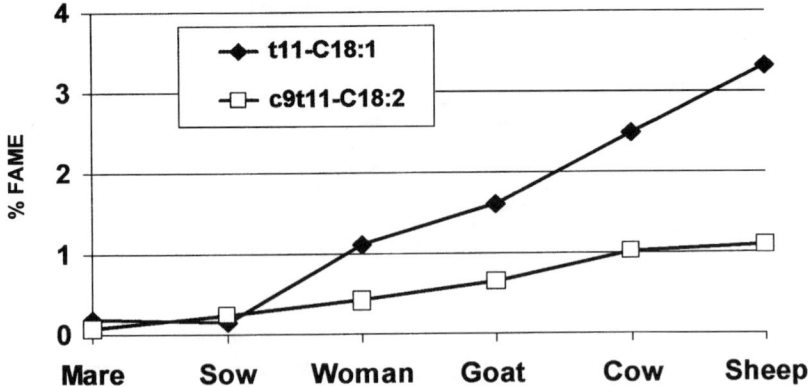

Fig. 16.4. Correlation between *trans* vaccenic acid and conjugated linoleic acid (CLA) concentration in the milk fat of different species. *Source:* Ref. 25.

Sunflower- rich diets (400 g/d) also increase the *trans*-10-/*trans*-11-18:1 ratio; buffer supplements up to 200 g/d decrease the ratio (16). The production of the intermediate *trans*-10-octadecenoic acid seems to be increased when highly acidic conditions occur in the rumen. More recently, evidence suggests that the *trans*-10-octadecenoic acid inhibits synthesis of the short- and medium-chain fatty acids in the mammary gland and induces milk fat depression (22). On the other hand, the percentage of CLA in the milk fat was not significantly influenced by pH conditions in cows fed buffer-supplemented PUFA-rich rations (16).

The milk fat CLA level can be affected by a number of factors as follows:

- Intake of pasture (PUFA-rich grass)
- Intake of polyunsaturated fatty acids, especially plant oils
- Dietary intake level and milk yield (CLA dilution in highly productive herds or individual animals)
- Ratio of forage to concentrate
- Stage of lactation (mobilization of body-fat stores and use of feed-derived fatty acids at the beginning of the lactation period; in the later stages of lactation, is predominantly the fat synthesis in the udder)
- Seasonal conditions
- Breed and individual conditions
- Production management (pasturing, silage feeding, composition and maturity stage of forage, indoor or outdoor farming)

References

1. Kelly, M.L., and Bauman, D.E. (1996) Conjugated Linoleic Acid: A Potent Anticarcinogen Found in Milk Fat, *Proc. Cornell Nutr. Conf. Feed Manuf.*, pp. 68–74, Cornell University, Ithaca, NY.
2. Kramer, J.K.G., Parodi, P.W., Jensen, R.G., Mossoba, M.M., Yurawecz, M.P., and Adlof, R.O.A. (1998) (Letter) Rumenic Acid: A Proposed Common Name for the Major Conjugated Linoleic Acid Isomer Found in Natural Products, *Lipids 33*, 835.
3. Pollard, M.R, Gunstone, F.D, James, A.T, and Morris, L.J. (1980) Desaturation of Positional and Geometric Isomers of Monoenoic Fatty Acids by Microsomal Preparations from Rat Liver, *Lipids 15*, 306–314.
4. Salminen, I., Mutanen, M., Jauhiainen, M., and Aro, A. (1998) Dietary *trans* Fatty Acids Increase Conjugated Linoleic Acid Levels in Human Serum, *Nutr. Biochem. 9*, 93–98.
5. Fogerty, A.C., Ford, G.L., and Svoronos, D. (1988) Octadeca-9,11-dienoic Acid in Foodstuffs and in the Lipids of Human Blood and Breast Milk, *Nutr. Rep. Int. 38*, 937–944.
6. Stanton, C., Lawless, F., Kjellmer, G., Harrington, D., Devery, R., Connolly, J.F., and Murphy, J. (1997) Dietary Influences on Bovine Milk *cis*-9, *trans*-11-Conjugated Linoleic Acid Content, *J. Food Sci. 62*, 1083–1086.
7. Lal, D., and Narayanan, K.M. (1984) Effect of Lactation Number on the Polyunsaturated Fatty Acids and Oxidative Stability of Milk Fats, *Indian J. Dairy Sci. 37*, 225–229.

8. Jiang, J., Bjoerck, L., Fondén, R., and Emanuelson, M. (1996) Occurrence of Conjugated *cis*-9, *trans*-11-Octadecadienoic Acid in Bovine Milk: Effects of Feed and Dietary Regimen, *J. Dairy Sci. 79,* 438–445.
9. Lawless, F., L'Escop, P., Devery, R., Connolly, B., Murphy, J., and Stanton, C. (1998) The Effect of Animal Breed on the Levels of CLA (*c*-9,*t*-11) in Bovine Milk, *Int. Dairy J. 8,* 578 (Abstr.).
10. Kelly, M.L., Kolver, E.S., Bauman, D.E., Van Amburgh, M.E., and Muller, L.D. (1998) Effect of Intake of Pasture on Concentrations of Conjugated Linoleic Acid in Milk of Lactating Cows, *J. Dairy Sci. 81,* 1630–1636.
11. Dhiman, T.R., Anand, G.R., Satter, L.D., and Pariza, M.W. (1996) Conjugated Linoleic Acid Content of Milk from Cows Fed Different Diets, *J. Dairy Sci. 79 (Suppl. 1),* 35.
12. Chouinard, P.Y., Corneau, L., Bauman, D.E., Butler, W.R., Chilliard, Y., and Drackley, J.K. (1998) Conjugated Linoleic Acid Content of Milk from Cows Fed Different Sources of Dietary Fat, *J. Anim. Sci. 76 (Suppl. 1),* 233 (Abstr. 906).
13. Griinari, J.M., Nurmela, K., Sairanen, A., Nousiainen, J.I., and Khalili, H. (1998) Effect of Dietary Sunflower Oil and Pasture Forage Maturity on Conjugated Linoleic Acid (CLA) Content in Milk Fat from Lactating Dairy Cows, *J. Anim. Sci. 76 (Suppl. 1),* 300 (Abstr. 1172).
14. Offer, N.W., Dixon, J., and Speake, B.K. (1998) Effect of Dietary Fat Supplements on Levels of *trans* Acids and CLA in Bovine Milk, CLA What's Going On. European Concerted Action No. 1, p. 4.
15. Kelly, M.L., Berry, J.R., Dwyer, D.A., Griinari, J.M., Chouinard, P.Y., Van Amburgh, M.E., and Bauman, D.E. (1998) Dietary Fatty Acid Sources Affect Conjugated Linoleic Acid Concentrations in Milk from Lactating Dairy Cows, *J. Nutr. 128,* 881–885.
16. Tischendorf, F., Jahreis, G., Möckel, P., and Schöne, F (1998) Influence of Dietary Buffers on Milk Fat Content and Fatty Acid Profile in Dairy Cows Fed Rations Rich in Mono- or Polyunsaturated Fatty Acids, *Proc. Soc. Nutr. Physiol. 8,* 743.
17. Thivierge, M.C., Chouinard, P.Y., Levesque, J., Girard, V., Seoane, J.R., and Brisson, G.J. (1998) Effect of Buffers on Milk Fatty Acids and Mammary Arteriovenous Differences in Dairy Cows Fed Ca Salts of Fatty Acids, *J. Dairy Sci. 81,* 2001–2010.
18. Chouinard, P.Y., Corneau, L., Kelly, M.L., Griinari, J.M., and Bauman, D.E. (1998) Effect of Dietary Manipulation on Milk Conjugated Linoleic Acid Concentrations, *J. Anim. Sci. 76 (Suppl. 1),* 233 (Abstr. 907).
19. Dhiman, T.R., Helmink, E.D., McMahon, D.J., Fife, R.L., and Pariza, M.W. (1998) Conjugated Linoleic Acid Content of Milk from Cows Fed Extruded Oilseeds, *J. Anim. Sci. 76 (Suppl. 1),* 353 (Abstr. 1386).
20. Hodgson, J.M., Wahlqvist, M.L., Boxall, J.A., and Balazs, N.D. (1996) Platelet *trans* Fatty Acids in Relation to Angiographically Assessed Coronary Artery Disease, *Atherosclerosis 120,* 147–154.
21. Corl, B.A., Chouinard, P.Y., Bauman, D.E., Dwyer, D.A., Griinari, J.M., and Nurmela, K.V. (1998) Conjugated Linoleic Acid in Milk Fat of Dairy Cows Originates in Part by Endogenous Synthesis from *trans*-11 Octadecenoic Acid, *J. Anim. Sci. 76 (Suppl. 1),* 233 (Abstr. 908).
22. Griinari, J.M., Dwyer, D.A., McGuire, M.A., Bauman, D.E., Palmquist, D.L., and Nurmela, K.V.V. (1998) *Trans*-Octadecenoic Acids and Milk Fat Depression in Lactating Dairy Cows, *J. Dairy Sci. 81,* 1251–1261.

23. Griinari, J.M., Dwyer, D.A., McGuire, M.A., and Bauman, D.E. (1996) Partially Hydrogenated Fatty Acids and Milk Fat Depression, *J. Dairy Sci. 79,* 177 (Abstr.).
24. Jahreis, G., Fritsche, J., and Steinhart, H. (1997) Conjugated Linoleic Acid in Milk Fat: High Variation Depending on Production System, *Nutr. Res. 17,* 1479–1484.
25. Jahreis, G., Fritsche, J., Schöne, F., and Steinhart, H. (1998) CLA in Milk of Different Species, CLA What's Going On. European Concerted Action No. 1, p. 4.
26. Jensen, R.G., Lammi-Keefe, C.J., Hill, D.W., Kind, A.J., and Henderson, R. (1998) The Anticarcinogenic Conjugated Fatty Acid, $9c,11t$-18:2, in Human Milk: Confirmation of Its Presence, *J. Hum. Lact. 14,* 23–27.
27. McGuire, M.K., Park, Y., Behre, R., Harrison, L.Y., Shultz, T.D., and McGuire, M.A. (1997) Conjugated Linoleic Concentrations of Human Milk and Infant Formula, *Nutr. Res. 17,* 1277–1283.
28. Dhiman, T.R., Satter, L.D., Pariza, M.W., Galli, M.P., and Albright, K. (1997) Conjugated Linoleic Acid (CLA) Content of Milk from Cows Offered Diets Rich in Linoleic and Linolenic Acid, *J. Dairy Sci. 80 (Suppl. 1),* 184 (P171).

Chapter 17

Dietary Control of Immune-Induced Cachexia: Conjugated Linoleic Acid and Immunity

Mark E. Cook[a,b,c,d], D. DeVoney[b], B. Drake[b], M.W. Pariza[b,c,d], L. Whigham[d], and M. Yang[b]

[a]Animal Sciences Department, [b]Environmental Toxicology Center, [c]Food Microbiology and Toxicology, and [d]Department of Nutritional Sciences, University of Wisconsin, Madison, WI 53706

Introduction

Our understanding of the profound immunomodulatory effects of conjugated linoleic acid (CLA) developed initially from research on animal growth and development in which nutritional, pharmacological, and management practices are used to optimize animal growth. Because infection and subsequent immune response classically result in weight loss through cachexia, these practices have included ways to reduce exposure or reduce the immune response itself. We were in search of an agent that could uncouple essential immune function from the negative wasting effects. We have found that agent in CLA.

Environmental Microbes and Animal Growth

Early in the 1950s, the role of the microbial environment in growth and animal development was beginning to be recognized. Studies with animals raised in gnotobiotic chambers showed an improvement in growth of >10% compared with animals raised in an environment in which microbes were present (1). Investigations of the role of environmental influences on animal growth continued through the 1950s. Numerous authors (2–4) showed that rearing young chicks in a "new" environment (not exposed to older birds) resulted in marked improvements in growth and feed efficiency. Chicks exposed to an environment that housed adult birds showed suppressed growth and feed efficiency. At this time, it was hypothesized that the cause of the growth suppression induced by the "old" environment was "infection" (2,3). Indeed, the exposure of gnotobiotic-reared chicks to normal bacterial flora suppressed growth and feed intake.

These early observations resulted in the following two monumental animal husbandry changes: (i) the use of antimicrobials and (ii) "all in, all out" rearing strategies. These management strategies have continued to this day and have provided selection pressures for the evolution of microbes and animal genetic lines for the last 50 years.

Methods to prevent the growth suppression caused by conventional environments have focused largely on elimination of "growth suppressing" microbes. Antimicrobials were effective in reducing the bacterial load on epithelial surfaces, and thus antibiotic use became the treatment of choice for enhancing growth rate and improving feed efficiency. Lev and Forbes (1) showed that nearly half of the 12% improvement in growth achieved by raising chicks in gnotobiotic facilities could be realized by feeding antibiotics. As a result, the use of antibiotics in animal feeds grew steadily throughout the 1950s, 1960s, and 1970s until, by the late 1970s, 50% of the total tonnage of antibiotics produced in the U.S. made its way into animal feeds, primarily as growth stimulants (5). Today, the use of antimicrobials in poultry and swine diets is commonplace. However, Europe has recently moved toward a ban of four widely used growth stimulants: bacitracin zinc, spiramycin, virginamycin, and tylosin phosphate (6). Trends toward the regulation of antibiotic use as growth stimulants suggest the need for alternative methods of protecting against growth suppression caused by microbes (or immune stimulation, explained later).

Although research is currently lacking, it is thought that genetic selection of animals for faster rates of gain and improved feed efficiency has affected the animal's response to its microbial environment. An animal that thrives in an immune-stimulating environment undoubtedly has lost the responsiveness described by Lev and Forbes in the 1950s (1). Over the last decade, we and others (unpublished data) have observed an increased resistance of the broiler to decreases in weight gain after endotoxin injection. Are genetic selection pressures resulting in immunocompromised animals?

Role of the Immune Response in Growth

Opportunities for developing new strategies to prevent growth suppression caused by environmental microbes have become available only recently. It is now well accepted that the animal's immunological reaction to immune stimulants is the cause of decreased growth and rate of development (7–11). Immune cells of the monocyte/macrophage lineage encountering non–self-immune stimulants release polypeptides (cytokines), which communicate with other cells of the immune system and the host. Interleukin-1 (IL-1) and tumor necrosis factor-α (TNF-α) are two cytokines that are essential in up-regulating the immune system. However, these cytokines also have pronounced effects on other nonlymphoidal target tissues (8).

Klasing et al. (12) showed that the anorexic and growth-suppressing effects of immune stimulation could be achieved by the direct injection of IL-1. Injection of TNF similarly suppressed food intake and growth. The anorexic and growth-suppressing effects of IL-1 and TNF after endotoxin exposure could be prevented or alleviated by administering antibodies reactive to the growth-suppressing cytokines or cytokine receptor antagonists in the immune-stimulated animal (13). Also, inhibition of cytokine activity could be viewed as causing immunosuppression because cytokines are extremely critical elements in normal immunological response. In fact, immunosuppressants tend to enhance animal growth in the absence of overt disease. Although antibodies and receptor antagonists could prove to be valuable tools in the treatment of acute endotoxic or septic shock, such therapy has little value in preventing immune-induced growth suppression in agricultural animals exposed to conventional environments. However, the ideal strategy for preventing growth suppression due to immune stimulation would be to block the response of nonlymphoidal tissues to cytokine action, thus preserving immune function while reducing negative effects on other tissues.

Although genetic selection for resistance to growth depression due to immune stimulation could have (and probably has) been applied inadvertently to farm animals, we have hypothesized that such selection pressure results in immune-suppressed animals. When animals are selected for faster rates of gain and improved feed efficiency in immune-stimulating environments, the capacity for immunologic response declines (14). Using elite lines of Pekin ducks heavily selected for growth over many generations, we have observed a significant negative correlation between growth rate and lymphocyte blastogenesis (unpublished data). As previously mentioned, the modern broiler has become increasingly more resistant to growth depression due to immune stimulation. In today's broiler, it is very difficult to induce the migration of macrophages into the peritoneal cavity. Although a definitive study linking genetic selection for growth and immune suppression is still lacking, the circumstantial observations made by several groups suggest that this link may actually exist.

Metabolic Basis for Cytokine-Induced Growth Suppression

Interleukin-1 and TNF (as well as several other cytokines) induce a host of metabolic changes (8,10). Cytokines can induce anorexia by stimulation of gastrointestinal peptides such as cholecystokinin (CCK) (15,16). Because gut peptides are released in the lumen of the gastrointestinal tract (17), we attempted to prevent anorexia in chickens by feeding antibodies reactive with selected gut peptides (18–21). Although conventionally reared animals (not intentionally immune stimulated) did not show increased food intake when fed antibody to CCK, enhanced growth and improved feed efficiency were realized (22). This biological alternative to antibiotics has been commercialized by using laying hens to generate large quantities of antibodies in the yolks of their eggs. We have continued to modify the role of gut peptides through antibody feeding with moderate success in stimulating animal performance. Additional methods of rendering nonlymphoidal tissue resistant to immune cytokines were warranted. Our goal, however, was to achieve resistance to the negative effects of the immune system, not by altering microbial ecology at the epithelial surfaces, nor by suppressing immunological function, but instead by altering the response of nonlymphoidal tissue to the surge of systemically released cytokines during the immune response.

Klasing *et al.* (12) showed that muscle wasting induced by immune stimulants could be achieved directly by the application of IL-1 to harvested muscle strips. When the gastrocnemius muscle was treated with IL-1, fractional muscle synthesis declined (decreased muscle synthesis and increased muscle degradation). Similar findings were reported by other groups in different animal species, involving the most potent catabolic cytokine, i.e., TNF.

A simple nutritional means of improving nonlymphoidal tissue resistance to cytokines began to emerge in the mid- to late 1980s. Dinarello (7) had reported that

IL-1–induced muscle catabolism was associated with prostaglandin E_2 (PGE_2) synthesis. When IL-1 was applied to muscle strips that were excised and placed in culture, the production of PGE_2 increased (23), whereas direct application of PGE_2 to muscle strips stimulated muscle degradation (24). In addition, it became apparent that PGE_2 could down-regulate the immune response (25). Hence, we hypothesized that during immune challenge, cytokines released from the immune-stimulated cells (i.e., TNF-α and IL-1) induced the release of phospholipid metabolites (arachadonic acid *via* phopholipases) which, when acted upon by lipoxygenase or cyclooxygenase, produced active lipid metabolites that caused muscle catabolism and hence weight loss (26). Modifying this principal metabolic pathway comprised the basis for research involving the lipid metabolite of linoleic acid known as conjugated linoleic acid (CLA).

Nonlymphoidal tissue production of eicosanoids is predicted to down-regulate the immune response and facilitate the response of nonlymphoidal tissues to cytokines. Therefore, modifying nonlymphoidal tissue response to immune cytokines *via* the eicosanoid pathway might allow improved growth in the immune-challenged animal by means other than immune suppression.

Conjugated Linoleic Acid

n-3 Fatty Acid Precedent

In 1993, we discussed the potential role of n-3 fatty acids in fish oil as a means of modifying the 2-series prostaglandin pathway (26). Metabolites of n-3 fatty acids result in the synthesis of 3-series eicosanoids. These fatty acids were shown to be competitive with arachadonic acid for cyclooxygenase. Because the 3-series prostaglandins have reduced or altered biological activity compared with 2-series prostaglandins, we predicted that they may be a useful nutrient in preventing immune-induced weight loss. Indeed, researchers had previously shown that anorexia caused by IL-1 and the immune response could be prevented by feeding high dietary levels of fish oils (8%) (27). The mechanism by which fish oils prevented immune-induced wasting appeared to be related to decreased release of catabolic cytokines from stimulated immune cells (27–29). At that time, we envisioned the following three problems associated with using fish oils to prevent the catabolic response to immune challenge in agricultural animals: (i) the level of fish oil necessary to prevent immune-induced weight loss was cost prohibitive; (ii) fish oils in animal feeds above a 1% inclusion level impart off-flavors to meat, milk, and eggs; and (iii) fish oil appeared to act by causing suppression of the immunological response (26). Our goal to protect nonlymphoidal tissue from immune-induced wasting would be achieved only if the method utilized were economical, did not adversely alter the quality of the food, and maintained the immune system's ability to respond to challenge. Because CLA was an altered metabolite of the precursor for 2-series prostaglandins (i.e., linoleic acid), its use in preventing the catabolic nature of the immune response became apparent.

Additional evidence that CLA may play a role in alleviating the catabolic nature of immune cytokines *via* the eicosanoid pathway was shown by Nugteren (30). In his work, the elongated and desaturated product of the 10-*trans*, 12-*cis* conjugated octadecadienoate isomer of CLA (5-*cis*, 8-*cis*, 12-*trans*, 14-*cis*-eicosatetraenoic acid) had the highest activity in competitive inhibition (competitive with arachadonic acid, 5-*cis*, 8-*cis*, 11-*cis*, 14-*cis*-eicosatetraenoic acid) of cyclooxygenase. Hence, at least one isomer of CLA, if elongated and desaturated, could potentially modify the eicosanoid pathway through enzymatic inhibition.

The anticarcinogenic effects of CLA (31–34) reported at the time suggested that CLA may not adversely affect the immune response. In a serendipitous street corner meeting, Pariza and I (M.E. Cook) decided to collaborate and study the role of CLA in immune function.

The Role of Conjugated Linoleic Acid in Preventing Immune-Induced Wasting

Using the model systems of Klasing *et al.* (12), we began our first test of CLA's ability to prevent immune-induced wasting (*ca.* 1990). Day-old chicks were fed linoleic acid (95% pure) or CLA (95%) synthesized from linoleic acid (43% 9-*cis*, 11-*trans*- and 44% 10-*trans*, 12-*cis*-octadecadienoate as the predominant CLA isomers). Half of the chicks in each dietary treatment group were injected with sterile-buffered saline; the other half received 1 mg/kg body weight endotoxin consisting of lipoplysaccharide (LPS) from *Escherichia coli*. The change in body weight over the next 24 h was monitored. Chicks fed CLA gained weight after endotoxin injection, whereas those fed the control diet either lost weight or failed to grow (26). Additional experiments were conducted in rats (26) and mice (35–37) to determine whether the protection against immune-induced wasting was conserved across several animal species. Both mice and rats were responsive to the protective effects of CLA against immune stress. Feeding fish oil (0.5%) at a dietary level similar to CLA (0.5%) provided no protection against the catabolic nature of endotoxin injection in mice (35–37).

Since then, numerous studies (unpublished data) that used a number of immune-related inducers of the catabolic response have been conducted and have supported these findings. In one of these models, the protective effects of CLA against wasting induced by the spontaneous autoimmune disorder in NZB/W F1 mice was tested (Yang *et al.*, unpublished data). Although mice fed CLA gained less weight than those fed a control diet, body weight changes after immune-induced proteinuria were significantly different between the two dietary groups. Autoimmune mice fed CLA lost significantly less weight than control-fed mice after developing immune-induced proteinuria, suggesting protection from immune-induced wasting.

Peritoneal injections of sephadex are also known to result in immune-induced wasting (12). When mice fed a control or CLA-supplemented diet were injected with sephadex, those fed CLA lost significantly less weight than the control-fed mice (38). Mice fed CLA also showed significantly less immune-related anorexia (35), which could partially explain reduced weight loss after immune stimulation.

Weight loss due to a direct injection of TNF was also reduced in CLA-fed mice compared with those fed linoleic acid (unpublished data). Although these findings were very encouraging, suppression of the immune response would achieve similar results. Strategies for preventing immune-induced cachexia in the past have focused on suppressing the inflammatory response or negating the effects of systematically released cytokines on nonlymphoidal tissue with the use of technologies involving anti-TNF or anti-IL-1 therapies. Does CLA prevent the catabolic effects of immune stimulation by suppressing the immunological response?

Role of Conjugated Linoleic Acid in the Immunological Response

In our early studies, we immediately hypothesized that CLA mediates protection against immune-induced wasting by suppressing the immune response. A series of studies were conducted to determine the influence of CLA on the immune response. Antibody response in rats or chicks to bovine serum albumin and sheep red blood cells was not affected adversely by feeding CLA (26). In a study involving autoimmune NZB/W F1 mice, we observed no delay in the expression of antinuclear antibodies (Yang et al., unpublished data). In contrast, expression of autoimmune antibodies appeared earlier in NZB/W F1 mice fed CLA. A more detailed presentation of the role of CLA on immunoglobulins was recently presented by Sugano et al. (39). They observed that CLA increased immunoglobulins (Ig)A, IgM, and IgG, but decreased IgE in serum. Immunoglobulin production by mesenteric lymphoid lymphocytes confirmed these findings in serum (with regard to IgA, IgG, and IgM). Lipopolysaccharide stimulation of mesenteric lymphoid lymphocytes showed that feeding rats CLA reduced IgE production. An immunoglobulin class switch from IgM to IgG then to IgE requires IL-4, a potent TH-2 cytokine. These results suggested that CLA may down-regulate the TH-2 immune response and/or up-regulate TH-1 (T-helper type 1) activities.

We reported (26) increased phagocytic activity of peritoneal macrophages from CLA-fed rats, which supported the previous findings of increased macrophage killing ability (40). However, when CLA was incubated with murine macrophages *in vitro*, phagocytosis was found to be reduced (41). In contrast, Chew et al. (41) reported that CLA added *in vitro* to murine macrophages had enhanced bacteriocidal activity. Differences found could result from methods used to study CLA effects on phagocytosis (*ex vivo* or *in vitro*). CLA has a pronounced effect on Δ-9 desaturase (42). The decrease in Δ-9 desaturase activity may be more evident *ex vivo* than *in vitro*. Also, it is unclear how CLA could on the one hand, decrease phagocytosis, and on the other hand, increase bacterial killing activity, unless CLA, under these *in vitro* conditions, directly affected bacterial survival.

Although incubation of porcine lymphocytes with CLA at $2-7 \times 10^{-5}$ mol/L CLA *in vitro* significantly decreased IL-2 production (41), lymphocytes isolated from CLA-fed mice have increased IL-2 production when stimulated *in vitro* (43,44). This apparent increase is consistent with CLA enhancing TH-1 cytokines in related immune function. We predict that this increase in IL-2 production may be due to inhibition of prostaglandin production as seen in several experimental models (44–46). However, other data show reductions in TNF and IL-6 measured *ex vivo* in lymphocytes from CLA-fed mice. Although these findings were unchanged with PGE_2 and leukotriene $(LT)B_4$, treatment with LPS reversed the suppressive effect of CLA on TNF, but not IL-6 production. There were no significant changes in IL-1 production under all conditions, although there was a consistent trend for IL-1 to be reduced by CLA. These data are not consistent with what would be predicted and was indeed found to some degree, with regard to the effects of CLA on eicosanoid production and with respect to eicosanoid effects on cytokine production. It has been reported that PGE_2 is a potent inhibitor of TNF production (47). Because desaturated and elongated metabolites of select isomers of CLA reduce the activity of cyclooxygenase (30), and it has been shown that CLA inhibits PGE_2 production (44,46), why does CLA down-regulate the production of TNF and IL-6?

Rats fed CLA were shown to have increased footpad swelling in response to phytohemagglutinin-P (26). This is a classic measure of delayed type II hypersensitivity, a complex immune response that involves elements of inflammation and lymphocyte proliferation. Similar effects have been reported in animals fed n-3 fatty acids (25). Because *in vivo* responses to phytohemagglutinin-P are T cell dependent, the effects of CLA on lymphocyte blastogenesis were determined. Splenocytes from mice fed CLA had increased proliferation when stimulated *in vitro* with phytohemagglutinin-P (35) or conconavalin A (Con A) (DeVoney, unpublished data). Similar findings were reported by others (41,43,48,49). Further studies performed with semipure isomers found that the 10-*trans*, 12-*cis* CLA isomer reproduces these findings in mice in Con A-stimulated lymphocyte proliferation assays. The 9-*cis*, 11-*trans* CLA isomer did not have similar activity, but there may have been interfering compounds in the isomer preparation used (DeVoney, unpublished data).

Interestingly, CLA-fed animals have larger spleens and increased splenic and blood lymphocyte populations both with and without immune stimulation (DeVoney *et al.*, unpublished data). Splenic lymphocytes from naïve mice fed CLA have a higher percentage of CD4 positive and a lower percentage of CD8 positive T cells, resulting in a higher CD4 to CD8 T-cell ratio (51). No significant changes were seen in the B-cell to T-cell ratios or in natural killer cell (NK) populations as a percentage of total lymphocytes (unpublished data, 49). Splenic lymphocyte subsets were also altered in CLA-fed chicks injected with LPS or phosphate buffered saline vehicle, indicating that dietary CLA increased the CD4/CD8 ratio, increased the ratio of IgM positive cells to T cells, greatly increased T lymphocytes, and reduced IgM positive cells after LPS injection (49).

We hypothesized that if CLA lowered PGE_2 levels and favored TH-1 type responses, then NK cell activity would be potentiated. However, *in vitro* measures of NK activity in splenocytes from mice fed CLA showed no increase in naïve NK activity. When mice were injected with Poly I-C to induce a greater NK cell response, lymphocytes from control-fed mice showed greater activity than those from CLA-fed mice (52). In contrast, CLA-fed vehicle-injected mice had greater NK activity than control-fed mice (52). The vehicle injection may model a low-level stimulation, indicating that CLA-fed animals are more sensitive to developing a NK cell response under these conditions. Also, NK activity induction resulting from Poly I-C injection is biphasic because high levels of the interferons actually reduce NK cell activity. Therefore, the reduction of NK activity at the higher level of immune stimulation may be due to higher cytokine levels as hypothesized. These results, taken together with findings on lymphocyte populations, indicate that CLA-fed animals have more total NK cells, which may be more sensitive to activation.

In light of the role that NK cells play against tumor cells, is it possible that enhanced low-level NK activity is responsible, in part, for the dramatic anticancer effects of CLA? SCID mice were used to test the role of immune function on the antitumorogenic properties of CLA (53,54). A single injection of an antimitotic agent was used to block natural immunity in the immunocompromised mice. The antitumor activity of CLA was preserved under these conditions, with tumors forming in all mice and dramatically shrinking away in the CLA-fed mice 6–8 wk after engraftment. Metastatic spread of the grafted tumors was gradually reduced or nonexistent in CLA-fed SCID mice. SCID mice do have inducible NK activity, and this researcher is not convinced that a single injection of the antimitotic agent is sufficient to block NK activity for the entire 14-wk course of the experiments. In fact, the late regression of the tumors, unlike the action of CLA in previous studies, may be due to a delay of the NK response.

Conjugated Linoleic Acid and Immune-Related Disorders

The up-regulation of immunological function suggested that CLA may exacerbate disorders associated with the immune response. The first hypothesis we sought to test was the following: Does CLA increase the severity of type I hypersensitivity? The model used was recently reported (46). Guinea pigs were fed either a control or CLA-95–supplemented diet for 3–4 wk. During this period, they were hyperimmunized with ovalbumen. Tracheas were suspended in a tissue chamber and superfused with Krebs buffer. After an equilibration period, superfusates were collected to assess basal mediator release. Tracheas were then superfused with ovalbumen to induce contraction and antigen-induced mediator release. Contractions were measured using a polygraph; superfusate fractions were collected and assayed for select mediators. In other experiments, trachea, bladder, and lungs were harvested from hyperimmunized guinea pigs and bathed in warmed buffer, then challenged with antigen after basal levels of buffer were collected to assess mediator release.

Although it was anticipated that the "immune enhancing" properties of CLA would exacerbate immune-related events associated with antigen challenge, this was not observed. Antigen-induced tracheal contractions of guinea pigs fed CLA were not enhanced. Eicosanoid mediators (e.g., PGE_2) were actually decreased in CLA-fed guinea pigs, findings that agree with those of other researchers (44,55). Similar findings were obtained in organ bath experiments. These results suggest that CLA may modulate the inflammatory events associated with immune challenge.

As previously mentioned, the influence of CLA on spontaneous antinuclear autoimmunity in NZB/W F1 mice was also tested. Although CLA shortened the time period for the development of proteinuria in NZB/W F1 mice, mean days of survival did not differ between control- and CLA-fed mice. Only at one time point in the study did the control mice significantly differ in percentage of survival compared with CLA-fed mice (Yang et al., unpublished data).

Age-related depression of mitogen-induced splenocyte blastogenesis appeared to be slightly alleviated by feeding CLA (43). Young mice fed CLA had increased basal and mitogen-induced IL-2 production compared with control-fed mice. However, these authors did not find enhanced delayed-type hypersensitivity as we previously reported (26). Differences could be a function of methodologies used. However, in general, the immune-modulating activities of CLA do not appear to adversely affect immune-related disorders.

Conclusions

Conjugated linoleic acid has consistently shown immune-modulating activities. It appears to enhance select immunological responses and possibly modulate the immune response toward a TH-1 type response. Although CLA enhances select immune responses, it does not appear to enhance immune-related consequences of immune disorders (i.e., Type 1 hypersensitivity and autoimmunity). In fact, CLA may actually mitigate the adverse effects of the immune response (i.e., immune-induced wasting). The mechanism by which CLA has an immune-enhancing effect, while simultaneously protecting nonlymphoid tissues from the "adverse" effects of immune cytokines, appears to be related to regulation of eicosanoid production (56,57). Data are soon forthcoming demonstrating specific isomers responsible for these biological effects. The studies of Nugteren (30) would strongly suggest that the 10-*trans*, 12-*cis* isomer is the biologically relevant isomer.

References

1. Lev, M., and Forbes, M. (1959) Growth Response to Dietary Penicillin of Germ-Free Chicks with a Defined Intestinal Flora, *Br. J. Nutr. 13*, 78–84.
2. Coates, M.E., Dickenson, C.D., Harrison, G.F., Kon, S.K., Cummins, S.H., and Cuthebertson, W.F.J. (1951) Mode of Action of Antibiotics in Stimulating Growth of Chicks, *Nature 168*, 332.
3. Coates, M.E., Dickinson, C.D., Harrison, G.F., Kon, S.K., Porter, J.W.G., Cummins, S.H., and Cuthbertson, W.F.J. (1952) A Mode of Action of Antibiotics in Chick Nutrition, *J. Sci. Food. Agric. 3*, 43–48.
4. Lillie, R.J., Sizemore, J.R., and Bird, H.R. (1952) Environment and Stimulation of Growth of Chicks by Antibiotics, *Poult. Sci. 31*, 466–475.
5. Von Houwelling, C.D. (1978) *Draft Environmental Impact Statement. Subtherapeutic Agents in Animal Feeds,* Food and Drug Administration, Washington, DC.
6. Muirhead, S. (1998) E.U. Ban of Antibiotics Draws Sharp Criticism, *Feedstuffs 70*, 1–4.
7. Dinarello, C.A. (1984) Interleukin-1, *Rev. Infect. Dis. 6*, 91–95.
8. Klasing, K.C., (1988) Nutritional Aspects of Leukocytic Cytokines, *J. Nutr. 118*, 1436–1446.
9. Klasing, K.C., and Johnstone, B.J. (1991) Monokines in Growth and Development, *Poult. Sci. 70*, 1781–1789.

10. Johnson, R.W. (1997) Inhibition of Growth by Proinflammatory Cytokines: An Integrated View, *J. Anim. Sci. 75,* 1244–1255.
11. Roubenoff, R. (1997) Inflammatory and Hormonal Mediators of Cachexia, *J. Nutr. 127,* 10145–10165.
12. Klasing, K.C., Laurin, D.E., Peng, R.K., and Fry, D.M. (1987) Immunologically Mediated Growth Depression in Chicks: Influence of Feed Intake, Corticosterone and Interleukin-1, *J. Nutr. 117,* 1629–1637.
13. Van Deuren, M., Dofferhoff, A.S.M., and Van Der Meer, J.W.M. (1992) Cytokines and the Response to Infection, *J. Pathol. 168,* 349–356.
14. Miller, C.C., Cook, M.E., Rodgers, G.E., and Kohl, H. (1992) Immune Response Difference in Different Strains of Ducks, *Poult. Sci. 71 (Suppl. 1),* 166.
15. Ohgo, S., Nakatsuro, K., Ishikawa, E., and Shigeru, M. (1992) Stimulation of Cholecystokinin (CCK) Release from Superfused Rat Hypothalamo-Neurohypophyseal Complexes by Interleukin-1 (IL-1), *Brain Res. 593,* 25–31.
16. Daun, J.M., and McCarthy, D.O. (1993) The Role of Cholecystokinin in Interleukin-1 Induced Anorexia, *Physiol. Behav. 54,* 237–241.
17. Rao, R.K. (1991) Biologically Active Peptides in the Gastrointestinal Lumen, *Life Sci. 48,* 1685–1704.
18. Cook, M.E., Jerome, D.L., Pimentel, J., and Miller, C. (1997) Feeding Egg Antibodies to Cholecystokinin (CCK) Improves Growth and Feed Conversion in Broilers, *Poult. Sci. 76 (Suppl. 1),* 95.
19. Cook, M.E., Jerome, D.L., Daley, M.J., Greenblatt, H., and Skarie, C. (1997) Performance Improvement Caused by Feeding Broilers Egg Yolk Antibodies to Cholecystokinin Is Correlated with Specific Antibody Dose Not the Mass of Egg Yolk, *Poult. Sci. 76 (Suppl. 1),* 95.
20. Cook, M.E., Pimentel, J.L., and Miller, C.C., U.S. Patent 5,827,517 (1998).
21. Cook, M.E., and Jerome, D.L., U.S. Patent 5,725,873 (1998).
22. Hooge, D.M. (1998) Studies Show Benefits of Cholecystokinin Antibodies, *Feedstuffs 70,* 12–16.
23. Goldberg, A.L., Baracos, V., Rodemann, H.P., Waxman, L., and Dinarello, C. (1984) Control of Protein Degradation in Muscle by Prostaglandins, Ca++, and Leukocytic Pyrogen (Interleukin-1), *Fed. Proc. 43,* 1301–1306.
24. Rodemann, H.P., and Goldberg, A.L. (1982) Arachidonic Acid, Prostaglandin E_2 and F_2 Influence Rates of Protein Turnover in Skeletal and Cardiac Muscle, *J. Biol. Chem. 257,* 1632–1638.
25. Hwang, D. (1989) Essential Fatty Acid and Immune Response, *FASEB J. 3,* 2052–2061.
26. Cook, M.E., Miller, C.C., Park, Y., and Pariza, M. (1993) Immune Modulation by Altered Nutrient Metabolism: Nutritional Control of Immune-Induced Growth Depression, *Poult. Sci. 72,* 1301–1305.
27. Hellerstein, M.K., Meydani, S.N., Meydani, M., Wu, K., and Dinarello, C.A. (1989) Interleukin-1-Induced Anorexia in the Rat. Influence of Prostaglandins, *J. Clin. Investig. 84,* 228–235.
28. Endres, S., Ghorbani, R., Kelley, V., Georgilis, K., Lonnemann, G., van der Meer, J.J.W.M., Cannon, J.G., Rogers, T.S., Klempner, M.S., Weber, P.C., Schaefer, E.J., Wolff, S.M., and Dinarello, C.A. (1989) The Effect of Dietary Supplementation with n-3 Polyunsaturated Fatty Acids on the Synthesis of Interleukin-1 and Tumor Necrosis Factor by Mononuclear Cells, *N. Engl. J. Med. 320,* 265–271.

29. Billiar, T.R., Bankey, P.E., Svingen, B.A., Curran, R.D., West, M.A., Holman, R.T., Simmons, R.L., and Cerra, F.B. (1988) Fatty Acid Intake and Kupffer Cell Function: Fish Oil Alters Eicosanoid and Monokine Production to Endotoxin Stimulation, *Surgery 104*, 343–349.
30. Nugteren, D.H. (1970) Inhibition of Prostaglandin Biosynthesis by 8-*cis*, 12-*trans*, 14-*cis*-Eicosatrienoic Acid and 5-*cis*, 8-*cis*, 12-*trans*, 14-*cis*-Eicosatetraenoic Acid, *Biochim. Biophys. Acta 210*, 171–176.
31. Pariza, M.W., and Hargraves, W.A. (1985) A Beef-Derived Mutagenesis Modulator Inhibits Initiation of Mouse Epidermal Tumors by 7,12-Dimethylbenz[a]anthracene, *Carcinogenesis 6*, 591–593.
32. Ha, Y.L., Grimm, N.K., and Pariza, M.W. (1987) Anticarcinogens from Fried Ground Beef: Heat Altered Derivatives of Linoleic Acid, *Carcinogenesis 8*, 1881–1887.
33. Ha., Y.L., Storkson, J., and Pariza, M.W. (1990) Inhibition of Benzo(α)pyrene-Induced Mouse Forestomach Neoplasia by Conjugated Dienoic Derivatives of Linoleic Acid, *Cancer Res. 50*, 1097–1101.
34. Ip, C., Chin, S.F., Scimeca, J.A., and Pariza, M.W. (1991) Mammary Cancer Prevention by Conjugated Dienoic Derivatives of Linoleic Acid, *Cancer Res. 51*, 6118–6124.
35. Miller, C.C., Park, Y., Pariza, M.W., and Cook, M.E. (1994) Feeding Conjugated Linoleic Acid to Animals Partially Overcomes Catabolic Response Due to Endotoxin Injection, *Biochem. Biophys. Res. Commun. 198*, 1107–1112.
36. Cook, M.E., and Pariza, M.W., U.S. Patent 5,430,066 (1995).
37. Cook, M.E., and Pariza, M.W., U.S. Patent 5,428,072 (1995).
38. Park, Y. (1996) Regulation of Energy Metabolism and the Catabolic Effects of Immune Stimulation by Conjugated Linoleic Acid, Ph.D. Thesis, University of Wisconsin, Madison.
39. Sugano, M., Tsujita, A., Yamasaki, M., Noguchi, M., and Yamada, K. (1998) Conjugated Linoleic Acid Modulates Tissue Levels of Chemical Mediators and Immunoglobulins in Rats, *Lipids 33*, 521–527.
40. Michal, J.J., Chew, B.P., Schultz, T.D., Wong, T.S., and Magnuson, N.S. (1992) Interaction of Conjugated Dienoic Derivatives of Linoleic Acid with β-Carotene on Cellular Host Defense, *FASEB J. 6*, A1102 (abstr.).
41. Chew, B.P., Wong, T.S., Shultz, T.D., and Magnuson, N.S. (1997) Effects of Conjugated Dienoic Derivatives of Linoleic Acid and β-Carotene in Modulating Lymphocyte and Macrophage Function, *Anticancer Res. 17*, 1099–1106.
42. Lee, K.N., Pariza, M.W., and Ntambi, J.M. (1998) Conjugated Linoleic Acid Decreases Hepatic Stearoyl-CoA Desaturase mRNA Expression, *Biochem. Biophys. Res. Commun. 248*, 817–821.
43. Hayek, M.G., Han, S.N., Wu, D., Watkins, B.A., Meydani, M., Dorsey, J.L., Smith, D.E., and Meydani, S.N. (1999) Dietary Conjugated Linoleic Acid Influences the Immune Response of Young and Old C57BL1/6NCr IBR Mice, *J. Nutr. 129*, 32–38.
44. Turek, J.J., Li, Y., Schoenlein, I.A., Allen, K.G.D., and Watkins, B.A. (1998) Modulation of Macrophage Cytokine Production by Conjugated Linoleic Acids Is Influenced by the Dietary n-6:n-3 Fatty Acid Ratio, *J. Nutr. Biochem. 9*, 258–266.
45. Wong, M., Boon, C., Wong, T., Hosick, H., Boylston, T., and Shultz, T. (1997) Effects of Dietary Conjugated Linoleic Acid on Lymphocyte Function and Growth of Mammary Tumors in Mice, *Anticancer Res. 17*, 987–994.

46. Whigham, L.D., Cook, E.B., Stahl, J.L., Saban, R., Pariza, M.W., and Cook, M.E. (1998) Conjugated Linoleic Acid Reduced Antigen-Induced Prostaglandin E_2 Release from Sensitized Tracheas, *FASEB J. 12,* A869 (abstr.).
47. Peters, T., Ulrich, K., and Decker, K. (1990) Interdependence of Tumor Necrosis Factor, Prostaglandin E_2 and Protein Synthesis in Lipopolysaccharide-Exposed Rat Kupffer Cells, *Eur. J. Biochem. 191,* 583–589.
48. Winchell, D.C., Clandinin, M.T., and Field, C.J. (1998) Conjugated Linoleic Acid Fed in a Low Polyunsaturated Fat Diet Increases Mitogen Responses in Rats, *FASEB J. 12,* A869 (abstr.).
49. DeVoney, D., Pariza, M.W., and Cook, M.E. (1997) Conjugated Linoleic Acid Increases Blood and Splenic T-Cell Response Post Lipopolysaccharide Injection, *FASEB J. 9,* 3355.
50. Cook, M.E., Cook, E.B., Stahl, J.L., Graziano, F.M., and Pariza, M.W., U.S. Patent 5,585,400 (1996).
51. Cook, M.E., Pariza, M.W., Yang, X., and DeVoney, D., U.S. Patent 5,674,901 (1997).
52. DeVoney, D., Pariza, M., and Cook, M.E. (1998) Conjugated Linoleic Acid (CLA) Enhances Murine Natural Killer Cell Activity in Naïve Animals but Attenuates Natural Killer Cell Activity After Stimulation with Polyinosinic-Polycytidylic Acid, *FASEB J. 12,* 5031.
53. Cesano, A., Visonneau, S., Scimeca, J., Kritchevsky, D., and Santolin, D. (1998) Opposite Effects of Linoleic Acid and Conjugated Linoleic Acid on Human Prostatic Cancer in SCID Mice, *Anticancer Res. 18,* 1429–1434.
54. Visonneau, S., Cesano, A., Tepper, S., Scimeca, J., Santoli, D., and Kritchevsky, D. (1997) Conjugated Linoleic Acid Suppresses the Growth of Human Breast Adenocacinoma Cells in SCID Mice, *Anticancer Res. 17,* 969–974.
55. Sugano, M., Tsujita, A., Yamasaki, M., Yamada, K., Ikeda, I., and Kritchevsky, D. (1997) Lymphatic Recovery, Tissue Distribution and Metabolic Effects of Conjugated Linoleic Acid in Rats, *J. Nutr. Biochem. 8,* 38–43.
56. Cook, M.E., and Pariza, M. (1998) The Role of Conjugated Linoleic Acid (CLA) in Health, *Int. Dairy J. 8,* 459–462.
57. Cook, M.E. (1999) Nutritional Effects of Vaccination, in *Veterinary Vaccines and Diagnostics,* (Schultz, R.D., ed.) Academic Press, San Diego.

Chapter 18
Incorporation of Conjugated Fatty Acid into Biological Matrices

Martin P. Yurawecz[a], John K.G. Kramer[b], Michael E.R. Dugan[c], Najibullah Sehat[a], Magdi M. Mossoba[a], Jun Jie Yin[a], and Yuoh Ku[a]

[a]Center for Food Safety and Applied Nutrition, U.S. Food and Drug Administration, Washington, DC 20204
[b]Southern Crop Protection, Food Research Center, Agriculture and Agri-Food Canada, Guelph, ON, Canada N1G 2W1
[c]Lacombe Research Center, Agriculture and Agri-Food Canada, Lacombe, AB, Canada, T4L 1W1

Introduction

There are two sources of conjugated fatty acids (CFA), natural and synthetic. Natural CFA are produced in the rumen as a result of partial biohydrogenation mainly of linoleic acid (1–3), and are found in milk fats and meats from ruminants (4–9). The main CFA isomer in these products was found to be *cis*-9, *trans*-11-octadecadienoic (18:2) acid (4), also known as rumenic acid (10). Rumenic acid can also be formed by Δ9 desaturation from *trans*-11-18:1 [(11) and Chapter 13, this volume], the monounsaturated fatty acid intermediate formed during biohydrogenation of linoleic acid in the rumen. Small amounts of *trans*-11-18:1 are also found in partially hydrogenated vegetable oils, which will give rise to rumenic acid (12). Rumenic acid represents ~80% of the CFA in natural products; the remainder consists of several minor CFA isomers of which *trans*-7, *cis*-9-18:2 (9) and *trans*-11, *cis*-13-18:2 (13) are the more prominent ones. On the other hand, synthetic CFA products, prepared by alkali isomerization of linoleic acid can contain either two or four major positional CFA isomers, each consisting of the prominent *cis/trans* isomers with minor amounts of *trans,trans* and *cis,cis* geometric isomers (14–16). The major *cis/trans* CFA isomers are *trans*-8, *cis*-10-18:2; *cis*-9, *trans*-11-18:2; *trans*-10, *cis*-12-18:2; and *cis*-11, *trans*-13-18:2 (13,15,17).

Largely in response to the favorable physiologic and pathologic reports described elsewhere in this book, human matrices (9,12,18–22), and human cancer tissues (20) were examined. In addition, numerous animal feeding studies have been conducted using the synthetic CFA mixtures to determine the etiology of these responses. One of the parameters often examined was the incorporation of CFA into tissue lipids in the hope of determining possible mechanisms of action. Human feeding studies of synthetic CFA are in progress, but the results of these studies have not yet been published.

To date, individual CFA isomers have not been evaluated for any biochemical or pathologic responses because pure isomers were not available in large enough amounts to undertake such studies. Such studies are necessary because it is becoming evident that CFA isomers have different responses (see Chapter 2).

In this chapter, we summarize and critically review published results of CFA incorporation into biological matrices from either natural sources or synthetic CFA mixtures. Each group will be further subdivided into studies showing CFA incorporation into total lipids vs. incorporation into individual lipid classes. References are generally ordered with increased developments in methodology.

Incorporation of CFA from Natural Sources

A number of human and animal matrices have been examined for their content of total CFA derived from dietary sources such as dairy products, meats, and partially hydrogenated vegetable oils, or from rumen matrices in which the rumen served as the source of the CFA.

Adipose Tissue. Ackman et al. (23) were the first to show by gas chromatography (GC) that small amounts of CFA, possibly arising from the diet, were present in human adipose tissue. Both *cis*-9, *trans*-11-18:2 and *trans*-9, *trans*-11-18:2 were detected. The relatively large amounts of the latter may be due to the BF_3-methylation procedure used (see Chapter 6). Rat and sheep adipose was examined after KOH hydrolysis of the total lipids and analysis of the free fatty acids (FFA) by reverse-phase high-performance liquid chromatography (HPLC) (8,24). Alkali hydrolysis prevented isomerization, and reverse-phase HPLC separated CFA plus some of their metabolites, but the HPLC method did not resolve individual CFA isomers. Fritsche et al. (20) examined the 4,4-dimethyloxazoline (DMOX) derivatives of the CFA isomers of human adipose tissue by GC-mass spectrometry (GC-MS) and found additional small amounts of *trans*-9, *cis*-11-18:2 and *cis*,*cis* CFA isomers. Analysis of human adipose tissue by silver-ion HPLC (Ag^+-HPLC) separated several *trans*,*trans* CFA isomers and identified *trans*-7, *cis*-9-18:2 (9) and *trans*-11, *cis*-13-18:2 (13,17,25) as the more abundant two among the minor CFA isomers.

Serum. Over a 10-year period, a group of researchers at the Whittington Hospital in London, UK, identified *cis*-9, *trans*-11-18:2 in total human serum lipids or phospholipids (18,26–28). They hydrolyzed the serum lipids using pancreatic or phopholipases followed by separation of the free CFA by reverse-phase HPLC. The CFA separated into two peaks by reverse-phase HPLC, which were identified as *cis*-9, *trans*-11-18:2 and *trans*-9, *trans*-11-18:2 (27). This method did not isomerize CFA, but it lacked the resolution to identify any of the minor CFA isomers. Huang et al. (19) used essentially the same method to determine the effect of cheese consumption on the CFA profile of plasma phospholipids in men. These authors reported only the presence of rumenic acid. Forgerty et al. (5) extracted total lipids from blood serum with chloroform/methanol (29), methylated the lipids with $NaOCH_3$, and identified *cis*-9, *trans*-11-18:2 by GC using a 25-m fused silica column. Alkali methylation did not isomerize the CFA, but the GC column used lacked resolution to identify even the *trans*-9, *trans*-11-18:2. Park and Pariza (30) found that calf and horse sera contained substantial amounts of *trans*-10, *cis*-12-18:2 in addition to *cis*-9, *trans*-11-18:2. Salminen *et al.* (12) showed an increase of *cis*-9, *trans*-11-18:2 in total serum of humans consuming a diet rich in dairy fats and partially hydrogenated vegetable oil. They used "acidic methanol" to methylate the lipids and identified only rumenic acid by GC.

Milk.   Parodi (4) was the first to characterize *cis*-9, *trans*-11-18:2 in cow milk by a combination of chemical and chromatographic procedures. A small amount of a 10,12-18:2 CFA isomer was detected, but was attributed to isomerization during chemical procedures used (31). By using GC, rumenic acid was subsequently identified in human milk as the only CFA isomer (5,22). McGuire *et al.* (20) reported additional small amounts of *trans*-10, *cis*-12-18:2 in human milk based on GC retention times. Banni *et al.* (8) found an 18:3 CFA in the milk of cows and sheep using reverse-phase HPLC, which was not further characterized. The use of a 100-m polar fused silica capillary column resolved three *cis*/*trans* isomers (*cis*-9, *trans*-11-18:2; *trans*-9, *cis*-11-18:2; *trans*-10, *cis*-12-18:2), a *cis,cis* isomer (*cis*-9, *cis*-11-18:2) and a *trans,trans* (*trans*-9, *trans*-11-18:2) CFA isomer in cow milk, which were identified by GC-MS (32). Ag^+-HPLC separated six *trans,trans*, seven minor *cis*/*trans* CFA isomers (*trans*-7, *cis*-9-18:2; *trans*-8, *cis*-10-18:2; *trans*-10, *cis*-12-18:2; *trans*-11, *cis*-13-18:2; *cis*-11, *trans*-13-18:2, and *cis*-12, *trans*-14-18:2) in addition to the major rumenic acid, as well as six *cis,cis* CFA isomers (9,13,17,25).

Cheese.   The first reports showed the following: (i) *trans,trans* isomers were the major CFA in cheese (33,34); (ii) the total CFA content increased during cheese formation (34); and (iii) the CFA isomer distribution was altered during cheese processing (35). Changes during cheese processing were not confirmed in subsequent reports (36,37). Cheese, like milk fat, has now been shown to contain ~20 CFA isomers, which were established by one group using a combination of chemical methods, GC and GC-MS (38), and by another group using Ag^+-HPLC, GC, and GC- MS methods (9,13,15,17,24). For a more complete list of the 20 CFA isomers found in cheese, see Chapter 18.7.

Meat of Ruminants.   In 1987, Ha *et al.* (39) isolated and identified by GC five CFA isomers in ground beef (*cis*-9, *trans*-11-18:2; *trans*-10, *cis*-12-18:2; *cis*-9, *cis*-11-18:2; *trans*-9, *trans*-11-18:2; and *trans*-10, *trans*-12-18:2), which were believed to be formed during the frying of ground beef. The spectrum of CFA isomers was similar to that obtained by alkali isomerization of linoleic acid. The high content of *trans,trans* CFA isomers found in beef was due to BF_3 methylation carried out at 100°C. By substituting a milder acid-catalyzed methylation procedure (4% HCl/methanol at 60°C), the same group found by GC the content of *cis*-9, *trans*-11-18:2 in total CFA to be between 79 and 92% in different meat products (6). Fogerty *et al.* (5), using $NaOCH_3$-catalyzed methylation, reported *cis*-9, *trans*-11-18:2 as the only CFA isomer in beef, lamb, and pork meat by GC. In addition to the major *cis*-9, *trans*-11-18:2, minor amounts of *trans*-9, *cis*-11-18:2, *cis*-9, *cis*-11-18:2, and *trans*-9, *trans*-11-18:2 were identified by GC retention times (40). Ag$^+$-HPLC separated several *trans,trans* CFA isomers, identified *trans*-7, *cis*-9-18:2 and *trans*-11, *cis*-13-18:2 as the two most abundant minor *cis/trans* CFA isomers among several others, and resolved a number of *cis,cis* CFA isomers (9,13,25).

Bile. Cow and sheep bile fluids were found to contain a relatively high amount of *cis*-9, *trans*-11-18:2 in the phosphatidylcholine (PC) fraction, the principal phospholipid in bile (41). By using chemical and chromatographic methods including Ag$^+$ thin-layer chromatography (Ag$^+$-TLC), the structure of the CFA was established as *cis*-9, *trans*-11-18:2. Furthermore, this acid was found to be present exclusively in the 2-position of PC. By monitoring the eluate at 234 nm during normal-phase HPLC separation of bile lipids, Cawood *et al.* (26) demonstrated that CFA in human bile lipids were associated mainly with PC. Reverse- phase HPLC analysis of the FFA produced after pancreatic lipase hydrolysis showed only one CFA peak. Smith *et al.* (42) clearly showed that the CFA in total human bile lipids was *cis*-9, *trans*-11-18:2. The FFA and the ester lipids were methylated with diazomethane and NaOCH$_3$, respectively. Proof for *cis*-9, *trans*-11-18:2 in bile lipids was obtained by using reverse-phase HPLC, GC with a nonpolar fused silica capillary column, and GC-MS of the 4-phenyl-1,2,4-triazoline-3,5-dione (PTAD) adduct of the CFA.

General Comment. In the past 10 years, there has been a general improvement in the analysis of CFA by the use of the following methods: (i) alkali- instead of acid-catalyzed methylation procedures to avoid isomerization of CFA isomers; (ii) long 100-m capillary GC columns; and (iii) Ag$^+$-HPLC that resolved individual CFA isomers. To date, the CFA content and isomer distribution of several natural products and some human matrices have been examined. However, there is no information on the positional distribution of CFA isomers on the glycerol moiety of triacylglycerols (TAG) or within molecular species of milk fats.

Incorporation of Synthetic CFA into Total Tissue Lipids

Cancer Studies. Several cancer studies in which CFA were evaluated reported either the concentration of CFA in total phospholipids of selected tissues (43–45), or the CFA content in total neutral lipids and phospholipids of selected tissues (46,47). The results of these feeding studies showed that incorporation of CFA was greater into the neutral lipid than into the phospholipid fraction (46,47). The only CFA isomer reported in the tissues examined was *cis*-9, *trans*-11-18:2 (43–47). The level of CFA in the tissue lipids was proportional to the amount of CFA fed to rats (44). The rate of depletion of CFA from a tissue, after discontinuation of the CFA diet, was greater in neutral lipids than in phospholipids (47).

For the first few studies (43–45), a laboratory-synthesized CFA mixture was fed to mice or rats. The CFA mixture consisted of two positional CFA isomers, 9,11-18:2 and 10,12-18:2, as determined by Ag$^+$-HPLC analysis of a representative sample supplied by Dr. Pariza (15). For subsequent cancer studies, a commercial CFA mixture was used (46,47); a mixture from the same commercial source that was later analyzed and found to consist of four positional CFA isomers (15). Regardless of the number of CFA isomers fed in these cancer studies, the GC columns used by these investigators would not have resolved more than two major *cis/trans* CFA peaks because 9,11-18:2 plus 8,10-18:2 and 10,12-18:2 plus 11,13-18:2 coelute on these relatively shorter and less polar GC columns. The general conclusions observed in these cancer studies (see above) remain the same regardless of the dietary CFA isomer composition, even though the identity of the CFA isomer responsible and the distribution of the minor CFA isomers remain unknown. It remains to be determined which of the four positional or geometric isomer(s) is (are) active and if their incorporation into cancer tissues is different.

Human cancer tissues are currently being investigated in several laboratories and the results of one group are presented in Chapter 20.

Liver. Belury and Kempa-Steczko (48) analyzed the liver lipids of mice fed a commercial CFA mixture that was believed to contain two positional CFA isomers (9,11- and 10,12-18:2) based on the GC column used, but this commercial product was subsequently shown to contain four major positional CFA isomers (8,10-, 9,11-, 10,12-, and 11,13-18:2) (15). The authors reported only two CFA peaks in the neutral lipid and phospholipid fractions from the liver lipids (48) because the GC column used combined 8,10- plus 9,11-18:2, and 10,12- plus 11,13-18:2. The concentration of the first eluting peak (labeled 9,11-18:2) was consistently greater than that of the second eluting peak (labeled 10,12-18:2). This result is similar to that observed for liver lipids of pigs fed a commercial CFA mixture with four positional isomers (49). In this latter study (49), the 9,11-18:2 isomer was preferentially accumulated in liver lipids.

Spleen. Turek *et al.* (50) fed rats a commercial CFA mixture that was believed to contain two positional CFA isomers (9,11- and 10,12-18:2) based on the GC column used, but this commercial product was subsequently shown to contain four major positional CFA isomers (8,10-, 9,11-, 10,12-, and 11,13-18:2) (15). The authors (50) reported two CFA peaks in total spleen lipids, *cis*-9, *trans*-11-18:2 plus *trans*-9, *cis*-11-18:2 [at ~0.7% of total fatty acid methyl esters (FAME)] and *trans*-10, *cis*-12-18:2 (at ~1.3% of total FAME). These results would suggest that *trans*-10, *cis*-12-18:2 was preferentially accumulated in spleen lipids (50). However, the GC column used did not separate 10,12- and 11,13-18:2; therefore, one can only conclude that the sum of these two CFA (10,12- plus 11,13-18:2) was greater than the sum of 8,10- plus 9,11-18:2. It appears that spleen lipids contained a significant amount of the 11,13-18:2 isomer, similar to heart lipids of pigs fed a four-positional CFA isomer mixture (49).

Incorporation of Synthetic CFA into Individual Tissue Lipid Classes

Sugano *et al.* (51) were the first to report the incorporation of dietary CFA isomers into the following five different liver phospholipids: PC, phosphatidylethanolamine (PE), phosphatidylinositol (PI), phosphatidylserine (PS), and diphosphatidylglycerol

(DPG, or cardiolipin). Rats were fed a commercial CFA mixture believed to contain two positional CFA isomers (9,11- and 10,12-18:2). The five eluting CFA peaks by GC were labeled in the order ($t9,c11/c9,t11$), ($t10,c12$), ($c9,c11$), ($c10,c12$), and ($t9,t11/t10,t12$). However, similar commercial CFA mixtures were shown to contain four major positional CFA isomers (15), and their own data (51) would support that conclusion. The first two CFA peaks were in fact mixtures of CFA isomers, i.e., 8,10- plus 9,11-18:2 and 10,12- plus 11,13-18:2. This lack of GC resolution led to misidentification of the second and third peaks in the DPG and to an exaggerated response attributed to the 10,12-18:2 isomer in most other lipid classes. In addition to the lack of resolution of the CFA isomers, the CFA were methylated using BF_3/methanol heated to an unspecified temperature, which resulted in excessive amounts of *cis/trans* to *trans,trans* isomerization of CFA of most phospholipids analyzed. The authors also reported the distribution of CFA in a number of other tissues, lymph, and serum (51). Again, the lack of GC resolution and the excessive amount of *trans,trans* CFA isomers formed during BF_3 methylation seriously distorted the CFA distribution.

Li and Watkins (52) examined the effect of feeding CFA on bone formation in rats and presented the total fatty acid composition of liver, muscle, femur cortical bone, femur bone marrow, and femur bone periosteum; no lipid subclasses were reported. They fed rats a commercial CFA mixture they believed contained only two major positional CFA isomers. On the basis of their GC separations, they reported three peaks labeled (c-9, t-11-18:2/ t-9, c-11-18:2), (t-10, c-12-18:2/ c-10, t-12-18:2), and (t-9, t-11-18:2/ t-10, t-12-18:2). However, the commercial CFA mixture fed appears to contain four major positional CFA isomers (15). Under the GC condition used, the first CFA peak is a mixture of 8,10- plus 9,11-18:2 and is generally similar to that of the concentration of 9,11-18:2 by itself. This is because the incorporation of the 8,10-18:2 isomers into tissue lipids was shown to be small when compared with the other CFA isomers (49). The second CFA peak is a mixture of 10,12- and 11,13-18:2; the latter CFA isomer was shown to accumulate extensively in some tissues and lipid classes (49). The results of this study showed an interesting comparison between the first (8,10- plus 9,11-18:2) and second (10,12- plus 11,13-18:2) eluting CFA peaks. In liver lipids, the first peak was more abundant; in muscle lipids the second peak was more prominent, and in the three bone matrices, the two peaks were generally of equal abundance. This result was consistent with our own data, which showed *cis*-9, *trans*-11-18:2 to be preferentially incorporated into liver lipids and *cis*-11, *trans*-13-18:2 preferentially incorporated into heart or muscular lipids (49). Except for the lack of resolution of CFA isomers (52), these data showed clear differences in CFA accumulation in different tissues. Furthermore, the authors used excellent techniques, such as pulverization of tissues at liquid N_2 temperature, extraction with chloroform/methanol, and methylation with $NAOCH_3$. An analysis of isolated lipid subclasses in the various tissues would be most informative.

Kramer *et al.* (49) presented the distribution of CFA isomers in the heart and liver of pigs fed a commercial CFA mixture which contained four major positional CFA isomers. The tissues were removed as quickly as possible from the animal, rinsed with water, frozen between two blocks of dry ice, and stored at −70°C until analyzed. The tissues were pulverized at dry ice temperature and the total lipids were extracted with 1:1 chloroform/methanol (53). The total lipids were separated by three-directional TLC using silica gel H plates; the spots were visualized under ultraviolet (UV) light after the plates were sprayed with 2′,7′-dichlorofluorescein (54). The advantage of using this TLC technique is that it separated all of the neutral lipids and phospholipids on one TLC plate for easy quantitation. The spots were removed, heptadecanoic acid (17:0) was added to each spot as internal standard for quantitation purposes, and the silica gel together with the isolated lipid classes were methylated using $NaOCH_3$/methanol, except for sphingomyelins (SM), which were methylated using HCl/methanol, and FFA, which were methylated using diazomethane (49).

All of the lipid fractions were then analyzed by using both GC on a 100-m CP Sil 88 column and by Ag^+-HPLC. The identity of all CFA isomers was confirmed by comparison to standards and analyzed by GC-Fourier transform infrared spectroscopy (FTIR) and GC-MS as their methyl esters and their DMOX derivatives. GC resulted in the resolution of 10 CFA peaks as shown in Figure 18.1A. The 100-m GC column separated the two previously unresolved *cis/trans* isomers, 11,13- and 10,12-18:2, but not the 8,10- and 9,11-18:2 isomers. In addition, all four *cis,cis* CFA isomers were separated, but three of the four *trans,trans* isomers were not resolved. Ag^+-HPLC served as a valuable complementary technique to GC because it allowed the resolution of all 12 possible geometric isomers of four positional CFA isomers (Fig. 18.1A′). Both of these methods were used to analyze all of the lipid classes of heart, liver, and adipose tissues from these pigs fed CFA.

The analysis of omental fat (Figs. 18.1B and 18.1B′), liver TAG (Figs. 18.1C and 18.1C′), and heart TAG (Figs. 18.1D and 18.1D′) showed that the distribution of CFA isomers in the dietary mixture (Figs. 18.1A and 18.1A′) was also available to the liver and heart. The liver phospholipids preferentially accumulated *cis*-9, *trans*-11-18:2 (Figs. 18.2A and 18.2A′; Fig. 18.3), except DPG, which showed a preference to accumulate *cis*-11, *trans*-13-18:2 (Fig. 18.3), whereas heart phospholipids showed a preference for incorporating *cis*-11, *trans*-13-18:2 (Figs. 18.2B and 18.2B′; Fig. 18.4), particularly into DPG (Figs. 18.2C and 18.2C′; and Fig. 18.4). The CFA isomers *trans*-8, *cis*-10-18:2 and *trans*-10, *cis*-12-18:2 did not accumulate extensively in any of the lipid classes of pig heart or liver. Positional analysis of CFA in TAG and the different phospholipids was not determined.

The accumulation of *cis*-11, *trans*-13-18:2 into heart lipids and in particular into DPG, the phospholipid found principally in inner mitochondrial membranes (55,56), may be viewed with concern. This CFA isomer (*cis*-11, *trans*-13-18:2) is not the major CFA found in natural products and it is not the minor 11,13-18:2 isomer found in milk and cheese, i.e., *trans*-11, *cis*-13-18:2 (13,17,25). The significance of this accumulation remains unknown.

Significance of CFA Inclusion into Membrane Lipids

Detailed analyses of CFA incorporation into different tissues and their subclasses are necessary to evaluate fully the physiologic and/or pathologic effect of different CFA

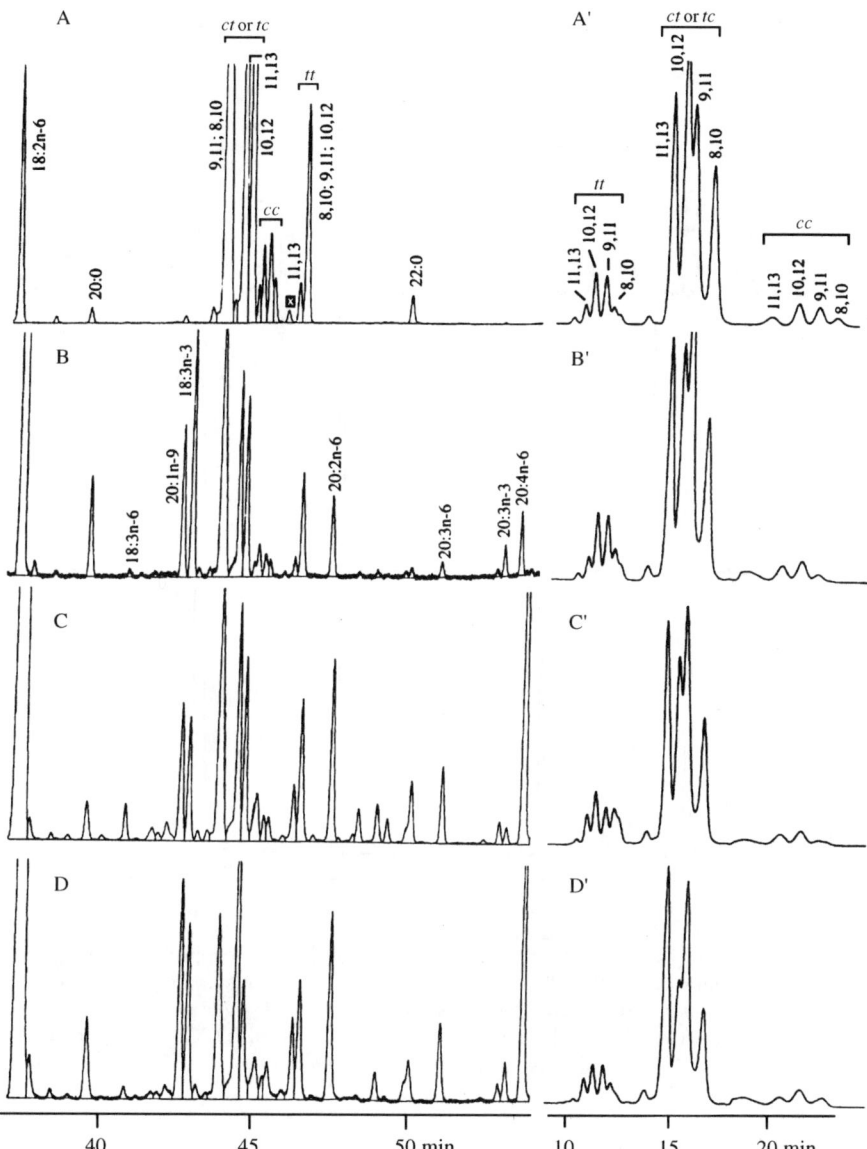

Fig. 18.1. Partial profiles obtained by gas chromatography (GC) (A– D) and silver-ion high-performance liquid chromatography (Ag⁺-HPLC) (A'–D') of the conjugated fatty acid (CFA) mixture fed to pigs (A,A'), and of CFA isomers found in omental fat (B,B'), liver triacylglycerols (TAG) (C,C') and heart TAG (D,D'). The GC region selected was between linoleic (18:2n-6) and arachidonic (20:4n-6) acids. "x" is an unknown in the GC chromatogram that was also found in the chromatograms of both the dietary CFA mixture and in the tissues of pigs fed the control and CFA diets; *c* and *t* refer to the *cis* and *trans* CFA isomers. The elution order of the *cis,cis* CFA isomers was as follows: 8,10-, 9,11-, 10,12-, and 11,13-18:2. (Published by permission of *Lipids*.)

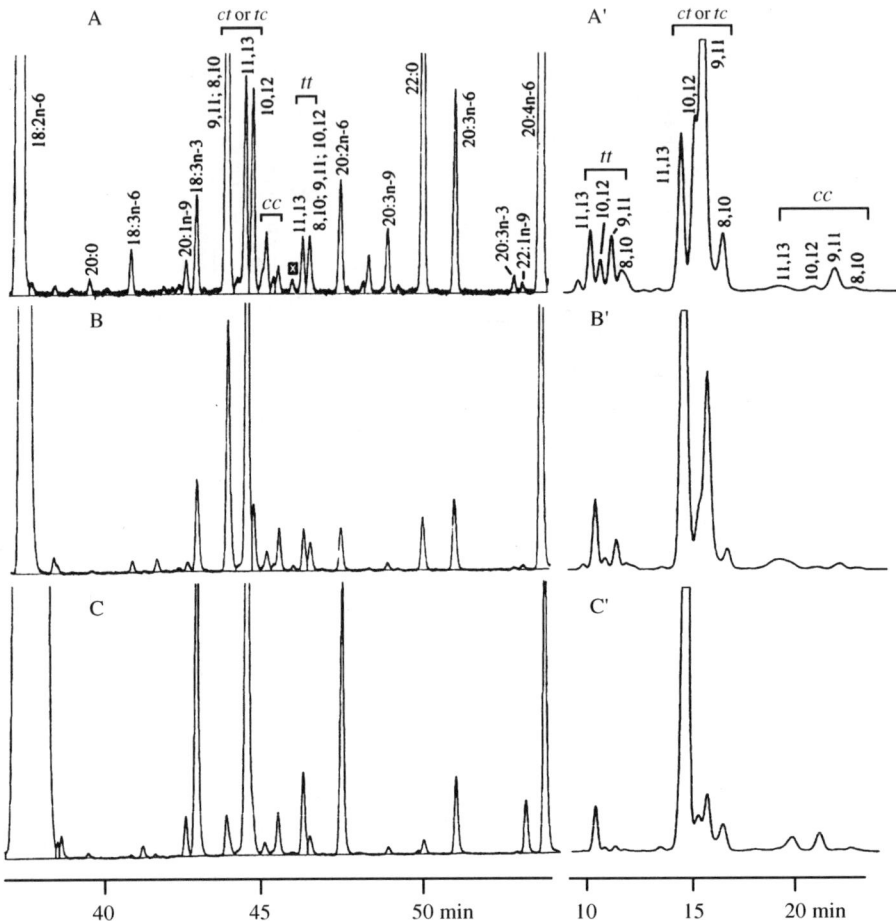

Fig. 18.2. Partial gas chromatography (GC) (A–C) and silver-ion high-performance liquid chromatography (Ag⁺-HPLC) (A′–C′) chromatograms of selected liver and heart lipid classes of pigs fed the conjugated fatty acid (CFA) diet: liver phosphatidylcholine (PC) (A,A′), heart PC (B,B′), and heart diphosphatidylglycerol (DPG) (C,C′). For further details see Figure 18.1. (Published by permission of *Lipids*.)

isomers, their effect in different tissues and membrane structures, and in determining the fate of individual CFA isomers in animals. However, there is a note of caution in interpreting incorporations of CFA isomers into tissue lipids.

First, a high accumulation of a CFA isomer(s) may represent selective uptake or slower rate of metabolism, or conversely, a low concentration of a CFA isomer(s) may signify discrimination or rapid metabolism. An example of the latter situation was recently observed by comparing the C_{20} CFA metabolites reported by Sébédio *et al.* (57) with the incorporation of C_{18} CFA isomers into pig tissue lipids (49).

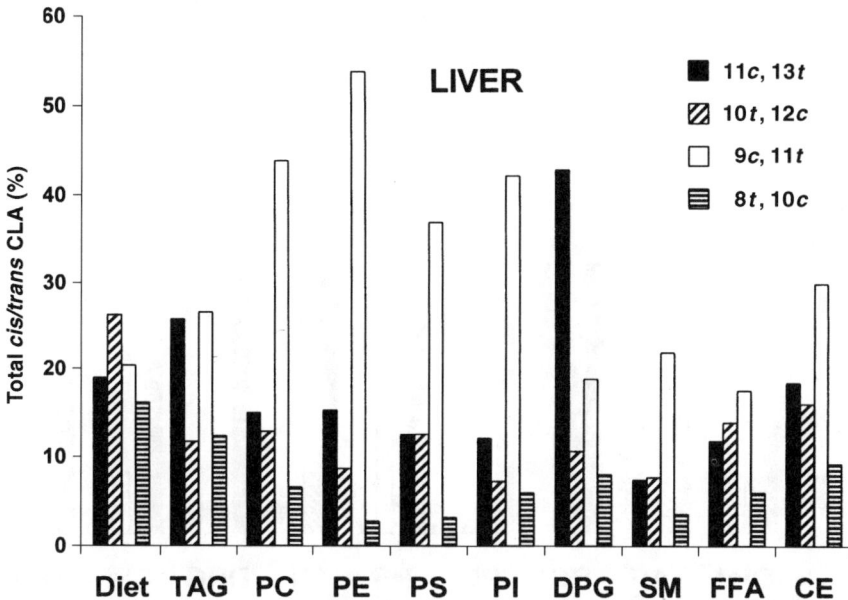

Fig. 18.3. Quantitative comparison of the four major *cis/trans* conjugated linoleic acid (CLA) isomers in the different liver lipid classes and the diet determined by silver-ion high-performance liquid chromatography (Ag$^+$-HPLC). Relative amounts (y-axis) are expressed as a percentage of total *cis/trans* CLA isomers. The distribution of the *cis/trans* CLA isomers in the diet are included for comparison. TAG, triacylglycerol; PC, phosphatidylcholine; PE, phosphatidylethanolamine; PS, phosphatidylserine; PI, phosphatidylinositol; DPG, diphosphatidylglycerol; SM, sphingomyelin; FFA, free fatty acids, CE, cholesteryl esters. (Published by permission of *Lipids*.)

Sébédio *et al.* (57) found and characterized three C_{20} metabolites in rat liver lipids after feeding a CFA mixture to essentially deficient rats. Two of the CFA metabolites found (8,12,14-20:3 and 5,8,12,14-20:4) appeared to be derived from 10,12-18:2 (57). But 10,12-18:2, one of the four positional isomers in the CFA mixture (15,57), accumulated very little in the liver lipids (49). This would suggest that this CFA isomer, 10,12-18:2, is actively metabolized and might explain the low concentration of 10,12-18:2 isomer in liver lipids (49), assuming that the two species behaved similarly.

Second, accumulation of a CFA isomer should not be considered as proof of a given physiologic or pathologic response. Several reports have shown that *cis*-9, *trans*-11-18:2 accumulated in the phospholipids of a given tissue. Accumulation was thought to provide evidence that *cis*-9, *trans*-11-18:2 was the active CFA isomer. By using the same logic, *cis*-11, *trans*-13-18:2 should be considered an active CFA isomer, because it accumulated in heart lipids and specifically DPG (49). However, no specific effect (negative or positive) has thus far been attributed to result from the accumulation of this CFA isomer. In all fairness, however, one may need to investigate the enzyme complexes of oxidative phosphorylation in mitochondria; they are

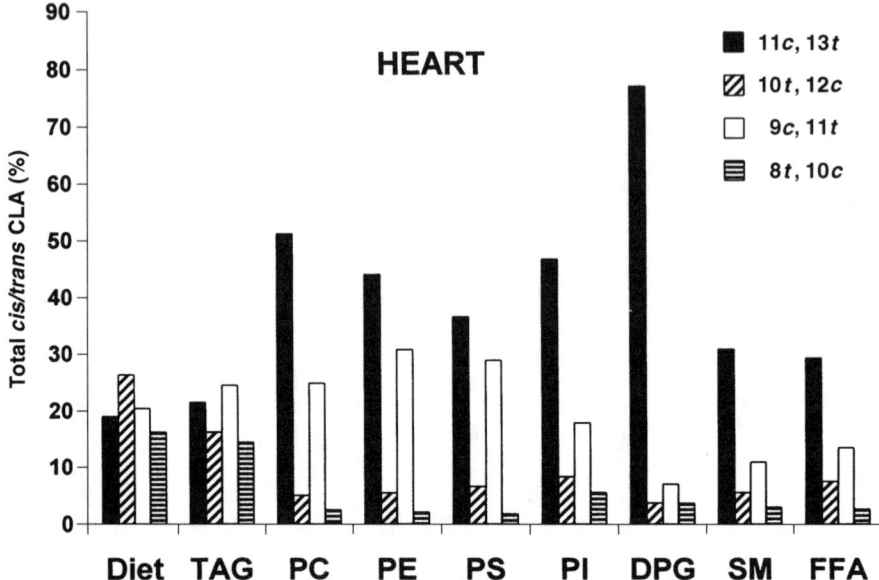

Fig. 18.4. Quantitative comparison of the four major *cis/trans* conjugated linoleic acid (CLA) isomers in the different heart lipid classes and the diet determined by silver-ion high-performance liquid chromatography (Ag$^+$-HPLC). For further details see Figure 18.3. (Published by permission of *Lipids*.)

closely associated with DPG, which contains this CFA isomer. A unique response due to the accumulation of docosahexaenoic acid into mitochondrial phospholipids was observed recently (58).

Third, incorporation of CFA isomer(s) or their metabolites may cause physical changes in membrane properties or protect the membrane. Some investigators have suggested that CFA are antioxidants (43,44), although this claim has been questioned (59,60). Preliminary evidence (J.J. Yin, unpublished results) indicates that inclusion of up to 5% 1-stearoyl-2-octadec-*cis*-9, *trans*-11-dienoyl-*sn*-glycero-3-phosphorylcholine (1-18:0, 2-CFA-PC) instead of 1-stearoyl- 2-octadec-*cis*-9, *cis*-12-dienoyl-*sn*-glycero-3-phosphorylcholine (1-18:0, 2-18:2 n-6-PC) into an artificial membrane consisting of soy or egg lecithins significantly affected oxygen transport within the membrane and affected the packing order of the alkyl side chains of the fatty acids in the membranes as demonstrated by electron spin resonance studies. These results are encouraging because they indicate that a small amount of CFA in membrane phospholipids can change the physical properties of a membrane, with possible physiologic consequences. It is interesting to note that the low concentration of 5% 1-18:0,2-CFA-PC in the oil was effective in causing structural changes in the membrane. Incidentally, maximum anticancer protection was also observed with a low concentration of conjugated linoleic acid (CLA) (1%) in the diet (44,46); 1% CLA in the diet is equivalent to 5% CLA in the oil of a diet containing 20% fat. Further studies are required to establish a mechanism between the structural effects on membranes and pathologic responses in biological systems.

Incorporation measurements should always be conducted in conjunction with metabolic interconversions. CFA metabolites and their subsequent transformation into physiologically active compounds constitute an area requiring further investigation. Limited studies have been reported on the effects of CFA on eicosanoid biosynthesis (52,61,62).

References

1. Kepler, C.R., Hirons, K.P., McNeill, J.J., and Tove, S.B. (1966) Intermediates and Products of the Biohydrogenation of Linoleic Acid by *Butyrivibrio fibrisolvens*, *J. Biol. Chem. 241*, 1350–1354.
2. Hughes, P.E., Hunter, W.J., and Tove, S.B. (1982) Biohydrogenation of Unsaturated Fatty Acids, *J. Biol. Chem. 257*, 3643–3649.
3. Harfoot, C.G., and Hazlewood, G.P. (1988) Lipid Metabolism in the Rumen, in *The Rumen Microbial Ecosystem*, (Hobson, P.N., ed.) pp. 285–322, Elsevier Applied Science, London.
4. Parodi, P.W. (1977) Conjugated Octadecadienoic Acids of Milk Fat, *J. Dairy Sci. 60*, 1550–1553.
5. Fogerty, A.C., Ford, G.L., and Svoronos, D. (1988) Octadeca-9,11-dienoic Acid in Foodstuffs and in the Lipids of Human Blood and Breast Milk, *Nutr. Rep. Int. 38*, 937–944.
6. Chin, S.F., Liu, W., Storkson, J.M., Ha, Y.L., and Pariza, M.W. (1992) Dietary Sources of Conjugated Dienoic Isomers of Linoleic Acid, a Newly Recognized Class of Anticarcinogens, *J. Food Comp. Anal. 5*, 185–197.
7. Lin, H., Boylston, T.D., Chang, M.J., Luedecke, L.O., and Shultz, T.D. (1995) Survey of the Conjugated Linoleic Acid Contents of Dairy Products, *J. Dairy Sci. 78*, 2358–2365.
8. Banni, S., Carta, G., Contini, M.S., Angioni, E., Deiana, M., Dessì, M.A., Melis, M.P., and Corongiu, F.P. (1996) Characterization of Conjugated Diene Fatty Acids in Milk, Dairy Products, and Lamb Tissues, *J. Nutr. Biochem. 7*, 150–155.
9. Yurawecz, M.P., Roach, J.A.G., Sehat, N., Mossoba, M.M., Kramer, J.K.G., Fritsche, J., Steinhart, H., and Ku, Y. (1998) A New Conjugated Linoleic Acid Isomer, 7 *trans*,9 *cis*-Octadecadienoic Acid, in Cow Milk, Cheese, Beef and Human Milk and Adipose Tissue, *Lipids 33*, 803–809.
10. Kramer, J.K.G., Parodi, P.W., Jensen, R.G., Mossoba, M.M., Yurawecz, M.P., and Adlof, R.O. (1998) Rumenic Acid: A Proposed Name for the Major Conjugated Linoleic Acid Isomer Found in Natural Products, *Lipids 33*, 835.
11. Pollard, M.R., Gunstone, F.D., James, A.T., and Morris, L.J. (1980) Desaturation of Positional and Geometric Isomers of Monoenoic Fatty Acids by Microsomal Preparations from Rat Liver, *Lipids 15*, 306–314.
12. Salminen, I., Mutanen, M., Jauhiainen, M., and Aro, A. (1998) Dietary *trans* Fatty Acids Increase Conjugated Linoleic Acid Levels in Human Serum, *J. Nutr. Biochem. 9*, 93–98.
13. Sehat, N., Kramer, J.K.G., Mossoba, M.M., Yurawecz, M.P., Roach, J.A.G., Eulitz, K., Morehouse, K.M., and Ku, Y. (1998) Identification of Conjugated Linoleic Acid Isomers in Cheese by Gas Chromatography, Silver-Ion High-Performance Liquid Chromatography and Mass Spectral Reconstructed Ion Profiles. Comparison of Chromatographic Elution Sequences, *Lipids 33*, 963–971.

14. Mounts, T.L., Dutton, H.J., and Glover, D. (1970) Conjugation of Polyunsaturated Acids, *Lipids 5*, 997–1005.
15. Sehat, N., Yurawecz, M.P., Roach, J.A.G., Mossoba, M.M., Kramer, J.K.G., and Ku, Y. (1998) Silver-Ion High-Performance Liquid Chromatographic Separation and Identification of Conjugated Linoleic Acid Isomers, *Lipids 33*, 217–221.
16. Yurawecz, M.P., Sehat, N., Mossoba, M.M., Roach, J.A.G., Kramer, J.K.G., and Ku, Y. (1999) Variations in Isomer Distribution in Commercially Available Conjugated Linoleic Acid, *Fett/Lipid 101* (in press).
17. Eulitz, K., Yurawecz, M.P., Sehat, N., Fritsche, J., Roach, J.A.G., Mossoba, M.M., Kramer, J.K.G., Adlof, R.O., and Ku, Y. (1999) Preparation, Separation and Confirmation of the Eight Geometric *cis/trans* Conjugated Linoleic Acid Isomers 8,10- Through 11,13-18:2, *Lipids* (in press).
18. Britton, M., Fong, C., Wickens, D., and Yudkin, J. (1992) Diet as a Source of Phospholipid Esterified 9,11-Octadecadienoic Acid in Humans, *Clin. Sci. 83*, 97–101.
19. Huang, Y.-C., Luedecke, L.O., and Shultz, T.D. (1994) Effect of Cheddar Cheese Consumption on Plasma Conjugated Linoleic Acid Concentrations in Men, *Nutr. Res. 14*, 373–386.
20. Fritsche, J., Mossoba, M.M., Yurawecz, M.P., Roach, J.A.G., Sehat, N., Ku, Y., and Steinhart, H. (1997) Conjugated Linoleic Acid (CLA) Isomers in Human Adipose Tissue, *Z. Lebensm. Unters. Forsch. A 205*, 415–418.
21. McGuire, M.K., Park, Y., Behre, R.A., Harrison, L.Y., Shultz, T.D., and McGuire, M.A. (1997) Conjugated Linoleic Acid Concentrations of Human Milk and Infant Formula, *Nutr. Res. 17*, 1277–1283.
22. Jensen, R.G., Lammi-Keefe, C.J., Hill, D.W., Kind, A.J., and Henderson, R. (1998) The Anticarcinogenic Conjugated Fatty Acid, 9c,11t-18:2, in Human Milk: Confirmation of Its Presence, *J. Hum. Lact. 14*, 23–27.
23. Ackman, R.G., Eaton, C.A., Sipos, J.C., and Crewe, N.F. (1981) Origin of *cis*-9, *trans*-11- and *trans*-9, *trans*-11-Octadecadienoic Acids in the Depot Fat of Primates Fed a Diet Rich in Lard and Corn Oil and Implications for the Human Diet, *Can. Inst. Food Sci. Technol. J. 14*, 103–107.
24. Banni, S., Day, B.W., Evans, R.W., Corongiu, F.P., and Lombardi, B. (1995) Detection of Conjugated Diene Isomers of Linoleic Acid in Liver Lipids of Rats Fed a Choline-Devoid Diet Indicates That the Diet Does Not Cause Lipoperoxidation, *J. Nutr. Biochem. 6*, 281–289.
25. Sehat, N., Rickert, R., Mossoba, M.M., Kramer, J.K.G., Yurawecz, M.P., Roach, J.A.G., Adlof, R.O., Morehouse, K.M., Fritsche, J., Eulitz, K.D., Steinhart, H., and Ku, Y. (1999) Improved Separation of Conjugated Fatty Acid Methyl Esters by Silver Ion–High Performance Liquid Chromatography, *Lipids 34*, (407–413).
26. Cawood, P., Wickens, D.G., Iversen, S.A., Braganza, J.M., and Dormandy, T.L. (1983) The Nature of Diene Conjugation in Human Serum, Bile and Duodenal Juice, *FEBS Lett. 162*, 239–243.
27. Iversen, S.A., Cawood, P., Madigan, M.J., Lawson, A.M., and Dormandy, T.L. (1984) Identification of a Diene Conjugated Component of Human Lipid as Octadeca-9,11-dienoic Acid, *FEBS Lett. 171*, 320–324.
28. Iversen, S.A., Cawood, P., and Dormandy, T.L. (1985) A Method for the Measurement of a Diene-Conjugated Derivative of Linoleic Acid, 18:2(9,11), in Serum Phospholipid, and Possible Origins, *Ann. Clin. Biochem. 22*, 137–140.

29. Bligh, E.G., and Dyer, W.J. (1959) A Rapid Method of Total Lipid Extraction and Purification, *Can. J. Biochem. Physiol. 37*, 911–917.
30. Park, Y., and Pariza, M.W. (1998) Evidence that Commercial Calf and Horse Sera Can Contain Substantial Amounts of *trans*-10,*cis*-12 Conjugated Linoleic Acid, *Lipids 33*, 817–819.
31. Emken, E.A., Scholfield, C.R., Davison, V.L., and Frankel, E.N. (1967) Separation of Conjugated Methyl Octadecadienoate and Trienoate Geometric Isomers by Silver-Resin Column and Preparative Gas–Liquid Chromatography, *J. Am. Oil Chem. Soc. 44*, 373–375.
32. Kramer, J.K.G., Fellner, V., Dugan, M.E.R., Sauer, F.D., Mossoba, M.M., and Yurawecz, M.P. (1997) Evaluating Acid and Base Catalysts in the Methylation of Milk and Rumen Fatty Acids with Special Emphasis on Conjugated Dienes and Total *trans* Fatty Acids, *Lipids 32*, 1219–1228.
33. Ha, Y.L., Grimm, N.K., and Pariza, M.W. (1989) Newly Recognized Anticarcinogenic Fatty Acids: Identification and Quantification in Natural and Processed Cheeses, *J. Agric. Food Chem. 37*, 75–81.
34. Shantha, N.C., Decker, E.A., and Ustunol, Z. (1992) Conjugated Linoleic Acid Concentration in Processed Cheese, *J. Am. Oil Chem. Soc. 69*, 425–428.
35. Werner, S.A., Luedecke, L.O., and Shultz, T.D. (1992) Determination of Conjugated Linoleic Acid Content and Isomer Distribution in Three Cheddar-Type Cheeses: Effects of Cheese Cultures, Processing, and Aging, *J. Agric. Food Chem. 40*, 1817–1821.
36. Jiang, J., Björck, L., and Fondén, R. (1997) Conjugated Linoleic Acid in Swedish Dairy Products with Special Reference to the Manufacture of Hard Cheeses, *Int. Dairy J. 7*, 863–867.
37. Lin, H., Boylston, T.D., Luedecke, L.O., and Shultz, T.D. (1998) Factors Affecting the Conjugated Linoleic Acid Content of Cheddar Cheese, *J. Agric. Food Chem. 46*, 801–807.
38. Lavillonnière, F., Martin, J.C., Bougnoux, P., and Sébédio, J.-L. (1998) Analysis of Conjugated Linoleic Acid Isomers and Content in French Cheeses, *J. Am. Oil Chem. Soc. 75*, 343–352.
39. Ha, Y.L., Grimm, N.K., and Pariza, M.W. (1987) Anticarcinogens from Fried Ground Beef: Heat-Altered Derivatives of Linoleic Acid, *Carcinogenesis 8*, 1881–1887.
40. Fritsche, S., and Fritsche, J. (1998) Occurrence of Conjugated Linoleic Acid Isomers in Beef, *J. Am. Oil Chem. Soc. 75*, 1449–1451.
41. Christie, W.W. (1973) The Structures of Bile Phosphatidylcholines, *Biochim. Biophys. Acta 316*, 204–211.
42. Smith, G.N., Taj, M., and Braganza, J.M. (1991) On the Identification of a Conjugated Diene Component of Duodenal Bile as 9Z,11E-Octadecadienoic Acid, *Free Radic. Biol. Med. 10*, 13–21.
43. Ha, Y.L., Storkson, J., and Pariza, M.W. (1990) Inhibition of Benzo(a)pyrene-Induced Mouse Forestomach Neoplasia by Conjugated Dienoic Derivatives of Linoleic Acid, *Can. Res. 50*, 1097–1101.
44. Ip, C., Chin, S.F., Scimeca, J.A., and Pariza, M.W. (1991) Mammary Cancer Prevention by Conjugated Dienoic Derivative of Linoleic Acid, *Can. Res. 51*, 6118–6124.
45. Liew, C., Schut, H.A.J., Chin, S.F., Pariza, M.W., and Dashwood, R.H. (1995) Protection of Conjugated Linoleic Acids Against 2-Amino-3-methylimidazo[4,5-*f*]quinoline-Induced Colon Carcinogenesis in the F344 Rat: A Study of Inhibitory Mechanisms, *Carcinogenesis 16*, 3037–3043.

46. Ip, C., Briggs, S.P., Haegele, A.D., Thompson, H.J., Storkson, J., and Scimeca, J.A. (1996) The Efficacy of Conjugated Linoleic Acid in Mammary Cancer Prevention Is Independent of the Level or Type of Fat in the Diet, Carcinogenesis 17, 1045–1050.
47. Ip, C., Jiang, C., Thompson, H.J., and Scimeca, J.A. (1997) Retention of Conjugated Linoleic Acid in the Mammary Gland is Associated with Tumor Inhibition During the Post-Initiation Phase of Carcinogenesis, *Carcinogenesis 18,* 755–759.
48. Belury, M.A., and Kempa-Steczko, A. (1997) Conjugated Linoleic Acid Modulates Hepatic Lipid Composition in Mice, *Lipids 32,* 199–204.
49. Kramer, J.K.G., Sehat, N., Dugan, M.E.R., Mossoba, M.M., Yurawecz, M.P., Roach, J.A.G., Eulitz, K., Aalhus, J.L., Schaefer, A.L., and Ku, Y. (1998) Distributions of Conjugated Linoleic Acid (CLA) Isomers in Tissue Lipid Classes of Pigs Fed a Commercial CLA Mixture Determined by Gas Chromatography and Silver-Ion High-Performance Liquid Chromatography, *Lipids 33,* 549–558.
50. Turek, J.J., Li, Y., Schoenlein, I.A., Allen, K.G.D., and Watkins, B.A. (1998) Modulation of Macrophage Cytokine Production by Cunjugated Linoleic Acids is Influenced by the Dietary n-6:n-3 Fatty Acid Ratio, *J. Nutr. Biochem. 9,* 258–266.
51. Sugano, M., Tsujita, A., Yamasaki, M., Yamada, K., Ikeda, I., and Kritchevsky, D. (1997) Lymphatic Recovery, Tissue Distribution, and Metabolic Effects of Conjugated Linoleic Acid in Rats, *J. Nutr. Biochem. 8,* 38–43.
52. Li, Y., and Watkins, B.A. (1998) Conjugated Linoleic Acids Alter Bone Fatty Acid Composition and Reduce *ex vivo* Bone Prostaglandin E_2 Biosynthesis in Rats Fed n-6 or n-3 Fatty Acids, *Lipids 33,* 417–425.
53. Kramer, J.K.G., and Hulan, H.W. (1978) A Comparison of Procedures to Determine Free Fatty Acids in Rat Heart, *J. Lipid Res. 19,* 103–106.
54. Kramer, J.K.G., Fouchard, R.C., and Farnworth, E.R. (1983) A Complete Separation of Lipids by Three-Directional Thin-Layer Chromatography, *Lipids 18,* 896–899.
55. Hoch, F.L. (1992) Cardiolipins and Biomembrane Function, *Biochim. Biophys. Acta 1113,* 71–133.
56. Shigenaga, M.K., Hagen, T.M., and Ames, B.N. (1994) Oxidative Damage and Mitochondrial Decay in Aging, *Proc. Natl. Acad. Sci. USA 91,* 10771–10778.
57. Sébédio, J.L., Juanéda, P., Dobson, G., Ramilison, I., Martin, J.C., Chardigny, J.M., and Christie, W.W. (1997) Metabolites of Conjugated Isomers of Linoleic Acid (CLA) in the Rat, *Biochim. Biophys. Acta 1345,* 5–10.
58. Watkins, S.M., Carter, L.C., and German, J.B. (1998) Dococahexaenoic Acid Acummulates in Cardiolipin and Enhances HT-29 Cell Oxidant Production, *J. Lipid Res. 39,* 1583–1588.
59. van den Berg, J.J.M., Cook, N.E., and Tribble, D.L. (1995) Reinvestigation of the Antioxidant Properties of Conjugated Linoleic Acid, *Lipids 30,* 599–605.
60. Chen, Z.Y., Chan, P.T., Kwan, K.Y., and Zhang, A. (1997) Reassessment of the Antioxidant Activity of Conjugated Linoleic Acids, *J. Am. Oil Chem. Soc. 74,* 749–753.
61. Liu, K.-L., and Belury, M.A. (1997) Conjugated Linoleic Acid Modulation of Phorbol Ester-Induced Events in Murine Keratinocytes, *Lipids 32,* 725–730.
62. Sugano, M., Tsujita, A., Yamasaki, M., Noguchi, M., and Yamada, K. (1998) Conjugated Linoleic Acid Modulates Tissue Levels of Chemical Mediators and Immunoglobulins in Rats, *Lipids 33,* 521–527.

Chapter 19

Bone Metabolism and Dietary Conjugated Linoleic Acid

Bruce A. Watkins[a,b], Yong Li[a], and Mark F. Seifert[b]

[a]Department of Food Science, Lipid Chemistry and Molecular Biology Laboratory, Purdue University, West Lafayette, IN
[b]Department of Anatomy, Indiana University School of Medicine, Indianapolis, IN

Introduction

Long bone growth and modeling are regulated by complex interactions among an individual's genetic potential, environmental influences, and nutrition. These interactions produce a bone architecture that balances functionally appropriate morphology with the skeleton's role in calcium and phosphorus homeostasis. Long bones of children increase in length and diameter by a process called modeling. Bone modeling represents an adaptive process of generalized and continuous growth and reshaping of bone, governed by the activities of osteoblasts and osteoclasts until the adult bone structure is attained. This growth requires that bone cells function normally. Bone modeling is distinct from bone remodeling, which describes the local, coupled process of bone resorption and formation that maintains skeletal mass and morphology in the adult. The numerous cell-derived growth regulatory factors present within skeletal tissues, e.g., prostaglandins, cytokines, and growth factors, exert local controls on skeletal metabolism. The prostaglandins are major players in bone metabolism but also have a role in joint disease. Many of the skeletal pathologies that afflict the adult, e.g., osteoporosis or rheumatoid arthritis, are the consequence of abnormal bone remodeling and metabolism, or constitute an inflammatory process. Data from recent studies suggest that the onset and severity of some of these pathologies may be delayed and lessened if bone modeling is optimized early in life or if diets are supplemented with nutrients that reduce tissue concentrations of factors that undermine skeletal health.

This chapter has the following goals: (i) to describe the types of cells located in bone; (ii) to explain the basic elements of bone modeling and remodeling; (iii) to discuss the roles of prostaglandins and cytokines involved in the local regulation of bone metabolism; and (iv) to document the role of lipids, including studies on conjugated linoleic acids, in bone biology and the relationship between these fatty acids and factors regulating skeletal metabolism.

Bone Cells and Bone Metabolism

Bone is a multifunctional organ that consists of a structural framework of mineralized matrix and contains heterogeneous populations of chondrocytes, osteoblasts, osteocytes, osteoclasts, endothelial cells, monocytes, macrophages, lymphocytes, and hemopoietic cells. This milieu of cells produces a variety of biological regulators that control local bone metabolism. Systemic calcitropic hormones [parathyroid hormone (PTH), estrogen, and 1,25(OH)$_2$vitamin D$_3$] and autocrine and paracrine factors, including prostaglandins, cytokines, and growth factors, orchestrate the cellular activities of bone modeling to increase the length and diameter, and properly shape long bones in children. Bone grows in size and shape through the collective activities of cells that produce, mineralize, and resorb bone matrix. Bone matrix is produced and mineralized through the activity of osteoblasts, whereas bone matrix resorption is accomplished by specialized multinucleated cells called osteoclasts (1). The combined and cooperative activities of osteoblasts and osteoclasts result in a bone architecture that provides mechanical support and protection for the body. In addition, it serves as a vital reservoir of minerals, principally calcium and phosphorus, necessary for maintaining normal cellular, neurologic, and vascular activities of the body.

Osteoblasts are mononucleated bone-forming cells that originate locally from mesenchymal stem cells. Osteoblasts are recruited to a site of bone formation where they actively synthesize and secrete an organic bone matrix called "osteoid," which is composed principally of type I collagen and other noncollagenous proteins. After its formation, osteoid normally undergoes rapid mineralization with hydroxyapatite. In addition to synthesizing bone matrix, osteoblasts maintain a high alkaline phosphatase activity and produce numerous regulatory factors including prostaglandins, cytokines, and growth factors. These locally produced compounds are reported to stimulate bone formation and/or bone resorption (2–5).

Osteoclasts are large multinucleated bone-resorbing cells. They form at skeletal sites from the fusion of mononuclear hemopoietic precursors that arrive *via* the vasculature. During bone resorption, osteoclasts produce and release lysosomal enzymes, hydrogen protons, and free radicals into a confined space or resorptive compartment next to bone, which dissolve the mineral and degrade bone matrix (6). Active osteoclasts are in contact with mineralized surfaces and produce distinctive resorptive cavities called Howship's lacunae. Thus, bone cells are under the regulatory control of systemic and local factors that modulate their activities and influence bone modeling and remodeling processes.

Bone Tissue and Growth

Bone is a dynamic connective tissue consisting of living cells embedded within or lining the surfaces of a mineralized organic matrix. Bone provides mechanical support for the body, and through attachment of muscles, allows for locomotive movement through space. Furthermore, skeletal tissue protects vital organs and serves as a metabolic reservoir of calcium and phosphate for the body. Anatomically, the bones of the skeleton can be classified according to their individual shapes: flat (bones forming the roof of the skull, scapula, and ilium), short (carpal and tarsal bones), irregular (vertebrae), and long (humerus, radius, ulna, femur, and tibia).

All bone is derived from mesenchymal tissue; however, two different histogenetic processes exist for producing bone, one direct and another indirect, through a temporary cartilage model. Intramembranous ossification occurs within presumptive flat bones by direct differentiation of mesenchymal cells into osteogenic cells. These osteoblasts deposit organic matrix within their embryonic connective tissue membrane, which becomes mineralized. Long bones are formed by endochondral ossification, a process in which embryonic mesenchymal cells differentiate into chondroblasts and chondrocytes, which secrete hyaline cartilage matrix and produce a cartilage model of the future bone. During maturation of this tissue, chondrocytes within the future shaft (diaphysis) and ends (epiphyses) of the bone hypertrophy, and the surrounding matrix undergoes mineralization. Invasion of these regions by the vasculature produces diaphyseal and epiphyseal centers of ossification. Mineralized cartilage matrix is removed by chondroclasts and replaced with bone by newly arrived osteogenic cells. These respective centers grow and expand toward one another but remain temporarily separate by a bar of cartilage, the epiphyseal growth plate. This plate of cartilage, interposed between epiphyseal and metaphyseal regions of a bone, provides the means for bones to grow in length. In this process, chondrocyte proliferation, matrix production, mineralization, and vascular invasion are balanced with removal of mineralized trabeculae from the metaphyseal side of the growth plate through osteoclastic activity. These primary trabeculae of mineralized cartilage enclosed by bone are progressively replaced by secondary trabeculae of bone produced by osteoblasts. This trabecular or cancellous bone constitutes a meshwork of tissue at the ends of long bones whose surfaces are lined by bone cells and whose intertrabecular spaces are filled with hemopoietic tissue. Trabecular bone is involved mainly with metabolic functions. Dense or cortical bone completely encases bones and is especially thick within the diaphyses or shafts of long bones. The diameters of bones increase *via* intramembranous ossification through apposition of bone matrix by osteoblasts located within the periosteum. Cortical bone serves primarily mechanical and protective functions.

Bone Modeling

Bone modeling describes the continuous changes in bone shape, length, and width throughout the growth of an individual until skeletal maturity is reached. In contrast to bone remodeling, bone modeling lacks local coupling of resorption with bone formation on bone surfaces. Resorption and formation occur on separate surfaces (periosteal or corticoendosteal); therefore, surface activation in modeling bone may be followed by either resorption or formation (Table 19.1).

Bone Remodeling

The skeletal morphology of the adult represents a sophisticated compromise between structural obligation and metabolic responsibility, serving the individual in support and locomotion while actively participating in the regulation of calcium homeostasis. This compromise is accomplished through the individual's genetic potential for

TABLE 19.1 Comparison of Bone Modeling and Remodeling Activities[a]

	Bone modeling	Bone remodeling
Local coupling	Formation and resorption are *not* coupled	Formation and resorption are coupled
Timing and sequence of activity	F and RS are continuous and occur on separate surfaces	Cyclical: A-RS-RV-F; formation always follows resorption
Extent of surface activity	100% of surfaces are active	20% of surfaces are active
Anatomical objectives	Gain in skeletal mass and change in skeletal morphology	Skeletal maintenance

[a]A, activation; RS, resorption; RV, reversal; F, formation.

growth and intricate interactions among nutrition, metabolism, endocrine factors, and transcription factors. Hormones and certain nutrients modulate the autocrine and paracrine influences (actions of prostaglandins, cytokines, and growth factors) on cells and are responsible for the maintenance of bone mass and architecture. In the adult skeleton, the coordination of bone-resorbing and bone-forming activities is termed the "bone remodeling cycle."

The regulation of bone remodeling, and its corresponding role in the maintenance of adult bone mass, is distinctly different from the processes that control skeletal growth and modeling in the young (Table 19.1). As the name implies, modeling is responsible for creating bone shape. Modeling of bone is an adaptive process, providing order and specificity to the generalized increase in bone mass that accompanies body growth.

Bone remodeling involves the removal and internal restructuring of previously existing bone and is responsible for the maintenance of tissue mass and architecture in the adult skeleton (7). Chemical and/or electrical stimuli activate local bone cell populations and coordinate their activities in removing and replacing discrete "packets" of bone at skeletal sites. These organized groups of cells are called "basic multicellular units," or BMU (8), which function within a defined remodeling cycle. Osteoblasts and osteoclasts are important members of the BMU. The cellular interactions occurring within a remodeling cycle are divided into the following four main events: activation, resorption, reversal, and formation (Table 19.1) (9). A remodeling cycle begins when a nonremodeling quiescent bone surface becomes "activated." The signals effecting activation are not fully understood, but it is believed that the bone lining cells covering inactive surfaces initiate this event. It is during this activation event that osteoclasts attach to the bone surface and resorb bone. This marks the resorption phase and results in the release of bone calcium. As the period of bone resorption subsides, the reversal phase represents a period of transition in which the bone surface is repopulated with newly formed osteoblasts.

The formation phase begins as osteoblasts commence deposition of new bone matrix (osteoid), which subsequently becomes mineralized. This phase, and the bone remodeling cycle, is complete when the osteoblasts refill the cavity created during the resorption phase. These events of resorption and formation are believed to be "coupled," such that resorption is always followed by bone formation. It is hypothesized that a reservoir of growth factors and cytokines, previously produced by osteoblasts and incorporated into the bone matrix, become available locally as autocrine/paracrine factors during osteoclastic bone resorption. These factors direct the proliferation, differentiation, and recruitment of new osteoblasts to the remodeling site and thus regulate the remodeling cycle (2–4) (Table 19.1).

Regulation of Bone Metabolism

Bone formation and bone resorption are regulated by systemic hormones and local factors produced in bone (2–4,10). Systemic hormones involved in stimulating bone formation include insulin, growth hormone (11), and estrogen (12), whereas those involved in stimulating bone resorption include 1,25-$(OH)_2$vitamin D_3 (13), PTH (14), thyroid hormone (15), and glucocorticoids (16). In addition, calcitonin (17) inhibits bone resorption. Even though several localized compounds act on bone cells, the prostaglandins (PG) seem to be the principal mediators of bone cell function because their biosynthesis and release from bone cells and associated tissues can be induced by several cytokines as well as systemic factors. PGE_2, which is synthesized from arachidonic acid, is a potent stimulator of bone resorption and is the primary PG affecting bone metabolism. The PG also influence insulin-like growth factors (IGF), which are major bone-derived growth factors (2). Once secreted and deposited in bone matrix, the IGF are released during osteoclastic bone resorptive activity, acting in an autocrine or paracrine fashion to stimulate new bone cell formation and matrix production. Thus, IGF, in concert with other bone growth factors, play an essential role in the efficiency with which bone formation is coupled to bone resorption. The relationship between PG and IGF is important to the maintenance of skeletal mass during aging, as well as being vital in optimizing acquisition of skeletal mass during critical stages of skeletal growth and development.

Cytokines

Cytokines are extracellular signaling proteins, secreted by effector cells, that act on nearby target cells. Cytokines exert their effects at low concentrations in an autocrine or paracrine fashion in cell-to-cell communications. Although generally stimulating anabolic processes in cells, cytokines also inhibit cell activity; hence, they could be called biological modifiers. One of the principal cellular responses produced by the action of cytokines is the synthesis and release of PGE_2.

The cytokines involved in bone modeling and remodeling include epidermal growth factor (EGF), fibroblast growth factor (FGF), interferon-γ (IFN-γ), interleukins (IL-1, IL-6), platelet-derived growth factor (PDGF), transforming growth factors (TGF-α, TGF-β), tumor necrosis factor-α (TNF-α), and insulin-like growth factors (IGF-I, IGF-II) (5,18,19) (Table 19.2). Most of these, such as IL-1 (20), IL-6

TABLE 19.2 Responses of Autocrine and Paracrine Factors in Bone[a]

Responses observed in bone	Cytokines, eicosanoids, or peptide growth factors[b]
Bone formation or matrix production	IGF, PGE$_2$, TGF-β
Bone resorption	EGF, FGF, IL, LT, PDGF, TGF-α, TNF-α
Collagen synthesis	FGF, IGF, TGF-β

[a]Source: Refs. 2–5,30.
[b]IGF, insulin-like growth factor; PGE$_2$, prostaglandin E$_2$; TGF-α/TGF-β, transforming growth factor; EGF, epidermal growth factor; FGF, fibroblast growth factor; IL, interleukin; LT, leukotriene; PDGF, platelet-derived growth factor; TNF-α, tumor necrosis factor-α.

(21), TNF (10), EGF (22), FGF (23), and PDGF (18), stimulate bone resorption. Some, such as IGF-I and IGF-II (24) and TGF-β (25), enhance bone formation, whereas others (FGF, PDGF, and TGF-β) also stimulate proliferation and differentiation of collagen-synthesizing cells.

Insulin-Like Growth Factors (IGF)

The IGF, also called somatomedins, are described as paracrine or autocrine regulatory polypeptides of cells. These compounds stimulate growth and synthesis of DNA, RNA, and proteins in cells. IGF are mitogenic and stimulate differentiation in a variety of cell types. Pituitary growth hormone (GH) controls tissue biosynthesis and secretion of IGF-I (or somatomedin C) postnatally, and it is through IGF-I that the tissue effects of GH are mediated. Serum concentrations of IGF-I are maintained by liver synthesis under the influence of GH. Much of the circulating IGF is bound to plasma IGF binding proteins (IGFBP).

Both IGF-I and IGF-II are conserved in skeletal tissue of vertebrates, including chickens, humans, and rats (14,24,26). The amount of IGF-I and IGF-II produced by bone cells is species dependent. In humans, neonatal mice, and chickens, more IGF-II than IGF-I is produced in the skeletal tissues, but adult mice and rats have more IGF-I than IGF-II in skeletal extracts (26). Although IGF-II is generally more abundant than IGF-I (27), IGF-I appears to be under greater regulatory control in bone (19,28). For example, PGE$_2$ (0.01–1 µmol/L) elevated IGF-I mRNA and polypeptide levels 1.9- to 4.7-fold; however, PGE$_2$ did not increase IGF-II mRNA or polypeptide levels in bone organ cultures of fetal rat calvaria (28).

Eicosanoids

In addition to the cytokines, which act as local modifiers of bone metabolism, certain eicosanoids [e.g., PG, leukotrienes (LT)], fatty acid-derived biological regulators, also exert stimulatory effects on bone formation and resorption. In 1970, Klein and Raisz (29) reported that PGE$_1$, PGE$_2$, PGA$_1$, and PGF$_{1α}$ increased the release of ^{45}Ca into the media from cultured fetal rat bone. Since then, numerous studies have demonstrated that PGE stimulates bone formation as well as bone resorption

(4,30,31). For example, PGE_2 stimulated collagen synthesis in cultured rat calvariae (32) and osteoblastic cells (33), and also increased the proliferation of osteoblasts in culture (34). PGE_2 was reported to increase cortical bone mass and intracortical bone remodeling in both intact and ovariectomized rats (35), and increased proximal tibial metaphyseal bone area in osteopenic ovariectomized rats (36). On the other hand, Raisz (4) reported that infusion of PGE_2 at a high concentration depressed osteogenesis in fetal rat calvariae. Thus, PGE_2 effects on bone formation may be dose related, i.e., stimulatory at low concentrations and inhibitory at high concentrations (32,37,38).

PGE also stimulates bone resorption. PGE_3, nearly equal in potency to PGE_2, caused increased resorption in cultures of fetal rat long bone (39). In addition, PGE_2 has been shown to mediate the effects of $1,25(OH)_2$vitamin D_3 (40), cytokines [TNF-α (41), IL-3 (40)], and growth factors [TGF-β (42), PDGF (43), and bFGF (44)] in enhancing bone resorption. Elevated production of PGE_2 has also been shown to be associated with several osteolytic disorders in humans, including bone loss associated with dental cysts, failing joint prostheses, chronic osteomyelitis, and certain neoplasms of bone (30).

Production of PG has been measured in human tissues, bone organ cultures, and osteoblasts (4,31). Physical stress (45) and systemic and local bone regulatory factors (PTH, EGF, PDGF, TGF, and IL) stimulate PG synthesis and release in osteoblast cell cultures or bone organ cultures (3,27,46).

Similar to the PG, the LT also play an important role in bone metabolism (47–49). Ren and Dziak (47) demonstrated that LTB_4 inhibited cell proliferation in cultured osteoblasts isolated from rat calvaria in a dose-dependent manner, but it may also interact with PG to regulate osteoblast activity. Others (48) report that LTC_4, LTD_4, and 5-hydroxyeicosatetranoic acid (HETE) stimulate isolated avian osteoclasts to resorb bone. Meghji *et al.* (49) found that LTB_4, LTC_4, LTD_4, 5-HETE, and 12-HETE stimulated bone resorption in calvariae of mice at picomolar concentrations, whereas PGE_2 produced this stimulatory effect at 10 nmol/L.

Conjugated Linoleic Acids

Conjugated linoleic acid (CLA) is the name used to describe a group of positional and geometric fatty acid isomers (octadecadienoic acids) derived from linoleic acid. The double bonds in CLA are conjugated, that is, not separated by a methylene group as in linoleic acid. Furthermore, CLA will not substitute for linoleic acid as an essential fatty acid. CLA occur naturally in ruminant food products (beef, lamb, and dairy) because of the process of bacterial biohydrogenation of linoleic acid in the rumen (50). The growing body of literature on CLA suggests that these isomeric conjugated fatty acids possess potent anticancer activity in the mammary gland of rats (51,52), and reduce proliferation of human breast cancer cell lines in culture (53) and in SCID mice (54). When fed to rats, CLA failed to produce any toxicological effect (55).

The positional isomers of CLA include Δ-7,Δ-9; Δ-8,Δ-10; Δ-9,Δ-11; Δ-10,Δ-12; Δ-11,Δ-13; and Δ-12,Δ-14 conjugated octadecadienoic acids (counting from the carboxyl end of the molecule). Each of the aforementioned positional conjugated diene isomers can occur in the following geometric configurations: *cis-,trans-*; *trans-,cis-*; *cis-,cis-*; and *trans-,trans-* (56). The most common CLA isomer found in food products is *cis*-9, *trans*-11-octadecadienoic acid (50), for which the name rumenic acid has been proposed (57).

The unique biological actions of CLA include anticarcinogenic, antiatherosclerotic, antioxidative, immunomodulative, and antibacterial activities. One of the earliest experiments on CLA indicated that these fatty acids, isolated from extracts of grilled ground beef, exhibited anticarcinogenic activity against chemically induced skin cancer in mice (58). The CLA isomers are potent anticancer nutrients for epidermal and mammary tumors in rodents; recent experiments on prostate cancer cell lines demonstrated that CLA isomers incorporate into cell lipids and, compared with linoleic acid, decrease cell proliferation (59). The mechanism of CLA action on cell function is not well described; however, our laboratory has shown in rats that these fatty acids reduce *ex vivo* PGE_2 production in bone organ culture (60), serum IGF-I binding proteins (61), and basal and lipopolysaccharide-induced levels of IL-6 by resident peritoneal macrophages (62). Other investigators have reported a significant reduction in serum PGE_2 (63,64) and splenic LTB_4 in rats given CLA (64).

Lipids in Skeletal Matrix Mineralization and Bone Growth

Matrix vesicles are small (100–200 nm diameter) structures that arise by budding from the plasma membranes of chondrocytes, osteoblasts, and odontoblasts (65). In growth plate cartilage, matrix vesicles are preferentially released from the lateral surfaces of maturing chondrocytes within the hypertrophic cell zone and enter the adjacent extracellular matrix, i.e., longitudinal septae. This process appears to represent a stage in the sequence of programmed cell death, or apoptosis, in chondrocytes, and the location of matrix vesicles coincides with site-specific initiation of matrix mineralization (66–68).

Matrix vesicles are lipid rich (69), and their membranes are especially concentrated in acidic phospholipids, such as phosphatidylserine, phosphatidylinositol, and phosphatidic acid, which are synthesized from polyunsaturated fatty acids (PUFA). The membrane-bound acidic phospholipids located within the matrix vesicle may act as non–energy-requiring Ca^{2+} traps that facilitate mineralization by forming lipid-Ca^{2+}-protein complexes (70,71). Matrix vesicles also contain several ion-transport proteins for calcium (annexin V, anchorin II, calbindin-D9k) and various phosphatases, especially alkaline phosphatase (68). Alkaline phosphatase is thought to act upon phosphoester molecules present in the extracellular fluid compartment, catalyzing the release of inorganic phosphate. Intravesicular enrichment of Ca^{2+} and PO^-_4 causes precipitation of an amorphous calcium phosphate that is converted to octacalcium phosphate and finally to hydroxyapatite crystals (71,72). The developing hydroxyapatite crystals grow and eventually rupture the matrix vesicle membrane to facilitate radial expansion of matrix mineralization by epitaxis.

Dietary Lipids Alter Local Factors Controlling Bone and Cartilage Metabolism

It is well documented that lipids play an important role in skeletal biology and bone health. For example, phospholipids facilitate cartilage mineralization in growth plate (73), and PG mediate messages from biomechanical forces (31) and aid in regulating anabolic factors, including IGF (2), to support bone formation. Emerging evidence from human and animal research supports the hypothesis that dietary lipids influence bone modeling and remodeling. Epidemiologic studies indicate that dietary fat intake is associated with reduced risk of vertebral and femoral fractures in adults, and saturated fat intake leads to increased bone density in children (74). Recent investigations with chicks and rats revealed that PUFA and CLA affect histomorphometric measurements of bone modeling (37,38,61).

Dietary fat may influence bone metabolism by altering PG biosynthesis (37,38,61). The PG, locally produced from 20-carbon essential fatty acid precursors (20:4n-6 and 20:5n-3) in osteogenic cells, regulate both bone formation and bone resorption (31). In support of the relationship among dietary PUFA, PG and bone metabolism, Watkins *et al.* (37,61) reported that dietary lipids (n-3 fatty acids and CLA) modulated the production of PGE_2, altered the concentration of IGF-I in bone tissues, and led to increased or decreased bone formation rates in growing chicks and rats. In these experiments, animals given n-3 fatty acids tended to show an increased rate of bone formation, suggesting a stimulatory effect on osteoblastic activity. The favorable effect of n-3 fatty acids on bone modeling in growing animals is supported by the observation of reduced bone mineral loss in ovariectomized rats supplemented with eicosapentaenoic acid (20:5n-3) (75); however, it is also possible that bone resorption is diminished with 20:5n-3.

The PGE_2 produced by osteoblasts may stimulate IGF-I synthesis or affect its action to support anabolic responses in bone (28). Studies with dairy fats revealed that butter fat blended with corn oil reduced *ex vivo* bone PGE_2, elevated bone IGF-I concentration, and increased bone formation rates in animals nearly 60% compared with those given diets higher in n-6 fatty acids (38). The responses observed in bone tissue suggest that moderating the action of n-6 fatty acids (linoleic acid) with n-3 fatty acids or CLA can benefit bone modeling. Others have reported health benefits attributed to moderating the dietary intake of n-6 fatty acids. For example, research on heart disease and dietary lipids shows a positive association between linoleic acid concentrations in adipose and coronary heart disease (76), and the need for decreasing the intake of PUFA to reduce oxidative modification of lipoproteins (77).

The studies on PUFA and bone modeling in animals suggest that dietary intakes of different PUFA families and CLA can affect bone modeling directly (by modulating PG biosynthesis) or indirectly (*via* IGF). The anabolic effects of PGE_2 may occur through stimulation of endogenous IGF-I production by osteoblasts (4) or by increasing bone cell responsiveness to IGF-I (78). Dietary fatty acids, depending upon the type (n-3 or CLA) and amount ingested, may therefore up-regulate or down-regulate IGF-I production in bone *via* their ability to modulate local concentrations of PGE_2 (37,38,61,79,80). The relationships between dietary PUFA and bone metabolism offer many promising opportunities for further biochemical and molecular research on local factors controlling bone cell function. The dietary factors that appear to influence bone cell activity to stimulate bone formation and bone resorption are illustrated in Figure 19.1.

Investigations with PUFA and epiphyseal (growth) cartilage and chondrocytes also indicate that dietary lipids affect cartilage metabolism to influence bone modeling. Experiments on growth cartilage demonstrated that this tissue selectively accumulates dietary fatty acids (82,83). These results suggest that chondrocytes are either sensitive to excess n-6 fatty acids or to an overproduction of PGE_2 (84). Growth cartilage in children and young animals contains small amounts of n-6 fatty acids but a relatively high concentration of 20:3n-9 (Mead acid) (85). The concentration of Mead acid is not reduced in growth cartilage of growing animals given diets adequate or enriched in linoleic acid (n-6) (82). Furthermore, supplementation of growth cartilage chondrocytes with linoleic or arachidonic acids depressed collagen synthesis (84), but these cells showed greater collagen synthesis when enriched with n-3 fatty acids or CLA (86,87).

In our laboratory, primary cultures of avian epiphyseal chondrocytes were studied to assess the effects of CLA (a mixture of CLA positional and geometric isomers) on fatty acid metabolism, collagen synthesis, and PGE_2 production (88). In a series of experiments, chondrocytes were enriched with CLA and linoleic acid (LA) (0, 50,

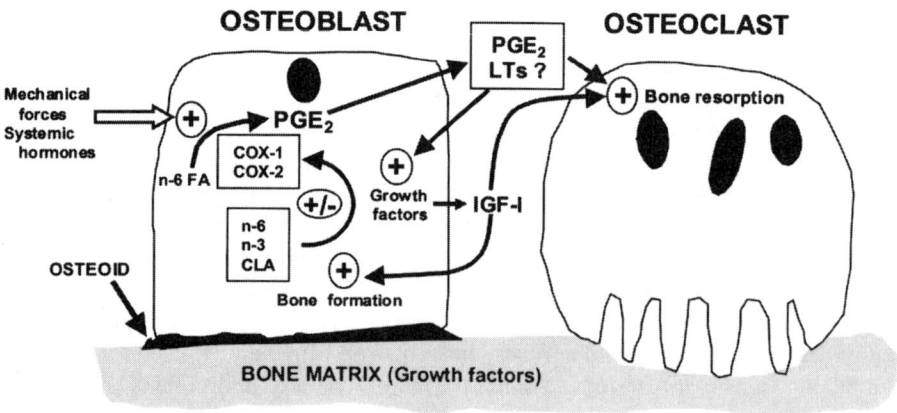

Fig. 19.1. Observed effects of dietary fatty acids and related compounds on osteoblastic and osteoclastic activity in bone (80). Excessive biosynthesis of prostaglandin E_2 (PGE_2) may depress bone formation and lead to
increased bone resorption. Altering the production of eicosanoids [PGE and leukotriene (LT)B] appears to optimize formation by osteoblasts perhaps by influencing insulin-like growth factor (IGF)-I production and action (81).

100, and 200 µmol/L) or arachidonic acid (AA) and eicosapentaenoic acid (EPA) (0 and 50 µmol/L). Chondrocytes enriched with CLA contained *cis*-9, *trans*-11 and *trans*-10, *cis*-12 18:2 as the primary CLA isomers. CLA decreased the concentrations of 16:1 and 18:1 in chondrocytes compared with the LA-treated cells. Enrichment of chondrocytes with CLA and LA affected collagen synthesis in a dose-dependent fashion. The LA and AA treatments reduced collagen synthesis as previously observed, but CLA and EPA appeared to stimulate its synthesis. Chondrocyte production of PGE_2 was reduced by CLA and EPA treatments, whereas LA and AA increased PGE_2 relative to the no fatty acid enrichment. These experiments suggest that CLA may positively influence growth plate cartilage function in the young and may reduce production of inflammatory PGE_2 in the adult. The effects of PUFA on collagen synthesis are summarized in Figure 19.2.

Dietary Conjugated Linoleic Acids Influence Bone Modeling in Animals

Watkins *et al.* (38) recently reported that butter fat led to a higher rate of bone formation in chicks compared with those given diets containing higher amounts of n-6 fatty acids (linoleic). The greater rate of bone formation in the butter-fed chicks was associated with reduced arachidonic acid level and *ex vivo* PGE_2 production in bone, and higher serum hexosamines and IGF-I in bone (38). Because the beneficial effects of butter fat on bone might be attributed to its CLA content (milk fat contained 1.5% CLA), studies with these fatty acids were conducted in rats.

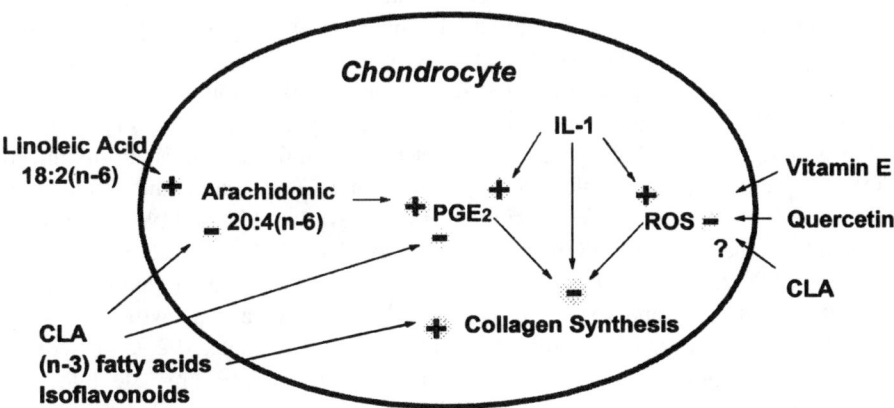

Fig. 19.2. The diagram illustrates the effects of dietary fatty acids [polyunsaturated fatty acids (PUFA) and conjugated linoleic acids (CLA)] and antioxidants on collagen synthesis in epiphyseal chondrocytes. Dietary fatty acids presumably affect collagen synthesis by modifying prostaglandin E_2 (PGE_2) production, whereas antioxidants prevent the inhibitory response of reactive oxygen species (ROS) on collagen synthesis. Interleukin-1 (IL-1) induces the production of both PGE_2 and ROS.

In our investigations with rats, dietary CLA led to differences in CLA enrichment of various organs and tissues; brain exhibited the lowest concentration of isomers, but bone tissues (periosteum and marrow) contained the highest (60). Moreover, CLA altered the fatty acid composition of rat tissues, reducing 18:1 in liver, skeletal muscle, heart, and bone marrow and periosteum (60). In a subsequent study with rats, feeding 0.5% CLA resulted in total CLA concentrations ranging from 0.27 to 0.43% in the polar lipid fraction and from 2.02 to 3.37% in the neutral lipids in liver, bone marrow, and bone periosteum. Our observation that CLA accumulated at a higher concentration in neutral lipids compared with polar lipids is consistent with that of Ip *et al.* (89) for rat mammary gland. In rats, the *trans*-10, *cis*-12 18:2 isomer was incorporated into the phospholipid fraction of tissue lipid extracts to the same extent as was the *cis*-9, *trans*-11 isomer (unpublished data). The ratio of *cis*-9, *trans*-11/*trans*-10, *cis*-12 roughly reflected the isomeric distribution of these CLA isomers in the diet given to rats. Moreover, the *cis*-9, *trans*-11 isomer of CLA was preferentially incorporated into the membrane phospholipids of rats (60), suggesting that *cis*-9, *trans*-11 might be a biologically active isomer.

It is hypothesized that CLA depresses arachidonate-derived eicosanoid biosynthesis based on reduced *ex vivo* PGE_2 production in rat bone organ culture and liver homogenate (61). The reduction in PGE_2 by CLA might be explained as a competitive inhibition of n-6 PUFA formation that results in lowered substrate availability for cyclooxygenase. Although there was a trend of reduced arachidonic acid concentration in bone tissues, the dramatic decrease in *ex vivo* PGE_2 production in bone organ culture could not be explained satisfactorily by a lack of substrate. The biosynthesis of PGE_2 in bone (cells of the osteoblast lineage) is highly regulated (90–92). A primary step for the regulation of PG formation is at the level of the enzyme cyclooxygenase (COX, also called prostaglandin G/H synthase). Fatty acids were shown to modulate the expression and activity of this key enzyme. For example, Nanji *et al.* (93) showed that saturated fat reduced peroxidation and decreased the levels of COX-2, the inducible form of COX, in rat liver. In a rat feeding study on colon tumorigenesis, a high-fat corn oil diet (rich in n-6 fatty acids) up-regulated COX-2 expression, but a high-fat fish oil diet (rich in n-3 fatty acids) inhibited it; however, expression of COX-1, the constitutive enzyme, was not affected (94). We speculate that CLA may influence PGE_2 production through the COX enzyme system, more likely on COX-2, to exert physiologic effects on bone (Fig. 19.3). Potentially, CLA may alter COX-2 action/expression to influence PTH (90) and growth factor (44) PG-dependent osteoclastic bone resorption, PGE receptor-mediated (92,95) actions on bone cells, and cytokine-induced extracellular release of PGE_2 (91) by osteoblasts.

Other possible mechanisms of action for CLA include reduced desaturation/ elongation of linoleic acid and inhibition of prostanoid biosynthesis by its isomeric analogs. Sébédio *et al.* (96) reported that CLA may be further desaturated and elongated to form conjugated 20:4 isomers, which might block the access of arachidonic acid to COX. The unusual 20:4 isomers derived from CLA might also affect the activity of the COX enzymes. Further study with CLA is required to confirm whether its isomeric analogs affect COX activity and expression.

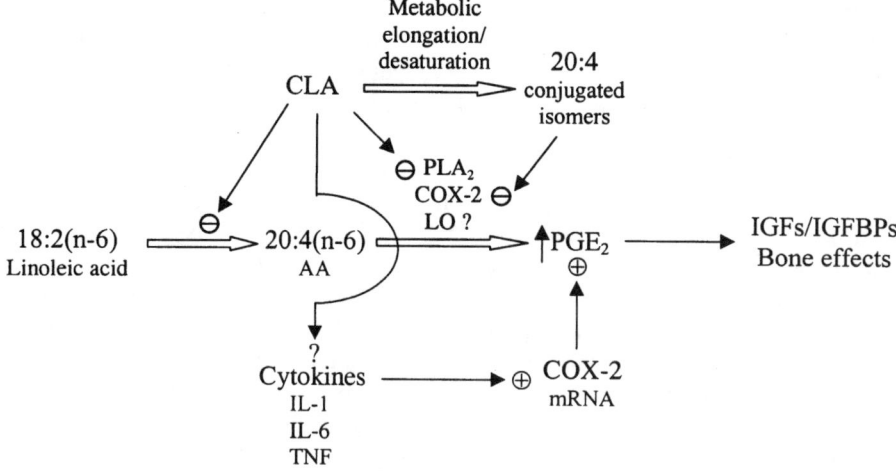

Fig. 19.3. Hypothetical mechanisms for the actions of conjugated linoleic acids (CLA) on prostaglandin E_2 (PGE_2) production and bone metabolism. Block arrows indicate biochemical reaction processes in which CLA and linoleic acid participate. Line arrows indicate possible effects of CLA and its metabolites on PGE_2 metabolism and their subsequent action on bone. *Source:* Ref. 61.

Dietary CLA effects on serum concentrations of IGF-I and IGF binding proteins (IGFBP) and their subsequent effect on bone modeling were examined in rats. Weanling male rats were fed AIN-93G diet containing 70 g/kg of added fat for 42 d. Treatments included 0 or 10 g/kg of CLA, and n-6 fatty acids in soybean oil (SBO) or n-3 fatty acids in menhaden oil + safflower oil (MSO) following a 2 × 2 factorial design (61). Serum IGFBP was influenced by n-6 and n-3 PUFA and CLA ($P = 0.01$ for 38–43 kDa bands corresponding to IGFBP-3). CLA increased IGFBP level in rats fed SBO ($P = 0.05$) but reduced it in those fed MSO ($P = 0.01$). Rats fed MSO had the highest serum IGFBP-3 level. This study also showed that CLA decreased liver IGF-I mRNA level in rats fed MSO supplemented with CLA ($P = 0.02$). Liver IGF-I mRNA expression was up-regulated by n-3 fatty acids and down-regulated by CLA in rats. In tibia, rats given CLA had reduced mineral apposition rate (3.69 vs. 2.79 μm/d) and bone formation rate [0.96 vs. 0.65 $\mu m^3/(\mu m^2 \cdot d)$]; however, bone formation rate tended to be higher with MSO. Dietary lipid treatments did not affect serum intact osteocalcin or bone mineral content. These results showed that dietary PUFA type and CLA modulate growth factors that regulate bone metabolism.

Dietary CLA supplementation lowered 18:1n-9 and total monounsaturated fatty acid in the polar fraction of liver, tibia bone marrow, and periosteum, whereas it increased 22:6n-3 and total n-3 in the polar fraction of liver and bone marrow. In the neutral fractions of tissue lipids, CLA treatment decreased 18:1n-9, 18:1n-7, 20:2n-6, 20:4n-6, 22:5n-3, 22:6n-3, total monounsaturated, total n-6, total n-3, and total PUFA, whereas CLA increased saturated fatty acids. In all tissues examined, except in the cortical bone of tibia, CLA decreased the level of 20:4n-6.

In rats given CLA, bone metabolism may be altered by modulating the production of the bone resorptive cytokines, such as interleukins (IL-1 and IL-6) and tumor necrosis factor (TNF), or the lipoxygenase product leukotriene B_4 (LTB_4) (Fig. 19.3). IL-1 and IL-6 have long been implicated in the pathophysiology of bone diseases such as rheumatoid arthritis (97) and postmenopausal osteoporosis (98,99). Dietary CLA was recently shown to lower basal and lipopolysaccharide (LPS)-stimulated IL-6 production and basal TNF production by resident peritoneal macrophages in rats (62). Furthermore, CLA reduced the release of LTB_4, (a lipoxygenase product of arachidonic acid), a strong bone resorption factor (100), from peritoneal exudate cells. Assuming that CLA has similar effects on these cytokines in bone, together with the fact that CLA reduced the production of PGE_2 in bone tissue, one could hypothesize that at a proper dietary level, the anti-inflammatory effects of CLA would be beneficial for the treatment of inflammatory bone disease.

This research is the first to show that CLA (naturally occurring in milk fat) affects bone metabolism in rats. The levels of CLA used in this study (1% dietary CLA), though higher than that found in conventional diets without supplementation, compare favorably with the ranges used (0.5–1.5%) in other investigations that examined anti-inflammatory and anticarcinogenic properties of CLA. Further work is required to evaluate more typical dietary levels of CLA as research elucidates the actions of these fatty acids on bone metabolism and health.

Degenerative Bone Diseases, Cyclooxygenases (COX-1 and COX-2) and Cancer

Osteoporosis is a significant health problem in the U.S., costing $13 billion each year to treat and convalesce adult patients (101). Osteoporosis is a condition of decreased bone mass that is prevalent in women and places them at risk for fractures. Although dietary calcium is believed to minimize bone mineral loss, calcium is not a singular treatment for osteoporosis. In addition, calcium intake above a daily requirement does not stimulate bone formation. Two theories have recently emerged to help explain the pathogenesis of osteoporosis. These theories focus on impaired coupling between bone formation and resorption, and a decline in the amount of bone growth factors deposited into bone matrix, leading to reduced bone formation (2,4). Therefore, a lack of control of PGE_2 production and decreased storage or action of IGF-I or other growth factors in bone may play roles in the etiology of osteoporosis. Dietary n-3 (20- and 22-carbon) fatty acids appear to up-regulate IGF-I, and CLA seem to up-regulate or down-regulate IGF-I action, depending on the fat source, in rat tissues to influence bone modeling (61). These observations are consistent with findings that PGE_2 stimulates IGF-I synthesis and IGF-II binding to type 2 IGF receptors in osteoblasts (81), and that dietary fatty acids modulate prostanoid biosynthesis, manipulating these autacoids to affect autocrine/paracrine actions on osteoblasts. Moreover, experiments demonstrating that basic fibroblast growth factor is an inducer of osteoclast formation by a mechanism requiring PGE_2 biosynthesis

(44) suggest that dietary fatty acids may influence bone resorption through prostanoid biosynthesis.

Because dietary fat constitutes 30–40% of the food calories in Western diets, the type of fat consumed can significantly influence the metabolic and physiologic processes controlling bone modeling in children and adults. The bioactive fatty acids, prostanoid precursors and inhibitors, likely modulate PGE_2 biosynthesis to influence anabolic actions (IGF-I and IGF-II) and inflammatory responses (cytokines) in bone and cartilage cells. Furthermore, the consumption of antioxidant nutrients and isoflavones (genistein) may contribute to better bone health by reducing the formation of free radicals and lipid peroxides that may depress osteoblastic activity (86,87).

Although osteoporosis is a major disease for postmenopausal women, osteoarthritis and inflammatory joint disease affect millions of people worldwide. Inflammatory cytokines (e.g. IL-1) are known to inhibit chondrocyte proliferation (97) and induce cartilage degradation, for which part of the response is mediated by PGE_2 (102). Excess production of PGE_2 is linked to joint pathology (rheumatoid arthritis), is known to exacerbate inflammatory responses, and results in a net loss of proteoglycan from articular cartilage (102). Because PGE_2 activation of the IGF-I/IGFBP axis may play an important role in cartilage biology and collagen and proteoglycan synthesis (103), dietary fatty acids may also be important for supporting joint repair. Investigations are warranted to explain the role of CLA and ratio of n-6/n-3 fatty acids on cartilage biology, joint disease, and ligament healing because dietary sources of these fatty acids exert potent effects on prostanoid biosynthesis to mediate cell activity.

The current knowledge of prostanoid formation and nonsteroidal anti-inflammatory drug action resides in understanding the regulation and expression of cyclooxygenase. Two isoforms of this enzyme exist, i.e., COX-1 and COX-2. The COX-1 is a constitutive enzyme responsible for generating PG that act physiologically, whereas COX-2 is an inducible enzyme that is stimulated by cytokines, growth factors, and tumor promoters. The COX-2 is responsible for production of PGE_2, which is associated with inflammatory reactions (arthritis), osteoporosis, and cancer (104). Because cancer is a leading cause of mortality, and the costs associated with the treatment and convalesce of osteoporosis and arthritis exceed $70 billion each year, research directed at controlling COX-2 is justified. Each year, cancer-related deaths due to breast and prostate cancer are at 88,000 (105), and arthritis afflicts 40 million persons in the U.S. (106).

Anti-inflammatory diets, including nutraceutical n-3 fatty acids, are associated with decreased pathogenesis of rheumatoid arthritis (secondary osteoporosis), reduced inflammatory diseases (107–109), and cancer risk (105). The common link among these diseases resides in the regulation/expression of COX-2. For example, multiple lines of evidence indicate that up-regulation of COX-2 contributes to tumorigenesis and inflammation, providing tissue levels of prostanoid precursors that influence formation of the proinflammatory PGE_2. In addition, chronic aspirin users (COX inhibitor) have reduced incidence of colorectal cancer. Both COX-1 and COX-2 inhibitors suppress experimental mouse skin carcinogenesis, and permanent overactivation of arachidonic acid metabolism appears to be the driving force for tumor development (110). Moreover, metastasis of cancer to bone is a frequent outcome of breast (about two thirds of patients with metastatic breast cancer have bone involvement) and prostate malignancies. The metastasis is often associated with significant morbidity (severe bone pain and pathologic fractures) due to osteolysis, and metastatic target bone is continually being remodelled under the influence of local and systemic factors (111).

Interestingly, recent investigations suggest that both COX-2 induction and an increase in the supply of arachidonic acid are required to greatly increase prostanoid production (112). Supplying arachidonic acid appears to increase prostanoids to reduce the effects of nonsteroidal anti-inflammatory drugs, including NS-398, a specific COX-2 inhibitor. Therefore, in our view, n-3 fatty acids and CLA isomers may act as potent anticancer nutrients because they not only directly/indirectly affect the activity and expression of COX-2 but may also reduce the supply of arachidonic acid to diminish prostanoid biosynthesis. In any case, one mode of action for CLA appears to be anti-inflammatory with respect to reducing PGE_2 production.

The data presented in this review describe consistent and reproducible effects of n-3 fatty acids and CLA on decreasing *ex vivo* PGE_2 production in bone organ cultures (37,61) and in various cell culture systems (87). The potent beneficial anticancer effect of CLA is likely linked, in part, to down-regulation of COX-2 activity. Future investigations on CLA should evaluate isomeric effects on COX-1 and COX-2 for which overexpression of the latter is associated with carcinogenesis and inflammation. This research would lead to the following: (i) important discoveries for bone modeling and remodeling, (ii) development of delivery systems in designed foods, and (iii) opportunities to identify a synergism between nutraceuticals and drug therapies to reduce cancer risk and control inflammatory bone/joint disease.

Acknowledgment

This paper was approved as Journal Paper Number 15953 of the Purdue Agricultural Experiment Station.

References

1. Baron, R. (1993) Anatomy and Ultrastructure of Bone, in *Primer on the Metabolic Bone Diseases and Disorders of Mineral Metabolism,* (Favus, M.J., ed.) 2nd edn., pp. 3–9, Raven Press, New York.
2. Baylink, D.J., Finkelman, R.D., and Mohan, S. (1993) Growth Factors to Stimulate Bone Formation, *J. Bone Miner. Res. 8,* S565–S572.
3. Mundy, G.R. (1993) Cytokines and Growth Factors in the Regulation of Bone Remodeling, *J. Bone Miner. Res. 8,* S505–S510.
4. Raisz, L.G. (1993) Bone Cell Biology: New Approaches and Unanswered Questions, *J. Bone Miner. Res. 8,* S457–S465.
5. Canalis, E. (1993) Regulation of Bone Remodeling, in *Primer on the Metabolic Bone Diseases and Disorders of Mineral Metabolism,* (Favus, M.J., ed.) 2nd edn., pp. 33–37, Raven Press, New York.

6. Blair, H.C., Schlesinger, P.H., Ross, F.P., and Teitelbaum, S.L. (1993) Recent Advances Toward Understanding Osteoclast Physiology, *Clin. Orthop. 294,* 7–22.
7. Frost, H.M. (1993) Bone Remodelling and Its Relationship to Metabolic Bone Disease, pp. 28–53, Charles C. Thomas, Springfield, IL.
8. Frost, H.M. (1963) Bone Remodelling Dynamics, Charles C. Thomas, Springfield, IL.
9. Parfitt, A.M. (1990) Progress in Basic and Clinical Pharmacology, (Kanis, J.A., ed.) pp. 1–27, Karger AG, Basel.
10. Watrous, D.A., and Andrews, B.S. (1989) The Metabolism and Immunology of Bone, *Semin. Arthritis Rheum. 19,* 45–65.
11. Nilsson, A., Ohlsson, C., Isaksson, O.G., Lindahl, A., and Isgaard, J. (1994) Hormonal Regulation of Longitudinal Bone Growth, *Eur. J. Clin. Nutr. 48,* S150–S160.
12. Chow, J., Tobias, J.H., Colston, K.W., and Chambers, T.J. (1992) Estrogen Maintains Trabecular Bone Volume in Rats Not Only by Suppression of Bone Resorption but Also by Stimulation of Bone Formation, *J. Clin. Investig. 89,* 74–78.
13. Raisz, L.G. (1990) Recent Advances in Bone Cell Biology: Interactions of Vitamin D with Other Local and Systemic Factors, *J. Bone Miner. Res. 9,* 191–197.
14. Kream, B.E., Petersen, D.N., and Raisz, L.G. (1990) Parathyroid Hormone Blocks the Stimulatory Effect of Insulin-Like Growth Factor-I on Collagen Synthesis in Cultured 21-Day Fetal Rat Calvariae, *Bone 11,* 411–415.
15. Klaushofer, K., Hoffmann, O., Gleispach, H., Leis, H.-J., Czerwenka, E., Koller, K., and Peterlik, M. (1989) Bone-Resorbing Activity of Thyroid Hormones Is Related to Prostaglandin Production in Cultured Neonatal Mouse Calvaria, *J. Bone Miner. Res. 4,* 305–312.
16. Lukert, B.P., and Raisz, L.G. (1990) Glucocorticoid-Induced Osteoporosis: Pathogenesis and Management, *Ann. Intern. Med. 112,* 352–364.
17. Lin, H.Y., Harris, T.L., Flannery, M.S., Aruffo, A., Kaji, E.H., Gorn, A., Kolakowski, L.F., Jr., Lodish, H.F., and Goldring, S.R. (1991) Expression Cloning of an Adenylate Cyclase-Coupled Calcitonin Receptor, *Science 254,* 1022–1024.
18. Canalis, E., McCarthy, T.L., and Centrella, M. (1989) Effects of Platelet-Derived Growth Factor on Bone Formation in vitro, *J. Cell Physiol. 140,* 530–537.
19. Canalis, E., Centrella, M., and McCarthy, T.L. (1991) Regulation of the Insulin-Like Growth Factor-II Production in Bone Cultures, *Endocrinology 129,* 2457–2462.
20. Marusic, A., and Raisz, L.G. (1991) Cortisol Modulates the Actions of Interleukin-1α on Bone Formation, Resorption, and Prostaglandin Production in Cultured Mouse Parietal Bones, *Endocrinology 129,* 2699–2706.
21. Ishimi, Y., Miyaura, C., Jin, C.H., Akatsu, T., Abe, T., Nakamura, Y., Yamaguchi, A., Yoshiki, S., Matsuda, T., Hirano, T., Kishimoto, T., and Suda, T. (1990) Il-6 Is Produced by Osteoblasts and Induces Bone Resorption, *J. Immunol. 145,* 3297–3303.
22. Lorenzo, J.A., Quinton, J., Sousa, S., and Raisz, L.G. (1986) Effects of DNA and Prostaglandin Synthesis Inhibitors on the Stimulation of Bone Resorption by Epidermal Growth Factor in Fetal Rat Long-Bone Cultures, *J. Clin. Investig. 77,* 1897–1902.
23. Simmons, H.A., and Raisz, L.G. (1991) Effects of Acid and Basic Fibroblast Growth Factor and Heparin on Resorption of Cultured Fetal Rat Long Bones, *J. Bone Miner. Res. 6,* 1301–1305.
24. McCarthy, T.L., Centrella, M., and Canalis, E. (1989) Regulatory Effect of Insulin-Like Growth Factors I and II on Bone Collagen Synthesis in Rat Calvarial Cultures, *Endocrinology 124,* 301–309.

25. Centrella, M., McCarthy, T.L., and Canalis, E. (1987) Transforming Growth Factor Beta Is a Bifunctional Regulator of Replication and Collagen Synthesis in Osteoblast-Enriched Cell Cultures from Fetal Rat Bone, *J. Biol. Chem. 262*, 2869–2874.
26. Bautista, C., Baylink, D.J., and Mohan, S. (1991) Isolation of a Novel Insulin-Like Growth Factor (IGF) Binding Protein from Human Bone: A Potential Candidate for Fixing IGF-II in Human Bone, *Biochem. Biophys. Res. Commun. 176*, 756–763.
27. Mohan, S., and Baylink, D.J. (1991) Bone Growth Factors, *Clin. Orthop. 263*, 30–48.
28. McCarthy, T.L., Centrella, M., Raisz, L.G., and Canalis, E. (1991) Prostaglandin E_2 Stimulates Insulin-Like Growth Factor I Synthesis in Osteoblast-Enriched Cultures from Fetal Rat Bone, *Endocrinology 128*, 2895–2900.
29. Klein, D.C., and Raisz, L.G. (1970) Prostaglandins: Stimulation of Bone Resorption in Tissue Culture, *Endocrinology 86*, 1436–1440.
30. Norrdin, R.W., Jee, W.S., and High, W.B. (1990) The Role of Prostaglandins in Bone in vivo, *Prostaglandins Leukot. Essent. Fatty Acids 41*, 139–149.
31. Marks, S.C., and Miller S.C. (1993) Prostaglandins and the Skeleton: The Legacy and Challenges of Two Decades of Research, *Endocr. J. 1*, 337–344.
32. Raisz, L.G., and Fall, P.M. (1990) Biphasic Effects of Prostaglandin E_2 on Bone Formation in Cultured Fetal Rat Calvaria: Interaction with Cortisol, *Endocrinology 126*, 1654–1659.
33. Hakeda, Y., Nakatani, Y., Kurihara, N., Ikeda, E., Maeda, N., and Kumegawa, M. (1985) Prostaglandin E_2 Stimulates Collagen and Non-Collagen Protein Synthesis and Prolyl Hydroxylase Activity in Osteoblastic Clone MC3T3-E1 Cells, *Biochem. Biophys. Res. Commun. 126*, 340–345.
34. Igarashi, K., Hirafuji, M., Adachi, H., Shinoda, H., and Mitani, H. (1994) Role of Endogenous PGE_2 in Osteoblastic Functions of a Clonal Osteoblast-Like Cell, MC3T3-E1, *Prostaglandins Leukot. Essent. Fatty Acids 50*, 169–172.
35. Jee, W.S., Mori, S., Li, X.J., and Chan, S. (1990) Prostaglandin E_2 Enhances Cortical Bone Mass and Activates Intracortical Bone Remodeling in Intact and Ovariectomized Female Rats, *Bone 11*, 253–266.
36. Mori, S., Jee, W.S., Li, X.J., Chan, S., and Kimmel, D.B. (1990) Effects of Prostaglandin E_2 on Production of New Cancellous Bone in the Axial Skeleton of Ovariectomized Rats, *Bone 11*, 103–113.
37. Watkins, B.A., Shen, C.-L., Allen, K.G.D., and Seifert, M.F. (1996) Dietary (n-3) and (n-6) Polyunsaturates and Acetylsalicylic Acid Alter *ex vivo* PGE_2 Biosynthesis, Tissue IGF-I Levels, and Bone Morphometry in Chicks, *J. Bone Miner. Res. 11*, 1321–1332.
38. Watkins, B.A., Shen, C-L., McMurtry, J.P., Xu, H., Bain, S.D., Allen, K.G.D., and Seifert, M.F. (1997) Dietary Lipids Modulate Bone Prostaglandin E_2 Production, Insulin-Like Growth Factor-I Concentration and Formation Rate in Chicks, *J. Nutr. 127*, 1084–1091.
39. Raisz, L.G., Alander, C.B., and Simmons, H.A. (1989) Effects of Prostaglandin E_3 and Eicosapentaenoic Acid on Rat Bone in Organ Culture, *Prostaglandins 37*, 615–625.
40. Collins, D.A., and Chambers, T.J. (1992) Prostaglandin E_2 Promotes Osteoclast Formation in Murine Hematopoietic Cultures Through an Action on Hematopoietic Cells, *J. Bone Miner. Res. 5*, 555–561.
41. Tashjian, A.H., Jr., Voelkel, E.F., Lazzaro, M., Goad, D., Bosma, T., and Levine, L. (1987) Tumor Necrosis Factor-α (Cachectin) Stimulates Bone Resorption in Mouse Calvaria *via* a Prostaglandin-Mediated Mechanism, *Endocrinology 120*, 2029–2036.

42. Tashjian, A.H., Jr., Voelkel, E.F., Lazzaro, M., Singer, F.R., Roberts, A.B., Derynck, R., Winkler, M.E., and Levine, L. (1985) Alpha and Beta Human Transforming Growth Factors Stimulate Prostaglandin Production and Bone Resorption in Cultured Mouse Calvaria, *Proc. Natl. Acad. Sci. USA 82*, 4535–4538.
43. Tashjian, A.H., Jr., Hohmann, E.L., Antoniades, H.N., and Levine, L. (1982) Platelet-Derived Growth Factor Stimulates Bone Resorption *via* a Prostaglandin-Mediated Mechanism, *Endocrinology 111*, 118–124.
44. Hurley, M.M., Lee, S.K., Raisz, L.G., Bernecker, P., and Lorenzo, J. (1998) Basic Fibroblast Growth Factor Induces Osteoclast Formation in Murine Bone Marrow Cultures, *Bone 22*, 309–316.
45. Somjen, D., Binderman, I., Berger, E., and Harell, A. (1980) Bone Remodelling Induced by Physical Stress Is Prostaglandin E_2 Mediated, *Biochim. Biophys. Acta 627*, 91–100.
46. Fermor, B., Gundle, R., Evans, M., Emerton, M., Pocock, A., and Murray, D. (1998) Primary Human Osteoblast Proliferation and Prostaglandin E_2 Release in Response to Mechanical Strain in vitro, *Bone 22*, 637–643.
47. Ren, W., and Dziak, R. (1991) Effects of Leukotrienes on Osteoblastic Cell Proliferation, *Calcif. Tissue Int. 49*, 197–201.
48. Gallwitz, W.E., Mundy, G.R., Lee, C.H., Qiao, M., Roodman, G.D., Raftery, M., Gaskell, S.J., and Bonewald, L.F. (1993) 5-Lipoxygenase Metabolites of Arachidonic Acid Stimulate Isolated Osteoclasts to Resorb Calcified Matrices, *J. Biol. Chem. 268*, 10087–10094.
49. Meghji, S., Sandy, J.R., Scutt, A.M., Harvey, W., and Harris, M. (1988) Stimulation of Bone Resorption by Lipoxygenase Metabolites of Arachidonic Acid, *Prostaglandins 36*, 139–149.
50. Parodi, P.W. (1997) Cows' Milk Fat Components as Possible Anticarcinogenic Agents, *J. Nutr. 127*, 1055–1060.
51. Ip, C. (1997) Review of the Effects of *trans* Fatty Acids, Oleic Acid, n-3 Polyunsaturated Fatty Acids, and Conjugated Linoleic Acid on Mammary Carcinogenesis in Animals, *Am. J. Clin. Nutr. 66*, 1523S–1529S.
52. Ip, C., Jiang, C., Thompson, H.J., and Scimeca, J.A. (1997) Retention of Conjugated Linoleic Acid in the Mammary Gland Is Associated with Tumor Inhibition During the Post-Initiation Phase of Carcinogenesis, *Carcinogenesis 18*, 755–759.
53. Cunningham, D.C., Harrison, L.Y., and Schultz, T.D. (1997) Proliferative Responses of Normal Human Mammary and MCF-7 Breast Cancer Cells to Linoleic Acid, Conjugated Linoleic Acid and Eicosanoid Synthesis Inhibitors in Culture, *Anticancer Res. 17*, 197–204.
54. Visonneau, S., Cesano, A., Tepper, S.A., Scimeca, J.A., Santoli, D., and Kritchevsky, D. (1997) Conjugated Linoleic Acid Suppresses the Growth of Human Breast Adenocarcinoma Cells in SCID Mice, *Anticancer Res. 17*, 969–974.
55. Scimeca, J.A. (1998) Toxicological Evaluation of Dietary Conjugated Linoleic Acid in Male Fisher 344 Rats, *Food Chem. Toxicol. 36*, 391–395.
56. Sehat, N., Yurawecz, M.P., Roach, J.A.G., Mossoba, M.M., Kramer, J.K.G., and Ku, Y. (1998) Silver-Ion High-Performance Liquid Chromatographic Separation and Identification of Conjugated Linoleic Acid Isomers, *Lipids 33*, 217–221.
57. Kramer, J.K.G., Parodi, P.W., Jensen, R.G., Mossoba, M.M., Yurawecz, M.P., and Adlof, R.O. (1998) Rumenic Acid: A Proposed Common Name for the Major Conjugated Linoleic Acid Isomer Found in Natural Products, *Lipids 33*, 835.

58. Ha, Y.L., Storkson, J., and Pariza, M.W. (1990) Inhibition of Benzo(a)pyrene-Induced Mouse Forestomach Neoplasia by Conjugated Dienoic Derivatives of Linoleic Acid, *Cancer Res. 50*, 1097–1101.
59. Cornell, K.K., Waters, D.J., Coffman, K.T., Robinson, J.P., and Watkins, B.A. (1997) Conjugated Linoleic Acid Inhibited the in vitro Proliferation of Canine Prostate Cancer Cells, *FASEB J. 11*, A579 (Abstr.).
60. Li, Y., and Watkins, B.A. (1998) Conjugated Linoleic Acids Alter Bone Fatty Acid Composition and Reduce *ex vivo* Prostaglandin E_2 Biosynthesis in Rats Fed n-6 or n-3 Fatty Acids, *Lipids 33*, 417–425.
61. Li, Y., Seifert, M.F., Ney, D.M., Grahn, M., Grant, A.L., Allen, K.D.G., and Watkins, B.A. (1999) Dietary Conjugated Linoleic Acids Alter Serum IGF-I and IGF Binding Protein Concentrations and Reduce Bone Formation in Rats Fed (n-6) or (n-3) Fatty Acids, *J. Bone Miner. Res., 14:* 1153–1162..
62. Turek, J.J., Li, Y., Schoenlein, I.A., Allen, K.G.D., and Watkins, B.A. (1998) Modulation of Macrophage Cytokine Production by Conjugated Linoleic Acids Is Influenced by the Dietary n-6:n-3 Fatty Acid Ratio, *J. Nutr. Biochem. 9*, 258–266.
63. Sugano, M., Tsujita, A., Yamasaki, M., Yamada, K., Ikeda, I., and Kritchevsky, D. (1997) Lymphatic Recovery, Tissue Distribution, and Metabolic Effects of Conjugated Linoleic Acid in Rats, *J. Nutr. Biochem. 8*, 38–43.
64. Sugano, M., Tsujita, A., Yamasaki, M., Noguchi, M., and Yamada, K. (1998) Conjugated Linoleic Acid Modulates Tissue Levels of Chemical Mediators and Immunoglobulins in Rats, *Lipids 33*, 521–527.
65. Anderson, H.C. (1967) Electron Microscopic Studies of Induced Cartilage Development and Calcification, *J. Cell Biol. 35*, 85–101.
66. Buckwalter, J.A., Mower, D., Ungar, R., Schaeffer, J., and Ginsberg, B. (1986) Morphometric Analysis of Chondrocyte Hypertrophy, *J. Bone Jt. Surg. 68*, 243–255.
67. Hunziker, E.B., Schenk, R.K., and Crus-Orive, L.M. (1987) Quantification of Chondrocyte Performance in Growth-Plate Cartilage During Longitudinal Bone Growth, *J. Bone Jt. Surg. 69*, 162–173.
68. Anderson, H.C. (1995) Molecular Biology of Matrix Vesicles, *Clin. Orthop. 314*, 266–280.
69. Gilder, H., and Boskey, A.L. (1990) Dietary Lipids and the Calcifying Tissues, in *Nutrition and Bone Development,* (Simmons, D.J., ed.) pp. 244–265, Oxford University Press, New York.
70. Genge, B.R., Xu, C., Wu, L.N.Y., Buzzi, W.R., Showman, R.W., Aresnault, A.L., Ishikawa, Y., and Wuthier, R.E. (1992) Establishment of the Primary Structure of the Major Lipid-Dependent Ca^{2+} Binding Proteins of Chicken Growth Plate Cartilage Matrix Vesicles: Identity with Anchorin CII (Annexin V) and Annexin II, *J. Bone Miner. Res. 7*, 807–818.
71. Sauer, G.R., and Wuthier, R.E. (1988) Fourier-Transform Infrared Characterization of Mineral Phases Formed During Induction of Mineralization by Collagenase-Released Matrix Vesicles in vitro, *J. Biol. Chem. 27*, 13718–13724.
72. Anderson, H.C. (1969) Vesicles Associated with Calcification in the Matrix of Epiphyseal Cartilage, *J. Cell Biol. 41*, 59–72.
73. Wuthier, R.E. (1993) Involvement of Cellular Metabolism of Calcium and Phosphate in Calcification of Avian Growth Plate Cartilage. *J. Nutr. 121*, 301–309.
74. Gunnes, M., and Lehmann, E.H. (1995) Dietary Calcium, Saturated Fat, Fiber and Vitamin C as Predictors of Forearm Cortical and Trabecular Bone Mineral Density in Healthy Children and Adolescents, *Acta Paediatr. 84*, 388–392.

75. Sakaguchi, K., Morita, I., and Murota, S. (1994) Eicosapentaenoic Acid Inhibits Bone Loss Due to Ovariectomy in Rats, *Prostaglandins Leukot. Essent. Fatty Acids 50*, 81–84.
76. Hodgson, J.M., Wahlqvist, M.L., Boxall, J.A., and Balazs, N.D. (1993) Can Linoleic Acid Contribute to Coronary Artery Disease? *Am. J. Clin. Nutr. 58*, 228–234.
77. Reaven, P., Parthasarathy, S., Grasse, B.J., Miller, E., Almazan, F., Mattson, F.H., Khoo, J.C., Steinberg, D., and Witztum, J.L. (1991) Feasibility of Using an Oleate-Rich Diet to Reduce the Susceptibility of Low-Density Lipoprotein to Oxidative Modification in Humans, *Am. J. Clin. Nutr. 54*, 701–706.
78. Hakeda, H.Z., Li, X.J., and Jee, W.S. (1991) Partial Loss of Anabolic Effect of Prostaglandin E_2 on Bone After Its Withdrawal in Rats, *Bone 12*, 173–183.
79. Watkins, B.A., and Seifert, M.F. (1996) Food Lipids and Bone Health, in *Food Lipids and Health*, (McDonald, R.E., and Min, B.D., eds.) pp. 71–116, Marcel Dekker, Inc., New York.
80. Watkins, B.A. (1998) Regulatory Effects of Polyunsaturates on Bone Modeling and Cartilage Function, *World Rev. Nutr. Diet. 83*, 38–51.
81. McCarthy, T.L., Ji, C., Casinghino, S., and Centrella, M. (1998) Alternate Signaling Pathways Selectively Regulate Binding of Insulin-Like Growth Factor I and II on Fetal Rat Bone Cells, *J. Cell. Biochem. 68*, 446–456.
82. Xu, H., Watkins, B.A., and Adkisson, H.D. (1994) Dietary Lipids Modify the Fatty Acid Composition of Cartilage, Isolated Chondrocytes and Matrix Vesicles, *Lipids 29*, 619–625.
83. Xu, H., Watkins, B.A., and Seifert, M.F. (1995) Vitamin E Stimulates Trabecular Bone Formation and Alters Epiphyseal Cartilage Morphometry, *Calcif. Tissue Int. 57*, 293–300.
84. Watkins, B.A., Xu, H., and Turek, J.J. (1996) Linoleate Impairs Collagen Synthesis in Primary Cultures of Avian Chondrocytes, *Proc. Soc. Exp. Biol. Med. 212*, 153–159.
85. Adkisson, H.D., Risener, F.S., Zarrinkar, P.P., Walla, M.D., Christie, W.W., and Wuthier, R.E. (1991) Unique Fatty Acid Composition of Normal Cartilage: Discovery of High n-9 Eicosatrienoic Acid and Low Levels of n-6 Polyunsaturated Fatty Acids, *FASEB J. 5*, 344–353.
86. Seifert, M.F., and Watkins, B.A. (1997) Role of Dietary Lipid and Antioxidants in Bone Metabolism, *Nutr. Res. 17*, 1209–1228.
87. Watkins, B.A., Seifert, M.F., and Allen, K.G.D. (1997) Importance of Dietary Fat in Modulating PGE_2 Responses and Influence of Vitamin E on Bone Morphometry, *World Rev. Nutr. Diet. 82*, 250–259.
88. Chen, Y. (1997) Modulatory Aspects of Polyunsaturated Fatty Acids on Epiphyseal Chondrocyte Function, M.S. Thesis, Purdue University, West Lafayette, IN.
89. Ip, C., Briggs, S.P., Haegele, A.D., Thompson, H.J., Storkson, J., and Scimeca, J.A. (1996) The Efficacy of Conjugated Linoleic Acid in Mammary Cancer Prevention Is Independent of the Level or Type of Fat in the Diet, *Carcinogenesis 17*, 1045–1050.
90. Maciel, F.M.B., Sarrazin, P., Morisset, S., Lora, M., Patry, C., Dumais, R., and Debrumfernandes, A.J. (1997) Induction of Cyclooxygenase-2 by Parathyroid Hormone in Human Osteoblasts in Culture, *J. Rheumatol. 24*, 2429–2435.
91. Pruzanski, W., Stefanski, E., Vadas, P., Kennedy, B.P., and van den Bosch, H. (1998) Regulation of the Cellular Expression of Secretory and Cytosolic Phospholipases A_2, and Cyclooxygenase-2 by Peptide Growth Factors, *Biochim. Biophys. Acta 1403*, 47–56.

92. Suda, M., Tanaka, K., Yasoda, A., Natsui, K., Sakuma, Y., Tanaka, I., Ushikubi, F., Narumiya, S., and Nakao, K. (1998) Prostaglandin E_2 (PGE_2) Autoamplifies Its Production Through EP1 Subtype of PGE Receptor in Mouse Osteoblastic MC3T3-E1 Cells, *Calcif. Tissue Int. 62*, 327–331.
93. Nanji, A.A., Zakim, D., Rahemtulla, A., Daly, T., Miao, L., Zhao, S.P., Khwaja, S., Tahan, S.R., and Dannenberg, A.J. (1997) Dietary Saturated Fatty Acids Down-Regulate Cyclooxygenase-2 and Tumor Necrosis Factor α and Reverse Fibrosis in Alcohol-Induced Liver Disease in the Rat, *Hepatology 26*, 1538–1545.
94. Singh, J., Hamid, R., and Reddy, B.S. (1997) Dietary Fat and Colon Cancer—Modulation of Cyclooxygenase-2 by Types and Amount of Dietary Fat During the Postinitiation Stage of Colon Carcinogenesis, *Cancer Res. 57*, 3465–3470.
95. Ono, K., Akatsu, T., Murakami, T., Nishikawa, M., Yamamoto, M., Kugai, N., Motoyoshi, K., and Nagata, N. (1998) Important Role of EP_4, a Subtype of Prostaglandin (PG) E Receptor, in Osteoclast-Like Cell Formation from Mouse Bone Marrow Cells Induced by PGE_2, *J. Endocrinol. 158*, R1–R5.
96. Sébédio, J.L., Juaneda, P., Dobson, G., Ramilison, I., Martin, J.C., Chardigny, J.M., and Christie, W.W. (1997) Metabolites of Conjugated Isomers of Linoleic Acid (CLA) in the Rat, *Biochim. Biophys. Acta 1345*, 5–10.
97. Arend, W.P., and Dayer, J.M. (1990) Cytokines and Cytokine Inhibitors or Antagonists in Rheumatoid Arthritis, *Arthritis Rheum. 33*, 305–315.
98. Pacifici, R., Rifas, L., McCracken, R., Vered, I., McMurty, C., Avioli, L.V., and Peck, W.A. (1989) Ovarian Steroid Treatment Blocks a Postmenopausal Increase in Blood Monocyte Interleukin 1 Release, *Proc. Natl. Acad. Sci. USA 86*, 2398–2402.
99. Girasole, G., Jilka, R.L., Passeri, G., Boswell, S., Boder, G., Williams, D.C., and Manolagas, S.C. (1992) 17β-Estradiol Inhibits Interleukin-6 Production by Bone Marrow-Derived Stromal Cells and Osteoblasts in vitro: A Potential Mechanism for the Antiosteoporotic Effect of Estrogens, *J. Clin. Investig. 89*, 883–891.
100. Garcia, C., Boyce, B.F., Gilles, J., Dallas, M., Qiao, M., Mundy, G.R., and Bonewald, L.F. (1996) Leukotriene B_4 Stimulates Osteoclastic Bone Resorption Both in vitro and in vivo, *J. Bone Miner. Res. 11*, 1619–1627.
101. Melton, L.J. (1997) Editorial: The Prevalence of Osteoporosis, *J. Bone Miner. Res. 12*, 1769–1771.
102. Fukuda, K., Dan, H., Saitoh, M., Takayama, M., and Tanaka, S. (1994) Superoxide Dismutase Inhibits Interleukin-1-Induced Degradation of Human Cartilage, *Agents Actions 42*, 71–73.
103. Di Battista, J.A., Dore, S., Morin, N., He, Y., Pelletier, J.P., and Martel-Pelletier, J. (1997) Prostaglandin E_2 Stimulates Insulin-Like Growth Factor Binding Protein-4 Expression and Synthesis in Cultured Human Articular Chondrocytes: Possible Mediation by Ca^{++}-Calmodulin Regulated Processes, *J. Cell. Biochem. 65*, 408–419.
104. Subbaramaiah, K., Zakim, D., Weksler, B.B., and Dannenberg, A.J. (1997) Inhibition of Cyclooxygenase: A Novel Approach to Cancer Prevention, *Proc. Soc. Exp. Biol. Med. 216*, 201–210.
105. Rose, D.P. (1997) Dietary Fatty Acids and Prevention of Hormone-Responsive Cancer. *Proc. Soc. Exp. Biol. Med. 216*, 224–233.
106. Lawrence, R.C., Helmick, C.G., Arnett, F.C., Deyo, R.A., Felson, D.T., Giannini, E.H., Heyse, S.P., Hirsch, R., Hochberg, M.C., Hunder, G.G., Liang, M.H., Pillemer, S.R.,

Steen, V.D., and Wolfe, F. (1998) Estimates of the Prevalence of Arthritis and Selected Musculoskeletal Disorders in the United States, *Arthritis Rheum. 41,* 778–799.
107. Kremer, J.M. (1996) Effects of Modulation of Inflammatory and Immune Parameters in Patients with Rheumatic and Inflammatory Disease Receiving Dietary Supplementation of n-3 and n-6 Fatty Acids, *Lipids 31,* S243–S247.
108. Geusens, P., Wouters, C., Nijs, J., Jiang, Y., and Dequeker, J. (1994) Long-Term Effect of Omega-3 Fatty Acid Supplementation in Active Rheumatoid Arthritis, *Arthritis Rheum. 37,* 824–829.
109. Shapiro, J.A., Koepsell, T.D., Voigt, L.F., Dugowson, C.E., Kestin, M., and Nelson, J.L. (1996) Diet and Rheumatoid Arthritis in Women: A Possible Protective Effect of Fish Consumption, *Epidemiology 7,* 256–263.
110. Marks, F., Furstenberger, G., and Muller-Decker, K. (1998) Arachidonic Acid Metabolism as a Reporter of Skin Irritancy and Target of Cancer Chemoprevention, *Toxicol. Lett. 96–97,* 111–118.
111. Orr, F.W., Kostenuik, P., Sanchez-Sweatman, O.H., and Singh, G. (1993) Mechanisms Involved in the Metastasis of Cancer to Bone, *Breast Cancer Res. Treat. 25,* 151–163.
112. Hamilton, L.C., Mitchell, J.A., Tomlinson, A.M., and Warner, T.D. (1999) Synergy Between Cyclo-Oxygenase-2 Induction and Arachidonic Acid Supply in vivo: Consequences for Nonsteroidal Anti-inflammatory Drug Efficacy, *FASEB J. 13,* 245–251.

Chapter 20

Conjugated Linoleic Acid (CLA) and the Risk of Breast Cancer

Flore Lavillonnière and Philippe Bougnoux

UPRES-EA 2103, University François-Rabelais, Tours, France

Introduction

Cancer is the leading cause of death for people <65 y old in Western countries; in contrast to death from heart diseases, there is no indication that mortality rates due to cancer will decrease. Many believe that cancer should be regarded more as a preventable rather than a treatable disease (1). Cancer is a multifactor disease in which environment contributes to a large part. Among preventable causes of cancer, estimates of dietary factors account for between 20 and 60%, according to the cancer sites (2,3). Therefore, much attention is focused on approaches that can provide new insights into the relation between dietary components of food and risk of cancer. Breast cancer is the leading cause of death among women <60 y old, and there is still no primary prevention of this type of cancer. Although early ecological studies on the relationship between diet, specifically between total fat intake and breast cancer, were promising, they have not led to precise recommendations because most of the subsequent analytical studies failed to demonstrate the existence of such a relationship (4). However, investigators in the field of cancer and nutrition have remained frustrated, mainly because they suspect that methodological issues (5) are largely responsible for premature negative conclusions. Specifically, insufficient precision and lack of sensitivity of dietary assessment methods, biases secondary to recall of diet, studies carried out in populations receiving dietary supplementation or with insufficient range of fat intake (6), lack of data on specific types of fatty acids or families of PUFA, lack of data taking into account interactions between PUFA and antioxidant lipids (7) are all potential faults that may have prevented nutritional epidemiologic studies from individualizing dietary components or profiles associated with breast cancer risk or protection.

Breast cancer is the clinical expression of the development of tumor tissue initially within a breast, then within different other sites during the metastatic evolution period. This tumor tissue results from a multistep carcinogenic process that probably extends over several decades. This process consists of an accumulation of genetic alterations acquired during a woman's life span, which leads to the neoplastic transformation of normal epithelial cells. Sometimes this phenomenon is accelerated by germinal genetic alterations that can be transmitted from parents to children. Such genetic predispositions represent an increased risk of breast cancer for which prevention through environmental factors such as diet would be highly desirable.

Dietary lipids can modulate mammary carcinogenesis at several stages; some of them may contribute to the prevention of acquired genetic alterations through their antimutagenic activity and therefore act at the very early stages of the initiation process. Lipids can also influence tumor promotion (6). Dietary lipids act on noncancer cells within tumor tissue (endothelial cells, fibroblasts, immune cells) and therefore affect angiogenesis and invasion processes at later stages of carcinogenesis (8). Moreover, lipids alter the immune response of the host (9). All of these steps are crucial to tumor progression, as well as growth and development of tumors.

Fatty acids, the main components of dietary lipids, have several properties that make them potential targets for use in a dietary prevention of breast cancer. For a few years, growing interest has focused on particular fatty acids, geometrical and positional isomers of linoleic acid, i.e., conjugated linoleic acids (CLA). These dietary conjugated dienes present interesting anticarcinogenic properties in animal studies of mammary carcinogenesis (10–15). Furthermore, these properties are unique in that CLA can act at very low concentration, close to that observed in the diet (16). No information is currently available on CLA and breast cancer; this is likely due to the difficulties in evaluating CLA intake in patients, which is inherent in the low sensitivity of dietary questionnaires. Insights into the relation of these natural fatty acids to breast cancer appear therefore to be highly desirable.

CLA and the Risk of Breast Cancer

We designed a case-control study to evaluate the relation between CLA and the risk of breast cancer (Lavillonnière et al., preliminary unpublished data). The CLA level in breast adipose tissue was used to reflect the individual's past dietary intake of CLA. The study involved 360 patients who were treated for breast tumors. These patients lived in Central France, an ethnically homogeneous region, with little migration flux. The group was comprised of patients who had localized breast cancer (n = 261, cases) and those treated for a benign breast tumor (n = 99, controls). At the time of surgery, a sample of breast adipose tissue was collected from each patient and used later for determination of CLA content. There was several differences among the two populations, i.e., the cases were older and there were more patients postmenopause in that population than in controls. Mean body mass index (BMI) was higher in cases than in controls, and age at menarche was lower in cases. Therefore, this population was representative of the group at risk for breast cancer because they expressed more risk factors than controls.

CLA in breast adipose tissues were analyzed by gas chromatography (GC) after derivatization with sodium methoxide and concentration by reversed-phase high-performance liquid chromatography (HPLC) (17). Identification of CLA peaks was made by comparison of retention times with that of an authentic standard and by mass spectrum analysis of methyl triazoline dione derivatives (MTAD). Study of the stereochemical distribution of CLA within the three positions of breast adipose tissue triacylglycerols was carried out by pancreatic lipase hydrolysis followed by analysis of the hydrolytic products.

The GC profiles of CLA between case and control tissues samples were very similar, with one CLA peak representing up to 85% of the total CLA isomers detected. This isomer was identified as the 18:2 $\Delta 9c,11t$ isomer of CLA. Mean CLA level in adipose tissues was 0.52% of total fatty acids. CLA level was higher in the breast adipose tissues from controls than in the tissues from cases. This suggests a protective effect of CLA against breast cancer.

The possible link between adipose tissue CLA and characteristics of the tumor was investigated in the population of breast cancer patients. There was no association between CLA level and tumor volume, extension of cancer cells to axillary lymph nodes, histopathologic or histoprognostic type, or estradiol or progesterone receptors content of the tumor, all parameters associated with breast cancer aggressiveness. This suggests that CLA levels are unlikely to be dependent on breast cancer characteristics.

Because the 18:2 $\Delta 9c,11t$ is the main isomer in breast tissue, the protective effect of CLA is likely to depend on this peculiar isomer. This hypothesis is consistent with results from a chemically-induced mammary carcinogenesis experiment in rats.

Dietary CLA Isomers and NMU-Induced Mammary Tumors in Rats

Rats were injected with N-methylnitrosourea (NMU) and then assigned to three experimental groups (n = 20 animals per group). They were fed a basal diet supplemented with either 1% by weight of a mixture of CLA isomers or 1% by weight of 18:2 $\Delta 9c,11t$ isomer, chemically prepared from castor oil. The control group was fed a sunflower oil–based diet. The number of tumors or the tumor mass was lower in the two rat groups assigned to either the CLA mixture diet or to the 18:2 $\Delta 9c,11t$ diet than in the control group (Table 20.1). These results demonstrate that the main dietary isomer, 18:2 $\Delta 9c,11t$, which is supposed to be the biologically active isomer of CLA, is potent as an anticarcinogenic agent (Lavillonnière et al., unpublished data).

The human data presented here offers potentially important and long-awaited information about CLA and breast cancer. As observed in several animal mammary tumor experiments, CLA may reduce the rate of breast cancer. The fact that 18:2 $\Delta 9c,11t$ is the main isomer in breast adipose tissue suggests that this isomer has a protective effect with regard to breast cancer. This observation may not extend to all types of cancer or to cancer at other sites. Such an inverse relationship was not observed by Griffin et al. (18) in patients with cervical neoplasia. They found CLA concentration to be higher in the group with precancerous lesions than in the control group. However, the methodology they applied provided the CLA concentration only in the phospholipids and was restricted further to the sn-2 position of this class of lipids. These differences may have led to conclusions different from those of our case-control study because we found that the proportion of CLA was not equivalent at each position of the triglyceride.

The significance of the difference in CLA level found between cases and controls deserves further investigation. Differences in CLA metabolism between cases

TABLE 20.1 Number and Growth of Mammary Tumors According to Enrichment of Diet with Conjugated Linoleic Acid (CLA) or 18:2 Δ9c,11t in Rats[a]

Dietary intervention[b]	n	Number of tumors	Tumor mass (g)
No CLA (control group)	20	46[c]	170[c]
Mixture of CLA isomers	20	31	94
18:2 Δ9c,11t	20	28	81

[a]20 wk after N-methylnitrosourea (MNU) injection; age at MNU administration, 52 d.
[b]Basal diet contained 5% sunflower oil and was supplemented with 1% by weight of free fatty acids depending on the experimental group.
[c]$P < 0.05$.

and controls cannot be ruled out. However, the lack of relationship between adipose tissue CLA level and the size of the tumor in breast cancer patients argues against the possibility that CLA would be lower in cases than in controls as a consequence of the presence of the disease (a possible flaw in any case-control study). Indeed, it is expected that the CLA content of breast adipose tissue reflects the past dietary intake of CLA, as has been observed for many PUFA (19–21). Additionally, the consumption of food rich in CLA leads to an increased CLA content in plasma or human milk (22–24), but no study has examined the incorporation of dietary CLA into adipose stores. Differences in dietary intake of specific fatty acids may influence their stereochemical position within the triacyglycerol molecule, either in human breast milk (25) or in rat (26) and ruminant adipose tissues (27). We actually found such positional differences for CLA between breast adipose tissues triacylglycerol in both the case and control populations (Table 20.2). This suggests that the intake or incorporation of CLA within triglycerides may be different between these two populations. Finally, the isomers of CLA in breast adipose tissue were similar to those found in many food items, as shown by the mass spectrum of the MTAD derivatives of CLA. Taken together, these observations would argue for differences in the dietary intake of CLA to at least partially explain the differences in breast adipose tissue CLA content between the cases and controls.

In conclusion, dietary CLA, and specifically the dietary 18:2 Δ9c,11t CLA isomer, seem to be protective against rat mammary tumors, and our data, although circumstantial, suggest that they are protective against breast cancer. Whether CLA may be considered as potential targets for use in a nutritional prevention of breast cancer remains an attractive issue. The amount of CLA needed, the duration of intervention, as well as the proper stage in life for such an intervention are not known at present. Paralleling a recent study carried out with dietary n-3 PUFA in breast cancer patients (28), a short-term dietary intervention with a diet high in CLA, such as dairy products, followed by an analysis of adipose tissue fatty acids would provide more insight into this issue.

TABLE 20.2 Positional Distribution of Conjugated Linoleic Acid (CLA) in Breast Adipose Tissue Triacylglycerol from Patients with Malignant (Cases) or Benign (Controls) Breast Tumors[a]

Population	Internal position (sn-2 position)	External positions (sn-1 and sn-3 positions)
	%	
Cases	30.4 ± 3.6[b]	69.6[b]
Controls	25 ± 5.1	75

[a]Determined after gas chromatography analysis of the products resulting from pancreatic lipase hydrolysis of breast adipose tissue triacylglycerol; mean value ± SEM, n = 15.
[b]$P < 0.05$ between case and control; t-test.

Acknowledgments

Claude Lhuillery is kindly acknowledged for the animal studies; Marie-Lise Jourdan for taking care of adipose tissue samples; and Gilles Body, Jacques Lansac, and the surgeons of the Onco-Gynecologic Unit for preparing adipose tissue during surgery. Olivier Le Floch provided patients' clinical files. Jean-Louis Sébédio and Jean-Charles Martin provided expertise for lipid biochemistry. FL was supported by CERIN and Région Centre.

References

1. Sporn, M.B. (1996) The War on Cancer, *Lancet 347,* 1377–1381.
2. Doll, R., and Peto, R. (1981) The Causes of Cancer: Quantitative Estimates of Avoidable Risks of Cancer in the United States Today, *J. Natl. Cancer Inst. 66,* 1191–1308.
3. Doll, R. (1992) The Lessons of Life: Keynote Address to the Nutrition and Cancer Conference, *Cancer Res. 52,* 2024s–2029s.
4. Hunter, D.J., Spigelman, D., Hadami, H.O., Beeson, L., Van der Brandt, P.A., Folsom, A.R., et al. (1996) Cohort Studies of Fat Intake and the Risk of Breast Cancer—A Pooled Analysis, *N. Engl. J. Med. 334,* 356–361.
5. Prentice, R., Pepe, M., and Self, S. (1989) Dietary Fat and Breast Cancer: A Quantitative Assessment of the Epidemiological Literature and a Discussion of Methodological Issues, *Cancer Res. 49,* 3147–3156.
6. Wynder, E.L., Cohen, L.A., Muscat, J.E., Winters, B., Dwyer, J.T., and Blackburn, G. (1997) Breast Cancer: Weighing the Evidence for a Promoting Role of Dietary Fat, *J. Natl. Cancer Inst. 89,* 766–775.
7. Bougnoux, P. (1999) n-3 PUFA and Cancer, in *Current Opinion in Clinical Nutrition and Metabolic Care, 2,* 121–126..
8. Jiang, W., Bryce, R., and Horrobin, D. (1998) Essential Fatty Acids: Molecular and Cellular Basis of Their Anti-Cancer Action and Clinical Implications, *Crit. Rev. Oncol. Hematol. 27,* 179–209.
9. Calder P. (1997) n-3 Polyunsaturated Fatty Acids and Immune Cell Function, *Adv. Enzyme Regul. 37,* 197–237.

10. Ip, C., Chin, S.F., Scimeca, J.A., and Pariza, M.W. (1991) Mammary Cancer Prevention by Conjugated Dienoic Derivative of Linoleic Acid, *Cancer Res. 51,* 6118–6124.
11. Ip, C., Singh, M., Thompson, H.J., and Scimeca, J.A. (1994) Conjugated Linoleic Acid Suppresses Mammary Carcinogenesis and Proliferative Activity of the Mammary Gland in the Rat, *Cancer Res. 54,* 1212–1215.
12. Ip, C., Scimeca, J.A., and Thompson, H.J. (1995) Effect of Timing and Duration of Dietary Conjugated Linoleic Acid on Mammary Cancer Prevention, *Nutr. Cancer 24,* 241–247.
13. Ip, C., Briggs, S.P., Haegele, A.D., Thompson, H.J., Storkson, J., and Scimeca, J.A. (1996) The Efficacy of Conjugated Linoleic Acid in Mammary Cancer Prevention Is Independent of the Level or the Type of Fat in the Diet, *Carcinogenesis 17,* 1045–1050.
14. Ip, C. (1997) Review of the Effects of *trans* Fatty Acids, Oleic Acid, n-3 Polyunsaturated Fatty Acids, and Conjugated Linoleic Acid on Mammary Carcinogenesis in Animals, *Am. J. Clin. Nutr.,* 1523S–1529S.
15. Ip, C., Jiang, C., Thompson, H.J., and Scimeca, J.A. (1997) Retention of Conjugated Linoleic Acid in the Mammary Gland Is Associated with Tumor Inhibition During the Post-Initiation Phase of Carcinogenesis, *Carcinogenesis 18,* 755–759.
16. Banni, S., and Martin, J.C. (1998) Conjugated Linoleic Acid and Metabolites, in *Trans Fatty Acids in Human Nutrition,* (Sébédio, J.L., ed.) pp. 261–302, The Oily Press, Dundee.
17. Lavillonnière, F., Martin, J.C., Bougnoux, P., and Sébédio, J.L. (1998) Analysis of Conjugated Linoleic Acid Isomers and Content in French Cheeses, *J. Am. Oil Chem. Soc. 75,* 343–352.
18. Griffin, J.F.A., Wickens, D.G., Tay, S.K., Singer, A., and Dormandy, T.L. (1987) Recognition of Cervical Neoplasia by the Estimation of a Free-Radical Reaction Product (Octadeca-9,11-dienoic Acid) in Biopsy Material, *Clin. Chim. Acta 163,* 143–148.
19. London, S.J., Sacks, F.M., Caesar, J., Stampfer, M.J., Siguel, E., and Willett, W.C. (1991) Fatty Acid Composition of Subcutaneous Adipose Tissue and Diet in Postmenopausal US Women, *Am. J. Clin. Nutr. 54,* 340–345.
20. Hunter, D.J., Rimm, E.B., Sacks, F.M., Stampfer, M.J., Colditz, G.A., Litin, L.B., Willet, W.C. (1992) Comparison of Measures of Fatty Acid Intake by Subcutaneous Fat Aspirate, Food Frequency Questionnaire, and Diet Records in a Free-Living Population of US Men, *Am. J. Epidemiol. 135,* 418–427.
21. Garland, M., Sacks, F.M., Colditz, G.A., Rimm, E.B., Sampson, L.A., Willett, W.C., Hunter, D.J. (1998) The Relation Between Dietary and Adipose Tissue Composition of Selected Fatty Acids in US Women, *Am. J. Clin. Nutr. 67,* 25–30.
22. Huang, T.C., Lloyd, O., Luedecke, D., Terry, D., and Shultz, D. (1994) Effect of Cheddar Consumption on Plasma Conjugated Linoleic Acid Concentrations in Men, *Nutr. Res. 14,* 373–386.
23. Britton, M., Fong, C., Wickens, D., and Yudkin, J. (1992) Diet as a Source of Phospholipid Esterified 9,11-Octadecadienoic Acid in Humans, *J. Clin. Sci. 83,* 97–101.
24. Park, Y.S., Behre, R.A., McGuire, M.A., Shultz, T.D., and McGuire, M.K. (1997) Dietary Conjugated Linoleic Acid (CLA) and CLA in Human Milk, *FASEB J. 11,* A239 (abstr.).
25. Jensen, R.G. (1996) The Lipids in Human Milk, *Prog. Lipid Res. 35,* 53–92.
26. Leray, C., Raclot, T., and Groscolas, R. (1993) Positional Distribution of n-3 Fatty Acids in Triacylglycerols from Rat Adipose Tissue During Fish Oil Feeding, *Lipids 28,* 279–284.

27. Smith, S.B., Yang, A., Larsen, T.W., and Tume, R.K. (1998) Positional Analysis of Triacylglycerols from Bovine Adipose Tissue Lipids Varying in Degree of Unsaturation, *Lipids 33*, 197–207.
28. Bagga, D., Capone, S., Wang, H.J., Heber, D., Lill, M., Chap, L., Glaspy, J.A. (1997) Dietary Modulation of Omega-3/Omega-6 Polyunsaturated Fatty Acid Ratios in Patients with Breast Cancer, *J. Natl. Cancer Inst. 89*, 1123–1131.

Chapter 21

Conjugated Linoleic Acid (CLA) in Lipids of Fish Tissues

Robert G. Ackman

Canadian Institute of Fisheries Technology, DalTech, Dalhousie University, Halifax, Nova Scotia B3J 2X4, Canada

Introduction

Before capillary gas–liquid chromatography (GLC), conjugated linoleic acids (CLA) were industrial materials and were examined by packed column GLC. However, in 1981, our capillary column GC showed an obvious *cis*-9,*trans*-11 isomer of linoleic acid in primate depot fats, including those of monkeys and humans (1). It could be traced in the one instance to the food items lard and corn oil, fed to the monkeys and intended to mimic a normal Canadian supply of dietary fatty acids. Figure 21.1 illustrates the GLC state of the art of that period, with two trace CLA peaks intermingled with human depot fat 18:3n-3 and C_{20} peaks. The analytical problems for CLA, common in dairy products (2), were then just manageable, usually benefitting from auxiliary fractionation by thin-layer chromatography (TLC) or similar enrichment. Little or no CLA has been found in margarines made from vegetable oils (3). After 1980, the rapid introduction of flexible fused silica columns for GLC

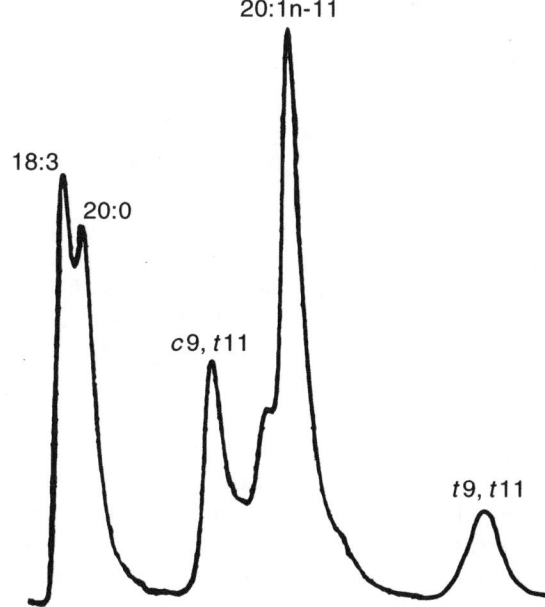

Fig. 1. Partial gas–liquid chromatography (GLC) analysis of the methyl esters of fatty acids of human depot fat sampled in 1980. Column: 46 m × 0.25 mm i.d., stainless steel, wall-coated with Silar-7 CP and operated at 170°C. The *cis*-9,*trans*-11-18:2 component follows the 20:0 peak and precedes the unknown component on the front of the main 20:1 peak, identified as 20:1n-9. *Source:* Ref. 1.

soon made the determination of the "essential" fatty acids linoleic and alpha-linolenic quite simple and accurate (4); such methods are now even more popular and also suitable for CLA analyses of vegetable oils and dairy products (5–7).

Once published, the CLA matter languished in our laboratory because these fatty acids were supposedly not of concern in marine lipids, our primary responsibility. The plant CLA content for several vegetable oils had in fact been included in our papers and the *cis*-9,*trans*-11-18:2 isomer could be explained by enzymes such as the soy lipoxygenase used at that time to identify "essential" polyunsaturated fatty acids, specifically those with methylene-interrupted *cis*,*cis* structure with the bonds in the n-6 and n-9 positions (AOCS Method Cd 15-78). This reaction also applied to appropriate n-3 fatty acids, including the major polyunsaturated fatty acids 20:5n-3 and 22:6n-3 [eicosapentaenoic (EPA) and docosahexaenoic (DHA), respectively] of fish oils and lipids (8). A similar lipoxygenase has even been isolated from fish tissue (9).

Consumption of linoleic acid in North America increased radically in the 1950–1970 era of belief in polyunsaturated fatty acids as the facile solution to heart disease (10). At the same time, consumption of dairy products decreased because of the overemphasized and misunderstood role of dietary saturated fatty acids and cholesterol. Consumption of fish in the U.S. has also declined steadily for decades except for a small increase when the epidemiologic results of a long-term study in Zutphen in The Netherlands (11) demonstrated the cardiac benefits of a regular intake of fish, a source of the long-chain n-3 fatty acids EPA and DHA. A small amount of the decrease in the consumption of fish in the U.S. as a source of EPA and DHA was offset by an increase in chicken consumption (12). The white meat of chicken is naturally a reasonable source of EPA and DHA, especially the latter, and can even be enriched in these fatty acids if desired (13). In those decades of upheaval in the dietary fatty acids available in the "Western" diet, there were also increasing numbers of published records of the fatty acid compositions of fish lipids. Most were compiled on lipids of whole fillets (14,15); shellfish were also included. A sharp demarcation in the linoleic acid content of fish showed that most marine fish typically had only 1–2% of linoleic acid, but some freshwater fish such as the channel catfish *Ictalurus punctatus*, popular in the U.S., could have as much as 12% of linoleic acid in the fatty acids of the edible fillet (16). This was higher than previously published for fatty acids of this fish in culture (15), and presumably reflected changes away from dietary marine fish meal (probably mostly menhaden in the case of U.S. farmed fish) to other protein materials such as soybean meal, or even to vegetable oils added to the diet. Wild and farmed catfish also differ, and USDA Handbook 8-15 cites ~3.5% linoleic acid in wild channel catfish (lipid 2.82%) and ~11.5% linoleic acid in farmed fish muscle (lipid 7.5%).

Recently, the cost of fish meal as a protein source and even that of fish oil rose temporarily to extraordinary heights (17). There has been considerable interest in replacing these sources of protein for growth and oil for energy in many species of farmed fish, for example brook charr *Salvelinus fontinalis* (18), Arctic charr *Salvelinus alpinus* (19), yellowtail *Seriola quinqueradiata* (20), rainbow trout *Oncorhynchus mykiss* (21,22), red drum *Sciaenops ocellatus* (23), white amur *Ctenopharyngodon idella* (24), Atlantic salmon *Salmo salar* (25–27), and striped bass *Morone chrysops* × *M. saxatilis* (28,29). Tilapia *Tilapia* sp., carp *Cyprinus carpio,* and many other warmwater fish feed naturally on vegetation; as these types move into the aquaculture industry, novel concentrated plant feed sources such as distillers grains, cottonseed meal, and canola protein will become more important (30). Today, seafoods are not considered to be important sources of CLA (31), but the term seafoods is so loosely used that it is often difficult to learn from brief references whether the source of most of the fat or lipid in the reference is from coldwater or warmwater fish, or from the fish alone or from the fish plus frying oil absorbed by batter in places such as the U.K. where deep-fried fish is popular (Table 21.1). In fact, interpretation of HPLC and GLC results (see below) adds to the confusion and should be corrected. Both *cis*-9,*trans*-11-18:2 and *trans*-10,*cis*-12-18:2 are likely to be absent from raw seafoods. This situation will also change for fried fish products because vegetable frying oils are changing in fatty acid composition (33,34), and the future position of CLA of plant origin in fried seafoods must be considered. This will vary widely with population habits (Table 21.1).

We now know that plant lipoxygenases are quite common (35); to prevent excessive destruction of the polyunsaturated fatty acids in seeds, dehydroperoxides must also be present and functional. Trace amounts of CLA, readily determined by ultraviolet (UV) absorption at 234 nm, are augmented in the refining of edible oils with acid-activated bleaching earths, especially at high temperatures (36). It is therefore inevitable that eventually some vegetable CLA will appear in farmed fish lipids proper if vegetable oils are fed. These may even be of poor quality for human consumption but can be tolerated by fish even if somewhat oxidized (37).

Analyses for CLA in food by GLC must be conducted with care. The used frying oil investigated by Sébédio *et al.* (38) with the highly polar GLC liquid phase Silar-10C (100% cyanosilicone) showed three CLA components following 18:3n-3, which itself coincided with 20:1, presumably 20:1n-9. In an analysis of the methyl esters of the fatty acids of fish, this region of the GLC analysis would be shared not only with 18:4n-3 and 18:4n-1 but also with 20:0, 20:1n-15, 20:1n-11, 20:1n-9, 20:1n-7, 20:1n-5, 20:2n-6, and so on (Fig. 21.2). The CLA peak proportions for

TABLE 21.1 Regional Consumption [g/(person·d)] of Fish Foods by Men and Women in the United Kingdom[a]

Seafood type	Scotland Male	Scotland Female	Northern Male	Northern Female	Midlands Male	Midlands Female	London, SE Male	London, SE Female
Deep-fried white fish	11	17	11	17	9	12	9	14
Other white fish	6	5	8	6	6	6	6	7
Shellfish	2	1	2	1	1	1	2	3
Oily fish	5	7	7	6	4	6	6	9
Total fish	24	30	28	25	20	25	23	33

[a]*Source:* Ref. 32.

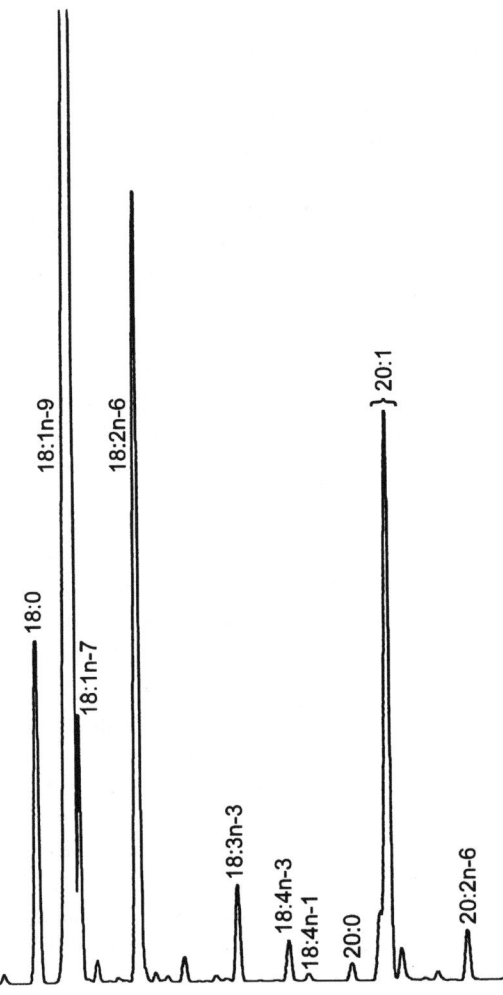

Fig. 21.2. Partial gas–liquid chromatography (GLC) analysis of the methyl esters of fatty acids from the kidney triacylglycerols of an experimental Atlantic salmon. Analysis on Omegawax 320 column, 30 m × 0.25 mm. The amount of linoleic acid is exaggerated over the norm for wild or even farmed commercial salmon by the major dietary fat (canola oil), but the positions of the 18:4n-3 and 18:4n-1 relative to 18:3n-3 and 20:0 are clearly illustrated.

heated sunflower and soybean oils are given by Sébédio *et al.* as 1:6 and 1:3 for (*cis,trans* + *trans,cis*) and *trans,trans* isomers, respectively. These are proportions similar to those reported in 1981 for four refined but unheated retail vegetable oils (1). Highly polar GLC columns may therefore not be more useful for CLA analyses than those of lower polarity based on polyglycols.

Unfortunately the C_{18}–C_{20} regions of GLC analyses of the methyl esters of fish lipids, even those of freshwater fish, are much more complex than those of most animal sources. For example, pork, lard, and mutton fatty acid methyl esters examined by Saeed *et al.* (39) on a FFAP (Carbowax type "free fatty acid phase") column show a good-sized retention time gap between 18:3n-3 and 20:0. However, all four of their published GLC charts, based on beef and mutton fatty acid analyses, show a peak immediately following that for 18:3n-3 and roughly equal to it in size. This was probably (see below) in the position for cis-9,trans-11 18:2. In analyses of fish lipids, this region of the chart is usually filled by 18:4n-3 and 18:4n-1 when the bonded forms of Carbowax from Supelco (SUPELCOWAX-10 or Omegawax 320, Bellefonte, PA), or other suppliers, are used (Fig. 21.2).

The late elution of CLA among the C_{18} fatty acids has led to a major misunderstanding in the CLA of seafoods in the food literature that could be exaggerated by the C_{18}–C_{20} overlap situation on cyanosilicones. At this time, it is sufficient to note that seafood was reported by Chin et al. (40), from HPLC analyses, to contain <0.1–0.8 mg CLA/g fat, but not to contain measurable cis-9,trans-11-18:2. Chin et al. (40), using a SUPELCOWAX-10 column, reported interfering substances eluting with the cis-9,trans-11 CLA. In all probability, this was in fact the 18:4n-3 peak common in marine fish lipids at the 1–2% level. As shown in Figure 21.2, the accompanying 18:4n-1 is usually one fifth to one tenth of the 18:4n-3 in marine lipids. These observations explain the n.d. = not detectable for cis-9,trans-12 entries in the tabulation of Chin et al. (40), repeated by O'Shea et al. (31). The trans-10,cis-12-18:2 should elute with or after 18:4n-1. The "total CLA" figures published (40) cannot now be identified from the information available, but both cis-9,trans-11-18:2 and trans-10,cis-12-18:2 are, from our other work, unlikely in common marine foods.

One recent objective of joint research into fish and human nutrition (41) was to investigate whether CLA would in fact readily deposit in fish lipids and to examine the GLC properties of CLA on Omegawax-320. Alkali-isomerized safflower seed oil was therefore fed to small, rapidly growing fish of three species. The first two, tilapia *(Tilapia nilotica)* and carp were freshwater, and the rockfish *(Sebastes schlegeli)* was marine. These are all fish that are popular in Korean aquaculture (41). Weight gain, feed efficiency, and specific growth rates were measured for triplicate groups of fish held in 40-L tanks with semicirculation, a biological filtration system, and a small and continuous amount of make-up water. The diets included, by weight, fish meal (45%), soybean meal (7.0%), wheat flour (30.0%), vitamin mixture (2.0%), mineral mixture (1.0%), and added lipid (15%). The latter was refined safflower oil or that oil replaced with isomerized oil to give 1.0, 2.5, 5.0, and 10.0% CLA. The extent of CLA in the isomerized safflower oil was 75%. Figure 21.3 shows part of the GLC pattern of the methyl esters of the fatty acids of this preparation.

Whole lipid was extracted from appropriate parts of experimental fish with chloroform/methanol (42) and separated into phospholipids and neutral lipids by the method of Juaneda and Rocquelin (43). Methyl esters of fatty acids were prepared by AOCS Method Ce 1b-89 and analysis by GLC in a Shimadzu (Kyoto, Japan) GC-14A gas chromatograph equipped with a flame ionization detector and a fused silica capillary column (Omegawax-320, 30 m × 0.32 mm i.d., Supelco). Identifications of fatty acids were confirmed by GC/MS, an Ion Trap Detector (Finnigan MAT 700, San Jose, CA) with data output controlled by a PC. The GC column (DB-1, 60 m × 0.25 mm, J&W Scientific, Folsom, CA) was passed directly into the Ion Trap Detector through a heated transfer line.

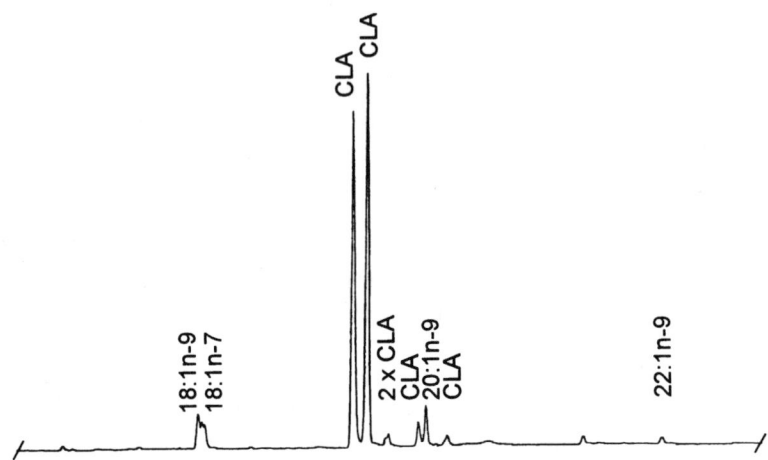

Fig. 21.3. Partial gas–liquid chromatography (GLC) analysis of an $AgNO_3$ thin-layer chromatography (TLC) isolate of the methyl esters of the isomerized safflower oil fatty acids, including CLA isomers. The two major CLA are from left to right, cis-9,trans-11-18:2 and trans-11,cis-12-18:2. Source: Ref. 41.

The six isomers of CLA that were prepared in the safflower oil and identified by GC analysis of the FAME of CLA are marked in Figure 21.3. Through the results of GC and GC/MS analysis, and supporting documentation (44), the two main peaks were identified as cis-9,trans-11-18:2 (eluting first) and trans-10,cis-12-18:2 (eluting later). The other results of the identification process suggest that two peaks were of the cis,cis type and two other peaks of the trans,trans type (41). Complete resolution of all CLA isomers by any chromatographic technology is very difficult (6,7). To assist in this identification process for the FAME from fish lipids, an $AgNO_3$-TLC plate (silver nitrate modified thin-layer chromatography) with a mobile development phase of benzene gave six bands (not shown). The CLA group fell cleanly below the band for methyl esters of saturated fatty acids and above the band for monoetheylenic fatty acids. The fish tissue FAME of CLA shown (Fig. 21.4) are from muscle phospholipids extracted from the group fed 5% CLA for 8 wk. The statement that the cis-9,trans-11 CLA isomer is "apparently" the only isomer incorporated into the phospholipid fraction of tissues of animals (40) is therefore incorrect with respect to fish.

The growth of the fish fed the diets including 1.0% CLA was normal compared with those receiving the control diet over the 8-wk period. The carp also tolerated the 2.5% CLA diet, but this dietary level did affect growth in the tilapia and rockfish (Table 21.2). The CLA content (mg/g lipid) in total muscle lipids of the three species after 8 wk of consuming the diets is shown in Table 21.3. For comparison, the average final weights are given, confirming the specific growth rate of Table 21.2. The proportions of the two major isomers in the fish muscles were all consistent, ranging from 81.0 to 89.8% of all CLA peaks on GLC, except for one value of 77.1% (carp, 1.0% diet). CLA can therefore be accumulated in the body lipids of both fresh-

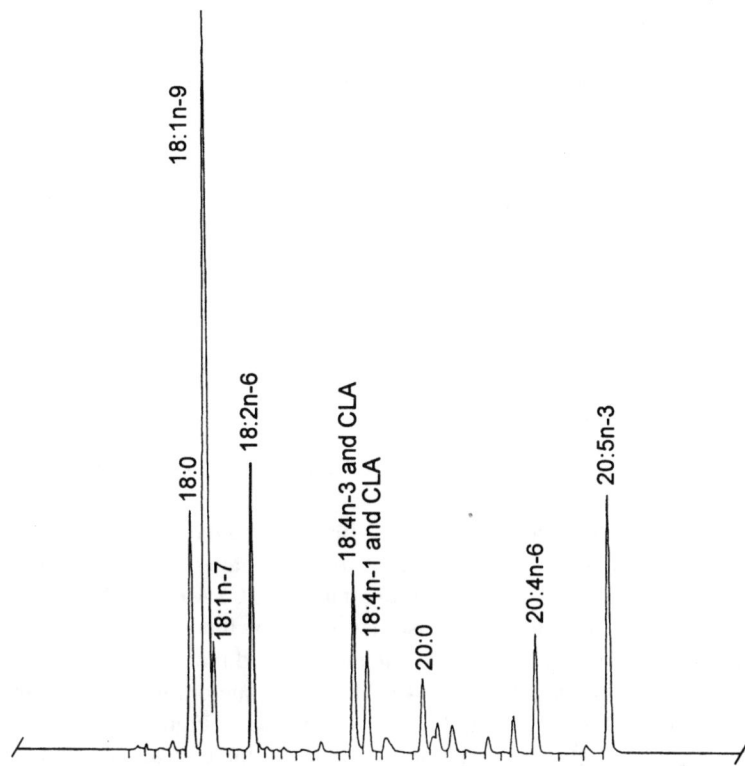

Fig. 21.4. Partial gas–liquid chromatography (GLC) analysis of the methyl esters of fatty acids of carp muscle phospholipids recovered from $AgNO_3$-TLC. *Source:* Ref. 41.

TABLE 21.2 Specific Growth Rates (%/day) of Three Fish Species Fed Diets with Different CLA Content Levels[a]

	Control diet	1% CLA	2.5% CLA	5.0% CLA	10.0% CLA
Carp	2.24	2.29	2.26	2.08	2.04
Tilapia	2.47	2.48	2.26	2.23	2.00
Rockfish	2.58	2.57	2.31	1.98	1.45

[a]*Source:* Ref. 41.

water and marine fish. This includes both major isomers in the phospholipids, contrary to an earlier opinion in which only the *cis*-9,*trans*-11 isomer was found in mammalian phospholipids(45).

That CLA can be included in the fatty acids of the phospholipids (Fig. 21.4) was unexpected, but Figure 21.4 also illustrates the point that the CLA fatty acids recovered from the carp phospholipids fall into the space on the GLC baseline normally

TABLE 21.3 CLA Contents (mg/g of fat) in Muscle Lipids of Fish Fed Different Levels of Dietary CLA and Mean Weights of Fish After 8 wk of Consuming Diets[a]

	Diet 1 Control	Diet 2 1.0%	Diet 3 2.5%	Diet 4 5.0%	Diet 5 10.0%
Start of growth trial					
Carp	0	130.1	206.9	214.5	218.4
Tilapia	0	41.3	84.0	118.1	180.9
Rockfish	0	51.0	106.1	155.3	126.2
Mean final weights (g) after 8 wk					
Carp	13.2	14.0	14.1	12.8	12.4
Tilapia	48.7	49.0	43.4	42.7	37.5
Rockfish	10.6	10.6	9.2	8.0	6.4

[a]Source: Ref. 41.

occupied by 18:4n-3 and 18:4n-1 on the polyglycol-based GLC column (cf. Fig. 21.2). Both Chin et al. (40) and Shantha et al. (46) investigated the loss of CLA due to several methods of preparing methyl esters of the conjugated fatty acids. In the latter work, the BF_3 esterification step, essentially that of AOCS Method Ce 1b- 89, caused a loss of 3% of CLA (the two major isomers); this method is not recommended in any event for CLA. However, it is convenient, provided the time for BF_3-MeOH exposure is short (cf. 47). Shantha et al. (46) illustrate a late-eluting and very minor artifact not noticed in Figure 21.3 and thus not expected in Figure 21.4.

Although 18:2n-6 (linoleic acid) and 18:3n-3 (α-linolenic acid) were targeted as "essential" for salmonids nearly 30 years ago by Castell et al. (48), results were confusing concerning the nature of the benefits of these acids in diets of both carp and U.S. channel catfish (49). Subsequently, for salmonids, actual marine fish oils became preferred because these were cheap and included the natural long-chain n-3 fatty acids. The European catfish *Silurus glanis,* however, accepted 18:3n-3 more readily than carp; >5% linoleic acid and 12% linolenic acid were deposited in the depot fat as well as good proportions of EPA (25%) and DHA (~10%) (50) when the fish were fed a suitable pelleted diet in which 50% vegetable matter replaced fish meal, but 14% total lipid derived in part from marine oils was also included. Many other studies point to vegetable proteins and oils becoming commonplace in the aquaculture industry raising warmwater fish (51).

The presence of conjugated linoleic acid (CLA) in the common vegetable oils is now a matter of public interest; however, the linoleic acid content of most marine fish lipids is so low as to continue to preclude speculation concerning whether the fatty acid of the muscle of this food would or would not contain CLA. The problem will be the new developments in farmed fish that accumulate fatty acids from a variety of dietary ingredients in their growth period. The reason is that the diets represent a balance between the cost and efficacy of the ingredients. The need is to provide good quality protein for growth and meet minimal specific needs for "essential" fatty acids, especially in salmonids (30,52), plus vitamins and minerals. For salmonids, replacement of the most common dietary fat, fish oil, with vegetable oils has become common lately. Several workers have experimented with feeding canola oil (26,53) to Atlantic salmon, *Salmo salar;* soybean and tallow have also been tested (25). Recently full-fat soybean has been evaluated as a salmon diet component (27). Some oilseed meals may be solvent extracted and still contain several percent of vegetable oil fatty acids. Experimental studies are often based on protein concentrates (54,55) or special meal stocks, but this is unlikely to be material for large-scale aquaculture. Increasingly, consumers want cold-pressed or similar solvent-free processed vegetable oils (56), increasing the possibility that natural CLA in the polar lipids of oil residues in oilseed meals may be passed from these residues in the protein source into farmed fish tissue.

Summary

CLA is conveniently prepared in quantity from any vegetable oil by alkali isomerization. This will produce two major isomers (*cis*-9,*trans*-11-18:2 and *trans*-10,*cis*-12-18:2) usually accompanied by other isomers (57), although this is not essential if moderate reaction conditions are employed (58). Deposition of CLA included in vegetable oil fed to fish took place without obvious discrimination among the isomers. Therefore, a potential source of CLA in fish will be the vegetable protein and/or vegetable oil used in fish farming, an industry that is increasingly including vegetable matter in feeds. Modern food preparation may also combine fish with various vegetable materials containing CLA, notably the oils used for deep frying that are absorbed by breading or batters. The GLC of methyl esters of CLA prepared with alkali transesterification followed by the brief BF_3-MeOH treatment for conversion of free fatty acids to esters (41) does not appear to discriminate among isomers or interfere with recovery and analysis. However, the GLC shows that for fish lipids, there could be an analytical problem because on our Omegawax-320 column, the most common CLA (*cis*-9,*trans*-11) coincides with the marine fatty acid 18:4n-3, and, in isomerized oils, the other major CLA (*trans*-10,*cis*-12) would coincide with the accompanying lesser 18:4n-1. The latter is formed by extension from algal 16:4n-1 (59); however, it normally comprises only about a tenth of the 18:4n-3 deposited if menhaden oil is fed, and may be barely visible in some other fish oils or lipids. Column to column variations, changes in "polarity" with aging or operating temperature, and the variety of modern GLC phases in use suggest that caution should be observed in reporting CLA in, or its absence from, fish and fish products unless any conflicts with the methyl esters of fatty acids common in fish, for example, 18:4n-3 and 18:4n-1 on polyglycol-based GLC columns and C_{20} fatty acids on more polar columns, are recognized. If straightforward GLC is not practical because of co-elution of the methyl ester of one or more of the fish fatty acids, $AgNO_3$-TLC (41) or HPLC (40) should effectively isolate the CLA for this analysis.

References

1. Ackman, R.G., Eaton, C.A., Sipos, J.C., and Crewe, N.F. (1981) Origin of *cis*-9,*trans*-11- and *trans*-9,*trans*-11-Octadecadienoic Acids in the Depot Fat of Primates Fed a Diet Rich in Lard and Corn Oil and Implications for the Human Diet, *Can. Inst. Food Sci. Technol. J. 14,* 103–107.
2. Parodi, P.W. (1977) Conjugated Octadecadienoic Acids of Milk Fat, *J. Dairy Sci. 60,* 1550–1553.
3. Precht, D., and Molkentin, J. (1997) Trans-Geometrical and Positional Isomers of Linoleic Acid Including Conjugated Linoleic Acid (CLA) in German Milk and Vegetable Fats, *Fett/Lipid 99,* 319–326.
4. Athnasios, A.K., Healy, E.J., Gross, A.F., and Templeman, G.J. (1986) Determination of *cis,cis*-Methylene Interrupted Polyunsaturated Fatty Acids in Fats and Oils by Capillary Gas Chromatography, *J. Assoc. Off. Anal. Chem. 69,* 65–67.
5. Wolff, R.L., Bayard, C.C., and Fabien, R.J. (1995) Evaluation of Sequential Methods for the Determination of Butterfat Fatty Acid Composition with Emphasis on *trans*-18:1 Acids. Application to the Study of Seasonal Variations in French Butters, *J. Am. Oil Chem. Soc. 72,* 1471–1483.
6. Lavillonnière, F., Martin, J.C., Bougnoux, P., and Sébédio, J.-L. (1998) Analysis of Conjugated Linoleic Acid Isomers and Content in French Cheeses, *J. Am. Oil Chem. Soc. 75,* 343–352.
7. Sehat, N., Yurawecz, M.P., Roach, J.A.G., Mossoba, M.M., Kramer, J.K.G., and Ku, Y. (1998) Silver Ion–High-Performance Liquid Chromatographic Separation and Identification of Conjugated Linoleic Acid Isomers, *Lipids 33,* 217–221.
8. Beare-Rogers, J.L., and Ackman, R.G. (1969) Reaction of Lipoxidase with Polyenoic Acids in Marine Oils, *Lipids 4,* 441–443.
9. Hsieh, R.J., and Kinsella, J.E. (1986) Lipoxygenase-Catalyzed Oxidation of n-6 and n-3 Polyunsaturated Fatty Acids: Relevance to and Activity in Fish Tissue, *J. Food Sci. 51,* 940–945, 996.
10. Lands, W.E.M. (1986) *Fish and Human Health,* Academic Press, Orlando, FL.
11. Kromhout, D., Bosschieter, E.B., and de Lezenne Coulander, C. (1985) The Inverse Relation Between Fish Consumption and 20-Year Mortality from Coronary Heart Disease, *N. Engl. J. Med. 312,* 1205–1209.
12. Raper, N.R., and Exler, J. (1991) ω3 Fatty Acids in the U.S. Food Supply, in *Health Effects of ω3 Polyunsaturated Fatty Acids in Seafoods,* (Simopoulos, A.P., Kifer, R.R., Martin, R.E., and Barlow, S.M., eds.) World Rev. Nutr. Diet., vol. 66, pp. 514–515, Karger, Basel (abstr.).
13. Ratnayake, W.M.N., Ackman, R.G., and Hulan, H.W. (1989) Effect of Redfish Meal Enriched Diets on the Taste and n-3 PUFA of 42-Day-Old Broiler Chickens. *J. Sci Food Agr. 49,* 59–74.
14. Kinsella, J.E. (1987) *Seafood and Fish Oils in Human Health and Disease,* Marcel Dekker, New York.
15. Ackman, R.G. (1989) Nutritional Composition of Fats in Seafoods, *Prog. Food. Nutr. Sci. 12,* 161–241.
16. Nettleton, J.A, Allen, W.H., Klatt, L.V., Ratnayake, W.M.N., and Ackman, R.G. (1990) Nutrients and Chemical Residues in One- to Two-Pound Mississippi Farm-Raised Channel Catfish *(Ictalurus punctatus), J. Food Sci. 55,* 954–958.

17. Anon. (1998) Meal Catch Set to Drop 10 m Tonnes, *Fishing News Int. 37*, 1–2.
18. Guillou, A., Soucy, P., Khalil, M., and Adambounou, L. (1995) Effects of Dietary Vegetable and Marine Lipid on Growth, Muscle Fatty Acid Composition and Organoleptic Quality of Flesh of Brook Charr *(Salvelinus fontinalis), Aquaculture 136,* 351–362.
19. Yang, X., and Dick, T.A. (1993) Effects of Dietary Fatty Acids on Growth, Feed Efficiency and Liver RNA and DNA Content of Arctic Charr, *Salvelinus alpinus* (L.), *Aquaculture 116,* 57–70.
20. Viyakarn, V., Watanabe, T., Aoki, H., Tsuda, H., Sakamoto, H., Okamoto, N., Iso, N., Satoh, S., and Takeuchi, T. (1992) Use of Soybean Meal as a Substitute for Fish Meal in a Newly Developed Soft-Dry Pellet for Yellowtail, *Nippon Suisan Gakkaishi 58,* 1991–2000.
21. Sanz, A., Morales, A.E., de la Higuera, M., and Cardenete, G. (1994) Sunflower Meal Compared with Soybean Meal as Partial Substitutes for Fish Meal in Rainbow Trout *(Oncorhynchus mykiss)* Diets: Protein and Energy Utilization, *Aquaculture 128,* 287–300.
22. Gomes, E.F., Rema, P., and Kaushik, S.J. (1995) Replacement of Fish Meal by Plant Proteins in the Diet of Rainbow Trout *(Oncorhynchus mykiss):* Digestibility and Growth Performance, *Aquaculture 130,* 177–186.
23. Gaylord, T.G., and Gatlin, D.M., III (1996) Determination of Digestibility Coefficients of Various Feedstuffs for Red Drum *(Sciaenops ocellatus), Aquaculture 139,* 303–314.
24. Bakir, H.M., Melton, S.L., and Wilson, J.L. (1993) Fatty Acid Composition, Lipids and Sensory Characteristics of White Amur *(Ctenopharyngodon idella)* Fed Different Diets, *J. Food Sci. 58,* 90–95.
25. Hardy, R.W., Scott, T.M., and Harrell, L.W. (1987) Replacement of Herring Oil with Menhaden Oil, Soybean Oil, or Tallow in the Diets of Atlantic Salmon Raised in Marine Net-Pens, *Aquaculture 65,* 267–277.
26. Polvi, S.M., and Ackman, R.G. (1992) Atlantic Salmon *(Salmo salar)* Muscle Lipids and Their Response to Alternative Dietary Fatty Acid Sources, *J. Agric. Food Chem. 40,* 1001–1007.
27. Bjerkeng, B., Refstie, S, Fjalestad, K.T., Storebakken, T., Rødbotten, M., and Roem, A.J. (1997) Quality Parameters of the Flesh of Atlantic Salmon *(Salmo salar)* as Affected by Dietary Fat Content and Full-Fat Soybean Meal as a Partial Substitute for Fish Meal in the Diet, *Aquaculture 157,* 297–309.
28. Nematipour, G.R., and Gatlin, D.M., III (1993) Requirement of Hybrid Striped Bass for Dietary (n-3) Highly Unsaturated Fatty Acids, *J. Nutr. 123,* 744–753.
29. Webster, C.D., Tiu, L.G., Tidwell, J.H., Van Wyk, P., and Howerton, R.D. (1995) Effects of Dietary Protein and Lipid Levels on Growth and Body Composition of Sunshine Bass *(Morone chrysops × M. saxatilis)* Reared in Cages, *Aquaculture 131,* 291–301.
30. Lim, C.E., and Sessa, D.J., (eds.) (1994) *Nutrition and Utilization Technology in Aquaculture,* AOCS Press, Champaign, IL.
31. O'Shea, M., Lawless, F., Stanton, D., and Devery, R. (1998) Conjugated Linoleic Acid in Bovine Milk Fat: A Food-Based Approach to Cancer Chemoprevention, *Trends Food Sci. Technol. 9,* 192–196.
32. U.K. Department of Health (1998) Report on Health and Social Subjects, # 48, Nutritional Aspects of the Development of Cancer. Report of the Working Group on Diet and Cancer of the Committee on Medical Aspects of Food and Nutrition Policy, HMSO, London.
33. Ackman, R.G. (1996) Fatty Acids in Newer Fats and Oils, in Edible Oils and Fat Products: General Applications, in *Bailey's Industrial Oil and Fat Products,* vol. 1, 5th edn., (Hui, Y.H., ed.) pp. 427–439, John Wiley and Sons, New York.

34. Perkins, E.G., and Erickson, M.D. (1996) *Deep Frying: Chemistry, Nutrition, and Practical Applications,* AOCS Press, Champaign, IL.
35. Martini, D., and Iacazio, G. (1995) Lipoxygenases and Associated Metabolic Paths, *OCL 2,* 374–385. (In French)
36. Brekke, O.L. (1980) Bleaching, in *Handbook of Soy Oil Processing and Utilization,* (Erickson, D.R., Pryde, E.H., Brekke, O.L., Mounts, T.L. and Falb, R.A., eds.) pp. 105–130, American Soybean Association and American Oil Chemists' Society, St. Louis and Champaign.
37. Koshio, S., Ackman, R.G., and Lall, S.P. (1994) Effects of Oxidized Herring and Canola Oils in Diets on Growth, Survival, and Flavor of Atlantic Salmon, *Salmo salar, J. Agric. Food Chem. 42,* 1164–1169.
38. Sébédio, J.-L., Grandgirard, A., Septier, Ch., and Prevost, J. (1987) Etat d'Altération de quelques Huiles de Friture Prélevées en Restauration, *Rev. Franç. Corps Gras 34,* 15–18.
39. Saeed, T., Abu-Dagga, F., and Rahma, H.A. (1986) Detection of Pork and Lard as Adulterants in Beef and Mutton Mixtures, *J. Assoc. Off. Anal. Chem. 69,* 999–1002.
40. Chin, S.F., Liu, W., Storkson, J.M., Ha, Y.L., and Pariza, M.W. (1992) Dietary Sources of Conjugatated Dienoic Isomers of Linoleic Acid, a Newly Recognized Class of Anticarcinogens, *J. Food Compos. Anal. 5,* 185–197.
41. Choi, B.-D., Kang, S.-J., Ha Y.-L., and Ackman, R.G. (1999) Accumulation of Conjugated Linoleic Acid (CLA) in Tissues of Fish Fed Diets Containing Various Levels of CLA, in *Quality Attributes of Muscle Foods,* (Xiong, Y.L., Ho, C.T., and Shahidi, F., eds.) Plenum Publishing, in press.
42. Bligh, E.G., and Dyer, W.J. (1959) A Rapid Method of Total Lipid Extraction and Purification, *Can. J. Biochem. Physiol. 37,* 911–917.
43. Juaneda, P., and Rocquelin, G. (1985) Rapid and Convenient Separation of Phospholipids and Non-Phosphorus Lipids from Rat Heart Using Silica Cartridges, *Lipids 20,* 40–45.
44. Ha, Y.L., Grimm, N.K., and Pariza, M.W. (1989) Newly Recognized Anticarcinogenic Fatty Acids: Identification and Quantification in Natural and Processed Cheeses, *J. Agric. Food Chem. 37,* 75–81.
45. Ha, Y.L., Strokson, J., and Pariza, M.W. (1990) Inhibition of Benzo(α)pyrene-Induced Mouse Forestomach Neoplasia by Conjugated Dienoic Derivatives of Linoleic Acid, *Cancer Res. 50,* 1097–1101.
46. Shantha, N.C., Decker, E.A., and Hennig, B. (1993) Comparison of Methylation Methods for the Quantitation of Conjugated Linoleic Acid Isomers, *J. Assoc. Off. Anal. Chem. Int. 76:* 644–649.
47. Ackman, R.G. (1998) Remarks on Official Methods Employing Boron Trifluoride in the Preparation of Methyl Esters of the Fatty Acids of Fish Oils, *J. Am. Oil Chem. Soc. 75,* 541–545.
48. Castell, J.D., Sinnhuber, R.O., Wales, J.H., and Lee, D.J. (1972) Essential Fatty Acids in the Diet of Rainbow Trout *(Salmo gairdneri):* Growth, Feed Conversion and Some Gross Deficiency Symptoms, *J. Nutr. 102,* 77–85.
49. Wilson, R.P. (1995) Lipid Nutrition of Finfish, in *Nutrition and Utilization Technology in Aquaculture,* (Lim, C.E., and Sessa, D.J., eds.) pp. 74–81, AOCS Press, Champaign, IL.
50. Füllner, G., and Wirth, M. (1996) The Influence of Nutrition on Fat Content and Fatty Acid Composition of European Catfish *(Silurus glanis), Fett/Lipid 98,* 300–304. (in German)
51. Ballestrazzi R., and Mion, A. (1993) Lipids and Teleostean Fish Feeding, *Riv. Ital. Acquacoltura 28,* 155–173. (In Italian)

52. Sargent, J., Henderson, R.J., and Tocher, D.R. (1989) The Lipids, in *Fish Nutrition*, (Halver, J.E., ed.) pp. 153–218, Academic Press, New York.
53. Dosanjh, B.S., Higgs, D.A., McKenzie, D.J., Randall, D.J., Eales, J.G., Rowshandeli, N., Rowshandeli, M., and Deacon, G. (1998) Influence of Dietary Blends of Menhaden Oil and Canola Oil on Growth, Muscle Lipid Composition, and Thyroidal Status of Atlantic Salmon *(Salmo salar)* in Sea Water, *Fish Physiol. Biochem. 19,* 123–134.
54. McCurdy, S.M., and March, B.E. (1992) Processing of Canola Meal for Incorporation in Trout and Salmon Diets, *J. Am. Oil Chem. Soc. 69,* 213–220.
55. Olvera-Novoa, M.A., Pereira-Pacheco, F., Olivera-Castillo, L., Pérez-Flores, V., Navarro, L., and Sámano, J.C. (1997) Cowpea *(Vigna unguiculata)* Protein Concentrate as Replacement for Fish Meal in Diets for Tilapia *(Oreochromis niloticus)* Fry, *Aquaculture 158,* 107–116.
56. Haumann, B.F. (1997) Mechanical Extraction: Capitalizing on Solvent-Free Processing, *INFORM 8,* 165–174.
57. Christie, W.W., Dobson, G., and Gunstone, F.D. (1997) Isomers in Commercial Samples of Conjugated Linoleic Acid, *J. Am. Oil Chem. Soc. 74,* 1231.
58. Ackman, R.G. (1998) Laboratory Preparation of Conjugated Linoleic Acids, *J. Am. Oil Chem. Soc. 75,* 1227.
59. Ackman, R.G., and Kean-Howie, J. (1995) Fatty Acids in Aquaculture: Are ω-3 Fatty Acids Always Important? in *Nutrition and Utilization Technology in Aquaculture*, (Lim, C.E. and Sessa, D.J., eds.) pp. 82–104, AOCS Press, Champaign, IL.

Chapter 22
Conjugated Linoleic Acids in Human Milk

Mark A. McGuire[a], Michelle K. McGuire[b], Peter W. Parodi[c], and Robert G. Jensen[d]

[a]Department of Animal and Veterinary Science, University of Idaho, Moscow, ID 83844–2330
[b]Department of Food Science and Human Nutrition, Washington State University, Pullman, WA 99164–6376
[c]Dairy Research and Development Corporation, Human Nutrition Program, Chernside, QLD, 4032 Australia
[d]Nutritional Sciences, University of Connecticut, Storrs, CT 06268–2637

Introduction

The purpose of this chapter is to describe the metabolic pathways by which conjugated linoleic acids (CLA) and the presumed biologically active isoform, rumenic acid or $c9,t11$-18:2 (RA), are conveyed to the mammary gland, incorporated into milk lipids therein, and potentially influence maternal and infant health. Thus, the origins and distribution of lipids in milk will first be described. Second, information concerning the amounts of CLA in human milk and formulas will be presented. Finally, mechanisms by which CLA in milk might potentially prevent disease in both mother and child will be discussed.

Origins and Distribution of the Lipids in Human Milk

Milk Fat Distribution

Milk is a very complex fluid, containing carbohydrates and salts in true solution, caseins in colloidal dispersion, cells and cellular debris, and lipids mainly in the form of emulsified globules. Lipid represents from 3 to 5% of milk and is composed mainly of triacylglycerols (TG, 98%), phospholipids (0.8%), and cholesterol (0.5%). Lipid occurs as globules emulsified in the aqueous phase (87%) of milk. Nonpolar lipids (e.g., TG, cholesterol esters and retinyl esters) are found in the core of these globules (1), which are covered with bipolar materials (e.g., phospholipids, proteins, mucopolysaccharides, cholesterol, and enzymes), organized into a loose layer called the milk lipid globule membrane (MLGM). The MLGM acts as an emulsion stabilizer. Milk fat globules range in size from 1 to 10 µm, with most of the globules smaller than 1 µm. However, milk fat globules of 4 µm account for the majority of the globules on a weight basis. Milk fat globules formed from microdroplets of lipid in the secreting cell can fuse and be extruded from the cell; these are surrounded by the MLGM.

Origins and Positional Placement of Milk Fatty Acids

In women consuming typical Western diets, fatty acids of 10, 12, and 14 carbons with no double bonds (10:0, 12:0 and 14:0; *de novo* fatty acids) are synthesized in the mammary gland and account for ~10% of milk fatty acids (2), whereas longer-chain fatty acids (e.g., 16:0, 18:0, 18:1, and 18:2; preformed fatty acids) total ~70% of the rest. Noteworthy is the fact that linoleic acid (18:2 n-6) and all *trans* isomers possibly including CLA originate from the maternal diet. Approximately 29% of milk fatty acids are transported to the mammary gland *via* chylomicrons and very low density lipoproteins (VLDL) that originate from dietary sources (3). The remaining 61% originate from mobilized adipose tissue or synthesis in other tissues. Oleic acid, a major dietary fatty acid, can also be synthesized by hepatic desaturation of 18:0 (4). It is possible that the desaturase that forms oleic acid from stearic acid could form RA from *trans*-11-18:1 (vaccenic acid).

Dietary fatty acids are transported to the mammary gland as TG by chylomicrons and VLDL where they are hydrolyzed by lipoprotein lipase and enter the secretory cells. The fatty acids mobilized in adipose tissue are transported to the mammary gland by serum albumin. Within mammary cells, fatty acids are activated to CoA esters which are, in turn, converted to TG and phospholipids by reactions with glycerol-3-phosphate (5). Interestingly, the positional placement of fatty acids into these lipids is species and organ specific. For example, 70% of the 16:0 in human milk TG is located at the *sn*-2 position (1). Currently, information about the positional distribution of RA in the triglyceride molecule is unknown. It is reasonable to assume that, even though RA has a unique structure, it would be located in a position in the triglyceride molecule similar to that of linoleic acid, which is found predominately at positions *sn*-2 and -3 (1). However, some fatty acids affect synthesis of other fatty acids (5) and may thus affect the typical synthesis of milk fat. CLA may be an example of such a fatty acid with unique effects on metabolism because of its structure. Several regulatory roles of CLA are described later in this chapter.

CLA in Human Milk and Infant Formula

Concentration of CLA

The concentrations of specific CLA isomers and total CLA in human milk have been reported by several investigators (6–9); these data are summarized in Table 22.1. It is noteworthy that Jensen *et al.* (8) confirmed the presence of RA in human milk using gas–liquid chromatography/mass spectrometry methodology. Further, Yurawecz *et al.* (10) found that milk produced by German women (n = 5) contained not only RA (80–84% of total CLA), but also the *t*7,*c*9-18:2 isoform (6–10% of total CLA). In summary, all samples of human milk analyzed to date have been found to contain RA in amounts surprisingly similar to those found in bovine milk.

TABLE 22.1 Concentrations of Total Conjugated Linoleic Acid (CLA) and Rumenic Acid (RA) in Human Milk[a]

Origin and description of milk	RA	Total CLA	Reference
	(mg/g lipid)		
Australia (n = 18; conventional diet)	5.8	NR[b]	Fogerty et al. (6)
Australia (n = 8; Hare Krishna diet)	11.2	NR	Fogerty et al. (6)
U.S. (n = 14)	3.6 ± 0.9	3.8 ± 0.9	McGuire et al. (7)
U.S. (n = 16; low dairy diet)	2.3 ± 0.1	NR	Park et al. (9)
U.S. (n = 16; high dairy diet)	3.8[c] ± 0.3	NR	Park et al. (9)
U.S. (n = 20)	2.1 ± 0.2	2.1 ± 0.2	Jensen et al. (8)

[a]Values are means ± SEM.
[b]NR, not reported.
[c]Value is greater ($P < 0.05$) than that reported for the same women when consuming a low dairy foods diet.

In contrast, data presented in the two publications (7,11) containing reports of the total CLA and RA contents of proprietary infant formulas suggest that formulas contain negligible amounts of any CLA isomer. The only exception to this appears to be formula (1.4 mg RA/g lipid) that is produced using ruminant fat as a major lipid source. Interestingly, Chin and colleagues (11) found that, although no RA was detected in the samples that they analyzed, there was a measurable amount of other CLA isomers (0.7 and 0.3 mg/g lipid in soy and bovine milk-based formulas, respectively). Further, it is evident that neither lot number nor preparation type (powdered vs. ready-to-feed) influenced the CLA content of the formula (7), suggesting that processing does not influence this parameter.

Maternal Diet and CLA Content of Human Milk

Fogerty et al. (6) reported values for human milk RA concentration of 3.1–8.5 mg/g among mothers eating conventional diets and 9.7–12.5 mg/g lipid among Hare Krishna mothers. These data suggest that diet may influence human milk RA concentration (and thus, infant RA intake), because members of the Hare Krishna faith consume large amounts of butter or ghee, as well as cheese (good sources of dietary CLA). However, maternal intake of RA was not reported. In response to this, a randomized, crossover study (9) was conducted in which women were asked to consume diets very low or very high in RA (low and high dairy intakes, respectively). Chronic maternal intake of RA was estimated by semiquantitative food-frequency questionnaire to be 227 ± 180 mg/d. Rumenic acid intakes estimated by 3-d written records were 34 ± 8 and 291 ± 75 mg/d during the low and high RA diet periods, respectively. Data suggest that the RA content of human milk was influenced strongly by maternal diet, such that women consuming diets high in dairy fat produced milk with greater concentrations of RA (see Table 22.1). This supports data collected previously in lactating rats showing that CLA intake can influence CLA content of milk (12).

However, multiple regression analyses suggested that both body mass index (a reflection of body fat content) and long-term intake of RA were more influential than current intake of RA in describing the variation in milk RA content. This is reflected by the fact that current dietary intake of RA increased eightfold with less than a twofold increase in milk RA and suggests that the majority of RA in milk is from mobilized adipose stores. As indicated earlier, ~30% of the fatty acids in milk is derived from the diet, whereas twice that amount comes from body stores. Thus, any substantial changes in the RA content of human milk would have to be achieved by more long-term changes in dietary RA intake.

Furthermore, we have recently conducted a human intervention trial to document the effect of maternal consumption of a commercially available CLA supplement on milk CLA content. In this experiment, women (n = 10) were enrolled in a double-blind, randomized, placebo-controlled trial and consumed CLA supplements for 5 d or a placebo for 5 d (separated by a 7-d washout period). Conjugated linoleic acid supplements contained ~50% RA and 50% $t10,c12$-CLA, whereas placebos contained olive oil. Data again suggest that maternal consumption of this CLA supplement resulted in significantly ($P < 0.01$) elevated milk CLA and RA concentrations (data not shown).

Summary

In conclusion, there exists ample evidence that human milk contains CLA, the majority of which is RA. As would be expected, either dietary or supplemental intake of CLA isoforms can influence their concentrations in human milk. Thus, breast-fed human infants consume significant amounts of RA and potentially other CLA isoforms throughout infancy. In contrast to this, infant formulas contain undetectable to negligible amounts of any CLA isoform. Thus, formula-fed infants consume no CLA before introduction of bovine milk or CLA-containing weaning foods.

CLA, Lactation and Breast Cancer

As described elsewhere in this publication, dietary CLA can produce a wide range of desirable physiologic responses. Of particular concern to us and others interested in the public health potential of these physiologic responses is whether the presence of CLA in human milk can benefit the mother and/or child. Specifically, the influence of maternal and infant dietary CLA on breast cancer, nutrient partitioning, immune response, and diabetes will be addressed here.

One outstanding aspect of CLA biology is its ability to suppress rat mammary tumorigenesis when incorporated into the diet at a level of 1% or even less (reviewed by Parodi in Refs. 13,14). Of particular interest is the observation that rats fed diets supplemented with 1% CLA, only during the period from weaning at 21 d of age until ~50 d when carcinogen was administered, were afforded protection from mammary cancer for life. On the other hand, when CLA feeding commenced after carcinogen administration, continuous supplementation was required to achieve equivalent protection from cancer (15–17).

Mammary Development and Breast Cancer Risk

In rats, the period from weaning until ~50 d of age coincides with the development of the mammary gland to adult stage morphology. At birth, the mammary glands are rudimentary and consist of one or two main lactiferous ducts arising from the nipple. With increasing age, the ducts develop profuse multiple branches that end in club-shaped terminal end buds (TEB). The number of TEB is highest at ~21 d of age, and each TEB is composed of 3–6 layers of actively proliferating epithelial cells. After 21 d of age, TEB progressively cleave into 3–5 smaller differentiated alveolar buds (AB) and then into lobules that represent a further stage of development and differentiation. Mammary gland development is uneven, and a number of TEB never differentiate, become atrophic, and form finger-shaped terminal ducts (TD). By 55 d of age, after progressive branching, the ducts reach the limits of the mesenchyme or mammary fat pads (18).

Chemically induced mammary tumors occur primarily in the undifferentiated, rapidly proliferating epithelium at the distal end of TEB and TD. Consequently, tumor development is greatest when TEB are actively differentiating to AB at ~40–46 d of age. Pregnancy and lactation protect the gland from chemically induced carcinogenesis due, at least in part, to complete and permanent differentiation of all TEB to lobules that have a secretory function during lactation. Differentiation of cells is associated not only with altered gland structure, but also with lengthening of the cell cycle (mainly the G_1 phase), decreased binding of carcinogen to DNA, and increased ability of the cells to remove carcinogen adducts from DNA (18).

There is now a growing body of literature that supports the thesis that events in early life, even during the prenatal period, influence the risk of breast cancer later in life (19,20). Confirmation that events at a young age are important for subsequent risk of breast cancer comes from studies of women exposed to high doses of ionizing radiation for medical conditions. For example, multiple chest fluoroscopies, irradiation of the breast for postpartum mastitis, and irradiation for enlarged thymus glands are associated with increased risk of breast cancer. This risk increases with decreasing age at exposure (21). Long-term follow-up studies of the incidence of breast cancer among atomic bomb survivors from Hiroshima and Nagasaki indicate an increased risk for women exposed before the age of 20 y. Infants and children <4 y old at the time of exposure had the highest (ninefold) risk of developing breast cancer (22).

Anbazhagan *et al.* (23) observed that mammary specimens from newborn babies and infants showed striking variations in the structural development of the breast, such that some had a well-developed ductal system with numerous branching ducts and terminal lobules, whereas others had a less developed ductal system without lobular development. Morphological variation in the newborn breast may result from differences in *in utero* hormone concentrations. Estrogen-induced cell proliferation can result in a greater number of epithelial cells becoming targets for carcinogens.

Breast-Feeding and Breast Cancer

Breast cancer incidence is low in countries in which breast-feeding is common. In developed countries, breast cancer incidence is increasing at a time when breast-feeding is declining (24). Thus, we wish to address the following questions: (i) Does breast-feeding protect the mother or infant from breast cancer? (ii) Can dietary CLA intake during lactation influence this relationship?

Several, but not all case-control studies have found that long duration of breast-feeding affords protection from breast cancer, especially among premenopausal women (25). Several mechanisms have been proposed to explain the beneficial effects of lactation. These include the following: (i) reduction in the number of lifetime ovulatory cycles, (ii) elimination of carcinogens from the breast to the milk, (iii) decrease in estrogen production or other hormones associated with cell growth, and (iv) alteration of breast morphology by induction of complete differentiation of ductal structures producing advanced stage lobules, thus making them less susceptible to carcinogens (25). We hypothesize that increased exposure to anticarcinogenic lipids, like CLA, may also help explain the protective effect of lactation on breast cancer.

The importance of a differentiated ductal structure is supported by the effect of parity. It is known that increased parity confers protection from breast cancer. Nevertheless, parous women still develop breast cancer. Russo *et al.* (26) studied the morphology of normal and cancerous breasts as influenced by parity and found that breasts of parous women with cancer were less developed and similar to the breasts of nulliparous women. Moreover, women in these two categories did not achieve the same degree of breast differentiation compared with parous women without malignancies.

Infant Feeding Mode, CLA and Breast Cancer

A limited number of case-control studies have examined the relationship between exposure to human milk during infancy and development of breast cancer later in life; results suggest that breast-feeding affords a modest reduction in breast cancer risk. Again, the benefit appears to be greatest for premenopausal women (24). Interestingly, intake of RA was suggested as a possible reason for the inverse association between cow's milk consumption and breast cancer risk in a recent Finnish prospective study (27). It is interesting to note that a case-control study in British Columbia showed that frequent childhood consumption of whole milk was more important than recent consumption for reducing the risk of breast cancer (28). As mentioned previously, it is clear that human milk contains RA, whereas most infant formulas do not. The question remains whether RA intake during infancy can influence breast cancer risk later in life.

Furthermore, one should consider the studies of Hilakivi-Clarke and colleagues (29,30) who showed that diets high in linoleic acid, fed during pregnancy only, increased the risk of developing carcinogen-induced mammary tumors in the mothers (29) and their female offspring (30). The high linoleic acid diet increased serum estradiol levels in the mother. In the female offspring, there was increased mammary fat pad area, numbers of TEB, and epithelial cell density. These increases probably resulted from the proliferative stimulus of increased hormone levels in the mother. Currently, we do not know whether exposure to CLA during fetal or neonatal life might have the opposite effect.

High dietary fat intake enhances mammary tumorigenesis in rats, and animal and vegetable fats are known to have different tumorigenic potential. Consumption of these fats either before or after tumor initiation can also influence tumor outcome (31). However, Ip et al. (32) found that the ability of 1% CLA to inhibit mammary tumor development was independent of the amount or type of fat in the diet. Although the mechanisms for the anticarcinogenic action of CLA are still being investigated, Thompson et al. (33) showed that feeding rats 1% CLA, from weaning for 1 mo, resulted in a reduction in the density of the mammary epithelium. This was accompanied by reduced proliferation in the TEB and lobulo-aveolar buds as indicated by reduced DNA synthesis. Again, the effect of high CLA intake before weaning has not been reported.

Direction of Future Research

Clearly, it will be difficult to establish the association between CLA intake during the nursing period and subsequent breast cancer risk by prospective epidemiologic studies because of the varying levels of CLA in dietary ruminant products and an uncertain knowledge of CLA in infant formula. Nevertheless, animal studies in which rats consume varied levels of CLA during pregnancy and lactation can and should be conducted. The different treatments can be evaluated for prevention of chemically induced mammary tumor incidence in the offspring. It will be more difficult to determine the influence of CLA intake during lactation on subsequent mammary tumor development in dams because pregnancy and lactation induce a protective effect from chemically induced carcinogenesis in rats (18).

Summary

Epidemiologic evidence suggests that some women are protected from breast cancer as a consequence of lactation. Moreover, some women are protected because they were breast-fed as infants. Is the RA content of human milk, at least in part, responsible for this protection? This proposition is biologically plausible because the RA content of human milk can approach the level found by Ip et al. (15–17,32) to exert an antitumorigenic response in rats.

CLA and Physiologic Programming in the Infant

Support for the concept that CLA intake during the nursing period may influence the risk of cancer and other diseases in later life comes from Lucas (34), a pioneer in the field of fetal and early childhood origins of disease. He proposed the idea of "biological programming," a process whereby a stimulus or insult (including nutrition) during a critical period of early development can have an immediate or long-term consequence on biological structure or physiologic functions. Indeed, studies with small mammals and primates show that nutrition during early life may have potentially important long-term effect on factors such as blood lipids, plasma insulin, obesity, atherosclerosis, behavior, and learning. In addition, the diet consumed by preterm infants during the first months of life can have a major effect on subsequent developmental attainment, growth, and allergic status in early childhood (34).

Moreover, a large number of epidemiologic studies suggest that interventions designed to alter the rate of growth of the fetus before birth or to alter body proportions at birth may change the risk for developing several common chronic adult diseases, such as coronary heart disease, stroke, hypertension, obesity, glucose intolerance, and adult-onset diabetes mellitus (35,36). We would like to explore here the idea that CLA intake during early life might impart additional long-lasting effects beyond those previously discussed for breast cancer. Included in these effects are those influencing the immune system, nutrient partitioning, glycemic control, and growth modulation.

CLA and Immune Modulation

Recent data suggest that CLA can regulate certain aspects of immune function, probably by modulation of prostaglandin and cytokine production (37,38). Rats fed 0.5 or 1.0% CLA had increased production of the antiallergenic immunoglobulins IgA and IgG, whereas IgE, which is associated with allergic reactions, was decreased (37). This suggests that human milk CLA may help prevent food allergies and induce oral tolerance. Interestingly, infants who are breast-fed for >6 mo have a lower risk of developing childhood lymphoma, a cancer whose risk is higher in children with immune deficiencies (39).

CLA and Nutrient Partitioning

Conjugated linoleic acid was first shown to be a growth factor in an interesting study by Chin *et al.* (12). Rats fed 0.5% CLA during gestation and lactation or lactation only, had improved postnatal body weight gain in their pups. Further, pups from dams fed CLA were leaner. Supplementation with CLA has also been shown to reduce body fat and enhance lean body mass in mice (40). More recent data suggest that CLA affects key enzymes and processes involved in lipid metabolism and storage. Whether CLA intake during human infancy influences growth and nutrient partitioning is unknown; however, the rat data from Chin *et al.* (12) suggest this to be the case. Because human milk but not formula contains CLA, the hypothesis that CLA affects growth and nutrient partitioning in human infants deserves further study.

CLA and Diabetes

Very recent evidence presented by Houseknecht *et al.* (41) suggests that CLA has antidiabetogenic properties; CLA consumption normalized impaired glucose tolerance and improved hyperinsulinemia in rats. Because several reports suggest that breast-fed infants have a lower risk of developing diabetes in later life (42,43), the relationship between CLA intake during the suckling period and later risk of developing this disease is of great public health interest.

CLA and Bone Growth

Work by Watkins *et al.* (44) indicates that milk fat stimulates bone formation rate in growing chicks. The effect was associated with lowering prostaglandin E_2 production.

This is possibly due to the presence of CLA in milk fat. Thus, one might hypothesize that CLA intake during periods of rapid bone development (e.g., infancy and adolescence) might influence long-term bone health.

Summary

Taken together, these physiologic effects suggest that CLA in human milk may program the suckling infant to resist obesity and obesity-related diseases such as diabetes in later life. While this is speculative, the hypothesis can and should be tested experimentally in animal models and statistically in large human data sets.

Overall Conclusions

Rumenic acid ($c9, t11$-18:2) found in foods from ruminants has many potential beneficial effects. Rumenic acid is found in human milk in amounts that depend upon the quantity and length of time that dairy products, beef, and related foods have been consumed. Currently, it is not known whether the benefits discussed throughout this book will be transferred to the breast-fed infant *via* milk, but RA and other CLA may help decrease the risks of several chronic diseases including breast cancer, diabetes, and obesity. Clearly, much additional research is required in this area to test this hypothesis.

References

1. Jensen, R.G. (1996) The Lipids in Human Milk, *Prog. Lipid. Res. 35,* 53–92.
2. Thompson, B.J., and Smith, S. (1985) Biosynthesis of Fatty Acids by Lactating Breast Epithelial Cells: An Evaluation of the Overall Contribution to the Overall Composition of Human Milk Fat, *Pediatr. Res. 19,* 139–143.
3. Hachey, D.L., Thomas, M.R., Emken, E.A., Garza, C., Brown-Booth, L., Adlof, R.O., and Klein, P.D. (1987) Human Lactation: Maternal Transfer of Dietary Triglycerides Labeled with Stable Isotopes, *J. Lipid Res. 28,* 1185–1192.
4. Kinsella, J.E. (1972) Stearyl CoA as a Precurser of Oleic Acid and Glycerolipids in Mammary Microsomes from a Lactating Bovine: Possible Regulatory Step in Milk Triglyceride Synthesis, *Lipids 7,* 349–355.
5. Barber, M.C., Clegg, R.A., Travers, M.T., and Vernon, R.G. (1997) Lipid Metabolism in the Lactating Mammary Gland, *Biochim. Biophys. Acta 1347,* 102–126.
6. Fogerty, A.C., Ford, G.L., and Svoronos, D. (1988) Octadeca-9,11-dienoic Acid in Foodstuffs and in the Lipids of Human Blood and Breast Milk, *Nutr. Rep. Int. 38,* 937–944.
7. McGuire, M.K., Park, Y, Behre, R.A., Harrison, L.Y., Shultz, T.D., and McGuire, M.A. (1997) Conjugated Linoleic Acid Concentrations of Human Milk and Infant Formula, *Nutr. Res. 17,* 1277–1283.
8. Jensen, R.G., Lammi-Keefe, C.J., Hill, D.W., Kind, A.J., and Henderson, R. (1998) The Anticarcinogenic Conjugated Fatty Acid, 9c,11t-18:2, in Human Milk: Confirmation of Its Presence, *J. Hum. Lact. 14,* 23–27.
9. Park, Y., McGuire, M.K., Behre, R., McGuire, M.A., Evans, M.A., and Shultz, T.D. (1999) High Fat Dairy Product Consumption Increases $\Delta 9c,11t$-18:2 (Rumenic Acid) and Total Lipid Concentrations of Human Milk, *Lipids.*

10. Yurawecz, M.P., Roach, J.A.G., Sehat, N, Mossoba, M.M., Kramer, J.K.G., Fritsche, J., Steinhart, H., and Ku, Y. (1998) A New Conjugated Linoleic Acid Isomer, 7 *trans*,9 *cis*-Octadecadienoic Acid, in Cow Milk, Cheese, Beef and Human Milk and Adipose Tissue, *Lipids 33*, 803–809.
11. Chin, S.F., Liu, W., Storkson, J.M., Ha, Y.L., and Pariza, M.W. (1992) Dietary Sources of Conjugated Dienoic Isomers of Linoleic Acid, a Newly Recognized Class of Anticarcinogens, *J. Food Comp. Anal. 5*, 185–197.
12. Chin, S.F., Storkson, J.M., Albright, K.J., Cook, M.E., and Pariza, M.W. (1994) Conjugated Linoleic Acid Is a Growth Factor for Rats as Shown by Enhanced Weight Gain and Improved Feed Efficiency, *J. Nutr. 124*, 2344–2349.
13. Parodi, P.W., (1997) Cows' Milk Fat Components as Potential Anticarcinogenic Agents, *J. Nutr. 127*, 1055–1060.
14. Parodi, P.W., (1997) Milk Fat Conjugated Linoleic Acid: Can it Help Prevent Breast Cancer? *Proc. Nutr. Soc. NZ 22*, 137–149.
15. Ip, C., Singh, M., Thompson, H.J., and Scimeca, J.A. (1994) Conjugated Linoleic Acid Suppresses Mammary Carcinogenesis and Proliferative Activity of the Mammary Gland in the Rat, *Cancer Res. 54*, 1212–1215.
16. Ip, C., Scimeca, J.A., and Thompson, H. (1995) Effect of Timing and Duration of Dietary Conjugated Linoleic Acid on Mammary Cancer Prevention, *Nutr. Cancer 24*, 241–247.
17. Ip, C., Jiang, C., Thompson, H.J., and Scimeca, J.A. (1997) Retention of Conjugated Linoleic Acid in the Mammary Gland Is Associated with Tumor Inhibition During the Post-Initiation Phase of Carcinogenesis, *Carcinogenesis 18*, 755–759.
18. Russo, J., and Russo, I.H. (1987) Biological and Molecular Basis of Mammary Carcinogenesis, *Lab. Investig. 57*, 112–137.
19. de Waard, F., and Trichopoulos, D. (1988) A Unifying Concept of the Aetiology of Breast Cancer, *Int. J. Cancer 41*, 666–669.
20. Anbazhagan, R., and Gusterson, B.A. (1994) Prenatal Factors May Influence Predisposition to Breast Cancer, *Eur. J. Cancer 30A*, 1–3.
21. Land, C.E., Boice, J.D., Shore, R.E., Norman, J.E., and Tokunaga, M. (1980) Breast Cancer Risk from Low-Dose Exposures to Ionizing Radiation: Results of Parallel Analysis of Three Exposed Populations of Women, *J. Natl. Cancer Inst. 65*, 353–376.
22. Tokunaga, M., Land, C.E., Yamamoto, T., Asano, M., Tokuoka, S., Ezaki, H., and Nishimori, I. (1987) Incidence of Female Breast Cancer Among Atomic Bomb Survivors, Hiroshima and Nagasaki, 1950–1980, *Radiation Res. 112*, 243–272.
23. Anbazhagan, R., Bartek, J., Monoghan, P., and Gusterson, B.A. (1991) Growth and Development of the Human Infant Breast, *Am. J. Anat. 192*, 407–417.
24. Titus-Ernstoff, L., Egan, K.M., Newcomb, P.A., Baron, J.A., Stampfer, M., Greenberg, E.R., Cole, B.F., Ding, J., Willett, W., and Trichopoulos, D. (1998) Exposure to Breast Milk in Infancy and Adult Breast Cancer Risk, *J. Natl. Cancer Inst. 90*, 921–924.
25. Enger, S.M., Ross, R.K., Henderson, B., and Bernstein, L. (1997) Breastfeeding History, Pregnancy Experience and Risk of Breast Cancer, *Br. J. Cancer 76*, 118–123.
26. Russo, J., Romero, A.L., and Russo, I.H. (1994) Architectural Pattern of the Normal and Cancerous Breast Under the Influence of Parity, *Cancer Epidemiol. Biomark. Prev. 3*, 219–224.
27. Knekt, P., Järvinen, R., Seppänen, R., Pukkala, E., and Aromaa, A. (1996) Intake of Dairy Products and the Risk of Breast Cancer, *Br. J. Cancer 73*, 687–691.

28. Hislop, T.G., Coldman, A.J., Elwood, J.M., Brauer, G., and Kan, L. (1986) Childhood and Recent Eating Patterns and Risk of Breast Cancer, *Cancer Detect. Prev. 9,* 47–58.
29. Hilakivi-Clarke, L., Onojafe, I., Raygada, M., Cho, E., Clarke, R., and Lippman, M.E. (1996) Breast Cancer Risk in Rats Fed a Diet High in n-6 Polyunsaturated Fatty Acids During Pregnancy, *J. Natl. Cancer Inst. 88,* 1821–1827.
30. Hilakivi-Clarke, L., Clarke, R., Onojafe, I., Raygada, M., Cho, E., and Lippman, M. (1997) A Maternal Diet High in n-6 Polyunsaturated Fats Alters Mammary Gland Development, Puberty Onset, and Breast Cancer Risk Among Female Rat Offspring, *Proc. Natl. Acad. Sci. USA 94,* 9372–9377.
31. Sylvester, P.W., Russell, M., Ip, M.M., and Ip, C. (1986) Comparative Effects of Different Animal and Vegetable Fats Fed Before and During Carcinogen Administration on Mammary Tumorigenesis, Sexual Maturation, and Endocrine Function in Rats, *Cancer Res. 46,* 757–762.
32. Ip, C., Briggs, S.P., Haegele, A.D., Thompson, H.J., Storkson, J., and Scimeca, J.A. (1996) The Efficacy of Conjugated Linoleic Acid in Mammary Cancer Prevention Is Independent of the Level or Type of Fat in the Diet, *Carcinogenesis 17,* 1045–1050.
33. Thompson, H., Zhu, Z., Banni, S., Darcy, K., Loftus, T., and Ip, C. (1997) Morphological and Biochemical Status of the Mammary Gland as Influenced by Conjugated Linoleic Acid : Implication for a Reduction in Mammary Cancer Risk, *Cancer Res. 57,* 5067–5072.
34. Lucas, A., (1991) in *The Childhood Environment and Adult Disease (Ciba Foundation Symposium 156)* pp. 38–55, Wiley, Chichester.
35. Barker, D.J.P. (1996) Growth *in utero* and Coronary Heart Disease, *Nutr. Rev. 54,* S1–S7.
36. Whincup, P.H. (1998) Fetal Origins of Cardiovascular Risk: Evidence from Studies in Children, *Proc. Nutr. Soc. 57,* 123–127.
37. Sugano, M., Tsujita, A., Yamasaki, M., Noguchi, M., and Yamada, K. (1998) Conjugated Linoleic Acid Modulates Tissue levels of Chemical Mediators and Immunoglobulins in Rats, *Lipids 33,* 521–527.
38. Turek, J.J., Li, Y., Schoenlein, I.A., Allen, K.G.D., and Watkins, B.A. (1998) Modulation of Macrophage Cytokine Production by Conjugated Linoleic Acids Is Influenced by the Dietary n-6:n-3 Fatty Acid Ratio, *J. Nutr. Biochem. 9,* 258–266.
39. Davis, M.K., Savitz, D.A., and Graubard, B.I. (1988) Infant Feeding and Childhood Cancer, *Lancet ii,* 365–368.
40. Park, Y., Albright, K.J., Liu, W., Storkson, J.M., Cook, M.E., and Pariza, M.W. (1997) Effect of Conjugated Linoleic Acid on Body Composition in Mice, *Lipids 32,* 853–858.
41. Houseknecht, K.L., Vanden Heuvel, J.P., Moya-Camarena, S.Y., Portocarrero, C.P., Peck, L.W., Nickel, K.P., and Belury, M.A. (1998) Dietary Conjugated Linoleic Acid Normalizes Impaired Glucose Tolerance in the Zucker Diabetic Fatty *fa/fa* Rat, *Biochem. Biophys. Res. Commun. 244,* 678–682.
42. Gerstein, H.C. (1994) Cow's Milk Exposure and Type 1 Diabetes Mellitus, *Diabetes Care 17,* 13–19.
43. Norris, J.M., Beaty, B., Klingensmith, G., Yu, L., Hoffman, M., Chase, H.P., Erlich, H.A., Hamman, R.F., Eisenbarth, G.S., and Rewers, M. (1996) Lack of Association Between Early Exposure to Cow's Milk Protein and β-Cell Autoimmunity, *J. Am. Med. Assoc. 276,* 609–614.
44. Watkins, B.A., Shen, C.-L., McMurty, J.P., Xu, H., Bain, S.D., Allen, K.G.D., and Seifert, M.F. (1997) Dietary Lipids Modulate Bone Prostaglandin E_2 Production, Insulin-Like Growth Factor-1 Concentration and Formation Rate in Chicks, *J. Nutr. 127,* 1084–1091.

Chapter 23

Influence of Dietary Conjugated Linoleic Acid on Lipid Metabolism in Relation to Its Anticarcinogenic Activity

Sebastiano Banni[a], Elisabetta Angioni[a], Gianfranca Carta[a], Viviana Casu[a], Monica Deiana[a], Maria Assunta Dessì[a], Leonardo Lucchi[b], Maria Paola Melis[a], Antonella Rosa[a], Silvana Vargiolu[a], and Francesco P. Corongiu[a]

[a]Dipartimento di Biologia Sperimentale, Sez. Patologia Sperimentale, Universita' di Cagliari, Cagliari, Italy
[b]Divisione di Nefrologia e Dialisi, Università degli Studi di Modena, Modena, Italy

Introduction

Even though known for decades (1), CLA became the object of intense research only after the discovery that it possesses antimutagenic activity (2). It has now been shown to possess anticarcinogenic (2–5) and antiatherogenic (6,7) actions as well, to affect fat partitioning in the body (8,9), and to normalize impaired glucose tolerance and improve hyperinsulinemia in the prediabetic ZDF rat (10). Much less is known, however, about the mechanism(s) whereby CLA exerts these effects. Its anticarcinogenic activity was originally thought to be due to antioxidant properties (3,11), but these properties have not been confirmed by results obtained to date in studies involving several biological models (12,13). Our current work is aimed at testing the hypothesis that the anticarcinogenic action of CLA stems from its metabolism and influence on tissue lipid metabolism.

Absorption and Metabolism of CLA in Animal Tissues

As a result of its peculiar structure with a conjugated diene double bond system, and in spite of two double bonds, CLA has a backbone that can be superimposed on that of oleic acid. As a consequence, its incorporation into rat liver lipids is similar to that of oleic acid and occurs preferentially into neutral lipids (14–16). Its assimilation into different tissues, therefore, mirrors their neutral lipid content, which is higher in adipose and mammary tissues than in liver and plasma (17) (Table 23.1). In a previous paper (16), we characterized different types of conjugated diene (CD) fatty acids in rat tissues by means of high-performance liquid chromatography (HPLC) with on-line diode array and mass spectrometer (MS) detectors. In liver lipids of rats fed a low level (0.04%) of CLA in partially hydrogenated oils, two more fatty acids were characterized as CD 18:3 and CD 20:3 (16). We concluded that elongation and desaturation of CLA occurs in rat liver without affecting the CD structure, even though CD 20:4, the expected end product of such processes, was not detected. However, similar analyses of liver phospholipids from lambs, which are naturally overexposed

TABLE 23.1 Tissue Accumulation of Conjugated Linoleic Acid (CLA) and Its Metabolites as a Function of CLA Intake[a]

	CLA	CD[b] 18:3	CD 20:3	CD 20:4
		(µg/mg of lipids)		
% of dietary CLA		Adipose tissue		
0.0	0.00 ± 0.00	0.00 ± 0.00	0.00 ± 0.00	
0.50	24.71 ± 3.22	0.47 ± 0.14	0.85 ± 0.21	
1.00	33.81 ± 8.49	0.63 ± 0.19	1.08 ± 0.38	
1.50	54.49 ± 17.76	1.02 ± 0.25	1.35 ± 0.40	
2.00	56.46 ± 8.21	1.05 ± 0.40	1.38 ± 0.35	
		Mammary tissue		
0.00	0.44 ± 0.18	0.04 ± 0.02	0.06 ± 0.03	
0.50	21.50 ± 7.35	0.44 ± 0.17	0.71 ± 0.37	
1.00	46.72 ± 2.83	0.78 ± 0.32	1.21 ± 0.05	
1.50	57.77 ± 14.63	0.84 ± 0.40	1.50 ± 0.47	
2.00	68.02 ± 16.79	0.99 ± 0.28	1.70 ± 0.58	
		Plasma		
0.00	0.00 ± 0.00	0.00 ± 0.00	0.00 ± 0.00	0.00 ± 0.00
0.50	12.41 ± 6.28	0.70 ± 0.23	0.47 ± 0.21	0.08 ± 0.05
1.00	23.85 ± 8.03	1.21 ± 0.57	0.70 ± 0.19	0.11 ± 0.06
1.50	27.26 ± 15.49	1.24 ± 0.62	0.82 ± 0.33	0.09 ± 0.05
2.00	46.48 ± 21.86	1.98 ± 1.03	1.34 ± 0.65	0.19 ± 0.12
		Liver		
0.00	0.20 ± 0.05	0.39 ± 0.21	0.00 ± 0.00	
0.50	7.55 ± 0.72	0.48 ± 0.12	0.31 ± 0.06	
1.00	15.94 ± 2.58	0.97 ± 0.30	0.46 ± 0.22	
1.50	17.76 ± 2.79	1.29 ± 0.25	0.66 ± 0.12	
2.00	29.00 ± 5.68	1.63 ± 0.70	0.87 ± 0.20	

[a]CLA and metabolites (mean ± SD) increased significantly in all tissues tested in proportion to the dietary intake ($P < 0.05$). *Source:* Ref. 17.
[b]CD, conjugated diene.

to CLA in their diet and its generation by their intestinal flora, showed the presence not only of CLA, CD 18:3, and CD 20:3, but also of CD 20:4 (18).

Recently Sébédio *et al.* (19) extended these findings while studying chain elongation and desaturation in the liver of rats reared on a fat-free diet and fed a commercial CLA triglyceride (180 mg/d). By using complementary HPLC and GC/MS techniques, they identified three higher CD metabolites of CLA, namely, 20:3Δ8,12,14, 20:4Δ5,8,12,14, and 20:4Δ5,8,11,13. They suggested that only two isomers of CLA, 18:2*t,c*10,12 and 18:2*c,t* 9,11, can be efficiently chain elongated and desaturated. In tissue lipids of rats fed either the *c,t* 9,11 or the *c,t* 10,12 CLA isomers, the same investigators detected a further CD fatty acid, CD 16:2 (20). The presence of this fatty acid indicates that CLA might undergo one round of peroxisomal β-oxidation.

Other evidence has been obtained (20) that CLA isomers may be handled differently by the desaturase and elongase enzymes as a result of differences in the position of the conjugated double bonds and the geometric isomerism. The question might be raised, therefore, whether a selective metabolization of the CLA isomers may lead to different biological activities.

More recent results indicate that generation of CD 20:4 may depend on the level of dietary linoleic acid because an inverse correlation was observed between tissue levels of CD 20:4 and the linoleic acid content of the diet (unpublished data).

Tissue Accumulation of CLA and Metabolites as a Function of CLA Intake

Progressively increasing levels of CLA were noted in rat mammary and peritoneal fat pads, liver, and plasma, as its content in the diet increased over a range of 0.5 to 2% ($P < 0.05$ in all three tissues) (17) (Table 23.1). The plasma levels seem actually to provide a good index of CLA intake. The mammary fat pads had higher CLA concentrations on a per milligram lipid basis than the liver. This result was as expected, given the ready and high incorporation of CLA into neutral lipids rather than phospholipids (14–16). It is apparent, therefore, that substantial amounts of CLA can be stored in the mammary gland, an organ that consists largely of adipocytes. The peritoneal pads had concentrations similar to those of the mammary pads, attesting that adipocyte neutral lipids constitute a major storage site and source of CLA in the body. In the same study, we found a gradual increase of the CD 18:3 and CD 20:3 metabolites in mammary tissue and liver, as a function of CLA intake, but their levels were much lower than those of CLA (Table 23.1), suggesting that only a small fraction of it was metabolized *via* the desaturation and elongation pathways. The total accumulation of metabolites was also proportional to the dietary intake of CLA, over the same range of 0.5 to 2% ($P < 0.05$) (Table 23.1).

Changes in Linoleic Acid Metabolites as a Function of CLA Intake

In rats fed a basal diet (AIN-76A) containing 5% corn oil, CLA, regardless of its intake level, did not interfere with the retention of linoleic acid (14). In mammary and adipose tissues, the content of linoleic acid metabolites dropped significantly ($P < 0.05$), up to a CLA dietary intake of 1% (17) (Fig. 23.1). Of particular interest was a 50% decrease in 20:4, the substrate for the cyclooxygenase (COX) and lipoxygenase (LPOX) pathways of eicosanoid biosynthesis. No further significant reductions occurred at CLA intakes >1%. Thus, the dose response of 20:4 corresponded closely to the dose response of the mammary gland cancer protection afforded by dietary CLA (17). Changes in linoleic acid metabolites (including 20:4) were not seen in liver and plasma (data not shown), suggesting that these effects may be associated with the greater availability of CLA in tissues in which neutral lipids are the predominant lipid. Interestingly, in mammary and adipose tissues, the loss of 20:4 and 20:3 polyunsaturated fatty acids (PUFA), which are substrates of the COX and LPOX pathways (21), seemed to be replaced by

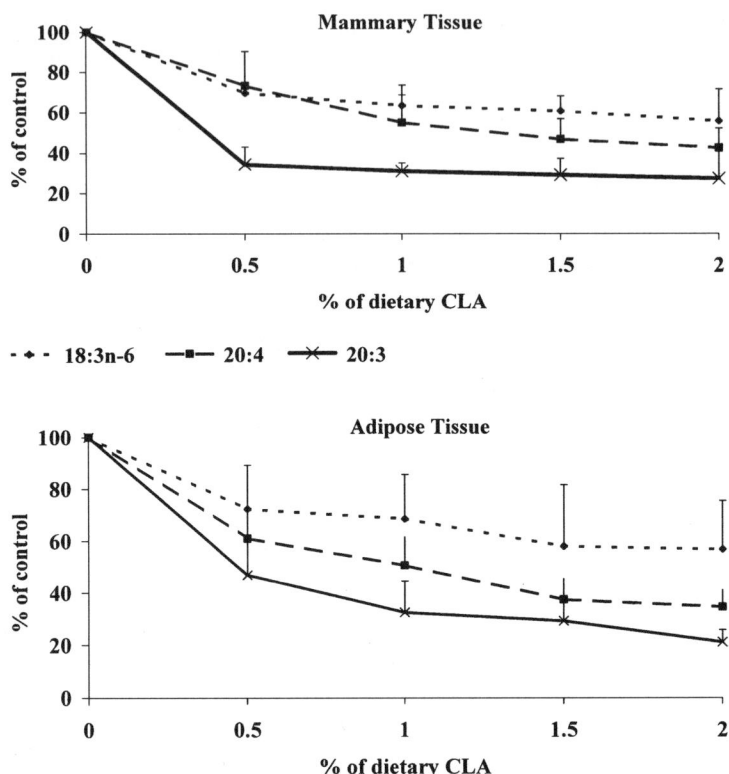

Fig. 23.1. Changes in linoleic acid metabolites as a function of CLA intake.

CD 18:3 and CD 20:3, which have been shown to inhibit those pathways (22,23) (Fig. 23.2). It would seem reasonable, therefore, to expect that CLA interferes with the biosynthesis of eicosanoids, an effect that could be involved not only in its anticarcinogenic activity, but also in its modulation of immune functions (24–26), its antiatherogenicity (6,7), and its induction of phorbol ester–mediated events in keratinocytes (27).

Terminal end buds (TEB) are the primary target site of chemical carcinogens inducing mammary carcinomas in rodents (28). CLA feeding was shown to decrease mammary epithelial cell branching and the overall density of TEB, as determined by digitized image analyses (28). These effects mirrored the 20:4 PUFA dose response; they were also evident at CLA dietary intakes of up to 1%, with no further decrements at higher intakes (17).

Retinol Content in Liver of Rats as a Function of CLA Intake

Increases in the peroxisomal β-oxidation of fatty acids are known to activate peroxisome proliferator-activated receptor (PPAR) α, and recently CLA was found to activate this receptor (29). Indications have been also obtained that PPAR α may be involved in

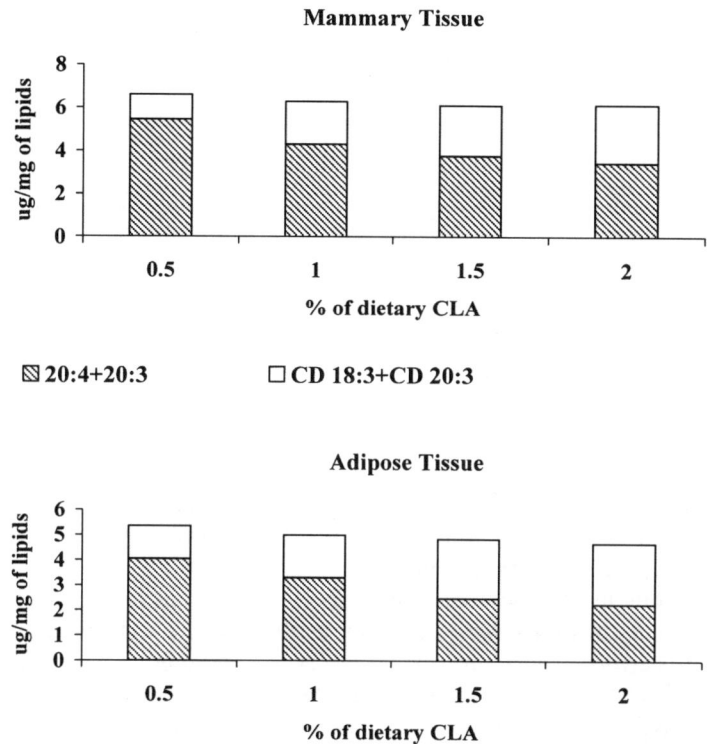

Fig. 23.2. Loss of 20:4 + 20:3 and its replacement by CD 18:3 + CD 20:3.

the expression of the cellular retinol binding protein (CRBP) gene of intestinal cells (30). It seems possible, therefore, that CLA may enhance the level of CRBP in these cells and thus lead to a sequestration of retinol in the lumen, increasing retinol absorption.

In a study originally designed to quantify CLA metabolites in the liver of rats fed up to 2% CLA, we noticed that an unidentified peak with a maximal absorption at 324 nm was consistently elevated (31). A literature search indicated that the peak might be retinol, and the indication was corroborated by introduction of retinol standards in the HPLC analyses. As shown in Table 23.2, retinol accumulated progressively in the liver as the intake of CLA increased, attaining a maximum elevation of about fivefold over control levels at a 2% intake. Total retinyl esters also increased, but to a maximum elevation of only about twofold or less; moreover, they displayed no dose response because they peaked between 0.5 and 1% CLA. The distribution of the various retinyl esters was not affected (data not shown), and the retinyl esters/retinol ratio decreased as dietary CLA intake increased, suggesting that CLA markedly alters the distribution of the retinol forms stored in the liver. In mammary tissue and plasma, only free retinol was detected, at modest but statistically significant elevations. A maximum was attained at 0.5% CLA and no dose-response effect

TABLE 23.2 Tissue Accumulation of Free and Esterified Retinol as a Function of Conjugated Linoleic Acid (CLA) Intake[a]

Dietary CLA (%)	Plasma (µg/mL)	Liver		Mammary tissue (µg/g of tissue)
		Free (µg/g of tissue)	Esters (µg/g of tissue)	
0	0.23 ± 0.04	1.97 ± 0.28	47.21 ± 8.30	0.45 ± 0.12
0.5	0.33 ± 0.10	3.31 ± 0.47	70.30 ± 14.66[b]	0.97 ± 0.33[c]
1	0.36 ± 0.10	5.75 ± 1.22[b]	84.11 ± 16.69[d]	1.09 ± 0.30[b]
1.5	0.37 ± 0.17[c]	7.51 ± 2.31[d]	91.28 ± 8.47[d]	1.14 ± 0.51[b]
2	0.40 ± 0.06[c]	10.71 ± 2.89[d]	85.21 ± 12.34[d]	1.15 ± 0.37[b]

[a]Values are means ± SD.
[b,c,d]Significantly different from control group (0% dietary CLA) at $P < 0.01$, $P < 0.05$, and $P < 0.001$, respectively. *Source:* Ref. 31.

was evident. The ratio of mammary tissue to plasma retinol was ~2 in control rats, and 3 in CLA-fed rats, suggesting that CLA may enhance the uptake and/or retention of retinol in the mammary gland.

At this point in time, however, only working hypotheses can be formulated concerning the mechanism by which CLA raises the vitamin A status in rats. Dietary retinol is reesterified in intestinal mucosa cells and absorbed in this form in association with lymph chylomicrons. During metabolism of the latter, all retinyl esters remain with chylomicron remnants and are removed similarly from circulation almost entirely by the parenchymal cells of the liver. After hydrolysis of the esters, retinol is transferred from the parenchymal to the stellate cells of the liver (32), where vitamin A is stored as retinyl palmitate in lipid droplets. Depending on the load of retinyl esters delivered to the liver, the regulatory control exerted by enzymes such as retinyl ester hydrolase and acyl CoA retinol acyltransferase is likely to determine the balance between retinol and retinyl esters stored in the liver. If CLA were to induce CRBP in liver cells, such an outcome could tip the scale in favor of retaining a higher proportion of retinol. This possibility could apply also to the increased level of retinol observed in the mammary tissue.

CLA and Its Metabolites in Human Tissues

A recent epidemiology study in Finland showed that habitual consumption of whole milk is associated with a reduced risk of breast cancer (33). This was a 25-y prospective study involving 4697 women with a mean age of 39 y at the time of recruitment and who were cancer free initially. Among individuals with the highest tertile of milk intake, there was a 60% decrease in relative risk. The adjustment for potential confounding factors, such as smoking, body mass index (BMI), number of childbirths, nutrients, and so on did not alter the results. A growing body of evidence has been gathered during the past decade indicating that milk fat may contain a number of components with anticancer activity, CLA among them (34). In another recent report (35), the ability of CLA to afford protection against breast cancer was assessed by measuring its level in breast adipose tissue, as a reflection of past CLA dietary intake. The tissue was obtained at the time of surgery from 360 patients with either carcinomas or benign tumors (controls). In all cases, the major isomer found was 18:2 $9c,11t$. An unconditional logistic regression model was used to obtain odds ratio estimates while adjusting for age, menopausal status, and BMI. An inverse association between CLA level and risk of breast carcinoma was found. The adjusted relative risk for women in the highest tertile of CLA level was 0.15 (95% confidence interval = 0.07–0.33) compared with women in the lowest tertile (P trend = 0.0001), suggesting a strong protective effect of CLA against breast cancer. Even though the above two studies were only prospective, the results obtained were at least consistent with the possibility that CLA may be effective also in humans.

We have measured the level of CLA in human plasma, adipose tissue, and red blood cell samples (unpublished data). Its tissue distribution mirrored that previously seen in rats, with a preferential accumulation in neutral lipid rich tissues (Table 23.3). The CD 18:3 and CD 20:3 metabolites, but not CD 20:4, were also detected. The failure to detect CD 20:4 may be explained by either a high dietary intake of linoleic acid or too low an intake of CLA. The latter possibility may also account for an observed difference in the CLA/metabolite ratio in human and rat tissues.

On comparing the CLA levels in human and rat plasma, it is apparent that in humans the level is too far below the optimal level to prevent cancer. With current everyday diets, therefore, it does not seem possible to increase the intake of CLA without increasing the intake of dietary animal fats beyond the recommended amount. An enrichment of animal fats with CLA could be an alternative. There are indeed many reports showing that it is possible, for example, to increase the CLA level in cow milk by manipulating the diet of the ruminants (36–39). With such an enriched milk and its derivatives, it might be possible to reach optimal CLA intakes. Another issue would have to be investigated, however, before implementing this approach. High tissue levels of CLA have been detected in certain human pathologic conditions (40–45); in patients with these conditions, much care would have to be exercised before supplying

TABLE 23.3 Distribution of Conjugated Linoleic Acid (CLA) and Its Metabolites in Human Tissues[a,b]

	Plasma	Adipose tissue (µg/mg of lipids)	RBC metabolites
CLA	1.070 ± 0.345	3.106 ± 0.860	0.695 ± 0.183
CD 18:3	0.087 ± 0.017	0.120 ± 0.032	0.169 ± 0.033
CD 20:3	0.090 ± 0.031	0.618 ± 0.203	0.063 ± 0.017

[a]Values are means ± SD.
[b]RBC, red blood cells; CD, conjugated diene.

them with CLA, as long as the mechanism underlying the pathologic increases has not been clarified.

Conclusions

CLA shows a unique activity in mammary cancer prevention in rats. When its consumption is limited to the period of pubescent mammary gland development, it confers a lasting protection against subsequent induction of mammary tumors (46). Furthermore, among the fatty acids known to influence mammary carcinogenesis, it is the only one able to down-regulate mammary epithelial growth during maturation, and therefore the size of the target cell population susceptible to carcinogenesis (28). CLA exerts these effects in a dose-dependent fashion; however, how its dietary intake and the tissue lipid metabolism changes it elicits influence mammary carcinogenesis has not yet been clarified.

This unusual PUFA is incorporated in tissue lipids as oleic acid, is metabolized as linoleic acid, and may undergo β-oxidation by peroxisomes with activation of PPAR α. It appears therefore to have three singular activities, each exerted usually by a different fatty acid. Its incorporation as oleic acid makes it more available in tissues such as the mammary gland in which neutral lipids are the predominant lipid, likely accounting for the fact that mammary carcinogenesis in the rat is prevented by very low dietary intakes. Its metabolism is similar to that of linoleic acid; thus it competes with the latter for the elongation and desaturation enzyme systems, leading on one hand to a decreased generation of eicosatrienoic and arachidonic acids, and on the other to a replacement of the natural substrates with CLA metabolites known to inhibit the COX and LPOX pathways (22,23). It seems reasonable, therefore, to anticipate that CLA may affect the biosynthesis of eicosanoids, an effect that has been documented to influence carcinogenesis (47–50). The anticarcinogenic action of CLA may be mediated also, at least in part, by an increased level of retinol in tissues that are the target sites of chemical carcinogens. A peroxisomal β-oxidation of CLA with consequent activation of PPAR α could account for the increase in the retinol content caused by CLA in various tissues. High doses of retinyl esters (51,52), as well as CLA feeding (4,28,53), have both been shown to suppress mammary gland development, to inhibit terminal duct and alveolar cell proliferation, and to block mammary tumorigenesis (4,53).

These three properties of CLA, as a fatty acid, might also explain the fact that its anticarcinogenic activity is independent of the type and amount of dietary fat (53) because the changes in lipid metabolism it brings about could work synergistically to compensate the effect(s) of dietary fat on mammary carcinogenesis.

The beneficial activities of CLA have been demonstrated to date only in animal experimental models. Even though there is some evidence in the literature linking CLA and mammary tumor prevention in humans, more studies are required to confirm this effect. The current CLA dietary intake in humans does not seem to be sufficient to exert beneficial effects. Dietary supplements are already widely available commercially as an alternate source. However, given that the supplements would have to be taken regularly, a more expeditious way could be an enhanced delivery of CLA through the food system. The introduction into the food system of CLA-enriched dairy fats and products could be a suitable and effective venue that would also not entail too high a dietary intake of animal fats. In ongoing studies, we are exploring whether CLA-enriched butter exerts the same preventive activity vs. mammary cancer in rodents as does synthetic CLA.

Acknowledgments

This work was supported by grants from the Regione Autonoma della Sardegna (PIC-Interreg II), Consiglio Nazionale delle Ricerche (CNR-Roma, Italy) Contract No. 96.0498.ST74, and MURST-programmi di rilevante interesse regionale No. 9805273874-014. The authors would like to thank Prof. Benito Lombardi for revising the manuscript.

References

1. Banni, S., and Martin, J.C. (1998) Conjugated Linoleic Acid and Metabolites, in Trans Fatty Acids in Human Nutrition, (Sébédio, J.L., and Christie, W.W., eds.) pp. 261–302, Oily Press, Dundee.
2. Pariza, M.W., and Hargraves, W.A. (1985) A Beef Derived Mutagenesis Modulator Inhibits Initiation of Mouse Epidermal Tumors by 7,12-Dimethylbenz(a)anthracene, Carcinogenesis 6, 591–593.
3. Ha, Y.L., Storkson, J., and Pariza, M.W. (1990) Inhibition of Benzo(a)pyrene-Induced Mouse Forestomach Neoplasia by Conjugated Dienoic Derivatives of Linoleic Acid, Cancer Res. 50, 1097–1101.
4. Ip, C., Singh, M., Thompson, H.J., and Scimeca, J.A. (1994) Conjugated Linoleic Acid Suppresses Mammary Carcinogenesis and Proliferative Activity of the Mammary Gland in the Rat, Cancer Res. 54, 1212–1215.
5. Liew, C., Schut, H.A.J., Chin, S.F., Pariza, M.W., and Dashwood, R.H. (1995) Protection of Conjugated Linoleic Acids Against 2-Amino-3-methylimidazo[4,5-f] quinoline-Induced Colon Carcinogenesis in the F344 Rat—a Study of Inhibitory Mechanisms, Carcinogenesis 16, 3037–3043.
6. Lee, K.N., Kritchevsky, D., and Pariza, M.W. (1994) Conjugated Linoleic Acid and Atherosclerosis in Rabbits, Atherosclerosis 108, 19–25.
7. Nicolosi, R.J., Courtemanche, K.V., Laitinen, L., Scimeca, J.A., and Huth, P.J. (1993) The Effect of Feeding Diets Enriched in Conjugated Linoleic Acid on Lipoproteins and Aortic Atherogenesis in Hamsters, Circulation 88, 2458.
8. Albright, K.J., Liu, W., Storkson, J.M., Hentges, E., Lofgren, P., Scimeca, J.A., Cook, M.E., and Pariza, M.W. (1996) Body Composition Repartitioning Following the Removal of Dietary Conjugated Linoleic Acid, J. Anim. Sci. 74, 152.
9. Dugan, M.E.R., Aalhus, J.L., Schaefer, A.L., and Kramer, J.K.G. (1997) The Effect of Conjugated Linoleic Acid on Fat to Lean Repartitioning and Feed Conversion in Pigs, Can. J. Anim. Sci. 77, 723–725.
10. Houseknecht, K.L., Vandenheuvel, J.P., Moyacamarena, S.Y., Portocarrero, C.P., Peck, L.W., Nickel, K.P., and Belury, M.A. (1998) Dietary Conjugated Linoleic Acid Normalizes Impaired Glucose Tolerance in the Zucker Diabetic Fatty fa/fa Rat, Biochem. Biophys. Res. Commun. 244, 678–682.

11. Ha, Y.L., Grimm, N.K., and Pariza, M.W. (1987) Anticarcinogens from Fried Ground Beef: Heat-Altered Derivatives of Linoleic Acid, *Carcinogenesis 8*, 1881–1887.
12. Banni, S., Angioni, E., Contini, M.S., Carta, G., Casu, V., Iengo, G.A., Melis, M.P., Deiana, M., Dessì, M.A., and Corongiu, F.P. (1997) Conjugated Linoleic Acid and Oxidative Stress, *J. Am. Oil Chem. Soc. 75*, 261–267.
13. van den Berg, J.J., Cook, N.E., and Tribble, D.L. (1995) Reinvestigation of the Antioxidant Properties of Conjugated Linoleic Acid, *Lipids 30*, 599–605.
14. Banni, S., Basford, R.E., Corongiu, F.P., and Lombardi, B. (1989) On the Question of Membrane-Lipid Peroxidation in the Liver of Rats Fed a Choline-Devoid Diet, *Adv. Biosci. 76*, 187–201.
15. Banni, S., Evans, R.W., Salgo, M.G., Corongiu, F.P., and Lombardi, B. (1990) Conjugated Diene and *trans* Fatty Acids in Tissue Lipids of Rats Fed an Hepatocarcinogenic Choline-Devoid Diet, *Carcinogenesis 11*, 2053–2057.
16. Banni, S., Day, B.W., Evans, R.W., Corongiu, F.P., and Lombardi, B. (1995) Detection of Conjugated Diene Isomers of Linoleic Acid in Liver Lipids of Rats Fed a Choline-Devoid Diet Indicates That the Diet Does Not Cause Lipoperoxidation, *J. Nutr. Biochem. 6*, 281–289.
17. Banni, S., Angioni, E., Casu, V., Melis, M.P., Carta, G., Corongiu, F.P., Thompson, H., and Ip, C. (1999) The Significance of a Decrease in Linoleic Acid Metabolites as a Potential Signaling Mechanism in Cancer Risk Reduction by Conjugated Linoleic Acid, *Carcinogenesis*.
18. Banni, S., Carta, G., Contini, M.S., Angioni, E., Deiana, M., Dessì, M.A., Melis, M.P., and Corongiu, F.P. (1996) Characterization of Conjugated Diene Fatty Acids in Milk, Dairy Products, and Lamb Tissues, *J. Nutr. Biochem. 7*, 150–155.
19. Sébédio, J.L., Juaneda, P., Dobson, G., Ramilison, I., Martin, J.C., Chardigny, J.M., and Christie, W.W. (1997) Metabolites of Conjugated Isomers of Linoleic Acid (CLA) in the Rat, *Biochim. Biophys. Acta 1345*, 5–10.
20. Angioni, E., Banni, S., Chardigny, J.M., Gregoire, S., Juaneda, P., Martin, J.C., and Sébédio, J.L. (1998) Metabolism of c,t 9,11 CLA and c,t 10,12 CLA in Rat Tissues, 2nd Meeting of the European Section of AOCS, Cagliari, 1–4 October (Abstr.).
21. Smith, W.L., Borgeat, P., and Fitzpatrick, F.A. (1991) The Eicosanoids: Cyclooxygenase, Lipoxygenase, and Epoxygenase Pathways, in *Biochemistry of Lipids, Lipoproteins and Membranes,* (Vance, D.E., and Vance, J., eds.) pp. 297–325, Elsevier, Amsterdam.
22. Nugteren, D.H. and Christ-Hazelhof, E. (1987) Naturally Occuring Conjugated Octadecatrienoic Acids Are Strong Inhibitors of Prostaglandin Biosynthesis, *Prostaglandins 33*, 403–417.
23. Nugteren, D.H. (1970) Inhibition of Prostaglandin Biosynthesis by 8*trans*, 12*trans*, 14*cis*-Eicosatrienoic Acid and 5*cis*, 8*cis*, 12*trans*, 14*cis*-Eicosatetraenoic Acid, *Biochim. Biophys. Acta 121*, 171–176.
24. Chew, B.P., Wong, T.S., Shultz, T.D., and Magnuson, N.S. (1997) Effects of Conjugated Dienoic Derivatives of Linoleic Acid and Beta-Carotene in Modulating Lymphocyte and Macrophage Function, *Anticancer Res. 17*, 1099–1106.
25. Miller, C.C., Park, Y., Pariza, M.W., and Cook, M.E. (1994) Feeding Conjugated Linoleic Acid to Animals Partially Overcomes Catabolic Responses Due to Endotoxin Injection, *Biochem. Biophys. Res. Commun. 198*, 1107–1112.

26. Wong, M.W., Chew, B.P., Wong, T.S., Hosick, H.L., Boylston, T.D., and Shultz, T.D. (1997) Effects of Dietary Conjugated Linoleic Acid on Lymphocyte Function and Growth of Mammary Tumors in Mice, *Anticancer Res. 17*, 987–993.
27. Liu, K.L. and Belury, M.A. (1997) Conjugated Linoleic Acid Modulation of Phorbol Ester-Induced Events in Murine Keratinocytes, *Lipids 32*, 725–730.
28. Thompson, H., Zhu, Z., Banni, S., Darcy, K., Loftus, T., and Ip, C. (1997) Morphological and Biochemical Status of the Mammary Gland as Influenced by Conjugated Linoleic Acid: Implication for a Reduction in Mammary Cancer Risk, *Cancer Res. 57*, 5067–5072.
29. Moya-Camarena, S.Y., Vanden Heuvel, J.P., and Belury, M.A. (1999) Conjugated Linoleic Acid Activates Peroxisome Proliferator-Activated Receptor α and β Subtypes but Does Not Induce Hepatic Peroxisome Proliferation in Sprague-Dawley Rats, *Biochim. Biophys. Acta, 1436*, 331–342.
30. Suzuki, R., Suruga, K., Goda, T., and Takase, S. (1998) Peroxisome Proliferator Enhances Gene Expression of Cellular Retinol-Binding Protein, Type II in Caco-2 Cells, *Life Sci. 62*, 861–871.
31. Banni, S., Angioni, E., Casu, V., Melis, M.P., Scrugli, S., Carta, G., Corongiu, F.P., and Ip, C. (1999) An Increase in Vitamin a Status by the Feeding of Conjugated Linoleic Acid, *Nutr. Cancer, 33*, 53–57.
32. Blomhoff, R., Green, M.H., and Norum, K.R. (1992) Vitamin A: Physiological and Biochemical Processing, *Annu. Rev. Nutr. 12*, 37–57.
33. Knekt, P., Jarvinen, R., Seppanen, R., Pukkala, E., and Aromaa, A. (1996) Intake of Dairy Products and the Risk of Breast Cancer, *Br. J. Cancer 73*, 687–691.
34. Parodi, P.W. (1997) Cow's Milk Fat Components as Potential Anticarcinogenic Agents, *J. Nutr. 127*, 1055–1060.
35. Lavillonière, F., Ferrari, P., Lhuillery, C., Martin, J.C., Sébédio, J.L., and Bougnoux, P. (1998) CLAs Containing High Proportion of the 18:2 Δ9*cis*, 11*trans* Isomer Exhibit a Protective Effect Both in Rat Mammary Carcinogenesis and in Human Breast Cancer, 2nd Meeting of the European Section of AOCS, Cagliari, 1–4 October (Abstr.).
36. Jiang, J., Bjoerck, L., Fonden, R., and Emanuelson, M. (1996) Occurrence of Conjugated *cis*-9,*trans*-11-Octadecadienoic Acid in Bovine Milk: Effects of Feed and Dietary Regimen, *J. Dairy Sci. 79*, 438–445.
37. Kelly, M.L., Berry, J.R., Dwyer, D.A., Griinari, J.M., Chouinard, P.Y., Vanamburgh, M.E., and Bauman, D.E. (1998) Dietary Fatty Acid Sources Affect Conjugated Linoleic Acid Concentrations in Milk from Lactating Dairy Cows. *J. Nutr. 128*, 881–885.
38. Oshea, M., Lawless, F., Stanton, C., and Devery, R. (1998) Conjugated Linoleic Acid in Bovine Milk Fat— A Food-Based Approach to Cancer Chemoprevention, *Trends Food Sci. Technol. 9*, 192–196.
39. Stanton, C., Lawless, F., Kjellmer, G., Harrington, D., Devery, R., Connolly, J.F., and Murphy, J. (1997) Dietary Influences on Bovine Milk *cis*-9,*trans*-11-Conjugated Linoleic Acid Content, *J. Food Sci. 62*, 1083–1086.
40. Banni, S., Angioni, E., Carta, G., Casu, V., Deiana, M., Dessì, M.A., Lucchi, L., Melis, M.P., Rosa, A., Scrugli, S., Sicbaldi, D., Solla, E., and Corongiu, F.P. (1997) Metabolism of Conjugated Linoleic Acid in Pathological States, 88th AOCS Annual Meeting 1994–1995 (Abstr.).

41. Banni, S., Lucchi, L., Baraldi, A., Botti, B., Cappelli, G., Corongiu, F., Dessì, M.A., Tomasi, A., and Lusvarghi, E. (1996) No Direct Evidence of Increased Lipid Peroxidation in Hemodialysis Patients, *Nephron 72,* 177–183.
42. Braganza, J.M., Wickens, D.G., Cawood, P., and Dormandy, T.L. (1983) Lipid Peroxidation (Free Radical Oxidation) Products in Bile from Patients with Pancreatic Disease, *Lancet, 2(8346):* 375–379.
43. Erskine, K.J., Iversen, S.A., and Davies, R. (1985) An Altered Ratio of 18:2 (9,11) to 18:2 (9,12) Linoleic Acid in Plasma Phospholipids as a Possible Predictor of Pre-Eclampsia, Lancet, *1(8428):* 554–555.
44. Fairbank, J., Hollingworth, A., Griffin, J., Ridgway, E., Wickens, D., Singer, A., and Dormandy, T. (1989) Octadeca-9,11-dienoic Acid in Cervical Intraepithelial Neoplasia: A Colposcopic Study, *Clin. Chim. Acta 186,* 53–58.
45. Situnayake, R.D., Crump, B.J., Thurnham, D.I., Davies, J.A., Gearty, J., and Davis, M. (1990) Lipid Peroxidation and Hepatic Antioxidant in Alcoholic Liver Disease, *Gut 31,* 1311–1317.
46. Ip, C., Scimeca, J.A., and Thompson, H. (1995) Effect of Timing and Duration of Dietary Conjugated Linoleic Acid on Mammary Cancer Prevention, *Nutr. Cancer 24,* 241–247.
47. Nakazawa, I., Iwaizumi, M., and Ohuchi, K. (1993) Some Features in Prostaglandin Synthesis of the Cancer Cells Which Metastasized into Liver from Intestinal Cancer Lesions, *Tohoku J. Exp. Med. 170,* 131–133.
48. Baxevanis, C.N., Reclos, G.J., Gritzapis, A.D., Dedousis, G.V., Missitzis, I., and Papamichail, M. (1993) Elevated Prostaglandin E_2 Production by Monocytes Is Responsible for the Depressed Levels of Natural Killer and Lymphokine-Activated Killer Cell Function in Patients with Breast Cancer, *Cancer 72,* 491–501.
49. Grubbs, C.J., Juliana, M.M., Eto, I., Casebolt, T., Whitaker, L.M., Canfield, G.J., Manczak, M., Steele, V.E., and Kelloff, G.J. (1993) Chemoprevention by Indomethacin of N- Butyl-*N*-(4-hydroxybutyl)-nitrosamine-Induced Urinary Bladder Tumors, *Anticancer Res. 13,* 33–36.
50. Rao, C.V., and Reddy, B.S. (1993) Modulating Effect of Amount and Types of Dietary Fat on Ornithine Decarboxylase, Tyrosine Protein Kinase and Prostaglandins Production During Colon Carcinogenesis in Male F344 Rats, *Carcinogenesis 14,* 1327–1333.
51. Moon, R.C., Grubbs, C.J., and Sporn, M.B. (1976) Inhibition of 7,12-Dimethylbenz(a)anthracene-Induced Mammary Carcinogenesis by Retinyl Acetate, *Cancer Res. 36,* 2626–2630.
52. McCormick, D.L., Burns, F.J., and Albert, R.E. (1981) Inhibition of Benzo[a]pyrene-Induced Mammary Carcinogenesis by Retinyl Acetate, *Cancer Res. 66(3):* 559–564.
53. Ip, C., Briggs, S.P., Haegele, A.D., Thompson, H.J., Storkson, J., and Scimeca, J.A. (1996) The Efficacy of Conjugated Linoleic Acid in Mammary Cancer Prevention Is Independent of the Level or Type of Fat in the Diet, *Carcinogenesis 17,* 1045–1050.

Chapter 24
Conjugated Linoleic Acid Metabolites in Rats

J.L. Sébédio

INRA, Unité de Nutrition Lipidique, 21034 DIJON Cédex, France

Introduction

Essential fatty acids are not only desaturated and elongated, leading to the formation of arachidonic acid from 18:2 n-6 and eicosapentaenoic and docosahexaenoic acids from 18:3 n-3, but also are elongated to "dead-end products," such as 20:2 n-6 from 18:2 n-6 and 20:3 n-3 from 18:3 n-3 (1–4).

Similarly, studies by Blank and Privett (5) showed that 9c,12t-18:2 could be converted into a *trans* isomer of eicosatetraenoic acid. Further studies demonstrated that at least one *trans* isomer of linoleic acid, 9c,12t-18:2, was converted into 5c,8c,11c,14t-20:4 (6–8) and that 9t,12c-18:2 was elongated into 11t,14c-20:2 (7,9). Some 18:3 geometrical isomers that result from the heat treatment of oils (10,11) can also be converted into *trans* isomers of 20:5 and 22:6 *in vivo* (12,13).

Conjugated 18:2 fatty acids (CLA) are also formed during heat treatment (frying) of oils (14). However, the main sources of CLA are ruminant fat and dairy products (15–20).

Although numerous studies have dealt with the desaturation and elongation of essential fatty acids as well as with their corresponding *trans* methylene interrupted isomers, very little has been published to date on the metabolism of these conjugated linoleic acid isomers. This review will focus on what is known about the metabolism of CLA in rats.

Metabolism of CLA in rats

In 1995, Banni *et al.* (21) showed the presence of 18:2 conjugated fatty acids in liver lipids of rats fed a partially hydrogenated oil that contained conjugated isomers of linoleic acid. They also detected the presence of 18:3 and 20:3 isomers having two conjugated double bonds as illustrated in Table 24.1. Because these fatty acids were not detected in the diet, they suggested that desaturation and elongation of CLA isomers might occur in rat liver. These isomers were not detected when the rats were fed a corn oil diet. Only the 18:2 isomers were detected in the adipose tissue when the diet was fed for 1 wk. Similarly, the same conjugated isomers were detected in lamb tissues (22).

Partially hydrogenated oils are known to contain mixtures of conjugated *cis*, *trans*, *trans,cis*, and *trans,trans* isomers (23). One major question was therefore to determine whether all of the conjugated isomers could be metabolized by the rat. For this reason, Sébédio *et al.* (24) fed rats a commercial mixture of CLA that has been widely used to date by other researchers (25,26).

TABLE 24.1 The 18:2, 18:3, and 20:3 Isomers with Two Conjugated Double Bonds Present in Liver Lipids of Rats Fed a Partially Hydrogenated Oil[a]

	18:2	18:3 (µg/mg)	20:3
Male rats			
CDPHO[b]	10.4 ± 0.97	1.82 ± 0.15	1.01 ± 0.31
CDCO[c]	ND	ND	ND
CSPHO[d]	3.91 ± 0.72	ND	0.37 ± 0.09
Female rats			
CDPHO	11.69 ± 1.21	1.40 ± 0.34	0.91 ± 0.31
CDCO	ND	ND	ND
CSPHO	4.48 ± 0.841	ND	0.40 ± 0.22

[a]*Source:* Adapted from Banni et al. (21).
[b]CDPHO, partially hydrogenated oil (choline-deficient diet).
[c]DCO, corn oil (choline-deficient diet).
[d]CSPHO, partially hydrogenated oil (choline-supplemented diet).
ND, not detected.

In a first set of experiments, the animals were force-fed a commercial CLA mixture for 6 d (180 mg/d). Analysis by high-performance liquid chromatography with ultraviolet (UV) detection revealed the presence of conjugated polyunsaturated fatty acids in the total liver lipid methyl esters (Fig. 24.1). After isolation, the metabolites were identified by gas–liquid chromatography (GLC) coupled with mass spectrometry (MS). Different derivatives such as dimethyoxazoline (27), picolinyl esters of the fully deuterated fatty acid (24), and the 4-methyl-1,2,4-triazoline-3,5-dione adducts (28) were utilized. The following four metabolites were identified: 20:3 Δ 8,12,14; 20:3 Δ 8,11,13; 20:4 Δ 5,8,12,14; and 20:4 Δ 5,8,11,13 (24,29).

Development of complex methodologies was necessary in order to elucidate the structures of these molecules because gas chromatography (GC)-MS of isolated 20:3 and 20:4 fractions revealed the presence of conjugated trienoic and tetraenoic isomers that were not separated even on highly polar columns (29). Only a combination of silver-ion and reversed-phase high-performance liquid chromatographic techniques (HPLC) as illustrated in Figure 24.2 permitted the separation of the four 20:3 and 20:4 isomers present in the mixture. Silver-ion HPLC is a powerful technique that is already being used for the conjugated dienes (30,31).

In most cases, mass spectra of the dimethyl oxazoline (DMOX) derivatives permitted elucidation of the structure of the molecule. As an example, we report in Figure 24.3 the mass spectra of the DMOX derivatives of the two 20:3 metabolites. Both spectra show intense peaks at m/z 113 and 126, typical of oxazoline derivatives, and an intense molecular ion at m/z 359, which indicates that both components are 20:3 isomers.

For compound X1, a mass interval of 12 units instead of 14 units for a saturated chain occurred between m/z 182 (C_7) and 194 (C_8), between 236 (C_{11}) and 248 (C_{12}), and between 262 (C_{13}) and 274 (C_{14}). This indicates the presence of three ethylenic bonds in the Δ8, Δ12, and Δ14 positions. Similarly, the ions at m/z 182 (C_7), 194

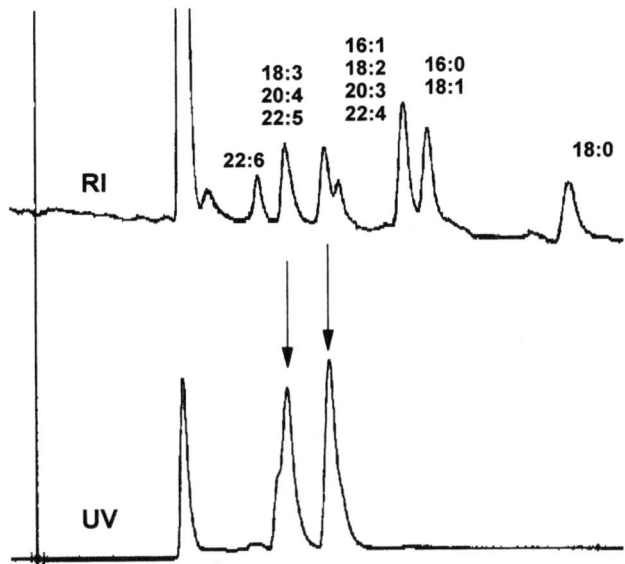

Fig. 24.1. Reversed-phase high-performance liquid chromatography (HPLC) analysis of fatty acid methyl esters from liver lipids of rats fed a mixture of conjugated linoleic acid (CLA) isomers. *Source:* Reproduced by permission from Ref. 25.

(C_8), 222 (C_{10}), 234 (C_{11}), 248 (C_{12}), and 260 (C_{13}) indicate the presence of three ethylenic bonds in the $\Delta 8$, $\Delta 11$, and $\Delta 13$ positions for the second 20:3 isomer (X2).

The present data show that some conjugated isomers of linoleic acid may be converted into unusual polyunsaturated fatty acid isomers (20:3 and 20:4), as has been demonstrated for mono-*trans* isomers of linoleic acid (6,7,9). If we assume that the normal pathways for fatty acids ($\Delta 6$ desaturation, elongation, $\Delta 5$ desaturation) may operate for 18:2 conjugated fatty acids, the 20:3 Δ 8,12,14 and the 20:4 Δ 5,8,12,14 are formed from the 18:2 Δ 10,12, whereas the 20:3 Δ 8,11,13 and the 20:4 Δ 5,8,11,13 would arise from the 18:2 Δ 9,11, the major conjugated acid in dairy products. However, the other conjugated acids, the Δ 8, 10 and Δ 11, 13 isomers present in the CLA mixture fed to the animals (Fig. 24.4), do not seem to be converted into significant amounts of long-chain polyunsaturated fatty acids under our experimental conditions.

However, because *cis,cis*, *cis,trans* and *trans,trans* isomers are present in commercial samples, it is not possible from the GC-MS studies described above to determine which isomers are the precursors of the 20:3 and 20:4 fatty acids. To do so, the collected 20:4 isomers were submitted to a hydrazine reduction to determine the geometry of each double bond (32). The resulting monoenes were identified and converted to the DMOX derivatives. The position of the ethylenic bond was determined by GC-MS. Two isomers were identified, i.e., the 20:4 Δ 5c,8c,11c,13t and the 20:4 Δ 5c,8c,12t,14c. Therefore, one could conclude that the two precursors are the 9c,11t, and the 10t,12c-18:2 isomers, the two major isomers of the commercial CLA mixture.

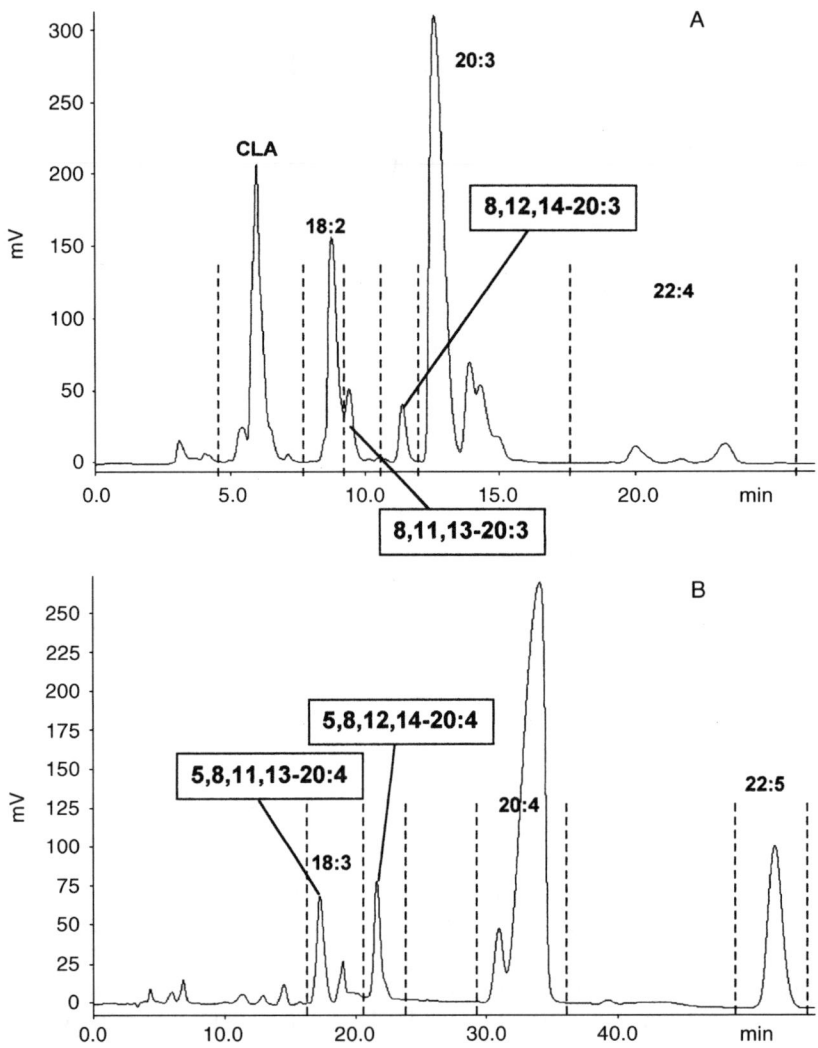

Fig. 24.2. Silver-ion high-performance liquid chromatography (Ag$^+$-HPLC) fractionation of (A) 20:3 and (B) 20:4 isomers isolated from liver lipids of rats fed a conjugated linoleic acid (CLA) mixture. *Source:* Adapted from Ref. 29.

It is important to point out that these two 20:4 isomers were observed when the rats were force-fed large quantities of CLA, whereas only 18:3 and 20:3 isomers were observed when CLA was directly introduced in a diet (21). To further study CLA metabolism, two isomers, the 9c,11t, and the 10t,12c-18:2 were synthesized (33,34) and fed to the animals for 6 wk (35).

After the 6-wk feeding period, analyses of the liver and adipose tissue lipids permitted identification of other C_{16} and C_{18} metabolites. These were the 16:2 Δ 8,10 and

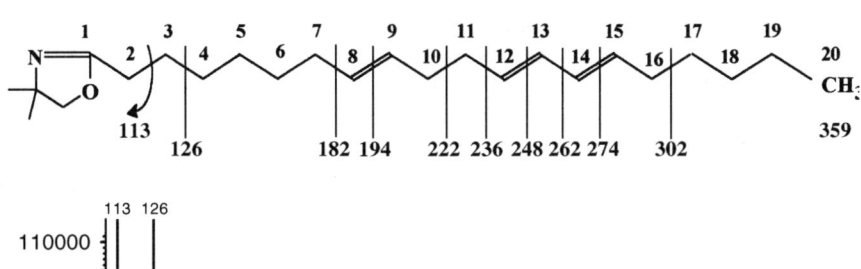
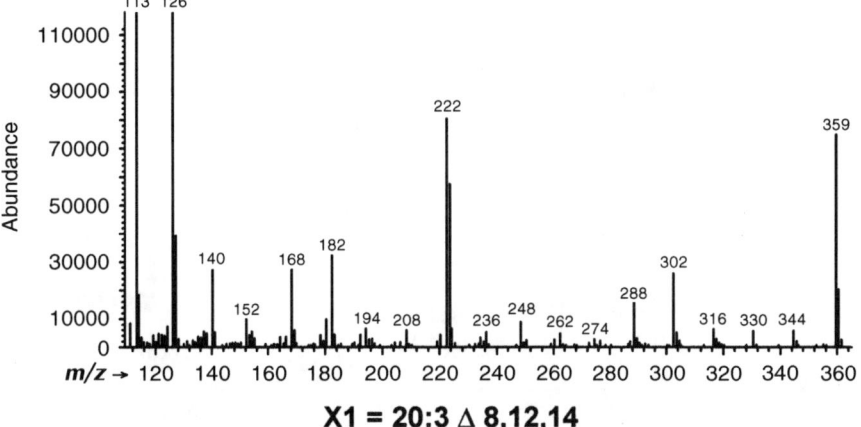
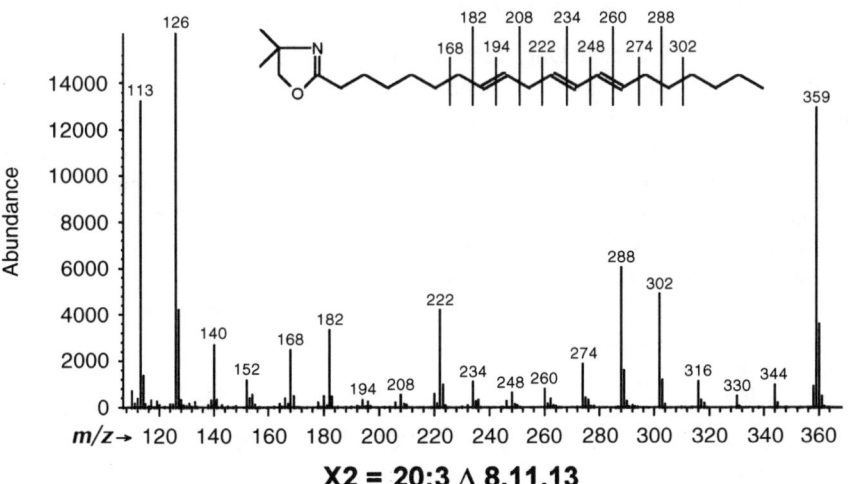

Fig. 24.3. Mass spectra of dimethyl oxazoline (DMOX) derivatives of two 20:3 conjugated isomers isolated from liver lipids of rats fed a conjugated linoleic acid (CLA) mixture. *Source:* Adapted from Ref. 24.

the 18:3 Δ 6,10,12. With the use of pure isomers, this study showed that the 9c,11t is converted into 18:3 and 20:3 conjugated fatty acids. The 10t,12c isomer is converted mainly into an 18:3 Δ 6,10,12, whereas only smaller quantities of the conjugated 20:3

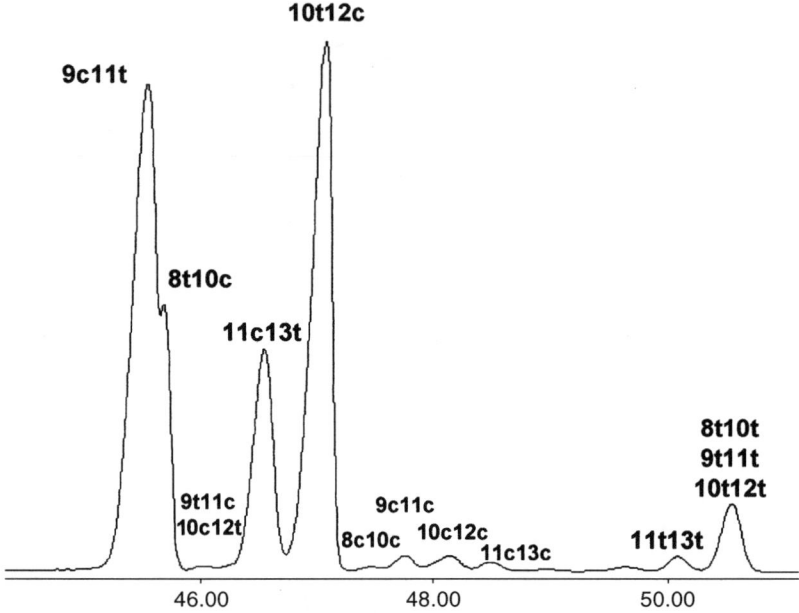

Fig. 24.4. Gas–liquid chromatography (GLC) analysis (CPSil 88, 100 m) of a commercial mixture of conjugated linoleic acid (CLA) isomers.

Δ 8,12,14 have been detected. Only a small quantity of 20:4 conjugated isomer, formed from the 10*t*,12*c* isomer, has been found in phospholipids. Furthermore, the 10*t*,12*c* isomer can also be converted into a conjugated 16:2 fatty acid having the ethylenic bonds in Δ 8 and Δ 10. Similar peroxisomal metabolic pathways have already been described by Luthria *et al.* (36) for linoleic acid which was transformed into a 7,10-16:2 and a 5,8-14:2. This 16:2 conjugated isomer was found only in liver lipids from animals fed the 10*t*,12*c*-18:2. This may suggest that 9*c*,11*t* and 10*t*,12*c* are metabolized differently through the peroxisomal β-oxidation pathway.

It would now be very interesting to study whether the higher metabolites could play any role in the eicosanoid synthesis.

Acknowledgment

These studies are part of a concerted action (FAIR PL 3671) funded by the Commission of the European Union.

References

1. Sprecher, H., (1981) Biochemistry of Essential Fatty Acids, *Prog. Lipid Res. 20,* 13–22.
2. Stearns, E.M., Ryvasy, J.A., and Privett, O.S. (1967), Metabolism of *cis*-11, *cis*-14- and *trans*-11, *trans*-14-Eicosadienoic Acids in the Rat, *J. Nutr. 93,* 486–490.
3. Bernert, J.T., and Sprecher, H. (1975) Studies to Determine the Role Rates of Chain Elongation and Desaturation Play in Regulating the Unsaturated Fatty Acid Composition of Rat Liver Lipids, *Biochim. Biophys. Acta 398,* 354–363.

4. Hagre, T.A., Christophersen, B.O., and Dannevig, B.H. (1986) Desaturation and Chain Elongation of Essential Fatty Acids in Isolated Liver Cells from Rat and Rainbow Trout, *Lipids 21*, 202–205.
5. Blank, M.L., and Privett, O.S. (1963) Studies on the Metabolism of *cis, trans* Isomers of Methyl Linoleate and Linolenate, *J. Lipid Res. 4*, 470–473.
6. Privett, O.S., Stearns, E.M., and Nickell, E.C. (1967) Metabolism of the Geometric Isomers of Linoleic Acid in the Rat, *J. Nutr. 92*, 303–310.
7. Berdeaux, O., Sébédio, J.L., Chardigny, J.M., Blond, J.P., Mairot, Th., Vatèle, J.M., Poullain, D., and Noël, J.P. (1996). Effects of *trans* n-6 Fatty Acids on the Fatty Acid Profile of Tissues and Liver Microsomal Desaturation in the Rat, *Grasas Aceites 47*, 86–99.
8. Ratnayake, W.M.N., Chen, Z.Y., Pelletier, G., and Weber, D. (1994) Occurrence of 5c,8c,11c,15t Eicosatetraenoic Acid and Other Unusual Polyunsaturated Fatty Acids in Rats Fed Partially Hydrogenated Canola Oil, *Lipids 29*, 707–714.
9. Beyers, E.C., and Emken, E.A. (1991) Metabolites of *cis,trans* and *trans,cis* Isomers of Linoleic Acid in Mice and Incorporation into Tissue Lipids, *Biochim. Biophys. Acta 1082*, 275–284.
10. Ackman, R.G., Hooper, S.N., and Hooper, D.L. (1974) Linoleic Acid Artifacts from the Deodorization of Oils, *J. Am. Oil Chem. Soc. 51*, 42–49.
11. Wolff, R.L. (1993) Further Studies on Artificial Geometrical Isomers of α-Linolenic Acid in Edible Linolenic Acid Containing Oils, *J. Am. Oil Chem. Soc. 70*, 219–224.
12. Grandgirard, A., Piconneaux, A., Sébédio, J.L., O'Keefe, S.F., Semon, E., and Le Quéré, J.L. (1989) Occurrence of Geometrical Isomers of Eicosapentaenoic and Docosahexaenoic Acids in Liver Lipids of Rats Fed Heated Linseed Oil, *Lipids 24*, 799–804.
13. Chardigny, J.M., Sébédio, J.L., Grandgirard, A., Martine, L., Berdeaux, O., and Vatèle, J.M. (1996) Identification of Novel *trans* Isomers of 20:5 n-3 in Liver Lipids of Rats Fed Heated Oil, *Lipids 31*, 165–168.
14. Sébédio, J.L., Grandgirard, A., and Prévost, J. (1988) Linoleic Acid Isomers in Heat Treated Sunflower Oils, *J. Am. Oil Chem. Soc. 65*, 362–366.
15. Parodi, P.W. (1977) Conjugated Octadecadienoic Acids of Milk Fat, *J. Dairy Sci. 60*, 1550–1553.
16. Fogerty, A.C., Ford, G.L., and Svoronos, D. (1988) Octadeca-9,11-dienoic Acid in Foodstuffs and in the Lipids of Human Blood and Breast Milk, *Nutr. Rep. Int. 35*, 937–944.
17. Ha, Y.L., Grimm, N.K., and Pariza, M.W. (1989) Newly Recognized Anticarcinogenic Fatty Acids: Identification and Quantification in Natural and Processed Cheeses, *J. Agric. Food Chem. 37*, 75–81.
18. Shantha, N.C., Decker, E.A., and Ustunol, Z. (1992) Conjugated Linoleic Acid Concentrations in Processed Cheese, *J. Am. Oil Chem. Soc. 69*, 425–428.
19. Jiang, J, Bjoerck, L., Fonden, R., and Emanuelson, M. (1996) Occurrence of Conjugated *cis*-9, *trans*-11-Octadecadienoic Acid in Bovine Milk: Effects of Feed and Dietary Regimen, *J. Dairy Sci. 79*, 438–445.
20. Shantha, N.C., Crum, A.D., and Decker, E.A. (1994) Evaluation of Conjugated Linoleic Acid Concentrations in Cooked Beef, *J. Agric. Food Chem. 42*, 1757–1760.
21. Banni, S., Day, B.W., Evans, R.W., Corongiu, F.P., and Lombardi, B. (1995) Detection of Conjugated Diene Isomers of Linoleic Acid in Liver Lipids of Rats Fed a Choline-Devoid Diet Indicates That the Diet Does Not Cause Lipoperoxidation, *J. Nutr. Biochem. 6*, 281–289.

22. Banni, S., Carta, G., Contini, M.S., Angioni, E., Deiana, M., Dessi, M.A., Melis, M.P., and Corongiu, F.P. (1996) Characterization of Conjugated Diene Fatty Acids in Milk, Dairy Products and Lamb Tissues, *J. Nutr. Biochem. 7,* 150–155.
23. Mossoba, M.M., McDonald, R.E., Armstrong, D.J., and Page, S.W. (1991) Identification of Minor C18 Triene and Conjugated Diene Isomers in Hydrogenated Soybean Oil and Margarine by GC-MI-FT-IR Spectroscopy, *J. Chromatogr. Sci. 29,* 324–330.
24. Sébédio, J.L., Juanéda, P., Dobson, G., Ramilison, I., Martin, J.C., Chardigny, J.M., and Christie, W.W. (1997) Metabolites of Conjugated Isomers of Linoleic Acid (CLA) in the Rat, *Biochim. Biophys. Acta 1345,* 5–10.
25. Banni, S., and Martin, J.C. (1998) in Trans *Fatty Acids in Human Nutrition,* (Sébédio, J.L., and Christie, W.W., eds.) pp. 261–302, The Oily Press, Dundee.
26. Christie, W.W., Dobson, G., and Gunstone, F.D. (1997) Isomers in Commercial Samples of Conjugated Linoleic Acid, *J. Am. Oil Chem. Soc. 74,* 1231.
27. Spitzer, V., Marx, F., and Pfeilsticker, K. (1994) Electron Impact Mass Spectra of the Oxazoline Derivatives of Some Conjugated Diene and Triene C_{18} Fatty Acids, *J. Am. Oil Chem. Soc. 71,* 873–876.
28. Dobson, G. (1998) Identification of Conjugated Fatty Acids by Gas Chromatography-Mass Spectrometry of 4-Methyl-1,2,4-triazoline-3,5-dione Adducts, *J. Am. Oil Chem. Soc. 75,* 137–142.
29. Juanéda, P., and Sébédio, J.L. (1999) Combining Silver-Ion and Reversed Phase High-Performance Liquid Chromatography for the Separation and the Identification of C_{20} Metabolites of Conjugated Linoleic Acid Isomers in Rat Liver Lipids, *J. Chromatogr. 724,* 213–219.
30. Sehat, N., Yurawecz, M.P., Roach, J.A.G., Mossoba, M.M., Kramer, J.K.G., and Ku., Y. (1998) Silver-Ion High-Performance Liquid Chromatographic Separation and Identification of Conjugated Linoleic Acid Isomers, *Lipids 33,* 217–221.
31. Adlof, R., and Lamm, T. (1998) Fractionation of *cis* and *trans* Oleic, Linoleic, and Conjugated Linoleic Fatty Acid Methyl Esters by Silver-Ion High-Performance Liquid Chromatography, *J. Chromatogr. 799,* 329–332.
32. Sébédio, J.L., Juanéda, P., Grégoire, S., Chardigny, J.M., Martin, J.C., and Ginies, C. (1998) Identification of *trans* C20:4 Conjugated Isomers in Liver Lipids of Rats Fed Conjugated Linoleic Acid (CLA), *Lipids,* submitted for publication.
33. Berdeaux, O., Christie, W.W., Gunstone, F.D., and Sébédio, J.L. (1997) Large Scale Synthesis of Methyl *cis*-9, *trans*-11-Octadecadienoate from Methyl Ricinoleate, *J. Am. Oil Chem. Soc. 74,* 1011–1015.
34. Berdeaux, O., Voinot, L., Angioni, E., Juanéda, P., and Sébédio, J.L. (1998) A Simple Method of Preparation of Methyl *trans*-10, *cis*-12- and *cis*-9, *trans*-11-Octadecadienoates from Methyl Linoleate, *J. Am. Oil Chem. Soc 75,* 1749–1755.
35. Sébédio, J.L., Angioni, E., Chardigny, J.M., Grégoire, S., Juanéda, P., Martin, J.C., and Berdeaux, O. (1999) 9*c*,11*t* and 10*t*, 12*c*-18:2 Do Not Influence Similarly Fatty Acid Composition of Rat Tissue Lipids, *Inform 10,* 543.
36. Luthria, D.L., Chen, Q., and Sprecher, H. (1997) Metabolites Produced During the Peroxisomal β-Oxidation of Linoleate and Arachidonate Move to Microsomes for Conversion Back to Linoleate, *Biochim. Biophys. Res. Commun. 233,* 438–441.

Chapter 25
Effect of Conjugated Linoleic Acid on Polyunsaturated Fatty Acid Metabolism and Immune Function

M. Sugano[a], M. Yamasaki[b], K. Yamada[b], and Y.-S. Huang[c]

[a]Prefectural University of Kumamoto, Kumamoto 862-8502, Japan
[b]Kyushu University School of Agriculture, Fukuoka 812-8581, Japan
[c]Ross Products Division, Abbott Laboratories, Columbus, OH

Introduction

Conjugated linoleic acid (CLA) is a collective term for the group of positional (between carbons 9 and 12) and geometrical isomers (*cis-cis, cis-trans, trans-cis, trans-trans*) derived from linoleic acid that have conjugated double bonds. CLA occurs naturally in a variety of foods, particularly dairy products and meat from ruminant animals (1,2). CLA is synthesized by microflora, *Butyrivibrio fibrisolvens,* in the rumen (3).

In recent years, CLA has gained a great deal of attention due to its purported protective properties against cancer and heart disease and evidence suggests that CLA can also regulate immune functions (4–17). The hypothesis stemmed from the analysis of the fatty acid composition of tissue phospholipids in animals fed CLA: there was a significant reduction in the proportion of arachidonic acid (18). The reduction of substrate availability may result in a decreased production of arachidonate-derived eicosanoids. Many of these eicosanoids are known to be potent regulators of immune functions. Furthermore, CLA may interfere with the enzymatic conversion of arachidonate to eicosanoids. In addition to these indirect effects, CLA, like linoleic acid, can be desaturated and elongated to more highly unsaturated molecules such as conjugated eicosatrienoic acid and conjugated eicosatetraenoic acid. Both conjugated fatty acids may serve as substrates for the synthesis of unknown eicosanoids (19–22). To date, knowledge is lacking concerning the active molecular form(s) of CLA and the mechanism by which CLA exerts its action.

This review will examine the metabolism of CLA and its effect on lipid and eicosanoid metabolism, and discuss the roles of prostaglandins and cytokines in the regulation of immune function.

Absorption and Metabolism of CLA

Dietary CLA is taken up and subsequently incorporated into body lipids (11). In rats with cannulated thoracic ducts, CLA was transported from the intestine as chylomicrons similarly to fatty acids such as linoleic acid (23). However, the recovery of CLA was lower than that of linoleic acid (Fig. 25.1). The absorption rates of individual CLA isomers differed as shown by differences in the amount of intragastrically administered

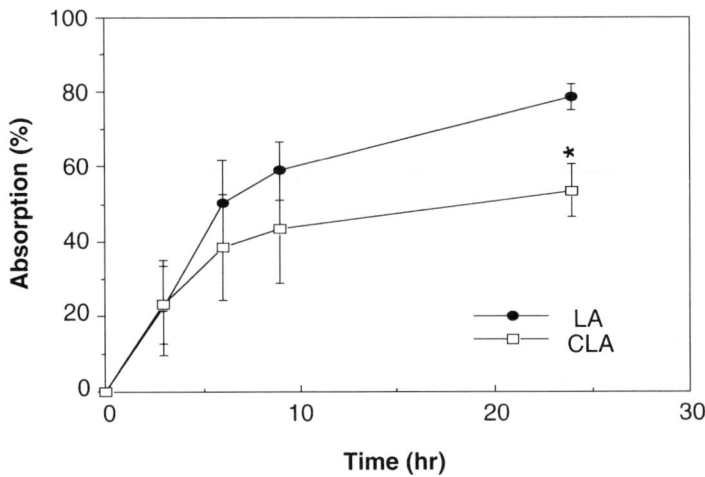

Fig. 25.1. Lymphatic absorption of linoleic (LA) and conjugated linoleic acids (CLA) in rats. Values are means ± SEM of 3 and 4 rats in the LA and CLA groups, respectively. *Significantly different ($P < 0.05$) compared with LA group. *Source:* Ref. 23.

CLA isomers appearing in the chylomicrons. More *trans-trans* than *cis-trans* and *trans-cis* isomers were detected in the chylomicrons. The distribution of these isomers in three stereospecific positions in the triglyceride molecules also differed. The *trans-cis* and *cis-trans* isomers were incorporated equally into all three positions (*sn*-1, *sn*-2, and *sn*-3) in triglyceride (TG) molecules, whereas the *trans-trans* isomers were incorporated preferentially into the *sn*-1 and *sn*-3 positions (23). The incorporation of CLA into different tissues also varied significantly (23) (Table 25.1). Notable accumulations were observed in lung and adipose tissue. The compositions of CLA isomers differed among tissues, i.e., *trans-trans* isomers accumulated significantly in adipose tissue, whereas they were not detected in brain. The distribution of CLA in different lipid fractions also differed. For example, all CLA isomers appeared in TG, whereas only the *c*9,*t*11-CLA isomer was found in liver phospholipids (11). The concentration of this isomer was highest in phosphatidylinositol and lowest in phosphatidylcholine and phosphatidlyserine (Table 25.1).

CLA has been identified in human tissues, such as blood, bile, adipose tissue, and milk (24–30). CLA was thought to be of dietary origin, through consumption of CLA-containing dairy products or meats from ruminant animals, or synthesized from linoleic acid through anaerobic microbial activity in the large bowel (24) or synthesized by dairy starter cultures (31). Because the amount of dietary linoleic acid intake or the level of linoleic acid in serum lipids has no significant effect on the concentration of CLA in human serum (32), CLA is unlikely to be synthesized endogenously. Evidence has suggested that CLA is formed from *trans*-vaccenic acid (Δ 11-18:1) by Δ 9-desaturation as shown in rat liver microsomal preparations (33,34).

TABLE 25.1 Effect of Dietary Conjugated Linoleic Acid (CLA) on Proportions of CLA in Various Tissues and Liver Phospholipids of Rats[a]

Total tissue lipids/ phospholipids	Tissue and dietary CLA level (%)	
	0 (wt%)	1.0
Adipose tissue	0.4 ± 0.3	9.6 ± 2.3
Serum	0.1 ± 0.1	4.0 ± 1.1
Liver	0.5 ± 0.3	3.9 ± 1.3
Lung	0.8 ± 0.4	8.9 ± 1.7
Heart	0.6 ± 0.3	3.5 ± 0.9
Kidney	0.5 ± 0.3	6.5 ± 0.8
Brain	0.2 ± 0.2	0.6 ± 0.1
Liver phospholipids		
Phosphatidylcholine	0.2 ± 0.1	0.8 ± 0.2
Phosphatidylethanolamine	0.2 ± 0.1	1.2 ± 0.4
Phosphatidylinositol	0.2 ± 0.2	4.5 ± 2.0
Phosphatidylserine	ND	0.8 ± 0.5
Cardiolipin	0.1 ± 0.1	2.3 ± 1.5

[a]Values are means ± SEM (n = 5); ND, not detected.

After absorption, dietary linoleic and α-linolenic acids are metabolized to long-chain polyunsaturated fatty acids (PUFA). Metabolism of dietary CLA, however, is less efficient as shown by the accumulation of CLA itself, but not of its derivatives (19–21), in various tissues (Table 25.1, Fig. 25.2). In mice, CLA isomers were incorporated more readily into neutral lipids than phospholipids when CLA was added to the diet. However, CLA disappeared more rapidly from neutral lipids than from phospholipids when supplementation was discontinued (11). Similarly, Belury and Kempa-Steczko (18) examined the role of CLA in lipid metabolism by determining the modulatory effect of CLA on fatty acid composition in mice fed different levels of CLA (0, 0.5, 1.0, and 1.5%) for 6 wk. They observed that dietary CLA was incorporated into liver lipids by displacing linoleic acid and arachidonic acid. In comparison with the control mice, the liver total lipids contained 60 and 30% less linoleic and arachidonic acid, respectively, in mice fed a 1.5% CLA diet. The effect of short-term feeding of CLA on the level of n-6 PUFA was more noticeable in neutral lipids than in phospholipids. This difference disappeared in mice fed CLA for longer periods. They concluded that CLA affects metabolic conversion of fatty acids in liver and ultimately results in modification of the n-6 fatty acid composition of phospholipids.

Liu and Belury (35) compared the incorporation of CLA with that of linoleic and arachidonic acids into murine keratinocytes, HEL-30 cells. After incubation with [^{14}C]linoleic acid, arachidonic acid, and CLA, they found that ~50% of linoleic and arachidonic acids, compared with only 30% of CLA, were incorporated into phosphatidylcholine of HEL-30 cells. More CLA was incorporated into phosphatidylserine,

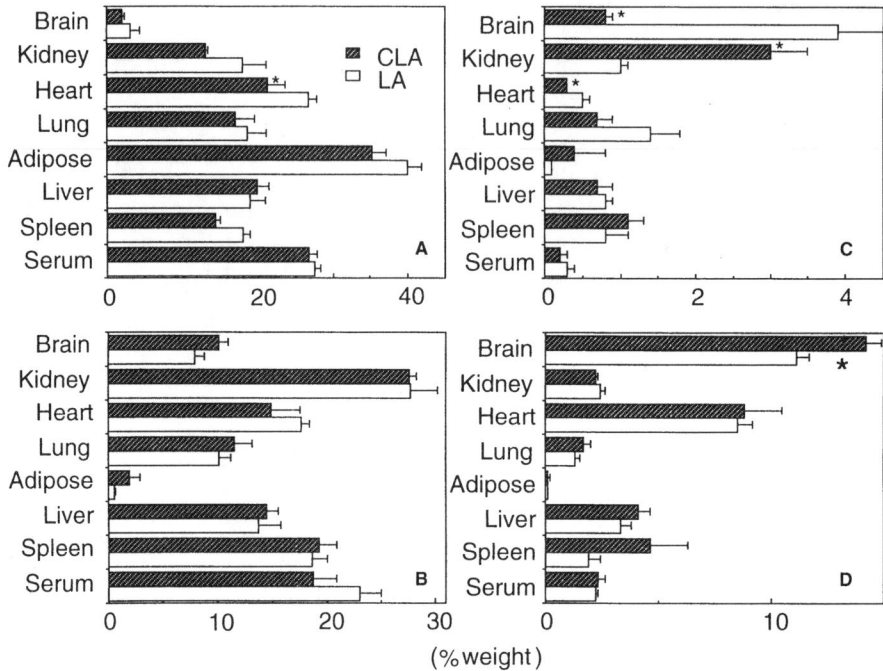

Fig. 25.2. Effect of conjugated linoleic acid (CLA) on fatty acid compositions of various tissues of rats. Values are means ± SEM of 3–6 rats. Significantly different from the corresponding linoleic acid (LA) group at *$P < 0.05$, **$P < 0.01$. (A) linoleic acid; (B) arachidonic acid; (C) dihomo-γ-linolenic acid; (D) docosahexaenoic acid. SD rats were fed 1% LA or 1% CLA diets for 2 wk. *Source:* unpublished data (Sugano et al.).

phosphatidylinositol, and phosphatidylethanolamine than into phosphatidylcholine. These observations indicate that the incorporation of CLA into phospholipids was significantly different from that of linoleic and arachidonic acids. Because phospholipids play a key role in cell membrane structure and function, it is possible that CLA exerts its biological effect through its effect on n-6 fatty acid composition of phospholipids.

How does CLA regulate the level of n-6 fatty acids in various tissues? One explanation is that CLA behaves like an analog of linoleic acid because of its structural similarity to linoleic acid. Belury and Kempa-Steczko (18) indicated that CLA could be desaturated to an unknown 18:3 product in a manner similar to the conversion of linoleic acid to γ-linolenic acid. Thus, CLA may compete with linoleic acid as a substrate for Δ 6 desaturase and competitively inhibit the desaturation of n-6 fatty acid. Consequently, this inhibition causes a reduction in the proportion of dihomo-γ-linolenic and arachidonic acids. Banni et al. (19–21) detected the presence of conjugated 18:3, conjugated 20:3, and conjugated 20:4 fatty acids, presumably the metabolites of CLA, in rat liver. These conjugated acids were also detected in mammary tissue of CLA-fed rats as well as in lamb liver (20). These findings suggest that both Δ 6- and Δ 5- desaturation and elongation of CLA can occur in liver.

Effect of CLA on Eicosanoid Production

The magnitude of eicosanoid synthesis is determined largely by the availability of substrate fatty acids in the membrane phospholipids. Eicosanoid production is also affected by the relative amounts of PUFA of different series, n-6 and n-3. For example, mammary and colon carcinogenesis is influenced differently by n-3 and n-6 PUFA; the former are protective, whereas the latter are promotive (36–38). It is believed by some that the effect of PUFA on carcinogenesis is modulated through the action of prostaglandins (PG) (39). Because CLA affects the metabolism of n-6 PUFA, it is plausible that CLA influences the production of arachidonic acid–derived eicosanoids.

Feeding 1.0% of CLA for 2 wk decreased serum PGE_2 levels in rats to 53% of that in control rats (22). Reports of another study showed that the effect of dietary CLA on serum PGE_2 level was dose dependent (17) (Table 25.2). When murine keratinocytes, HEL-30 cells preincubated with 0.2/7.4KBq/mL of [^{14}C]arachidonic acid and either linoleic acid or CLA were challenged with 12-O-tetradecanoyl-13-acetate (TPA), the level of PGE_2 in cells pretreated with CLA was significantly lower than that in cells pretreated with linoleic acid (40). At present, no information exists concerning effects of CLA on the production of 2-series PG other than PGE_2 and thromboxane. Isomers of CLA could be metabolized to one or more biologically active products, such as CLA-derived eicosanoids, which could regulate the synthesis of arachidonic acid metabolites such as PGE_2. It is also possible that CLA-derived eicosanoids could exhibit biological activity independently (19,41–44).

Clinical manifestations of hypersensitive immune reactions such as allergic reactions are induced by the production of chemical mediators, histamine, and

TABLE 25.2 Effect of Dietary Conjugated Linoleic Acid (CLA) on Levels of Prostaglandin E_2 in Spleen and Serum of Rats[a,b]

	Dietary CLA level (%)		
Tissue	0	0.5	1.0
Spleen (pg/g)	13.4 ± 2.6	14.6 ± 3.5	16.0 ± 3.7
Serum (pg/mL)	23.2 ± 0.9[a]	19.9 ± 1.0[a,b]	17.7 ± 0.8[b]

[a]Values are means ± SEM (n = 5); values not sharing a common superscript letter (a,b) are significantly different, $P < 0.05$.
[b]Source: Ref. 17.

leukotriene, and also prostaglandin triggered by the allergen-immunoglobulin, IgE. Leukotrienes (LT) are the lipoxygenation products of 20-carbon fatty acids. Because the allergic reaction can be modulated by long-chain n-3 and n-6 PUFA (45,46), and CLA can modulate the synthesis of prostaglandins, one could raise the question of whether CLA regulates the production of LT. We have previously reported that CLA feeding in rats lowered the level of LTB_4 but not LTC_4 in spleen. In contrast, a decrease in LTC_4 production in lung was observed (17) (Table 25.3). Because the levels of arachidonic acid, a substrate for LT synthesis, in spleen and lung phospholipids were comparable in all dietary groups (unpublished data), the different modulatory effect of CLA on LT production in these two tissues could not be attributed solely to substrate availability. In the same experiment, we also measured the release of LT by calcium ionophore A23187-stimulated peritoneal exudate cells (PEC). The results showed that CLA inhibited the release of LTB_4 (17). CLA feeding has also been shown to lower the proportions of n-6 fatty acids in PEC, suggesting that reduced LT formation was due to decreased substrate availability. On the other hand, when PEC collected from Wistar rats were stimulated *in vitro* with the calcium ionophore A23187, the presence of CLA in the medium had no effect on LTB_4 production (unpublished data). CLA feeding did not affect the release of histamine by PEC or the intracellular level of histamine (17), suggesting that the inhibitory effect of CLA on the production of particular LT is not exerted *via* lipoxygenase activity, but is likely modulated by multiple factors.

Effect of CLA on Immunoglobulin Production

B lymphocytes, antibody-producing cells, play a major role in humoral immunity. In type I allergy, induction of the production of allergen-specific IgE is a primary step (47). When the IgE receptor (FceR) on the mast cells or basophils is bridged by an

TABLE 25.3 Effect of Dietary Conjugated Linoleic Acid (CLA) on Levels of Leukotriene (LT)B_4 and LTC_4 in Lung and Spleen of Rats[a,b]

	Dietary CLA level (%)		
Tissue	0	0.5	1.0
Spleen		ng/g	
LTB_4	47.3 ± 2.4[a]	43.0 ± 2.7[a,b]	38.2 ± 2.6[b]
LTC_4	16.9 ± 1.2	18.7 ± 1.9	15.4 ± 1.8
Lung			
LTB_4	37.6 ± 11.2	29.5 ± 3.6	24.4 ± 3.7
LTC_4	34.7 ± 3.2[a]	15.3 ± 2.9[b]	11.1 ± 3.0[b]

[a]Values are means ± SEM (n = 5); values not sharing a common superscript letter (a,b) are significantly different, $P < 0.05$.
[b]*Source:* Ref. 17.

IgE-specific allergen, Ca^{2+} flows into the cells. This triggers the release of chemical mediators such as histamine and LT, subsequently inducing allergic disorders. IgA, in contrast, serves as an antiallergic factor by interfering with intestinal absorption of allergen. IgG also functions as an antiallergic factor through competition with the allergen-specific IgE to bind the receptor (FceR) on the surface of the target cells such as mast cells and basophils. Thus, both immunoglobulin IgA and IgG exert a preventive effect on type I allergy.

To evaluate whether CLA feeding affects immunoglobulin production, SD rats were fed a diet supplemented with CLA at levels of 0.5 and 1.0% for 3 wk. The spleen and mesenteric lymph node (MLN) were excised and lymphocytes prepared (17). The lymphocytes were then cultured in RPMI1640 medium containing 10% fetal bovine serum. After 24 h incubation, the concentrations of different antibodies in the culture supernatant were measured by enzyme-linked immunosorbent assay (ELISA). Results showed that the concentrations of IgA and IgM in MLN lymphocytes from rats fed 1.0% CLA were 3 and 50 times higher, respectively, than those in lymphocytes from control rats. IgG was detected in the CLA-fed group but not in the control rats. The level of IgG in the 1.0% CLA-fed group was nine times higher than that in the 0.5% CLA group. Production of IgG in spleen lymphocytes stimulated by lipopolysaccharide (LPS) was also enhanced by CLA feeding. The level of IgG was 1.3 times higher in the 1.0% CLA group than in the control group. CLA feeding significantly raised the levels of immunoglobulins in serum, such as IgA, IgM, and IgG, but decreased levels of IgE, particularly in rats fed 1.0% CLA.

The effects of CLA on the balance of $CD4^+$ and $CD8^+$ T lymphocytes in the spleen lymphocytes have also been measured. The $CD4^+$ cells are the group of helper T lymphocytes and CD8+ cells are the group of suppressor T lymphocytes. Results showed that CLA feeding had no significant effect on the balance of CD4+ and CD8+ T lymphocytes, suggesting that CLA enhances immunoglobulin production without influencing T-cell subsets.

A more detailed study examining the effect of feeding different levels of CLA on immunoglobulin production has also been conducted. In this study, SD rats were fed 0.05, 0.1, 0.25, 0.5, 1.0, or 2.0% CLA for 3 wk, following the same experimental protocol as that of the preceding experiment (unpublished data). Results are shown in Tables 25.4 and 25.5. In spleen lymphocytes, the levels of IgA, IgG, and IgM in the culture supernatant were significantly higher even in rats fed the lowest dose of CLA than in the control group. The increases were dose dependent. In MLN lymphocytes, the production of IgA, IgG, and IgM was not changed when the dietary CLA level was <0.25%, but significant increases were demonstrated when the CLA level exceeded 0.5%. These findings suggest that sensitivity to dietary CLA varied among tissues. MLN lymphocytes might have acquired resistance to CLA because of their presence at the site of CLA absorption and consequent exposure to a higher concentration of CLA. In spleen lymphocytes, no CLA was detected at dietary CLA levels <0.25%, and the cellular fatty acid composition was not affected, Thus, the regulation of immunoglobulin production by dietary CLA could not be attributed to

TABLE 25.4 Effect of Dietary Conjugated Linoleic Acid (CLA) on Immunoglobulin (Ig) Production in Spleen Lymphocytes of Rats[a,b]

	Dietary CLA level (%)		
Immunoglobulin	0	0.05	0.1
			ng/mL
IgA	ND	9.8 ± 0.7^b	14.3 ± 1.1^c
IgG	18.5 ± 1.0^a	38.9 ± 1.5^b	48.5 ± 3.1^c
IgM	30.2 ± 1.4^a	52.7 ± 1.2^b	$62.9 \pm 1.6^{c,d}$

[a]Values are means ± SEM (n = 5); ND, not detected. Values not sharing a common superscript letter (a–e) are significantly different, $P < 0.05$.
[b]Source: Yamasaki et al. (unpublished data).

TABLE 25.5 Effect of Dietary Conjugated Linoleic Acid (CLA) on Immunoglobulin (Ig) Production in Mesenteric Lymph Node Lymphocytes of Rats[a,b]

	Dietary CLA level (%)		
Immunoglobulin	0	0.05	0.10
			ng/mL
IgA	6.6 ± 0.2^a	2.6 ± 0.3^b	1.6 ± 0.2^b
IgG	25.1 ± 0.6^a	20.0 ± 0.3^b	24.6 ± 0.7^a
IgM	$9.8 \pm 0.3^{a,e}$	8.0 ± 0.3^b	$9.3 \pm 0.2^{a,e}$

[a]Values are means ± SEM (n = 5); values not sharing a common superscript letter (a–e) are significantly different, $P < 0.05$.
[b]Source: Yamasaki et al. (unpublished data).

the change of fatty acid composition or to the level of CLA incorporated into the cell membranes.

The effect of CLA on immunoglobulin production *in vitro* differed from that observed *in vivo*. There are many compounds that can regulate immunoglobulin production *in vitro*. For example, at the level of 0.1 mmol/L, PUFA can accelerate the production of IgE while inhibiting the production of IgG and IgM (48,49). The regulatory function of PUFA may be associated with lipid peroxidation because the thiobarbituric acid (TBA) value has correlated with this function. Hydrogen peroxide is also known to regulate immunoglobulin production *in vitro* (48). In a preliminary study, we cultured lymphocytes from SD rats in the presence of CLA. Unlike unsaturated fatty acids, CLA did not show specificity for immunoglobulin production. We have also used the U-266 cell line (human IgE productive B lymphoma) to evaluate the antiallergic effect of CLA on IgE production. However, no effects were observed.

Effect of CLA on Cytokine Production

To date, information about the effect of CLA on the production of cytokines, the potent regulators of the immunoglobulin production, is scarce. As discussed above, CLA can regulate the production of LTB_4, which in turn can stimulate the production of in-

	0.25	0.5	1.0	2.0
	17.3 ± 1.6[c,d]	18.6 ± 1.1[d]	16.5 ± 0.6[c,d]	26.8 ± 1.0[e]
	57.2 ± 0.6[c]	50.3 ± 0.4[c]	49.3 ± 0.9[c]	74.4 ± 6.3[d]
	65.6 ± 0.3[c]	64.5 ± 0.3[c,d]	62.0 ± 0.7[d]	75.1 ± 0.9[e]

	0.25	0.5	1.0	2.0
	2.0 ± 0.3[b]	8.7 ± 0.6[c]	1.8 ± 0.3[b]	7.6 ± 0.4[d]
	24.4 ± 0.4[a]	42.2 ± 0.7[c]	12.6 ± 2.1[d]	33.4 ± 0.7[e]
	9.1 ± 0.2[a]	14.2 ± 0.3[c]	1.7 ± 0.6[d]	10.4 ± 0.4[e]

terleukin (IL)-1, IL-2, and interferon (IFN) (50). IL-4, IL-5, tumor necrosis factor (TNF) and IFN induce class-specific immunoglobulin production of lymphocytes (51–54). The dietary n-3 PUFA stimulate immunoglobulin production (55,56) and regulate the level of cytokines such as IL-1α, IL-1β, and TNF (57). More studies examining the interaction between CLA and cytokine production are warranted to elucidate the mechanism by which CLA regulates immunoglobulin production.

Effect of CLA on Cellular Immunity

CLA also influences cellular immunity. First, CLA and β-carotene in combination stimulated proliferation *in vitro* of lymphocytes stimulated by pokeweed mitogen (PWM), concanavalin A (ConA), or phytohemagglutinin (PHA) (58). On the other hand, CLA inhibited the production of IL-2 by lymphocytes and also phagocytotic activity of the macrophages. When CLA and β-carotene were added together, spontaneous proliferation and cytotoxic activity of lymphocytes were both enhanced. Hence, CLA and β-carotene enhanced cellular immunity system synergistically. *In vivo* investigations of effects of CLA on cellular immunity have also been reported. After 3 or 6 wk of feeding of 0.1, 0.3, or 0.9% CLA in Balb/c mice, CLA stimulated the proliferation of spleen lymphocytes previously stimulated by PWM and enhanced the production of IL-2 (59).

In experiments on both humoral and cellular immunity, the mode of action of CLA was quite different *in vitro* and *in vivo*. Before promoting or recommending the use of CLA for enhancement of the immune system, elucidation of the mechanism of action is imperative.

Concluding Remarks

In this chapter, we reviewed the effects of CLA on PUFA metabolism, eicosanoid production and immune indices. The diversity of the functions and efficacy of CLA is remarkable; no other natural fatty acids exert such diverse functions. CLA may be found to have various clinical uses. In a recent report, Belury *et al.* (16,60) indicated that CLA affects lipid metabolism through activation of the peroxisome proliferator–activated receptor (PPAR). Studies on novel physiologic functions such as the antidiabetic effect and the regulation of β-oxidation enzyme activity are in progress. Overall, it is reasonable to believe that CLA may be used as an antiallergic factor through its enhancement of immunoglobulin production and regulation of eicosanoid production. However, mechanism(s) for these effects have not been elucidated. More studies examining the molecular mechanism of action of CLA are warranted.

References

1. Chin, S.F., Liu, W., Storkson, J.M., and Pariza, M.W. (1992) Dietary Sources of Conjugated Diene Isomers of Linoleic Acid, a Novelly Recognized Class of Anticarcinogens, *J. Food Comp. Anal. 5,* 185–207.
2. Ha, Y.L., Grimm, N.K., and Pariza, M.W. (1989) Newly Recognized Anticarcinogenic Fatty Acids: Identification and Quantitation in Natural and Processed Cheese, *J. Agric. Food Chem. 37,* 75–81.
3. Kepler, C.R., Hirons, K.P., McNeill, J.J., and Tove, S.B. (1966) Intermediates and Products of the Biohydrogenation of Linoleic Acid by *Butyrivibrio fibrisolvens, J. Biol. Chem. 241,* 1350–1354.
4. Ha, Y.L., Grimm, N.K., and Pariza, M.W. (1987) Anticarcinogens from Fried Ground Beef: Heated-Altered Derivatives of Linoleic Acid, *Carcinogenesis 8,* 1881–1887.
5. Belury, M.A., Nickel, K.P., Bird, C.E., and Wu, Y. (1996) Dietary Conjugated Linoleic Acid Modulation of Phorbol Ester Skin Tumor Promotion, *Nutr. Cancer 26,* 149–157.
6. Ip, C., Chin, S.F., Scimeca, J.A., and Pariza, M.W. (1991) Mammary Cancer Prevention by Conjugated Dienoic Derivatives of Linoleic Acid, *Cancer Res. 51,* 6118–6124.
7. Ip, C., Singh, M., Thompson, H.J., and Scimeca, J.A. (1994) Conjugated Linoleic Acid Suppresses Mammary Carcinogenesis and Proliferative Activity of the Mammary Gland in the Rat, *Cancer Res. 54,* 1212–1215.
8. Ip, C., Chin, S.F., Scimeca, J.A., and Thompson, H.J. (1995) Effect of Timing and Duration of Dietary Conjugated Linoleic Acid on Mammary Cancer Prevention, *Nutr. Cancer 24,* 241–247.
9. Ip, C., Jiang, C., Thompson, H.J., and Scimeca, J.A. (1997) Retention of Conjugated Linoleic Acid in the Mammary Gland Is Associated with Tumor Inhibition During the Post-Initiation Phase of Carcinogenesis, *Carcinogenesis 18,* 755–759.

10. Ip, C., Briggs, S.P., Haegele, A.D., Thompson, H.J., Storkson, J., and Scimeca, J.A. (1996) The Efficacy of Conjugated Linoleic Acid in Mammary Cancer Prevention Is Independent of the Level or Type of Fat in the Diet, *Carcinogenesis 17,* 1045–1050.
11. Ha, Y.L., Storkson, J., and Pariza, M.W. (1990) Inhibition of Benzo[a]pyrene-Induced Mouse Forestomach Neoplasia by Conjugated Dienoic Derivatives of Linoleic Acid, *Cancer Res. 50,* 1097–1101.
12. Cesano, A., Visonneau, S., Scimeca, J.A., Kritchevsky, D., and Santoli, D. (1998) Opposite Effects of Linoleic Acid and Conjugated Linoleic Acid on Human Prostatic Cancer in SCID Mice, *Anticancer Res. 18,* 833–838.
13. Lee, N., Kritchevsky, D., and Pariza, M.W. (1994) Conjugated Linoleic Acid and Atherosclerosis in Rabbit, *Atherosclerosis 124,* 19–25.
14. Nicolosi, R.J., Rogers, E.J., Kritchevsky, D., Scimeca, J.A., Huth, P.J. (1997) Dietary Conjugated Linoleic Acid Reduces Plasma Lipoproteins and Early Aortic Atherosclerosis in Hypercholesterolemic Hamsters, *Artery 22,* 266–277.
15. Park, Y., Albright, K.J., Liu, W., Storkson, J.M., Cook, M.E., and Pariza, M.W. (1997) Effect of Conjugated Linoleic Acid on Body Composition in Mice, *Lipids 32,* 853–858.
16. Houseknecht, K.L., Heuvel, J.P.V., Moya-Camarena, S.Y., Portocarrero, C.P., Peck, L.W., Nickel, K.P., and Belury, M.A. (1998) Dietary Conjugated Linoleic Acid Normalizes Impaired Glucose Tolerance in the Zucker Diabetic Fatty *fa/fa* Rat, *Biochem. Biophys. Res. Commun. 244,* 678–682.
17. Sugano, M., Tsujita, A., Yamasaki, M., Noguchi, M., and Yamada, K. (1998) Conjugated Linoleic Acid Modulates Tissue Levels of Chemical Mediators and Immunoglobulin in Rats, *Lipids 33,* 521–527.
18. Belury, M.A., and Kempa-Steczko, A. (1997) Conjugated Linoleic Acid Modulates Hepatic Lipid Composition in Mice, *Lipids 32,* 199–204.
19. Banni, S., Day, B.W., Evans, R.W., Corongiu, F.P., and Lombari, B. (1995) Detection of Conjugated Diene Isomers of Linoleic Acid in Liver Lipids of Rats Fed a Choline-Devoid Diet Indicates That the Diet Does Not Cause Lipoperoxidation, *J. Nutr. Biochem. 6,* 281–289.
20. Banni, S., Carta, G., Contini, M.S., Angioni, E., Deiana, M., Dessi, M.A., Melis, M.P., and Corongiu, F.P. (1996) Characterization of Conjugated Diene Fatty Acids in Milk, Dairy Products, and Lamb Tissues, *J. Nutr. Biochem. 7,* 150–155.
21. Sébédio, J.L., Juaneda, P., Dobson, G., Ramilison, I., Martin, J.C., Chardigny, J.M., and Christine, W.W. (1997) Metabolites of Conjugated Isomers of Linoleic Acid (CLA) in the Rat, *Biochim. Biophys. Acta 1345,* 5–10.
22. Thompson, H., Zhu, Z., Banni, S., Darcy, K., Loftus, T., and Ip, C. (1997) Morphological and Biochemical Status of the Mammary Gland as Influenced by Conjugated Linoleic Acid: Implication for a Reduction in Mammary Cancer Risk, *Cancer Res. 57,* 5067–5072.
23. Sugano, M., Tsujita, A., Yamasaki, M., Yamada, K., Ikeda., I., and Kritchevsky, D. (1997) Lymphatic Recovery, Tissue Distribution, and Metabolic Effects of Conjugated Linoleic Acid in Rats, *J. Nutr. Biochem. 8,* 38–43.
24. Fogerty, A.C., Ford, G.L., and Svoronos, D. (1988) Octadeca-9,11-dienoic Acid in Foodstuffs and in the Lipids of Human Blood and Breast Milk, *Nutr. Rep. Int. 38,* 937–944.
25. Cawood, P., Wickens, D.G., Iversen, S.A., Braganza, J.M., and Dormandy, T.L. (1983) The Nature of Diene Conjugation in Human Serum, Bile and Duodenal Juice, *FEBS Lett. 162,* 239–243.

26. Iversen, S.A., Cawood, P., Madigan, M.J., Lawson, A.M., and Dormandy, T.L. (1984) Identification of Diene Conjugated Component of Human Lipid as Octadeca-9,11-dienoic Acid, *FEBS Lett. 171,* 320–324.
27. Smith, G.N., Tai, M., and Braganza, J.M. (1991) On the Identification of a Conjugated Diene Component of Duogenal Bile as 9Z,11E-Octadecadienoic Acid, *Free Radic. Biol. Med. 10,* 13–21.
28. Harrison, K., Cawood, P., Iverson, A., and Dormandy, T. (1985) Diene Conjugation Patterns in Normal Human Serum, *Life Chem. Rep. 3,* 41–44.
29. Situnayake, R.D., Crump, B.J., Zezulka, A.V., Daris, M., McConkey, B., and Thurnham, D.I. (1990) Measurement of Conjugated Diene Lipids by Derivative Spectroscopy in Heptane Extracts of Plasma, *Ann. Clin. Biochem. 27,* 258–266.
30. Yurawecz, M.P., Roach, J.A.G., Sehat, N., Mossoba, M.M., Kramer, J.K.G., Fritsche, J., Steinhart, H., and Ku, Y. (1998) A New Conjugated Linoleic Acid Isomer, 7 *trans*, 9 *cis*-Octadecadienoic Acid, in Cow Milk, Cheese, Beef and Human Milk and Adipose Tissue, *Lipids 33,* 803–809.
31. Jiang, J., Bjorck, L., and Fonden, R. (1998) Production of Conjugated Linoleic Acid by Dairy Starter Cultures, *J. Appl. Microbiol. 85,* 95–102.
32. Herbel, B.K., McGuire, M.K., McGuire, M.A., and Shultz, T.D. (1998) Safflower Oil Consumption Does Not Increase Plasma Conjugated Linoleic Acid Concentrations in Human Plasma, *Am. J. Clin. Nutr. 67,* 332–337.
33. Pollard, M.R., Gunstone, F.D., James, A.T., and Morris, L.J. (1980) Desaturation of Positional and Geometric Isomers of Monoenoic Fatty Acid by Microsomal Preparations from Rat Liver, *Lipids 15,* 306–314.
34. Salminen, I., Mutanen, M., Jauhiainen, M., and Aro, A. (1998) Dietary *trans* Fatty Acids Increase Conjugated Linoleic Acid Levels in Human Serum, *J. Nutr. Biochem. 9,* 93–98.
35. Liu, K.L., and Belury, M.A. (1997) Conjugated Linoleic Acid Modulation of Phorbol Ester-Induced Events in Murine Keratinocytes, *Lipids 32,* 725–730.
36. Wynder, E.L. (1975) The Epidemiology of Large Bowel Cancer, *Cancer Res. 35,* 3388–3394.
37. Caygill, C.P.J., and Hill, M.J. (1995) Fish, n-3 Fatty Acids and Human Colorectal and Breast Cancer Mortality, *Eur. J. Cancer 4,* 329.
38. Caygill, C.P.J., and Hill, M.J. (1996) Fat, Fish, Fish Oil and Cancer, *Br. J. Cancer 74,* 159–164.
39. Carter, C.A., Milholland, R.J., Shea, W., and Ip, M.M. (1983) Effect of the Prostaglandin Synthetase Inhibitor Indomethacin on 7,12-Dimethylbenz[a]-anthracene-Induced Mammary Tumorigenesis in Rats Fed Different Levels of Fat, *Cancer Res. 43,* 3559–3562.
40. Liu, K.L., and Belury, M.A. (1998) Conjugated Linoleic Acid Reduces Arachidonic Acid Content and PGE_2 Synthesis in Murine Keratinocytes, *Cancer Lett. 127,* 15–22.
41. Cook, M.E., Miller, C.C., Park, Y., and Pariza, M.W. (1993) Immune Modulation by Altered Nutrient Metabolisms: Nutritional Control of Immune-Induced Growth Depression, *Poult. Sci. 72,* 1301–1305.
42. Miller, C.C., Park, Y., Pariza, M.W., and Cook, M.E. (1994) Feeding Conjugated Linoleic Acid to Animals Partially Overcomes Catabolic Responses Due to Endotoxin Injection, *Biochem. Biophys. Res. Commun. 198,* 1107–1112.
43. Liew, C., Schutt, H.A.J., Chin, S.F., Pariza, M.W., and Dashwood, R.H. (1995) Protection of Conjugated Linoleic Acids Against 2-Amino-3-methylimidazol[4,5-*f*]

quinoline-Induced Colon Carcinogenesis in the F344 Rat: A Study of Inhibitory Mechanisms, *Carcinogenesis 16,* 3037–3043.
44. Belury, M.A. (1995) Conjugated Dienoic Acid Linoleate: A Polyunsaturated Fatty Acid with Unique Chemoprotective Properties, *Nutr. Rev. 53,* 83–89.
45. Zurier, R.B. (1993) Fatty Acids, Inflammation and Immune Responses, *Prostaglandins Leukot. Essent. Fatty Acids 48,* 57–62.
46. Calder, P.C. (1995) Fatty Acid, Dietary Lipids and Lymphocyte Functions, *Biochem. Soc. Trans. 23,* 302–309.
47. Metcalfe, D.D. (1991) Food Allergy, *Curr. Opin. Immunol. 3,* 881–886.
48. Yamada, K., Hung, P., Yoshimura, K., Taniguchi, S., Lim, B.O., and Sugano, M. (1996) Effect of Unsaturated Fatty Acids and Antioxidants on Immunoglobulin Production by Mesenteric Lymph Node Lymphocytes of Sprague-Dawley Rats, *J. Biochem. 120,* 138–144.
49. Hung, P., Yamada, K., Lim, B.O., Mori, M., Yuki, T., and Sugano, M. (1997) Effect of Unsaturated Fatty Acids and α-Tocopherol on Immunoglobulin Levels in Culture Medium of Rat Mesenteric Lymph Node and Spleen Lymphocytes, *J. Biochem. 121,* 1054–1060.
50. Rola-Pleszczynski, M., and Lamoire, I. (1985) Leukotrienes Augment Interleukin 1 Production by Human Monocytes, *J. Immunol. 135,* 3958.
51. Elson, C.O., and Beagley, K.W. (1994) in *Physiology of the Gastrointestinal Tract,* (Johnson, L.R., ed.) pp. 243–265, Raven Press, New York.
52. Gauchat, J.F., Gascan, H., Roncarolo, M.G., Rousset, F., Pene, J., and de Vries, J.E. (1991) Regulation of Human IgE Synthesis: The Role of CD4+ and CD8+ T-Cells and the Inhibitory Effects of Interferon-α, *Eur. Respir. J. 4,* 31s–38s.
53. Ochel, M., Vohr, H.W., Pfeiffer, I., and Gleichmann, E. (1991) IL-4 Is Required for the IgE and IgG1 Increase and IgG1 Autoantibody Formation in Mice Treated with Mercuric Chloride, *J. Immunol. 146,* 3006–3011.
54. Pene, J., Rousset, F., Briere, F., Chretien, I., Wideman, J., Bonnefoy, J.Y., and de Vries, J.E. (1988) Interleukin 5 Enhances Interleukin 4-Induced IgE Production by Normal Human B Cells. The Role of Soluble CD23 Antigen, *Eur. J. Immunol. 18,* 929–935.
55. Prickett, J.D., Robinson, D.R., and Bloch, K.J. (1982) Enhanced Production of IgE and IgG Antibodies Associated with a Diet Enriched in Eicosapentaenoic Acid, *Immunology 46,* 819–826.
56. Watanabe, S., Sakai, N., Yasui, Y., Kimura, Y., Kobayashi, T., Mizutani, T., and Okayama, H. (1994) A High α-Linoleate Diet Suppresses Antigen-Induced Immunoglobulin E Response and Anaphylactic Shock in Mice, *J. Nutr. 124,* 1566–1573.
57. Endres, S., Ghorbani, R., Kelley, V.E., Georgilis, K., Lonnemann, G., van der Meer, J.W., Cannon, J.G., Rogers, T.S., Klempner, M.S., and Weber, P.C. (1989) The Effect of Dietary Supplementation with n-3 Polyunsaturated Fatty Acid on the Synthesis of Interleukin-1 and Tumor Necrosis Factor by Mononuclear Cells, *N. Engl. J. Med. 320,* 265–271.
58. Chew, B.P., Wong, T.S., Shultz, T.D., and Magnuson, N.S. (1997) Effects of Conjugated Dienoic Derivatives of Linoleic Acid and β-Carotene in Modulating Lymphocyte and Macrophage Function, *Anticancer Res. 17,* 1099–1106.
59. Wong, M.W., Chew, B.P., Wong, T.S., Hosick, H.L., Boylston, T.D., and Shultz, T.D. (1997) Effects of Dietary Conjugated Linoleic Acid on Lymphocyte Function and Growth of Mammary Tumors in Mice, *Anticancer Res. 17,* 987–994.
60. Belury, M.A., Moya-Camarena, S.Y., Liu, K.L., and Vander Heuvel, J.P. (1997) Dietary Conjugated Linoleic Acid Induces Peroxisome-Specific Enzyme Accumulation and Ornithime Decarboxylase Activity in Mouse Liver, *J. Nutr. Biochem. 8,* 579–584.

Chapter 26

Regulation of Stearoyl-CoA Desaturase by Conjugated Linoleic Acid

James M. Ntambi[a,b], Youngjin Choi[a], and Young-Cheul Kim[a]

[a]Department of Biochemistry
[b]Department of Nutritional Sciences, University of Wisconsin-Madison, Madison, WI 53706

Introduction

Conjugated dienoic derivatives of linoleic acid (CLA) is a collective term for a group of positional and geometric isomers of linoleic acid. The *cis*-9,*trans*-11 and, to a lesser extent, the *trans*-10,*cis*-12 isomers occur naturally in many foods, with higher concentrations found in products from ruminant animals. CLA has been clearly shown to have a number of biological actions including the following: It inhibits carcinogenesis and atherosclerosis in several animal models (1–5); it reduces body fat content and increases lean body mass in mice, rats, and pigs; and it has the ability to normalize the impaired glucose tolerance of Zucker diabetic fatty *fa/fa* rats (6–8). One of the effects of CLA that has been observed consistently regarding lipid metabolism is its ability to alter the fatty acid composition of tissues by reducing the levels of monounsaturated fatty acids (9). Monounsaturated fatty acids are synthesized by the stearoyl-CoA desaturase (SCD) enzyme, which catalyzes the Δ^9-*cis* desaturation of fatty acyl-CoA substrates. The preferred substrates are palmitoyl- and stearoyl-CoA (10), which are converted into palmitoleoyl- and oleoyl-CoA, respectively. Changes in the activity of stearoyl-CoA desaturase in tissues are reflected in cell membrane phospholipid and triglyceride composition; therefore, regulation of stearoyl-CoA desaturase has the potential to regulate a variety of key physiologic variables including insulin sensitivity, metabolic rate, adiposity, atherosclerosis, cancer, and obesity. These physiologic variables have all been shown to be influenced by CLA. The purpose of this chapter is to review the known effects of CLA on the regulation of the stearoyl-CoA desaturase in liver and during fat cell differentiation.

Effect of CLA on Stearoyl-CoA Desaturase Enzyme Activity

Studies of the effects of CLA on the fatty acid profile in chicken eggs, rabbit aorta, mouse and rat liver, human HepG2, and mouse liver H2.35 cells have shown that CLA acts by decreasing the levels of monounsaturated fatty acids, mainly palmitoleic (16:1n-7) and oleic acids (18:1n-9) (11). We examined the levels of monounsaturated fatty acids in livers of mice that had been maintained on a fat-free, high-carbohydrate diet (CHO) or a 5% corn oil diet (CO), each supplemented with 0.5% CLA for 2 wk (9). Fatty acid analysis of liver microsomes showed that livers of mice that were maintained on diets supplemented with CLA contained significantly lower amounts of monounsaturated fatty acids compared with those maintained on CHO or 5.0% corn oil (Table 26.1). Thus the ratio of 16:1 to 16:0 or 18:1 to 18:0 (Δ^9 desaturation index) was much higher in CHO- or CO-fed mice than in mice fed diets supplemented with CLA. The decrease in desaturation index indicates a decrease in stearoyl-CoA desaturase enzyme activity. The effect of CLA on the levels of monounsaturated fatty acids was also examined *in vitro* using H2.35 liver cells. The results showed that the ratios of 16:1 to 16:0 and 18:1 to 18:0 were higher in the control cells compared with the treated cells. Taken together, the *in vivo* and *in vitro* results indicate that CLA reduces the levels of monounsaturated fatty acids by reducing the activity of stearoyl-CoA desaturase. Polyunsaturated fatty acids (PUFA), mainly the n-3 and n-6 series, including linoleic acid, have also been shown to inhibit SCD activity in liver, adipocytes, and lymphocytes (12). However, CLA seems to be more potent than n-3 and n-6 PUFA in reducing the activity of stearoyl-CoA desaturase in mouse liver and mouse cell lines (9).

The consequences of the inhibition of stearoyl-CoA desaturase by CLA may be relevant to lipoprotein metabolism. The hepatic packaging and secretion of very low density lipoprotein (VLDL) require synthesis of apolipoprotein B-100 (apoB-100) as well as sufficient amounts of oleic acid, originating from the diet or synthesized by stearoyl-CoA desaturase (13). CLA has been shown to decrease plasma triglyceride (TG) as well as VLDL and LDL levels in rabbits and apoB-100 synthesis and TG secretion in HepG2 cells (14). Recent studies using cultured hepatocytes showed that oleic acid stimulated both the synthesis and secretion of VLDL and apoB-100 (15). These studies suggest that the decrease in VLDL and apoB-100 secretion by CLA could be due to its inhibitory effects on the stearoyl-CoA desaturase enzyme activity.

CLA has clearly been shown to be an anticarcinogenic nutrient, but the mechanism by which CLA inhibits neoplasia is not completely understood. Cancer cells appear to have an altered balance of saturated to monounsaturated fatty acids. For example, the saturated fatty acid to monounsaturated fatty acid ratio was found to be reduced in a

TABLE 26.1 Effect of Conjugated Linoleic Acid on 16:1/16:0 and 18:0/18:1 Ratios in Mouse Liver[a,b]

	Fatty acid (%)				Ratio (Δ^9 desaturation index)	
	16:1	16:0	18:1	18:0	16:1/16:0	18:1/18:0
CHO	4.3	27.1	42	12.8	0.16	3.3
CHO + 0.5% CLA	2.1[c]	30.2[c]	36.9[c]	19.7[c]	0.07[c]	1.9[c]
CO	2.1	32.3	18.1	19.7	0.07	0.9
CO + 0.5% CLA	0.8[c]	30.4	17.9	24.6[c]	0.03[c]	0.72[c]

[a]Values are means of pooled samples. SEM are <5% of mean.
[b]Fatty acid composition is expressed as a percentage of total recovered fatty acids. Mice were fed either fat-free, high-carbohydrate diet (CHO) or 5.0% corn oil diet (CO) with 0.5% CLA supplementation for 2 wk. Hepatic microsomal fractions were isolated. Total lipids were extracted and analyzed for fatty acid composition. *Source:* Ref. 9.
[c]The difference is significant at $P < 0.05$ by Student's *t*-test.

number of tumor cell lines or experimental tumors (16), indicating a higher activity of the stearoyl-CoA desaturase activity in these cells. Malignant cells from patients with leukemia were found to have a lower stearic to oleic acid ratio compared with normal white cells (17). Thus, one possible mechanism of the anticancer activity of CLA is to alter the ratio of saturated to monounsaturated fatty acids by reducing the activity of the stearoyl-CoA desaturase. Inhibition of SCD by CLA therefore may inhibit cancer cell growth. Further work is required to elucidate the role of SCD in cancer and the mechanism of action of CLA on SCD in this disease condition.

Many mechanisms may exist for regulating the activity of SCD and other enzymes, such as lipoprotein lipase and carnitine palmitoyltransferase, by CLA (6). Some evidence has been provided that CLA could have a direct inhibitory effect on these enzymes. In the case of the liver studies and adipocyte differentiation, the most convincing evidence, however, indicates that the effects of CLA on stearoyl-CoA desaturase enzyme activity are at the level of gene expression as discussed below.

CLA Decreases Stearoyl-CoA Desaturase mRNA Expression in Liver

Two mouse and rat stearoyl-CoA desaturase genes (SCD1 and SCD2) have been isolated and characterized (18–20). Dietary PUFA have been shown to suppress the expression of the SCD1 mRNA in liver (21). Very low levels of hepatic SCD1 mRNA are expressed when rats or mice are fed a regular semipurified diet that contains fat. When mice or rats are switched from that diet to a fat-free, high-carbohydrate diet, the expression of the SCD1 gene increases (21). We used this dietary manipulation to study the effect of CLA on the expression of the SCD1 mRNA in liver. We examined SCD1 mRNA levels in livers of mice that had been maintained on a fat-free, high-carbohydrate diet (CHO) or a 5% corn oil (CO) diet, each supplemented with 0.5% CLA for 2 wk (9). Mice fed the CHO diet expressed a high level of SCD1 mRNA in liver, consistent with previous results (21). However, the addition of 0.5% CLA to this diet decreased the expression of stearoyl-CoA desaturase mRNA by 45% (Fig. 26.1). Similarly, the addition of 0.5% of CLA to the CO diet decreased stearoyl-CoA desaturase mRNA by 75%. Fatty acid analysis in liver microsomes showed that the livers of mice that were maintained on diets supplemented with CLA contained significantly lower monounsaturated fatty acids compared with those of CHO- or CO-fed mice (Table 26.1). The ratio of 16:1 to 16:0 or 18:1 to 18:0 (Δ^9 desaturation index) was much higher in CHO- or CO-fed mice than in those fed diets supplemented with CLA. The decrease in desaturation index indicates a decrease in stearoyl-CoA desaturase enzyme activity. *In vitro*, treatment of H2.35 cultured liver cells with 150 µmol/L CLA resulted in a 50% decrease in SCD1 mRNA. Similar results have been obtained by treating human HepG2 cells with CLA (T. Drews and J. Ntambi, unpublished results). Taken together, the *in vivo* and *in vitro* results indicate that CLA reduces stearoyl-CoA desaturase enzyme activity by suppressing its mRNA expression.

Synthetically prepared CLA, which has been used in numerous experiments, consists of several different isomers (1). The two major isomers are *cis*-9,*trans*-11

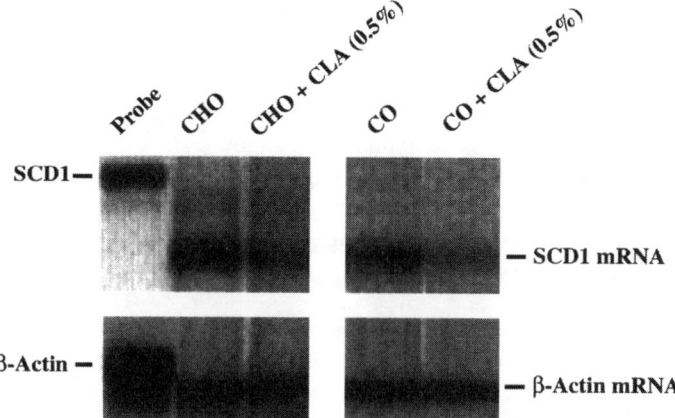

Fig. 26.1. Effect of conjugated linoleic acid (CLA) on hepatic stearoyl-CoA desaturase (SCD)1 mRNA content in mice. Mice were fed either a fat-free, high-carbohydrate diet (CHO) or a 5.0% corn oil diet (CO), each with or without 0.5% CLA supplementation, for 2 wk. Total RNA was isolated and analyzed by RNase protection assay for SCD1 and β-actin mRNA contents. The autoradiogram is representative of three separate experiments in which identical results were obtained. *Source:* Ref. 9.

CLA (42%) and *trans*-10,*cis*-12 CLA (44%). It has been suggested that each CLA isomer has a different biological activity (1). The results on the SCD1 gene expression in liver cells indicate that the *trans*-10,*cis*-12 CLA is the isomer that reduces both the enzyme activity and mRNA expression (9). The amount of *trans*-10,*cis*-12 CLA isomer that reduced SCD1 mRNA expression by >80% in H2.35 cells was approximately one half the amount of arachidonic acid (20:4 n-6) that was used to repress SCD mRNA in H2.35 cells. This would suggest that CLA is more potent than arachidonic acid in suppressing SCD mRNA.

In general, long-chain PUFA, e.g., 18:2n-6, 18:3n-3, 18:3n-6, 20:4n-6, 20:5n-3, and 22:6n-3, repress the expression of the genes that encode lipogenic enzymes including SCD1 (12). It was proposed that the basic requirements for a dietary fatty acid to inhibit expression of lipogenic genes were to contain 18 carbons and possess at least two methylene interrupted double bonds in the 9- and 12-positions (22). However, it should be noted that the *trans*-10,*cis*-12 CLA isomer, which is implicated in reducing SCD1 mRNA, does not contain a double bond at position 9 but contains double bonds at positions 6 and 8 from the ω-carbon. The only double bond position that CLA and other PUFA share is at position 6. The inhibitory effects that CLA and other PUFA have on the expression of the SCD1 mRNA levels may be related to the position and orientation of just one of the double bonds present in all of these PUFA.

The mechanisms of CLA action on SCD1 gene expression possibly involve decreased SCD mRNA stability and/or gene transcription. We have shown previously that PUFA repress the expression of the SCD1 gene in liver at the level of gene transcription, and we have localized a PUFA responsive element (PUFA-RE) in the promoter of stearoyl-CoA desaturase genes (23). Whether this region mediates the action of CLA on the transcription of the SCD1 gene awaits further investigation. Recently, it has been shown that dietary CLA is associated with elevated expression of a number of genes such as acyl-CoA oxidase that are responsive in part to the peroxisome proliferator-activated receptor-α (PPARα) in liver (24). Because peroxisome proliferators activate SCD1 gene expression in mouse liver (25), it is unlikely that CLA represses SCD1 gene expression through a PPARα-dependent mechanism.

CLA Regulates Stearoyl-CoA Desaturase Gene Expression During Adipocyte Differentiation

Adipose tissue plays an important role in the metabolism of lipids. Fat cells differentiate from fat cell precursors or preadipocytes. Studies from several laboratories that used the 3T3-L1 and 3T3-F442A preadipocyte mouse adipogenic cell lines have shown that the mechanism of adipocyte differentiation involves the interplay between growth factors and transcription factors such as C/EBPα and PPARγ2 (26,27). The induction of C/EBPα and PPARγ2 in the early stages of adipocyte differentiation leads to the transcriptional activation of adipocyte genes including stearoyl-CoA desaturase. Stearoyl-CoA desaturase enzyme then synthesizes palmitoleic acid and oleic acids, which constitute ~58% of the monounsaturated fatty acids present in differentiated 3T3-L1 preadipocytes and also in mouse adipose tissue *in vivo*. We and others (28) have found that CLA inhibits the differentiation of 3T3-L1 preadipocytes into adipocytes. In the presence of 100 µmol/L CLA, the cells do not accumulate lipid droplets and fail to express SCD1 mRNA. The expression of other adipocyte genes such as adipocyte P2 (aP2) is also reduced but not to the same extent as SCD1 (Fig. 26.2). CLA also inhibits the proliferation of 3T3-L1 preadipocytes (28). The mechanism by which CLA inhibits adipocyte differentiation and adipocyte gene expression is presently unknown. However, studies from this laboratory have shown that prostaglandin $F_{2\alpha}$ (PGF$_{2\alpha}$), a product of linoleic acid metabolism, inhibits adipocyte differentiation through the prostanoid FP^2 (FP) receptor (29). Activation of the FP receptor results in activation of the mitogen-activated protein (MAP) kinase pathway, which results in phosphorylation of PPARγ2 with consequent inhibition of adipogenesis and adipocyte gene expression (30). Whether CLA uses a similar mechanism to inhibit adipocyte differentiation remains to be determined. In contrast to the *in vitro* studies with 3T3-L1 cells, *in vivo* studies in *fa/fa* rats indicate that CLA feeding results in enhanced adipocyte differentiation with enhanced adipocyte gene expression (7). In these studies, the CLA is thought to mimic the effects of thiazolidinediones (TZD) by acting as a ligand for PPARγ2. More directly, it has been shown that CLA is able to activate the PPARγ2 *in vitro* (7). The CLA probably binds directly to PPARγ2, thereby activating the receptor. The activated receptor then interacts with a specific DNA sequence in responsive genes such as aP2 and subsequently activates adipocyte gene transcription. Recently, we have found that the *trans*-10,*cis*-12-isomer of CLA is the inhibitor of the differentiation of 3T3-L1 preadipocytes into adipocytes, whereas the *cis*-9,*trans*-11 CLA isomer has no inhibitory effects and in fact seems to enhance adipocyte differentiation (Y. Choi and J. M. Ntambi, unpublished results). On the basis of both the *in vitro* and *in vivo* observations, it is pos-

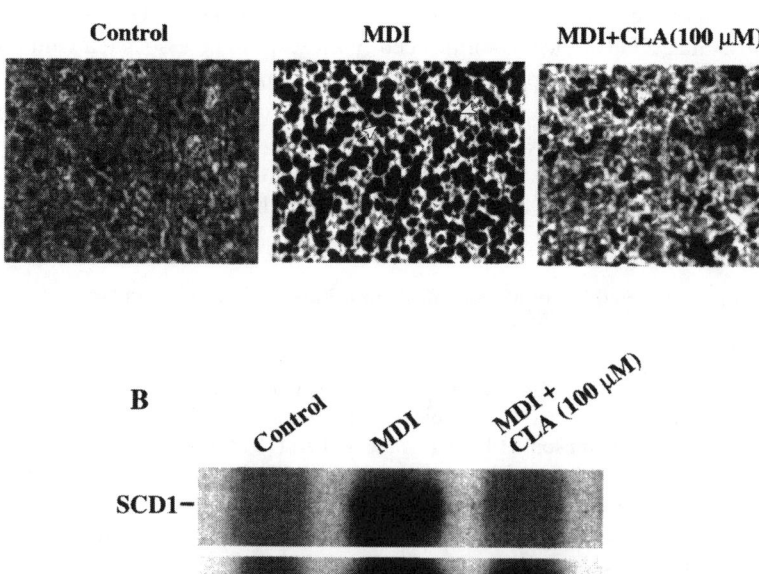

Fig. 26.2. Conjugated linoleic acid (CLA) inhibits preadipocyte differentiation and expression of stearoyl-CoA desaturase (SCD)1 mRNA. Two-day postconfluent 3T3-L1 preadipocytes were stimulated to differentiate with the standard differentiation cocktail (MDI) as previously described (18) or MDI supplemented with 100 µmol/L CLA. (A) After 7 d, Oil Red O staining was performed on the preadipocytes (control), MDI and MDI + 100 µmol/L CLA cells. Arrow points to a fat droplet in a differentiated cell. (B) Total RNA was isolated and analyzed by Northern blot using SCD1 or aP2 (28) specific cDNA radioactive probes. The lower pannel indicates ethidium bromide-stained gel, showing 28S and 18S ribosomal RNAs for equal loading of RNA.

sible that fat reduction caused by CLA treatment in animals is due to the *trans*-10,*cis*-12 isomer acting as the inhibitor, whereas the *cis*-9,*trans*-11 mimics the antidiabetic and adipogenic actions of TZD by activating PPAγ2. More research is required in this area.

Summary

Although CLA has demonstrated a range of beneficial biological activities in various biological models, its mechanism of action is not yet understood completely. The data reported in the past few years have confirmed that CLA is a nutrient that can act as a cellular regulator. The evidence presented in this chapter shows that CLA plays a role in lipid metabolism by regulating the expression of the stearoyl-CoA desaturase gene in liver and during adipocyte differentiation. Regulation of stearoyl-CoA desaturase results in changes in the levels of monounsaturated fatty acids in tissues. Changes in the levels of monounsaturated fatty acids in various cell types seem to be associated with insulin sensitivity, metabolic rate, adiposity, atherosclerosis, cancer, and obesity, all of which have been shown to be influenced by CLA. The studies on the regulation of SCD gene expression have begun to address the molecular action of CLA. Our knowledge of the mechanism of action of CLA is therefore expanding; continued research should provide insight into the role of CLA in diseases such as cancer, diabetes, and atherosclerosis.

Acknowledgments

We thank Henry Bene and Victor Munsen for their critical review of this chapter.

References

1. Ha, Y.L., Grimm, N.K., and Pariza, M.W. (1987) Anticarcinogens from Fried Ground Beef: Heat-Altered Derivatives of Linoleic Acid, *Carcinogenesis 8,* 1881–1887.
2. Ip, C., Singh, M., Thompson, H.J., and Scimeca, J.A. (1994) Conjugated Linoleic Acid Suppresses Mammary Carcinogenesis and Proliferative Activity of the Mammary Gland in the Rat, *Cancer Res. 54,* 1212–1215.
3. Lee, K.N., Kritchevsky, D., and Pariza, M.W. (1994) Conjugated Linoleic Acid and Atherosclerosis in Rabbits, *Atherosclerosis 108,* 19–25.
4. Ha, Y.L., Storkson, J., and Pariza, M.W. (1990) Inhibition of Benzo(a)pyrene-Induced Mouse Forestomach Neoplasia by Conjugated Dienoic Derivatives of Linoleic Acid, *Cancer Res. 50,* 1097–1101.
5. Ip, C., Chin, S.F., Scimeca, J.A., and Pariza, M.W. (1991) Mammary Cancer Prevention by Conjugated Dienoic Derivative of Linoleic Acid, *Cancer Res. 51,* 6118–6124.
6. Park, Y., Albright, K.J., Liu, W., Storkson, J.M., Cook, M.E., and Pariza, M.W. (1997) Effect of Conjugated Linoleic Acid on Body Composition in Mice, *Lipids 32,* 853–858.
7. Houseknecht, K.L., Vanden Heuvel, J.P., Moya-Camarena, S.Y., Portocarrero, C.P., Peck, L.W., Nickel, K.P., and Belury, M.A. (1998) Dietary Conjugated Linoleic Acid Normalizes Impaired Glucose Tolerance in the Zucker Diabetic Fatty *fa/fa* Rat, *Biochem. Biophys. Res. Commun. 244,* 678–682.
8. Dugan, M.E.R., Aalhus, J.L., Schaefer, A.L., and Kramer, J.K.G. (1997) The Effect of Conjugated Linoleic Acid on Fat to Lean Repartitioning and Feed Conversion in Pigs, *Can. J. Anim. Sci. 77,* 723–725.
9. Lee, K.N., Pariza, M.W., and Ntambi, J.M. (1998) Conjugated Linoleic Acid Decreases Hepatic Stearoyl-CoA Desaturase mRNA Expression, *Biochem. Biophys. Res. Commun. 248,* 817–821.
10. Enoch, H.G., Catala, A., and Stritmatter, P. (1976) Mechanism of Rat Liver Microsomal Stearyl-CoA Desaturase. Studies of the Substrate Specificity, Enzyme-Substrate Interactions and Function of the Lipid, *J Biol. Chem. 251,* 5095–5103.
11. Lee, K.N., Storkson, J.M., and Pariza, M.W. (1995) Dietary Conjugated Linoleic Acid Changes Fatty Acid Composition in Different Tissues by Decreasing Monounsaturated Fatty Acids. IFT Annual Meeting, *Book of Abstracts,* p. 183.
12. Sessler, A.M., and Ntambi, J.M. (1998) Polyunsaturated Fatty Acid Regulation of Gene Expression, *J. Nutr. 128,* 923–926.
13. Gibbins, G.F. (1990) Assembly and Secretion of Hepatic Very Low Density Lipoprotein, *Biochem. J. 268,* 1–13.

14. Lee, K.N. (1996) Conjugated Linoleic Acid and Lipid Metabolism, Ph.D. Thesis, University of Wisconsin, Madison.
15. Moberly, J.B., Cole, T.G., Alpers, D.H., and Schonfeld, G. (1990) Oleic Acid Stimulation of Apolipoprotein B Secretion from HepG2 Cells and Caco2 Cells Occurs Post Transcriptionally, *Biochem Biophys. Acta 1042,* 70–80.
16. Calorini, L., Fallani, A., Tombaccini, D., Mugnai, G., and Ruggieri, S. (1987) Lipid Composition of Cultured B16 Melanoma Cell Variants with Different Lung-Colonizing Potential, *Lipids 22,* 651–656.
17. Apostolov, K., Barker, W., Catovsky, D., Goldman, J., and Matutes, E. (1985) Reduction in the Stearic to Oleic Acid Ratio in Leukaemic Cells—A Possible Chemical Marker of Malignancy, *Blut 50,* 349–354.
18. Ntambi, J.M., Buhrow, S.A., Kaestner, K.H., Christy, R.J., Sibley, E., Kelly, T.J., and Lane, M.D. (1988) Differentiation-Induced Gene Expression in 3T3-L1 Preadipocytes: Characterization of a Differentially Expressed Gene Encoding Stearoyl-CoA Desaturase, *J. Biol. Chem. 263,* 17291–17300.
19. Kaestner, K.H., Ntambi, J.M., Kelly, T.J., and Lane M.D. (1989) Differentiation-Induced Gene Expression: A Second Differentially Expressed Gene Encoding Stearoyl-CoA Desaturase, *J. Biol. Chem. 264,* 14755–14761.
20. Mihara, K. (1990) Stucture and Regulation of of Rat Liver Microsomal Stearoyl-CoA Desaturase Gene, *J. Biochem. 108,* 1022–1029.
21. Ntambi, J.M. (1992) Dietary Regulation of the Stearoyl-CoA Desaturase 1 Gene Expression in Mouse Liver, *J. Biol. Chem. 267,* 10925–10930.
22. Clarke, S.D., and Jump, D.B. (1993) Regulation of Gene Transcription by Polyunsaturated Fatty Acids, *Prog. Lipid Res. 32,* 139–149.
23. Waters, K.M., Miller, C.M., and Ntambi, J.M. (1997). Localization of a Polyunsaturated Fatty Acid Response Region in Stearoyl-CoA Desaturase Gene 1, *Biophys. Biochem. Acta 1349,* 33–42.
24. Belury, M.A., Moyacamarena, S.Y., Liu, K.L., and Heuvel, J.P.V. (1997) Dietary Conjugated Linoleic Acid Induces Peroxisome-Specific Enzyme Accumulation and Ornithine Decarboxylase Activity in Mouse Liver, *J. Nutr. Biochem. 8,* 579–584.
25. Miller, C.W., and Ntambi, J.M. (1996) Peroxisome Proliferators Induce Mouse Liver Stearoyl-CoA Desaturase 1 Gene Expression, *Proc. Natl. Acad. Sci. USA 93,* 9443–9448.
26. Lin, F.T., and Lane, M.D. (1994) CCAAT/Enhancer Binding Protein Alpha Is Sufficient to Initiate the 3T3-L1 Adipocyte Differentiation Program, *Proc. Natl. Acad. Sci. USA 91,* 8757–8761.
27. Tontonoz, P., Hu, E., Graves, R.A., Budavari, A.I., and Spiegelman, B.M. (1994) mPPAR gamma 2: Tissue-Specific Regulator of an Adipocyte Enhancer, *Genes Dev. 8,* 1224–1234.
28. Brodie, A.E., Manning, V.A., Ferguson, K.R., Jewell, D.E., and Hu, C.Y. (1998) Conjugated Linoleic Acid Inhibits Differentiation of Pre- and Post-Confluent 3T3-L1 Adipocytes but Inhibits Cell Proliferation Only in Preconfluent Cells, *J. Nutr.,* in press.
29. Casimir, D.A., Miller, C.W., and Ntambi, J.M. (1996). Preadipocyte Differentiation Blocked by Prostaglandin Stimulation of Prostanoid FP^2 Receptor in Murine 3T3-L1 Cells, *Differentiation 60,* 203–210.
30. Reginato, M.J., Krakow, S.L., Bailey, S.T., and Lazar, M.A. (1998) Prostaglandins Promote and Block Adipogenesis Through Opposing Effects on Peroxisome Proliferator-Activated Receptor γ, *J. Biol. Chem. 273,* 1855–1858.

Chapter 27

Conjugated Linoleic Acid for Altering Body Composition and Treating Obesity

Richard L. Atkinson

Departments of Medicine and Nutritional Sciences, University of Wisconsin, Madison, WI

Introduction

Conjugated linoleic acid (CLA) is currently marketed in the United States as a dietary supplement. Except under special circumstances, current Federal law does not allow direct marketing claims for dietary supplements, but the focus of marketing efforts is directed toward the implications that CLA will reduce fat mass and increase lean body mass in humans. The basis for these clinical impressions arises from animal studies that demonstrate remarkable alterations in body fat and lean body mass (1–4). At present, few studies exist that have evaluated the mechanisms of action of CLA in altering body composition, and there are no peer-reviewed studies in the medical literature that have documented the effects of CLA on humans. This chapter will discuss some of the potential mechanisms of action of CLA, briefly review the animal studies, and discuss the very limited data on the effects of CLA in humans.

Sources of CLA in the Human Diet

CLA comes predominantly from animal products in the human diet (5). As reviewed elsewhere in this volume, CLA is produced by bacteria in the ruminant stomach of animals used for food. The major sources in the American diet are diary products and meat from cattle, sheep, and goats. Few other ruminant animals are eaten in the United States, but may be a more significant source in certain other countries. Fat contained in milk, butter, and cheese contains the most concentrated amounts of CLA. For reasons that are not entirely clear, turkey meat has a modest amount of CLA. Finally, hydrogenated fats, such as margarines and cooking oils, may have modest amounts of CLA.

The usual intake of CLA for the average American is ~100–300 mg/d. It is difficult to obtain more than ~400–500 mg/d from the usual dietary sources. It has been observed that "range fed" ruminant animals have more CLA than do animals raised in a feed lot. This is apparently due to a greater tendency to form CLA from the vegetation eaten by ruminant animals in the fields compared with the production of CLA from grains fed in the feedlot. However, there are seasonable variations in the amount of CLA produced and differences among different areas of the country, even in "range fed" animals.

From the animal studies described below, it is clear that large amounts of CLA must be ingested to produce the desired alterations in body composition. The only way to obtain quantities sufficient for this purpose is to ingest additional amounts of purified CLA as a dietary supplement or enriched food.

Potential Mechanisms of Action of CLA

There are numerous theories on the mechanisms of alterations in body composition seen in animals ingesting CLA, but few studies have been carried out to address this question. Table 27.1 lists some of the potential mechanisms by which CLA may reduce fat mass and/or increase lean body mass. Alteration of membrane structure is an overarching potential mechanism that might affect a number of activities such as enzyme and hormonal response, permeability, membrane fluidity, and receptor number and/or function. CLA has a different spatial configuration than the parent linoleic acid molecule, which could produce an altered membrane structure.

CLA may be metabolized to different compounds in the eicosanoid pathway, thus affecting the amounts of arachidonic acid and prostaglandins (6). Phinney *et al.* (7) postulated that altered concentrations and metabolism of arachidonic acid are involved in the etiology of obesity of Zucker fatty rats.

CLA is known to alter cytokines (8), and tumor necrosis factor (TNF)-α and other cytokines may play major roles in fat metabolism and the etiology of at least some types of obesity. For example, Hotamisligil *et al.* (9) showed that TNF-α may be involved in the insulin resistance of enlarged adipocytes.

Abnormalities in the genetic regulation of peroxisome proliferator–activated receptor (PPAR) activity have been implicated in the etiology of obesity (10). PPAR is a critical enzyme in the regulation of adipocyte differentiation, and alterations in PPAR might affect the ability of an animal or human to form additional adipocytes. CLA is known to alter PPAR metabolism (11).

Increased amounts of CLA in the diet reduce fat mass and increase lean body mass. This is the picture seen with administration of β-3 adrenergic agonists to animals (12) and is postulated to occur in humans (13). β-3 agonists increase both the metabolic rate and the proportion of energy derived from fat oxidation. CLA has been shown to increase metabolic rate and specifically to increase fat oxidation.

Finally, it is possible that CLA might alter the secretion or action of growth hormone, insulin-like growth factor, or other growth factors. Administration of growth hormone results in an increase in accumulation of lean body mass and a decrease in

TABLE 27.1 Potential Mechanisms of Action of CLA to Alter Body Composition

1.	Alter membrane structure or function
2.	Alter eicosanoid, arachidonic, and/or prostaglandin metabolism
3.	Alter cytokine production or response
4.	Alter peroxisome proliferator–activated receptor (PPAR) activity
5.	Alter sympathetic nervous system activity
6.	Alter growth hormone or growth factors

fat mass due to increased lipolysis (14). Chin *et al.* (1) suggested that CLA acts like a growth factor, at least in rats, because it enhanced weight gain and improved feed efficiency.

Animal Studies: Effects of CLA on Body Composition

The effects of CLA on body composition in several species of animals will be discussed in other chapters in this volume; thus, the topic will be only summarized here. CLA has been shown to reduce body fat and increase lean body mass in several species of animals including chickens, mice, rats, and pigs (1–5,15). In virtually all of the studies in animals, growing animals were used. Body composition in animals fed CLA was compared with that of animals fed a control diet, usually with linoleic acid or vegetable oil. In many of these studies, the actual amount of fat increased as the animal grew, but the percentage as well as the total amount of body fat was lower in the CLA-fed animals compared with the controls.

Mice are the most responsive of animals to CLA (2,4,5). In some studies, mice fed CLA had 70% less body fat, but the average reduction compared with controls was ~50%. Rats were less responsive and had only ~15% less body fat than the controls. Pigs are especially interesting because they are probably the closest model for humans that has been tested. Pariza *et al.* (16, unpublished data) studied pigs starting at about 25 kg and followed their ingestion of CLA or control diet for almost 4 mo, or to an end weight of ~140 kg. Pigs were fed 0.48 and 0.95% CLA by weight of the diet. Because pigs eat ~3–5 kg of food each day, the amount of pure CLA was ~40 g/d for some animals. There was a statistically significant reduction in backfat thickness of 27% compared with control pigs fed vegetable oil. Conversely, lean body mass increased by ~5%. Feed efficiency was increased in the CLA-fed pigs, who gained more weight per kilogram of food fed. An extensive evaluation for side effects of CLA was conducted before and after killing. Laboratory tests revealed an elevated white blood cell count in CLA-fed pigs compared with controls. At autopsy, there were no abnormalities in any of the organs tested, including brain, heart, lungs, liver, spleen, kidneys, and gastrointestinal tract.

Human Studies: Effects of CLA on Body Composition

Only three studies performed in humans to evaluate body composition exist, and these have been published only in abstract form. The first study was performed in Norway on 20 subjects who were of normal weight to mildly overweight (E. Thom, unpublished data). This was a randomized, double-blind, placebo-controlled trial of CLA, 1.8 g/d, vs. placebo for a 3-mo period. Body composition measurements were performed using a near infrared apparatus (Futrex 5000A). The initial body weights and percentage body fat measurements for the CLA and placebo subjects were 70.9 vs. 71.8 kg and 21.3 vs 22.0%, respectively. After 3 mo, the placebo group had a body weight of 72.4 kg and body fat of 22.4%. The CLA group had a nonsignificant decrease in body weight to 70.2 kg, but the percentage of body fat decreased by 4.3% to 17.0% ($P < 0.05$).

Ferreira et al. (17) evaluated 24 experienced resistance-trained males who were matched into two groups for body weight and total training volume. In a randomized study design, the CLA group was given six capsules per day of a 60% pure CLA preparation and the placebo group was given olive oil capsules for a total period of 28 d. Body composition was determined by dual-energy X-ray absorptiometry (DEXA). Subjects also were tested for strength performance with bench press and leg press exercises at the beginning and end of the 28-d period. There were no significant differences in body mass, percentage of body fat, or lean body mass between placebo and CLA subjects. Also, there were no significant changes in strength performance on the exercise tests. However, the trends for both strength exercises favored an improved performance in the CLA group; thus it is possible that a Type II error was present.

The length of time that the CLA preparation was given was short compared with the time it was administered in animals, and the doses of CLA given to the humans were much lower on a per kilogram basis. Additional studies would be required with a larger number of subjects given CLA over a longer time to conclude that it is not effective in enhancing muscle strength and lean body mass.

We evaluated CLA for the treatment of obesity (18). Eighty obese subjects were randomized to CLA vs. placebo in a double-blind fashion over 6 mo. All subjects were asked to follow a standardized diet and exercise regimen with a modest reduction in calories and a modest increase in exercise. Subjects took 2.7 g of CLA daily and body composition was assessed by underwater weights. Mean body weight at baseline in the CLA and placebo groups was 97.1 kg and 96.9 kg, respectively. The percentage of body fat was 38.8 and 35.7%, respectively. There were no significant changes in either body weight or body fat after 6 mo. Both groups lost ~2.5 kg of weight and ~1 kg of body fat. *Post-hoc* analyses of the data revealed a subpopulation of individuals who gained lean body mass. There were twice as many subjects in the CLA group than in the placebo group who gained lean body mass, but the CLA subjects lost body fat on average, whereas the placebo group gained. The difference in body fat change was statistically significant. There were no major side effects or adverse events noted in either group, with the exception of one person in the CLA group who exhibited significant edema and weight gain for a period of several weeks. This resolved when CLA was discontinued, but the subject restarted CLA without the knowledge of the investigators and had no further problems over the last ~3 mo of the study.

This study had several limitations, including the fact that only one dose of CLA was studied, and that all subjects were following a weight-losing diet and exercise program. In virtually all of the animal studies, growing animals were used. It seems possible that CLA may act to prevent the accumulation of body fat and promote the accumulation of lean body mass during the weight gain process. Thus, this protocol may not have been the appropriate design to determine an effect of CLA on body composition.

In all animal species tested, CLA produced a reduction in body fat and an increase in lean body mass compared with the control group. Because almost all of the studies were done in growing animals, it is difficult to determine the effects of CLA on body composition of adult animals. The three human studies were performed on adults, and two of the three did not show any effect on body fat or lean body mass. However, the doses of CLA used in humans were much lower than those used in animals. Animals were fed ~1% of the diet by weight. For 150-g rats consuming ~20 g of food per day, this represented ~0.2 g of CLA, or ~1.3 g/kg of CLA per day. Pigs ate about 4 kg of food per day, and received ~40 g of 60% pure CLA, or ~24 g/d. A 100-kg pig received 240 mg/(kg·d) of CLA. In contrast, our 97-kg obese humans received 2.7 g of pure CLA per day, or ~27 mg/(kg·d) of CLA. Studies with higher doses of CLA are warranted.

Summary and Conclusions

CLA is a metabolically active compound that dramatically reduced body fat and increased lean body mass in growing animals in which it has been tested. The potential mechanisms of action are not known, but because CLA has been demonstrated to affect a wide variety of enzymes and hormones in the body, multiple mechanisms may come into play. Priority areas for future research include attention to the changes CLA produces in membrane structure and function, the effects on prostaglandin and cytokine function, and the effects on sympathetic nervous system activity.

Human studies of CLA have been disappointing; two of three studies did not demonstrate an effect of CLA on body composition. However, uncertainty about the appropriate dose for humans, coupled with the possibility that the effects may be best demonstrated during a weight accumulation phase, suggests that dose-response studies are critical. Studies of the usefulness of CLA in the prevention of weight gain would be of major interest. In addition, studies in growing children may be considered in the future when questions of dose are answered and more information on safety is available.

Acknowledgments

Supported in part with funds from Natural Nutrition, Volda, Norway, and the Beers-Murphy Clinical Nutrition Center, University of Wisconsin, Madison, WI.

References

1. Chin, S.F., Storkson, J.M., Albright, K.J., Cook, M.E., and Pariza M.W. (1994) Conjugated Linoleic Acid Is a Growth Factor for Rats as Shown by Enhanced Weight Gain and Improved Feed Efficiency, *J. Nutr. 124,* 2344–2349.
2. Park, Y., Albright, K.J., Liu, W., Storkson, J.M., Cook, M.E., and Pariza, M.W. (1997) Effect of Conjugated Linoleic Acid on Body Composition in Mice, *Lipids 32,* 853–858.
3. Pariza, M.W., Park, Y., Kim, S., Sugimoto, K., Albright, K., Liu, W., Storkson, J., and Cook, M. (1997) Mechanism of Body Fat Reduction by Conjugated Linoleic Acid, *FASEB J. 11,* A139 (Abstr.).

4. West, D.B., Delany, J.P., Camet, P.M., Blohm, F., Truett, A.A., and Scimeca, J. (1998) Effects of Conjugated Linoleic Acid on Body Fat and Energy Metabolism in the Mouse, *Am. J. Physiol. 44*, R667–R672.
5. Pariza, M.W. (1997) Conjugated Linoleic Acid, a Newly Recognized Nutrient, *Chemistry and Industry,* June 16, pp. 464–466.
6. Liu, K.L., and Belury, M.A. (1998) Conjugated Linoleic Acid Reduces Arachidonic Acid Content and PGE(2) Synthesis in Murine Keratinocytes, *Cancer Lett. 127,* 15–22.
7. Phinney, S.D., Davis, P.G., Johnson, S.B., and Holman, R.T. (1991) Obesity and Weight Loss Alter Serum Polyunsaturated Lipids in Humans, *Am. J. Clin. Nutr. 53,* 8318.
8. Turek, J.J., Li, Y., Schoenlein, I.A., Allen, K.G.D., and Watkins, B.A. (1998) Modulation of Macrophage Cytokine Production by Conjugated Linoleic Acids Is Influenced by the Dietary n-6–n-3 Fatty Acid Ratio, *J. Nutr. Biochem. 9,* 258–266.
9. Hotamisligil, G.S., Arner, P., Caro, J.F., Atkinson, R.L., and Spiegelman, B.M. (1995) Increased Adipose Tissue Expression of Tumor Necrosis Factor-Alpha in Human Obesity and Insulin Resistance, *J. Clin. Investig. 95,* 2409–2415.
10. Ristow, M., Muller-Wieland, D., Pfeiffer, A., Krone, W., and Kahn, C.R. (1998) Obesity Associated with a Mutation in a Genetic Regulator of Adipocyte Differentiation, *N. Engl. J. Med. 339,* 953–959.
11. Belury, M.A., Moyacamarena, S.Y., Liu, K.L., and Heuvel, J.P.V. (1997) Dietary Conjugated Linoleic Acid Induces Peroxisome-Specific Enzyme Accumulation and Ornithine Decarboxylase Activity in Mouse Liver, *J. Nutr. Biochem. 8,* 579–584.
12. Sasaki, N., Uchida, E., Niiyama, M., Yoshida, T., and Saito, M. (1998) Anti-Obesity Effects of Selective Agonists to the Beta 3-Adrenergic Receptor in Dogs. II. Recruitment of Thermogenic Brown Adipocytes and Reduction of Adiposity after Chronic Treatment with a Beta 3-Adrenergic Agonist, *J. Vet. Med. Sci. 60,* 465–469.
13. Liu, Y.L., Toubro, S., Astrup, A., and Stock, M.J. (1995) Contribution of β_3-Adrenoceptor Activation to Ephedrine-Induced Thermogenesis in Humans, *Int. J. Obes. 19,* 678–685.
14. Rudman, D., Feller, A.G., Nagraj, H.S., Gergans, G.A., Lalitha, P.Y., Goldbery, A.F., Schlenker, R.A., Cohn, L., Rudman, I.W., and Mattson, D.E. (1990) Effects of Human Growth Hormone in Men over 60 Years Old, *N. Engl. J. Med. 323,* 1–6.
15. Dugan, M.E.R., Aalhus, J.L., Schaefer, A.L., and Kramer, J.K.G. (1997) The Effect of Conjugated Linoleic Acid on Fat to Lean Repartitioning and Feed Conversion in Pigs, *Can. J. Anim. Sci. 77,* 723–725.
16. Doyle, E. (1998) Scientific Forum Explores CLA Knowledge, *INFORM 9,* 69–72.
17. Ferreira, M., Kreider, R., Wilson, M., and Almada, A. (1999) Effects of Conjugated Linoleic Acid (CLA) Supplementation During Resistance Training on Body Composition and Strength, *J. Strength Conditioning Res.,* in press.
18. Atkinson, R.L., Gomez, T., Clark, R.L., and Pariza, M.W. (1998) Clinical Implications for CLA in the Treatment of Obesity, Program of the Annual Meeting, National Nutritional Foods Association, July 15–16, San Antonio, TX.

Chapter 28

Feeding CLA to Pigs: Effects on Feed Conversion, Carcass Composition, Meat Quality, and Palatability

Michael E.R. Dugan and Jennifer L. Aalhus

Meat Research Section, Lacombe Research Centre, Agriculture and Agri-Food Canada, Lacombe, Alberta, Canada T4L 1W1

Introduction

Studying the effects of feeding conjugated linoleic acid (CLA) to meat animals is an exciting and rapidly developing field of research. In the past few years, CLA consumption has been found to have many potential benefits, as outlined in other chapters, including protection against many types of cancer, atherosclerosis, and the negative effects associated with immune stimulation. Meat animals rarely succumb to cancer or atherosclerosis, for obvious reasons, but reductions in feed intake and body weight due to immune stimulation can represent substantial economic losses. Interest in feeding CLA to livestock was further enhanced by reports that CLA improved feed efficiency in rats (1) and repartitioned body fat to lean in mice (2).

Lacombe's Interest in CLA

The Research Branch of Agriculture & Agri-Food Canada has 18 specialized Centres, and the Lacombe Research Centre holds the national mandate for fresh meat research. Part of this mandate includes finding ways to improve feed utilization and lean production efficiencies. In the past, improvements in efficiencies have been realized through genetic selection and, more recently, by treatment with exogenous growth promotants. However, the use of exogenous promotants (e.g., β-adrenergic agonists or growth hormone) to improve animal growth and feed efficiency has been thwarted because of their effects on meat quality and consumer worries over potential residues. Injecting pigs with porcine somatotropin (pST) dramatically increases lean yield and improves feed utilization efficiency (3), but pST has been found to negatively affect pork quality by reducing marbling fat levels (4). Feeding the β-adrenergic agonist ractopamine also increases lean yield but has been reported to increase pork shear force (5). Therefore, testing CLA at Lacombe was of interest to determine whether CLA could improve production efficiencies without adversely affecting meat quality or palatability. In addition, in contrast to worries over residues associated with other exogenous growth promotants, CLA-enriched meats could potentially be marketed as value-added products.

Early CLA research was limited to feeding small animals, but large-animal feeding trials are now feasible as a result of large-scale commercial CLA production.

Studies on the effects of feeding CLA to pigs are at various stages of completion at a number of institutions, and most results are presently available only in abstract form. Paper length publications are limited at present to results from a single feeding trial conducted at Lacombe (6–8). This chapter will, therefore, concentrate on published results from Lacombe and rely on later editions of this book to comprehensively compare results pending from Lacombe and other institutions. It must be reemphasized that results from Lacombe are the product of a single feeding trial, and data from several more trials are needed to enable a critical assessment of the potential merits of adding CLA to animal feeds.

Feeding CLA at Lacombe

At Lacombe, we were limited to feeding CLA to pigs or cattle, and we chose to feed pigs for our initial trial. CLA is naturally abundant in ruminant meat and dairy products, as outlined in other chapters, but we fed pigs because of their lower potential for gastric fatty acid hydrogenation before intestinal absorption. Feeding pigs would, therefore, give CLA a better opportunity for physiologic activity and/or incorporation into tissue lipids. The objectives of the trial were as follows:

- Determine if CLA could improve feed utilization efficiency, growth performance and carcass composition in finishing pigs.
- Determine if CLA had any effect on pork quality or palatability.
- Determine to what extent pigs would be able to concentrate CLA in their tissues (Chapter 18).

Lacombe Methods

Animals and Diets

The pigs used for this study were offspring from industry-typical lean genetic stock (Landrace boars and Large-White × Landrace sows). Animals were raised and slaughtered in accordance with the principles and guidelines set out by the Canadian Council on Animal Care (9). A total of 108 pigs was used for this study (54 gilts and 54 barrows). Pigs were housed three to a pen with *ad libitum* access to feed and water. Three blocks of 12 pens were used, with pigs of the same gender and similar weights penned together. Pigs were acclimated to their pens for 1–2 wk, and experimental diets were fed when block average weights approached 60 kg (actual average ± SEM, 61.5 ± 0.45 kg). Experimental diets consisted of a basal diet (Table 28.1) to which was added either 2% CLA or 2% sunflower oil (control). The isomeric composition of dietary CLA is given in Chapter 18. The CLA was synthesized from sunflower oil, and both oils were purchased in free fatty acid form from Natural Lipids (Hovdebygda, Norway). The feeding level of CLA that would be physiologically active in pigs was not known; therefore, 2% CLA oil was selected to provide levels over and above those shown to be effective in rodents (1,2). Diets were formulated

TABLE 28.1 Diet Composition[a]

Ingredient	%
Wheat	20.0
Barley	55.7
Soybean meal (48% CP)	11.7
Canola meal	7.5
CLA or sunflower oil	2.0
Calcium carbonate	1.1
Dicalcium phosphate	0.9
Sodium chloride	0.4
Lysine·HCl	0.25
Threonine	0.05
Vitamin and mineral premix[b]	0.4
Energy and nutrient content[c]	
Moisture (%)	13.4
GE (MJ/kg)	16.2
CP (%)	16.3
Fat (%)	3.2
Lysine (%)	0.89
Threonine (%)	0.50
Ca (%)	0.707
P (%)	0.601

[a]Abbreviations: CP, crude protein; CLA, conjugated linoleic acid; GE, gross energy.
[b]The vitamin/mineral premix supplied per kilogram of complete feed: vitamin A, 8000 IU; vitamin E, 100 IU; vitamin D, 1000 IU; vitamin K, 0.097 mg; vitamin B_{12}, 20 µg; niacin, 80 mg; biotin, 1.61 mg; thiamine, 3.67 mg; choline, 1323 mg; riboflavin, 5.52 mg; pentothenate, 23.2 mg; magnesium, 2.03 g; potassium, 7.1 g; iron, 235 mg; zinc, 184 mg; manganese, 100 mg; copper, 28.6 mg; selenium, 0.045 mg; iodine, 1.1 mg.
[c]As-fed basis, analyzed concentrations.

commercially to meet or exceed National Research Council (10) nutrient requirements. Protein, lysine, and threonine were intentionally added in excess of requirements to support any CLA-induced fat-to-lean repartitioning. Pigs were slaughtered when pen average weights approached 105 kg (actual average, 106 ± 0.54 kg). Pig weights and feed consumptions per pen were measured weekly.

Animal Slaughter and Carcass Quality Assessment

At slaughter, pigs were stunned (400 V for 3 s), shackled, exsanguinated, scalded, dehaired, and eviscerated under simulated commercial conditions. The right loin (longissimus thoracis) temperature and pH were measured at 45 min, 3 h, and 24 h *post-mortem* using a hand-held Fisher Scientific Accumet 1002 pH meter equipped with a Xerolyt spear-type electrode (Ingold Messtechnik AG, Urdorf, Switzerland). Cores from the right loin were also collected at these times using a 19-mm stainless steel cork bore. The cores were flash-frozen in liquid nitrogen and stored at −80°C for glycogen and lactate determinations as described by Yambayamba et al. (11). Samples from the liver, heart, subcutaneous fat, and omental fat were also collected at 45 min *post-mortem* and frozen (−80°C) for lipid class analyses (results reported in Chapter 18). At 24 h *post-mortem*, the two carcasses nearest the pen average weight were used for determination of carcass composition [division of the left side of the carcass into primal cuts including picnic (lower shoulder), butt (upper shoulder), hock, ham, loin, and belly with further dissection into fat, bone, and lean] according to the procedures described by Martin et al. (12). Left loins were collected at 24 h from all carcasses for meat quality analyses and a 25-mm chop was removed at the 12th rib for determination of drip loss. Drip loss was measured as the amount of purge resulting during the storage of the chop in a polyethylene bag for 48 h at 2°C. At 48 h postslaughter, the left loin was subjectively evaluated as a consensus of three experienced panelists for color and structure of the longitudinal surface between the 4th and 12th ribs using the Agriculture Canada Quality Standards 5-Point Scale (13). Color standards range from 1 = extremely pale to 5 = extremely dark. Structure was rated from 1 = extremely soft, exudative, dough-like, usually with an open grainy texture, to 5 = extremely firm, dry, sticky, with closed and grainy texture. After subjective scoring, two 25-mm chops were removed from the left loin adjacent to the 11th rib; one chop was used to measure light reflectance, marbling, and shear force; the second chop was vacuum packaged and frozen (−40°C) for sensory evaluation.

Marbling was scored using the American Meat Science Association (14) recommended procedures for beef carcass evaluation pictorial standards (range 100–1100; 100 = devoid of marbling and 1100 = very abundant marbling). After a 20 min bloom, Commission Internationale de l'Eclairage (15) light reflectance coordinates [L* (brightness), a* (red-green axis), and b* (yellow-blue axis)] of the chop surface were measured in duplicate using a Chroma-Meter II (Minolta Canada, Mississauga, Canada); the results were converted to hue angle (H_{ab} = arctan[b*/a*]) and chroma ($C_{ab} = [a*^2 + b*^2]^{0.5}$) before averaging. The chops were then cooked and used to objectively measure tenderness (shear force). Specifically, the chops were first placed in individual beakers and immersed in 150 mL of room temperature saline (0.9% NaCl). The beakers were placed in an 80°C shaking water bath, and chop internal temperatures were monitored with thermocouples. When chop temperatures reached 72°C, chops were immediately transferred to polyethylene bags and cooled in a water/ice bath to arrest further cooking. After overnight cooling at 4°C, two cores (19 mm) were removed from the cooked chop perpendicular to the chop surface. Cores were sheared once parallel to the chop surface using an Instron Model 4301 Materials Testing System (Burlington, Canada) equipped with a Warner-Bratzler cell, and peak shear forces for each chop were averaged.

The remaining left loin was ground twice through a 3.2-mm grinding plate for the determination of moisture and soluble protein. Moisture content was calculated as the weight lost after heating 100 g of ground tissue at 105°C for 24 h. Soluble protein was measured as described by Barton-Gade (16), except that the result was expressed as grams soluble protein per kilogram of lean muscle instead of optical density (17). Dried loin was analyzed for both crude protein (AOAC Method 39.1.15) and crude intramuscular (IM) fat (AOAC Method 39.1.05)(18).

Sensory Evaluation

Loin chops were removed from the freezer and placed in a refrigerator at 4°C to thaw for 24 h. Fifteen minutes before broiling, the chops were removed from the refrigerator and vacuum packaging. Thermocouples were inserted horizontally to the midpoint along the long axis of the chop. Chops were placed in a commercial electric oven (174°C) on a broiling pan (at a distance of 10 cm from the chop surface to the broiling element) and heated to an internal temperature of 40°C, turned, and cooked to a final internal temperature of 72°C. Each chop was cut into 1.3-cm cubes, avoiding connective tissue and large areas of fat. Six cubes from each chop were randomly assigned to a six-member semi-trained sensory panel, screened according to established American Meat Science Association guidelines (19). Samples were placed in glass jars in a circulating water bath and allowed to equilibrate to 70°C before evaluation. Each sample was evaluated for initial and overall tenderness, amount of perceptible connective tissue, juiciness, and flavor intensity, using a 9-point descriptive scale (9 = extremely tender, no perceptible connective tissue, extremely juicy, and extremely intense pork flavor; 1 = extremely tough, abundant connective tissue, extremely dry, and extremely bland pork flavor). In addition, each sample was evaluated for flavor desirability and overall palatability using a 9-point hedonic scale (9 = extremely desirable; 1 = extremely undesirable). All panel evaluations were conducted in well-ventilated, partitioned booths, under 1076 lx of incandescent and fluorescent white light. Distilled water and unsalted soda crackers were provided to cleanse the palate of residual flavors between samples (20).

Statistical Analyses

Data were analyzed using the general linear model procedures of SAS (21). For data consisting of single measurements, the statistical model included the main effects of block, diet, gender, and diet by gender interaction. For data consisting of serial measurements, time was added to the model as a repeated measure together with its two- and three-way interactions with diet and gender. Animal performance data were analyzed using pen as the experimental unit, and carcass and meat quality measurements were analyzed on an individual animal basis. Performance data for one out of 18 pens fed the sunflower oil-containing diet were excluded due to the lameness of one pig. The subjective scores for color and structure were discrete in nature, and their frequencies were subjected to chi-square analysis to determine their probability of being influenced by diet, gender, or diet by gender interaction.

Lacombe Results

Growth Performance and Feed Conversion Efficiency

Pigs fed the CLA diet tended to have decreased feed intakes (–5.2%, $P = 0.07$), improved feed conversion efficiencies (FCE; –5.9%, $P = 0.06$), and similar average daily gains (ADG; $P = 0.84$) compared with pigs fed the sunflower oil diet (Table 28.2). Gender effects on performance were as expected, with gilts having lower ADG (–7.6%, $P = 0.01$), feed intakes (–12.5%, $P = 0.01$), and a tendency for lower FCE (–4.6%, $P = 0.06$). Feeding CLA did not interact with gender for any performance parameter, indicating a consistent effect of CLA across gender.

Gross Carcass Composition

CLA feeding did not affect carcass weight but did increase lean (+2.3%, $P = 0.02$) and reduce subcutaneous fat (–6.8%, $P = 0.01$) in total saleable cuts (Table 28.3). Effects of gender were as expected, with gilts having more lean (+7.8%, $P = 0.01$) and less fat (–27% subcutaneous fat, $P = 0.01$; –17.6% intermuscular fat, $P = 0.01$) per kilogram of total cuts. CLA feeding did not interact with gender for any component of total or individual cuts. Within individual cuts, the only component consistently reduced by CLA feeding was subcutaneous fat (–11.1% picnic, $P = 0.01$; –12.2% butt, $P = 0.01$; –6.1% loin, $P = 0.04$; –5.8% ham, $P = 0.05$), and the loin was the only individual cut exhibiting significantly increased lean (+3.6%, $P = 0.02$) (Table 28.4). On an individual animal basis, CLA feeding increased lean in saleable cuts by ~800 g and decreased subcutaneous fat in saleable cuts by ~850 g. To put this in perspective, with 18 million pigs slaughtered in Canada per year, the feeding of 2% CLA during the last 45 d of finishing could increase saleable lean yield by 14.2 million kg, reduce fat by 15.2 million kg, and save ~115,000 t of feed per year.

At present, the exact mechanism of CLA-induced fat-to-lean repartitioning is not known. Park *et al.* (2) indicated that CLA may act in rats by increasing lipolysis while reducing lipoprotein lipase activity in adipose tissue and enhancing fatty acid oxidation in both muscle and adipose tissue. Park *et al.* (2) also indicated that CLA

TABLE 28.2 Animal Performance[a]

Parameter	Diet			Gender		
	CLA	Sunflower	SEM	Gilt	Barrow	SEM
Avg. daily gain (kg/d)	1.01	1.01	0.02	0.97[b]	1.05[c]	0.02
Feed intake[d] (kg/d)	2.92	3.08	0.05	2.79[b]	3.19[c]	0.05
Feed conversion efficiency	2.89	3.07	0.06	2.91	3.05	0.06

[a]Values are means with SEM. For diet, n = 17 for control, n = 18 for conjugated linoleic acid (CLA). For gender, n = 17 for gilt, n = 18 for barrow.
[b,c]Means with different superscripts within the same row are different ($P < 0.01$).
[d]As-fed basis.

TABLE 28.3 Gross Carcass Composition[a]

	Diet			Gender		
	CLA	Sunflower	SEM	Gilt	Barrow	SEM
Total cuts per carcass (kg)	56.6	56.8	0.22	56.5	57.0	0.22
Total cut composition (g/kg)						
Lean	618[b]	604[c]	4	634[d]	588[e]	4
Bone	106	107	1	107	106	1
Subcutaneous fat	206[d]	221[e]	4	169[d]	232[e]	4
Intermuscular fat	48.0	46.5	0.4	42.7[d]	51.8[e]	1.1

[a]Values are means with SEM. For diet, n = 36 for control and n = 35 for conjugated linoleic acid (CLA), and for gender, n = 35 for gilts and n = 36 for barrows.
[b,c]Means with different superscripts within the same row are different ($P < 0.05$).
[d,e]Means with different superscripts within the same row are different ($P < 0.01$).

TABLE 28.4 Composition of Principal Lean Cuts[a]

Cut composition (g/kg total cuts)	Diet			Gender		
	CLA	Sunflower	SEM	Gilt	Barrow	SEM
Picnic (upper shoulder)						
Lean	91.3	89.5	1.3	92.8[d]	88[e]	1.3
Bone	11.4	11.8	0.3	11.6	11.6	0.3
Subcutaneous fat	17.5[d]	19.7[e]	0.6	15.8[d]	21.4[e]	0.6
Intermuscular fat	10.1	9.95	0.38	8.9[d]	11.2[e]	—
Butt (lower shoulder)						
Lean	90.9	89.5	1.0	93[d]	87.3[e]	1.0
Bone	6.52	5.85	0.56	5.76	6.61	0.56
Subcutaneous fat	30.1[d]	34.3[e]	1.0	30.2[d]	34.2[e]	1.0
Intermuscular fat	12.5	12.3	0.4	10.7[d]	14.1[e]	0.4
Loin						
Lean	201[b]	194[c]	2	206[d]	189[e]	2
Bone	46.9	47.2	0.6	47.9	46.3	0.6
Subcutaneous fat	85.6[b]	91.2[c]	1.9	79.8[d]	97.0[e]	1.9
Intermuscular fat	12.4	12.1	0.5	10.9[d]	13.5[e]	0.5
Ham						
Lean	223	219	2	230[d]	212[e]	2
Bone	30.2	30.1	0.4	30.2	30.1	0.4
Subcutaneous fat	66.3[b]	70.4[c]	1.4	63.6[d]	73.0[e]	1.4
Intermuscular fat	11.0	10.5	0.3	10.3	11.1	0.3
Belly						
Skinless/trimmed	31.5	28.8	1.4	31.7	28.6	1.4

[a]Values are means with SEM. For diet, n = 36 for control and n = 35 for conjugated linoleic acid (CLA), and for gender, n = 35 for gilts and n = 36 for barrows.
[b,c]Means with different superscripts within the same row are different ($P < 0.05$).
[d,e]Means with different superscripts within the same row are different ($P < 0.01$).

may act by affecting eicosanoid synthesis from other fatty acids or that CLA-derived eicosanoids may have their own biological activity. More comprehensive details on the effects of CLA on fat partitioning are presented in Chapter 27.

Glycogen, Lactate, pH and Temperature

The *post-mortem* rate and extent of muscle glycogen utilization are primary determinants of pork quality. Glycogen is metabolized to lactate in the *post-mortem* period *via* glycogenolysis and anaerobic glycolysis. The rate and extent of muscle pH and temperature decline subsequently influence pork quality. If the pH decline is rapid in the early *post-mortem* period while the carcass temperature is elevated, pork can become pale, soft, and exudative (PSE). Alternatively, if the pH decline is slow and remains relatively high due to a lack of glycogen, pork can become dark, firm, and dry (DFD).

In this study, loin glycogen and lactate concentrations were unaffected by diet, gender, or diet by gender interaction ($P > 0.05$). As expected, *post-mortem* glycogen concentrations decreased ($P < 0.01$) and lactate concentrations increased over time ($P < 0.01$, Fig. 28.1). These changes, however, were well within values reported for normal pork, and no time by diet, time by gender, or time by diet by gender interactions were found.

Post-mortem loin pH was not affected by diet or gender ($P > 0.05$), but barrow loins were warmer than gilt loins (+0.7°C; $P < 0.01$) and loins from CLA-fed pigs

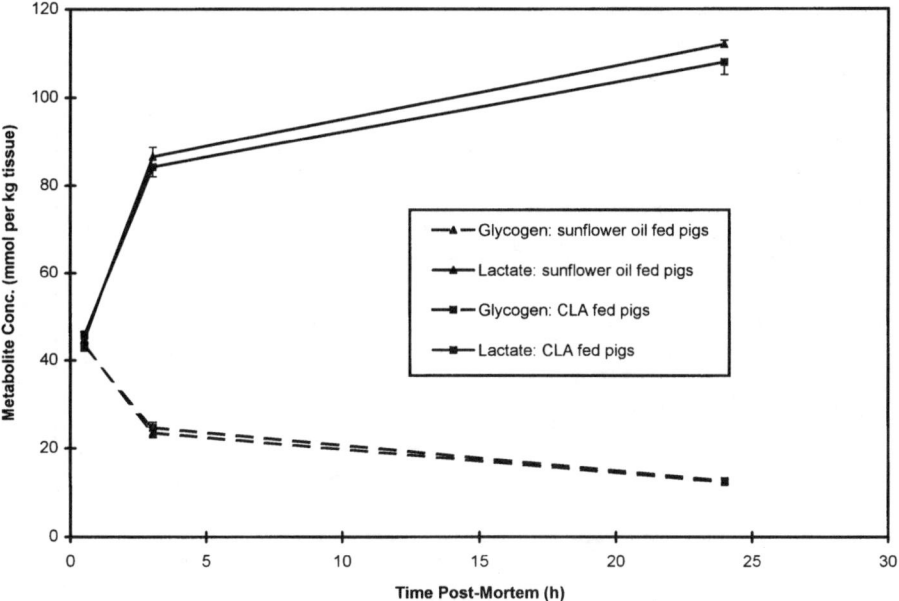

Fig. 28.1. *Post-mortem* changes in longissimus thoracis glycogen and lactate concentrations (106-kg pigs; n = 54; values are means ± SEM).

were warmer than loins from sunflower oil–fed pigs (+0.3°C; $P < 0.01$). As expected, both pH and temperature declined *post-mortem* ($P < 0.01$, Fig. 28.2), and rates of decline were within values required for normal pork quality development. The main effects of diet and gender on temperature were further defined by their significant interactions with time ($P < 0.01$). At 3 h, barrow loins were 2.7°C warmer than gilt loins, and loins from CLA-fed pigs were 1.15°C warmer than loins from sunflower oil–fed pigs. The higher barrow loin temperatures at 3 h coincided with a time by gender interaction effect on pH ($P < 0.01$). Barrow loin pH values at 3 h were 0.09 pH units lower than those of gilts. Gender differences in pH and temperature were likely due to barrows having more subcutaneous fat than gilts, with increased subcutaneous fat helping to maintain loin temperature and glycolytic activity. Pigs fed CLA, however, had less subcutaneous fat than sunflower oil–fed pigs. Loins from pigs fed CLA may have stayed warmer because of their larger lean (metabolically active) mass. Overall, feeding CLA did not effect loin glycogen, lactate, or pH, and had only a minor effect on carcass temperature.

Loin Objective Color and Subjective Scoring

The L* (lightness) values for chops were unaffected by diet or gender (Table 28.5). Hue angle was slightly increased (a shift from red toward the yellow; $P < 0.05$) in chops from barrows vs. gilts, and CLA-fed pigs had slightly higher chroma values (color

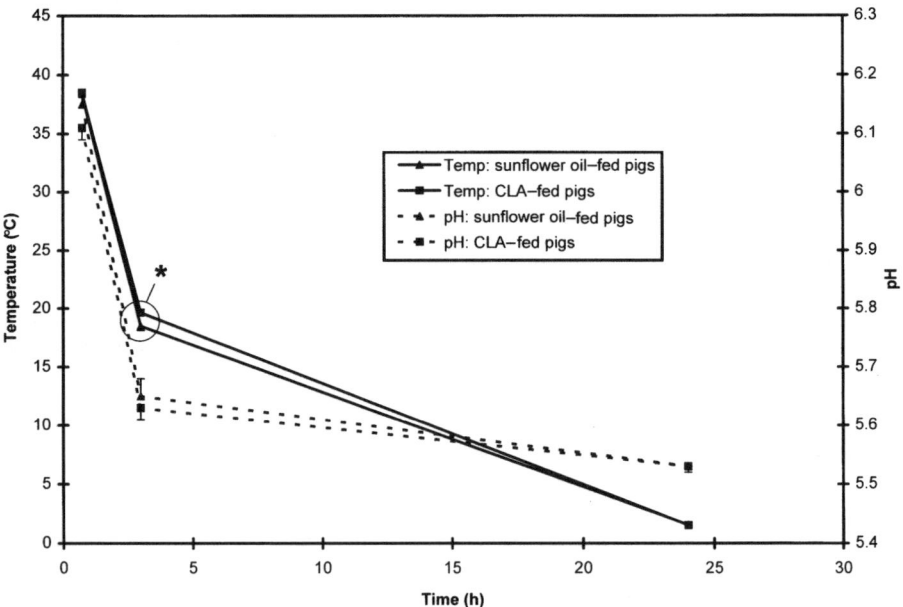

Fig. 28.2. *Post-mortem* changes in longissimus thoracis pH and temperature (*$P < 0.05$ vs. time matched comparison)(106-kg pigs; n = 54; values are means ± SEM).

TABLE 28.5 Loin Objective Color and Subjective Color, Structure and Marbling Scores[a]

Parameter	Diet			Gender		
	CLA	Sunflower	SEM	Gilt	Barrow	SEM
L* (lightness)	53.2	52.8	0.3	52.2	53.8	0.3
Hue	41.0	41.1	0.3	40.6[b]	41.5[c]	0.3
Chroma	9.05[b]	8.21[c]	0.23	8.40	8.85	0.2
Color score	2.96	2.94	0.02	2.97	2.94	0.02
Structure score	2.97	2.95	0.02	2.96	2.96	0.02
Marbling score[d]	434[e]	390[f]	9.2	388[e]	436[f]	9

[a]Values are means with SEM, n = 54 for both diet and gender; CLA, conjugated linoleic acid.
[b,c]Means with different superscripts within the same row are different ($P < 0.05$).
[d]Represents an interaction effect between diet and gender at $P < 0.01$.
[e,f]Means with different superscripts within the same row are different ($P < 0.01$).

saturation; $P < 0.05$). Subjective color and structure scores were unaffected by diet or gender. Marbling scores were elevated in chops from CLA-fed pigs over sunflower oil–fed pigs ($P < 0.01$), and in barrows over gilts ($P < 0.01$). In addition, a diet by gender interaction ($P < 0.01$) was found for marbling. Gilt marbling scores were 18.8% higher when CLA was fed vs. sunflower oil, whereas barrow marbling scores were increased only 3.7% by feeding CLA. Increases in barrow loin marbling may have been obscured slightly by their higher hue angle.

Loin Physical Characteristics

Loin shear force, drip loss, and soluble protein levels were not affected by diet, gender, or diet by gender interaction ($P > 0.05$, Table 28.6). Loins from CLA-fed pigs had significantly less moisture (–5 g/kg; $P < 0.05$) than loins from sunflower-fed pigs. Reduced moisture levels in loins from CLA-fed pigs were due in part to a 22.1% increase in intramuscular (IM) fat in dried loin ($P < 0.01$), equivalent to a 3.77 g increase per kilogram fresh loin. As expected, barrows also had greater loin IM fat than gilts (+25.7% on a dry matter basis; $P < 0.01$). A diet by gender interaction for IM fat was not found, indicating that CLA feeding accentuated loin IM fat deposition equally across genders. Barrows relative to gilts, and pigs fed the CLA diet vs. the sunflower oil diet had lower loin crude protein (CP) levels when expressed on a dry matter basis ($P < 0.01$). These effects were not found when CP was expressed on a wet weight basis ($P > 0.05$), indicating the maintenance of a protein to muscle mass ratio. When combined with gross carcass composition results, the overall effects of CLA feeding on loin composition (per kg wet weight) include a 17.3 g decrease in subcutaneous fat, an 18.0 g increase in lean and the increased lean was due in part to a 3.77 g increase in IM fat.

Increases in loin IM fat due to CLA were consistent with increases seen in subjective marbling scores. Diet and gender, however, were found to interact for subjective marbling scores, but not for (solvent extracted) IM fat. The diet by gender interaction

TABLE 28.6 Loin Composition and Quality Measurements[a]

Parameter	Diet			Gender		
	CLA	Sunflower	SEM	Gilt	Barrow	SEM
Moisture (g/kg)	741[b]	746[c]	1	744	742	1
Intramusclular fat (g/kg dry)	73.9[d]	60.5[e]	2.6	59.5[d]	74.8[e]	2.6
Crude protein (g/kg dry)	874[d]	888[e]	2	886[d]	875[e]	2
Shear force (kg/cm^2)	5.88	5.95	0.12	5.98	5.86	0.12
Drip loss (g/kg)	50.3	45.1	2.1	45.9	49.6	2
Soluble protein (mg/g)	174	175	2.6	175	175	2.6

[a]Values are means with SEM, n = 54 for both diet and gender; CLA, conjugated linoleic acid.
[b,c]Means with different superscripts within the same row are different ($P < 0.05$).
[d,e]Means with different superscripts within the same row are different ($P < 0.01$).

found for marbling scores may, therefore, be due to the inability to discern marbling fat subjectively as hue angle increases or the possibility that as IM fat increases, more triacylglycerol is stored in the myofibers and left undetectable to the subjective scorer. Increased IM fat has been previously noted when fat-to-lean repartitioning was induced with ractopamine (5), but the mechanism for increasing highly valued IM fat is not presently known. Potentially, adipocyte size and maturity in different fat depots may play a role in determining how each depot reacts when exposed to CLA.

Loin Sensory Evaluation

No detectable differences in any palatability characteristics were noted between loins from pigs fed the CLA and those fed the sunflower oil diet (Table 28.7). In addition, no differences in palatability characteristics were found between genders, with one exception. Overall palatability of barrow loin was slightly higher than gilt loin (+0.43 unit; $P < 0.05$). This effect may have been associated with the aforementioned gender effect on IM fat, but the diet and gender effects on IM fat were of similar magnitude. Gender differences in overall palatability, therefore, may be related only partially to IM fat levels.

A minor diet by gender interaction was noted for loin chop juiciness ($P < 0.05$). The gender by diet interaction was due to a nonsignificant 0.31 unit increase in barrow loin juiciness and a nonsignificant 0.29 unit decrease in gilt loin juiciness when pigs were fed the CLA vs. the sunflower oil diet.

Lacombe Conclusions

Feeding a diet containing 2% CLA tended to reduce feed intake (6.4 kg/pig over the finishing phase) and improve feed conversion efficiency (5.9%), but did not affect average daily gain. Feeding the CLA diet reduced subcutaneous fat (850 g/carcass) while increasing both lean (800 g/carcass) and intramuscular fat (3.8 g/kg fresh loin). CLA consumption did not adversely affect any measured meat quality or

TABLE 28.7 Loin Palatability[a]

	Diet			Gender		
Characteristic	CLA	Sunflower	SEM	Gilt	Barrow	SEM
Initial tenderness	5.34	5.38	0.11	5.25	5.47	0.11
Juiciness[b]	5.53	5.52	0.10	5.50	5.54	0.10
Flavor	5.67	5.57	0.06	5.56	5.68	0.06
Flavor intensity	5.51	5.51	0.06	5.58	5.44	0.06
Connective tissue	6.41	6.47	0.06	6.37	6.51	0.06
Overall tenderness	5.75	5.59	0.10	5.59	5.74	0.10
Overall palatability	5.46	5.46	0.07	5.35[c]	5.57[d]	0.07

[a]Values are means with SEM, n = 54 for both diet and gender; CLA, conjugated linoleic acid.
[b]Represents an interaction effect between diet and gender ($P < 0.05$).
[c,d]Means with different superscripts within the same row are different ($P < 0.05$).

palatability parameter, and increased intramuscular fat levels indicate that CLA has the potential to improve pork quality. Feeding CLA to pigs, therefore, appears to be a way to repartition subcutaneous fat to lean in pigs without adversely affecting meat quality.

CLA Research at Other Institutions

As indicated previously, results from other centers on the effects of feeding CLA to pigs are limited and in abstract form. Lack of experimental details at this point make it difficult to compare and contrast results with data presented thus far. However, some trends are evident and are summarized below.

Work at the University of Wisconsin (22) indicates that feeding a diet containing 0.48% or 0.96% CLA to pigs from 26 kg through to finishing reduced backfat thickness by 24% as measured by ultrasound. Eight pigs were fed per diet, and comparisons were made relative to pigs fed a 0% CLA control ration. Direct carcass measurements confirmed reduced backfat thickness, and lean yields were increased, but no effects on carcass weight or longissimus muscle area were observed. These results were in close agreement with the Lacombe results. Interestingly, during the first 49 d of feeding, the CLA-fed pigs had reduced feed intakes and weight gains. From 49 to 84 d, the CLA-fed pigs exhibited compensatory growth and weighed the same as control pigs at the end of the trial. In addition, feed efficiency was noted to improve from 49 to 84 d, and this was sufficient to result in a significant improvement in feed efficiency during the entire feeding period.

Work at Iowa State University (23) also indicates backfat thickness (ruler measured) decreases when CLA is fed. In this study, 8 barrows were fed per diet, and diets contained 0, 0.12, 0.25, 0.50, or 1.0% CLA. Pigs were fed from 26.3 kg to 116 kg, and average daily gains increased with increasing levels of dietary CLA. Average daily feed intake was not affected by the CLA content of the diet; hence a linear increase in feed-to-gain ratio was noted with increasing dietary CLA. Improvements in feed to gain were in agreement with Lacombe and Wisconsin results. Additionally, differences in loin eye area were not detected, in agreement with Wisconsin data. An interesting observation was that belly hardness increased with increasing dietary CLA and may thus help firm-up problematic soft bellies.

A group at Kansas State University is also researching the effects of feeding CLA to pigs (24,25). Commercial preparations of CLA to date have been the product of alkaline isomerization of oils rich in linoleic acid. O'Quinn et al. (24) reported feeding a novel source of CLA produced during the Kraft (sulfate) paper process (modified tall oil; MTO). In this study, three diets were fed, with 12 barrows per diet. The diets included a control (no CLA), a 0.5% CLA (manufactured from sunflower oil), and a 0.5% MTO diet. Pigs were fed in two phases, from 36.3 to 72.6 kg and from 72.6 to 104.3 kg. The basal diet consisted of corn and soybean meal with 1% lysine for the initial phase and 0.75% lysine for the final phase. Pigs fed the CLA diet had reduced weight gains from 36.3 to 72.6 kg, and neither the CLA- nor MTO-fed pigs had weight gains different from the control animals for the remainder of the study. Weight gains for the CLA fed pigs over the entire experiment were, however, less than those for MTO-fed pigs. In a second experiment (25), feeding graded levels of MTO (0, 0.25, 0.5, and 1%) did not affect average daily gain, average daily fed intake, or feed to gain. Feeding MTO did, however, decrease tenth and last rib backfat levels, and longissimus muscle area tended to increase. In addition, loin drip loss was found to decrease when MTO was fed, in contrast to the lack of effect shown when CLA was fed in the Lacombe study. Belly firmness was also increased when feeding MTO and this agrees well with Iowa results (23).

The majority of data available supports Lacombe study conclusions. Kansas results, however, appear to contradict those from other centers indicating CLA induces improvements in feed conversion efficiency. Explanations for this discrepancy will fully be forthcoming once full experimental details are published. One possible reason for the difference may be the two-phase feeding system used by the Kansas researchers. If the pigs fed CLA in the Kansas study were lighter when the second diet was introduced, and the second diet had a lower nutrient density than the first, the pigs may not have had an adequate nutrient supply to fuel compensatory growth (noted in the Wisconsin trial). Two-phase feeding systems are, however, a common industry practice and this suggests a definite need to determine if, and how, pig nutrient requirements change when feeding CLA.

Future CLA Research

Making suggestions for CLA research will be easier once more baseline data are available. Interpretation and integration of forthcoming results will require close scrutiny of several factors including: the dietary CLA content and its isomeric composition, chemical form (free acid, salt, ester) and digestibility; the duration of CLA feeding; the composition of the diet and the number, breed and gender of pigs fed per diet.

Given present limitations some research priorities are, however, clear. Current research emphasis is currently on grower and finisher pigs and further need in this area exists to help define otimal levels of CLA and other nutrients. Additionally the growth and potential reproductive effects of CLA in the breeding herd will be of interest, as will any effects on the pre- and post-weaning piglet. Fundamental work is still also needed to determine the mechanisms and activities of individual CLA isomers and how their concentrations might be preferentially increased during CLA synthesis. Further editions of this book will, therefore, be necessary to chronical evolving research and aid in identifying future research needs.

References

1. Chin, S.F., Storkson, J.M., Albright, K.J., Cook, M.E., and Pariza, M.W. (1994) Conjugated Linoleic Acid Is a Growth Factor for Rats as Shown by Enhanced Weight Gain and Improved Feed Efficiency, *J. Nutr. 124*, 2344–2349.
2. Park, Y., Albright, K.J., Liu, W., Storkson, J.M., Cook, M.E., and Pariza, M.W. (1997) Effect of Conjugated Linoleic Acid on Body Composition in Mice, *Lipids 32*, 853–858.
3. Dugan, M.E.R., Tong, A.K.W., Carlson, J.P., Schricker, B.R., Aalhus, J.L., Schaefer, A.L., Sather, A.P., Murray, A.C., and Jones, S.D.M. (1997) The Effects of Porcine Somatotropin, Gender and Porcine Stress Syndrome on Growth, Carcass Composition and Pork Quality, *Can. J. Anim. Sci. 77*, 233–240.
4. Beermann, D.H. (1990) Implications of Biotechnologies for Control of Animal Growth, in *Proceedings of the 43rd Reciprocal Meat Conference*, pp. 87–96, National Livestock Meat Board, Chicago.
5. Aalhus, J.L., Jones, S.D.M., Schaefer, A.L., Tong, A.K.W., Robertson, W.M., Merrill, J.K., and Murray, A.C. (1990) The Effect of Ractopamine on Performance, Carcass Composition and Meat Quality of Finishing Pigs, *Can. J. Anim. Sci. 70*, 943–952.
6. Dugan, M.E.R., Aalhus, J.L., Schaefer, A.L., and Kramer, J.K.G. (1997) The Effect of Feeding Conjugated Linoleic Acid on Fat to Lean Repartitioning and Feed Conversion in Pigs, *Can. J. Anim. Sci. 77*, 723–725.
7. Kramer, J.K.G., Sehat, N., Dugan, M.E.R., Mossoba, M.M., Yurawecz, M.P., Roach, J.A.G., Eulitz, K., Aalhus, J.L., Schaefer, A.L., and Ku, Y. (1998) Distributions of Conjugated Linoleic Acid (CLA) Isomers in Tissue Lipid Classes of Pigs Fed a Commercial CLA Mixture Determined by Gas Chromatography and Silver-Ion High-Performance Liquid Chromatography, *Lipids 33*, 549–558.
8. Dugan, M.E.R., Aalhus, J.L., Jeremiah, L.E., Kramer, J.K.G., and Schaefer, A.L. (1999) The Effects of Feeding Conjugated Linoleic Acid on Subsequent Pork Quality, *Can. J. Anim. Sci. 79*, 45–51.
9. CCAC (1993) *A Guide to the Care and Use of Experimental Animals*, 2nd edn., Canadian Council on Animal Care, Ottawa, Canada.
10. NRC (1988) Nutrient Requirements of Swine, 9th edn., National Academy Press, Washington, DC.
11. Yambayamba, E.S.K., Aalhus, J.L., Price, M.A., and Jones, S.D.M. (1996) Glycogen Metabolites and Meat Quality in Feed Restricted-Refed Heifers, *Can. J. Anim. Sci. 76*, 517–522.

12. Martin, A.H., Fredeen, H.T., Weiss, G.M., Fortin, A., and Sim, D. (1981) Yield of Trimmed Pork Product in Relation to Weight and Backfat Thickness of the Carcass, *Can. J. Anim. Sci. 61,* 299–310.
13. Agriculture Canada (1984) Pork Quality—A Guide to Understanding Colour and Structure of Pork Muscles, Agriculture Canada, Publication 5180/B, Ottawa, Canada.
14. AMSA (1990) Recommended Procedures for Beef Carcass Evaluation and Carcass Contests, American Meat Science Association and the National Livestock and Meat Board, Chicago, IL.
15. CIE (1976) Commission International de l'Eclairage, 18th Session, CIE Publication 36, London.
16. Barton-Gade, P.A. (1984) Method of Estimating Soluble Sarcoplasmic and Myofibrillar Proteins in Pig Meat, *Slagterienes Forskninginstitut,* 30.
17. Murray, A.C., Jones, S.D.M., and Sather, A.P. (1989) The Effect of Preslaughter Feed Restriction and Genotype for Stress Susceptibility on Pork Quality and Composition, *Can. J. Anim. Sci. 69,* 83–91.
18. AOAC (1995) Official Methods of AOAC International, 16th edn., AOAC, Washington DC.
19. AMSA (1995) Research Guidelines for Cookery, Sensory Evaluation and Instrumental Tenderness Measurements of Fresh Meat, American Meat Science Association and the National Livestock and Meat Board, Chicago, IL.
20. Larmond, E. (1977) Laboratory Methods for Sensory Evaluation of Foods, Agriculture Canada Publication 1637, Ottawa, Canada.
21. SAS (1985) SAS User's Guide: Statistics, 5th edn., SAS Institute, Cary, NC.
22. Cook, M.E., Jerome, D.L., Crenshaw, T.D., Buege, D.R., Pariza, M.W., Albright, K.J., Schmidt, S.P., Scimeca, J.A., Lofgren, P.A., and Hentges, E.J. (1998) Feeding Conjugated Linoleic Acid Improves Feed Efficiency and Reduces Carcass Fat in Pigs, *FASEB J. 12,* A836.
23. Thiel, R.L., Sparks, J.C., Wiegand, B.R., Parrish, F.C., Jr., and Ewan, R.C. (1998) Conjugated Linoleic Acid Improves Performance and Body Composition in Swine, in *Abstracts of the American Society of Animal Science,* p. 61, Midwestern Sectional Meetings, Des Moines, IA.
24. O'Quinn, P.R., Nelssen, J.L., Tokach, M.D., Goodbrand, R.D., Woodworth, J.C., Unruh, J.A., and Owen, K.Q. (1998) A Comparison of Modified Tall Oil and Conjugated Linoleic Acid on Growing-Finishing Pig Growth Performance and Carcass Characteristics, in *Abstracts of the American Society of Animal Science,* p. 61, Midwestern Sectional Meetings, Des Moines, IA.
25. O'Quinn, P.R., Smith, J.W., II, Nelssen, J.L., Tokach, M.D., Goodbrand, R.D., and Owen, K.Q. (1998) Effects of Increasing Modified Tall Oil on Growing-Finishing Pig Growth Performance and Carcass Characteristics, in *Abstracts of the American Society of Animal Science,* p. 61, Midwestern Sectional Meetings, Des Moines, IA.

Chapter 29

Dietary Sources and Intakes of Conjugated Linoleic Acid Intake in Humans

Michelle K. McGuire[a], Mark A. McGuire[b], Kristin Ritzenthaler[a], and Terry D. Shultz[a]

[a]Department of Food Science and Human Nutrition, Washington State University, Pullman, WA 99164–6376
[b]Department of Animal and Veterinary Science, University of Idaho, Moscow, ID 83844–2330

Introduction

Although it is assumed that there are only two essential fatty acids required by adult humans (1), it is becoming clear that the inclusion of other fatty acids in the diet can promote improved health and well-being in some individuals. Conjugated linoleic acid (CLA), a natural dietary component found primarily in dairy, beef, and lamb products, is one such fatty acid that possesses an impressive range of promising health benefits. This claim is supported by a growing number of experimental studies utilizing laboratory animals, as well as human and cell culture systems, suggesting that CLA may prevent or inhibit the growth of various types of cancers (2–6), heart disease (7,8), and diabetes (9). In addition, CLA may affect growth (10,11), the immune system (12), and bone metabolism (13,14). These findings are reviewed in depth within other sections of this publication and therefore will not be evaluated further here. Nonetheless, this profusion of recently published data provides undeniable evidence that dietary CLA may influence human health.

The term CLA generally refers to a mixture of conjugated positional and geometric isomers of linoleic acid (18:2n-6). However, the *cis*-9,*trans*-11-octadecadienoic acid, also referred to as rumenic acid (RA; 15), is considered likely to be the major biologically active form (16) and is the most prominent of the CLA isomers found naturally in the diet of humans. Currently, conflicting published data exist concerning whether humans can synthesize RA from other fatty acids (17,18). However, it is probable that the majority of RA found in the human body originates from the diet, and increased consumption of foods containing high levels of RA has been shown to increase the RA concentrations of plasma (18,19) and human milk (20,21). In particular, animal foods of ruminant origin are thought to contribute most of the total CLA and RA in human diets, and numerous publications have focused on the concentrations of these fatty acids in foods (e.g., 20–25). However, it is recognized that the CLA and RA contents of most foods are somewhat variable, mainly as a result of differences in environmental conditions and diet of the originating ruminant species (26,27). Thus, the estimation of CLA and RA intakes in humans, compared with intakes of other fatty acids, may be especially difficult. In fact, there have been only a limited number of publications that have included estimates of human CLA or RA consumption. Thus, the purpose of this review is to outline what is currently known about CLA and RA food sources as well as their intakes in various subpopulations of humans and to make specific recommendations concerning the focus of future research in this area.

Sources of CLA and RA

Two decades ago, Parodi (28) reported that milk fat contained conjugated octadecadienoic acids. Since then, numerous investigators have studied and documented the CLA and RA concentrations of many foods. A partial listing of these publications is presented in Table 29.1. In general, these data show clearly that foods from ruminant species contribute significant amounts of CLA, the majority of which is in the form of RA. Variation in the RA content of similar foods appears to result primarily from differences in the composition of the raw product; factors affecting the biosynthesis of RA and its incorporation into ruminant tissues and milk are reviewed elsewhere in this publication. Additionally, human milk but not infant formula has been shown to contain RA (21,31,38). In summary, breast-fed infants and individuals consuming foods of ruminant origin obtain significant amounts of RA in their daily diets.

Estimates of CLA and RA Intakes in Humans

Dietary CLA intake in humans has been estimated by using limited nutrient databases in conjunction with 3-d dietary records (17,19,39), semiquantitative food-frequency questionnaires (20), and data from a German national food intake survey (25). Further, there has been one report (18) of RA intakes estimated by using data

TABLE 29.1 Published Reports Containing Conjugated Linoleic Acid Contents of Foods

Origin of food	Foods analyzed	Reference
Australia	Milk	28
Australia	Beef, lamb, pork, lard, chicken, eggs, butter, margarine, human milk	21
U.S.	Milk, cheese, processed cheese, beef	29
India	Ghee	30
U.S.	Milk, dairy products, processed cheese and meat, beef, lamb, pork, poultry, eggs, canned foods, seafood, oils, fats, processed foods, infant foods	31
U.S.	Processed cheeses	32
U.S.	Cheddar cheeses	23
U.S.	Processed cheeses	33
U.S.	Beef (raw, cooked, fried, baked, broiled and microwaved)	34
U.S.	Milk, dairy products, processed cheese spread	23
U.S.	Dairy products	35
Italy	Milk, dairy products, lamb	24
U.S.	Human milk, infant formula	36
U.S.	Wildlife and game	37
U.S.	Human milk	38
Germany	Milk, dairy products, meat, processed meat, seafood, snack foods, chocolates, cakes, pastries	25

collected from analyses of food duplicate portions in conjunction with dietary records. Finally, there have been two additional estimates of human CLA intake (40,41), but the methodologies used to obtain these estimates were not described in detail. A summary of these published data is presented in Table 29.2. It is noteworthy that some of these estimates are for usual intakes (17,25,39), whereas others estimate CLA or RA intake during specific dietary intervention periods (19,20).

Although earlier preliminary estimates of CLA intake suggested that humans may consume between 0.5 and 1 g/d (40,41), more rigorously obtained estimates suggest that CLA and RA intakes are substantially below this amount. Herbel *et al.* (17) utilized data collected from 3-d dietary records in conjunction with a nutrient database containing a limited number of published values (23,31) for the CLA contents of foods. These investigators reported that young men and women (age: 28 ± 1 y) living

TABLE 29.2 Published Estimates (means ± SEM) of Rumenic Acid (RA) Intakes in Humans

Description of populations	RA intake (mg/d)	Method used	Reference
U.S.	1000[a]	NR[b]	40
Australia	500–1000[a]	NR	41
Washington/Idaho, U.S. Men and women combined (n = 12)	127[a]	3-d dietary records in conjunction with published values for CLA contents of foods	17
Germany Males Females	430 350	National food intake survey in conjunction with values for RA contents of German foods	25
Finland Men and women combined High dairy foods diet (n = 80) High *trans* fatty acid diet (n = 40) High stearic acid diet (n = 40)	310 ± 41 40 ± 6 90 ± 16	Biochemical analyses of food duplicate portions in conjunction with dietary records	18
Washington/Idaho, U.S. College-aged subjects Males (n = 19) Females (n = 18)	137 ± 84 52 ± 44	3-d dietary records in conjunction with published values for RA contents of foods	39
Washington/Idaho, U.S. Lactating women (n = 16)	227 ± 180	Semiquantitative food-frequency questionnaire in conjunction with published values for RA contents of foods	20
Washington/Idaho, U.S. Lactating women (n = 16) Low dairy foods diet High dairy foods diet	15 ± 24 291 ± 75	3-d dietary records in conjunction with published values for RA contents of foods	20

[a]Values represent intakes of all isomers of conjugated linoleic acid; other values represent intakes of RA only.
[b]NR, not reported.

in the United States consume ~127 mg CLA/d. Somewhat similar values were obtained by Ritzenthaler *et al.* (39) who found that young men and women consumed 137 and 52 mg RA/d, respectively. It is important to note the extremely low usual RA intakes by these college-age women. On the basis of food consumption data collected in the West German National Consumption Survey (42), which included 10,985 households with 23,209 7-d dietary records, Fritsche and Steinhart (25) estimated dietary RA intake to be 430 and 350 mg/d for German men and women, respectively. Several reasons might explain the higher intakes estimated for Germans compared with those for the U.S. population. First, Germans consume relatively more fat than Americans (42), thus increasing the likelihood that Germans would consume more RA. Second, Fritsche and Steinhart (25) included the RA content of an impressive number (n = 139) of locally obtained German foods in their nutrient database, and the values presented for these foods are substantially greater than values previously published for similar foods obtained in the United States. It is probable that this resulted in a higher (and possibly more accurate) estimate of RA for the German population compared with that reported by Herbel *et al.* (17) and Ritzenthaler *et al.* (39). Finally, both Herbel *et al.* (17) and Ritzenthaler *et al.* (39) studied relatively young men and women in contrast to the overall German population studied by Fritsche and Steinhart (25), suggesting that age may influence CLA intake in humans.

We have investigated this hypothesis by estimating RA intakes for U.S. and German populations of various ages. To do this, we utilized data describing the RA contents of American and German foods (containing beef and dairy products) with information concerning national consumption of RA-containing foods in these two countries (42,43) in various age ranges (Fig. 29.1). Using this methodology, we have estimated mean RA intakes over a broad range of ages to be 95 and 89% for males and females, respectively, of those estimated by Fritsche and Steinhart (25). These data

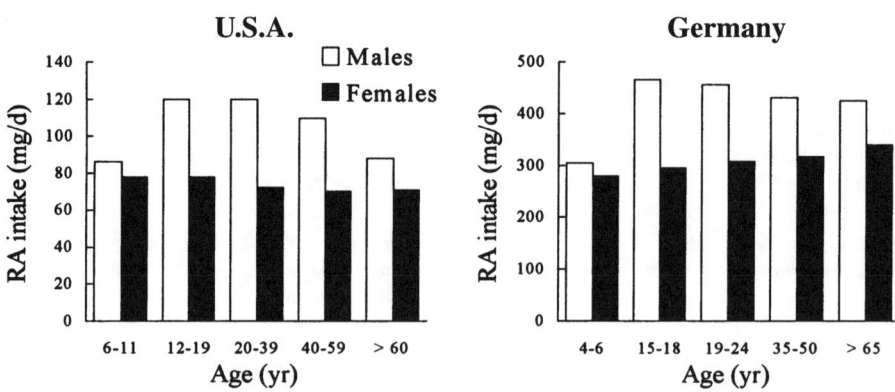

Fig. 29.1. Estimated intakes of rumenic acid (RA) in males and females living in the U.S. and Germany. Data were calculated using published values for the RA contents of German and U.S. foods (containing beef or dairy products) in conjunction with national dietary intake data.

suggest that, compared with the U.S. population and regardless of gender, RA consumption is greater for Germans in every age category. However, gender, age, and nationality interact to influence RA consumption. For example, RA consumption for German females continually increases with age, whereas that for U.S. females decreases. Clearly, further investigation should consider the interaction among age, gender, and ethnicity when studying RA intake and its putative physiologic effects.

Park *et al.* (20) used both semiquantitative food-frequency questionnaires and 3-d dietary records to estimate chronic as well as current RA intake of lactating women. Baseline data collected from the food-frequency questionnaires suggested that chronic RA intake of these women was 227 mg/d. In this intervention study with low and high dairy product consumption, 3-d dietary records estimated RA intakes to be 15 and 291 mg/d, respectively. Although the food-frequency questionnaire used in this study was not validated with a more direct measure of RA intake, data indicated that chronic RA intake estimated by the food-frequency questionnaire related significantly and positively to human milk RA concentration (Park *et al.*, unpublished data), suggesting that RA intakes estimated by this method did indeed reflect RA content of adipose stores. These estimates of RA intake are substantially higher than those reported by Herbel *et al.* (17) and Ritzenthaler *et al.* (39), which may be explained by the increase in fat intake that would be anticipated in lactating women, the use of a more complete database than that utilized by Herbel *et al.* (17), or a combination of these factors. However, apart from cultural differences, the reasons why RA intake estimates for lactating U.S. women were lower than those for nonlactating German women (25) are not obvious. Clearly, differences in utilization of these indirect methodologies to approximate RA intake can influence the obtained estimates, and both standardization and validation of these methods are warranted. Further, the effect of physiologic state (e.g., lactation) may influence RA intake.

Finally, Salminen *et al.* (18) used food duplicate methodology, a more direct assessment of dietary intake, in conjunction with 5-wk dietary records to estimate RA intake in Finnish men and women consuming experimental diets that differed in fatty acid content. These more rigorously obtained data suggest that RA intake can range from ~40 to 310 mg/d for a combined population of Finnish men and women. Although this study provides relatively more accurate and reliable estimates of RA intake compared with the previously mentioned reports, these values do not represent typical intakes of this population. Nonetheless, the similarity of the data obtained by Salminen *et al.* (18) and Park *et al.* (20) for maximal RA intakes on a high dairy fat diet suggests that human consumption of RA is certainly lower than the originally proposed estimates (40,41), and that the use of more indirect methodologies (i.e., food-frequency questionnaire and 3-d dietary record) can predict group RA intakes with some degree of accuracy.

Summary

Rigorous documentation of dietary CLA and RA intake in any population is not available currently. However, several investigators have utilized indirect methodologies to estimate both typical and extreme intakes of CLA and RA in a limited number of populations. Using 3-d dietary records in conjunction with published CLA or RA contents of foods, typical CLA intake has been estimated to be between 52 and 137 mg/d for young men and women in the United States (17,39) and 430 and 350 mg/d for German men and women, respectively (25). These values were supported and extended here by our estimation of RA intakes across the lifespan in German and U.S. populations (see Fig. 29.1).

The use of a semiquantitative food-frequency questionnaire to estimate typical RA intake in lactating women suggests that RA intake in this population is 227 mg/d (20). When individuals were asked to consume diets either low or high in dairy foods, minimum and maximum RA intakes were reported to be 15 (20) and 430 mg/d (25), respectively. These values are supported by data collected in the single study published to date that utilized biochemical analyses of food duplicate portions to estimate RA intake (18). Together, these data indicate that typical RA intake is well below that previously suggested (40,41) at least in individuals living in the United States, Germany, and Finland.

Recommendations for Future Research

The National Research Council (44) stated that CLA has been proven unequivocally to inhibit carcinogenesis in animal models. Since that time, an abundance of data has indicated that CLA, or specific isomers thereof, may influence human health. Thus, determination of human CLA and RA intakes appears warranted. Presently, there are no studies published that have documented typical CLA or RA intake using food duplicate methodology; only methodologies relying on the accuracy and completeness of available CLA and RA contents of foods have been used. Because CLA and RA concentrations within similar foods can vary greatly depending on the initial composition of raw food products, the accuracy and adequacy of these databases are questionable. Thus, it is recommended that rigorous validation of these nutrient databases be conducted using food duplicate methodology as the gold standard. Moreover, other potential indicators of CLA and RA intake, such as fasting plasma or adipose tissue concentrations of RA, should be investigated. Additionally, documentation of the differential contribution of CLA and RA from different food sources in different subpopulations would be of interest. Although we suspect that the major food sources (i.e., dairy and meat products) of CLA and RA remain constant, their relative contributions to dietary intake may vary by gender, age, physiologic state, and ethnicity.

Furthermore, estimations of the dietary intakes of CLA isomeric forms by infants, children, and adolescents have not been documented. McGuire et al. (36) have shown that human milk, but not infant formulas, contains significant amounts of RA. Consequently, it is clear that breast-fed infants consume more RA than do formula-fed infants. This could be especially important due to the published effects of CLA consumption on body weight of the suckling neonate (10). However, neither RA intake in infants nor its relationship with body composition has been documented. Furthermore, Ip et al. (45) have shown that the timing of dietary CLA intake influences its anticarcinogenic capacity in mammary tissues, such that it is more effective when consumed during periods of active mammary gland development. Thus, age and dietary CLA intake may interact, such that increased CLA intake during adolescence might preferentially decrease the risk of cancer in women. The findings of Ritzenthaler *et al.* (39) that young, college-aged women typically consume very low amounts of RA further support our hypothesis that CLA intake is influenced by age. Thus, it appears prudent to document CLA and RA intakes of various subpopulations of humans, including infants, children, and adolescents. Further, physiologic state (e.g., lactation) may influence RA intake.

Finally, examination of the relationships among dietary intake of CLA isomers, their concentrations in adipose tissue and plasma, and risk of various chronic degenerative diseases (e.g., cancer, diabetes, and osteoporosis) is essential for scientists and public health officials to draw conclusions concerning the importance of dietary CLA isomers in human health. Thus, enhancing our knowledge concerning isomeric CLA intake in various populations must remain a primary focus for research in this area.

References

1. National Research Council (1989) *Recommended Dietary Allowances*, 10th edn., National Academic of Science, Washington, DC.
2. Ha, Y.L., Grimm, N.K., and Pariza, M.W. (1987) Anticarcinogens from Fried Ground Beef: Heat-Altered Derivatives of Linoleic Acid, *Carcinogenesis 8*, 1881–1887.
3. Ip, C., Chin, S.F., Scimeca, J.A., and Pariza, M.W. (1991) Mammary Cancer Prevention by Conjugated Dienoic Derivative of Linoleic Acid, *Cancer Res. 51*, 6118–6124.
4. Shultz, T.D., Chew, B.P., Seaman, W.R., and Luedecke, L.O. (1992) Inhibitory Effect of Conjugated Dienoic Derivatives of Linoleic Acid and β-Carotene on the *in vitro* Growth of Human Cancer Cells, *Cancer Lett. 63*, 125–133.
5. Belury, M.A., Nickel, K.P., Bird, C.E., and Wu, Y. (1996) Dietary Conjugated Linoleic Acid Modulation of Phorbol Ester Skin Tumor Promotion, *Nutr. Cancer 26*, 149–157.
6. Cunningham, D.C., Harrison, L.Y., and Shultz, T.D. (1997) Proliferative Responses of Normal Human Mammary and MCF-7 Breast Cancer Cells to Linoleic Acid, Conjugated Linoleic Acid and Eicosanoid Synthesis Inhibitors in Culture, *Anticancer Res. 17*, 197–204.
7. Lee, K.N., Kritchevsky, D., and Pariza, M.W. (1994) Conjugated Linoleic Acid and Atherosclerosis in Rabbits, *Atherosclerosis 108*, 19–25.
8. Nicolosi, R.J., Roger, E.J., Kritchevsky, D., Scimeca, J.A., and Huth, P.J. (1997) Dietary Conjugated Linoleic Acid Reduces Plasma Lipoproteins and Early Aortic Atherosclerosis in Hypercholesterolemic Hamsters, *Artery 22*, 266–277.
9. Houseknecht, K.L., Vanden Heuvel, J.P., Moya-Camarena, S.Y., Portocarrero, C.P., Peck, L.W., Kwangok, P.N., and Belury, M.A. (1998) Dietary Conjugated Linoleic Acid Normalizes Impaired Tolerance in the Zucker Diabetic Fatty *fa/fa* Rat, *Biochem. Biophys. Res. Commun. 244*, 678–682.
10. Chin, S.F., Storkson, J.M., Albright, K.J., Cook, M.E., and Pariza, M.W. (1994) Conjugated Linoleic Acid Is a Growth Factor for Rats as Shown by Enhanced Weight Gain and Improved Feed Efficiency, *J. Nutr. 124*, 2344–2349.
11. Cook, M.E., Jerome, D.L., Crenshaw, T.D., Buege, D.R., Pariza, M.W., Schmidt, S.P., Scimeca, J.A., Lofgren, P.A., and Hentges, E.J. (1998) Feeding Conjugated Linoleic Acid Improves Feed Efficiency and Reduces Carcass Fat in Pigs, *FASEB J. 12*, Abstr. 4843.

12. Cook, M.E., Miller, C.C., Park, Y., and Pariza, M. (1993) Immune Modulation by Altered Nutrient Metabolism: Nutritional Control of Immune-Induced Growth Depression, *Poult. Sci. 72*, 1301–1305.
13. Li, Y., and Watkins, B.A. (1998) Conjugated Linoleic Acids Alter Bone Fatty Acid Composition and Reduce *ex vivo* Prostaglandin E_2 Biosynthesis in Rats Fed n-6 or n-3 Fatty Acids, *Lipids 33*, 417–425.
14. Watkins, B.A., Shen, C.L., McMurtry, J.P., Xu, H., Bain, S.D., Allen, K.G.D., and Seifert, M.F. (1997) Dietary Lipids Modulate Bone Prostaglandin E_2 Production, Insulin-Like Growth Factor-I Concentration and Formation Rate in Chicks, *J. Nutr. 127*, 1084–1091.
15. Kramer, J.K.G., Parodi, P.W., Jensen, R.G., Mossoba, M.M., Yurawecz, M.P., and Adlof, R.O. (1998) Rumenic Acid: A Proposed Common Name for the Major Conjugated Linoleic Acid Isomer Found in Natural Products, *Lipids 33*, 835.
16. Ha, Y.L., Storkson, J., and Pariza, M.W. (1990) Inhibition of Benzo(a)pyrene-Induced Mouse Forestomach Neoplasia by Conjugated Dienoic Derivatives of Linoleic Acid, *Cancer Res. 50*, 1097–1101.
17. Herbel, B.K., McGuire, M.K., McGuire, M.A., and Shultz, T.D. (1998) Safflower Oil Consumption Does Not Increase Plasma Conjugated Linoleic Acid Concentrations in Humans, *Am. J. Clin. Nutr. 67*, 332–337.
18. Salminen, I., Mutanen, M., Jauhiainen, M., and Aro, A. (1998) Dietary *trans* Fatty Acids Increase Conjugated Linoleic Acid Levels in Human Serum, *J. Nutr. Biochem. 9*, 93–98.
19. Huang, Y.C., Luedecke, L.O., and Shultz, T.D. (1994) Effect of Cheddar Cheese Consumption on Plasma Conjugated Linoleic Acid Concentration in Men, *Nutr. Res. 14*, 373–386.
20. Park, Y.S., Behre, R.A., McGuire, M.A., Shultz, T.D., and McGuire, M.K. (1997) Dietary Conjugated Linoleic Acid (CLA) and CLA in Human Milk, *FASEB J. 11*, Abst. 1387.
21. Fogerty, A.C., Ford, G.L., and Svoronos, D. (1988) Octadeca-9,11-dienoic Acid in Foodstuffs and in the Lipids of Human Blood and Breast Milk, *Nutr. Rep. Int. 38*, 937–944.
22. Werner, S.A., Luedecke, L.O., and Shultz, T.D. (1992) Determination of Conjugated Linoleic Acid Content and Isomer Distribution in Three Cheddar-Type Cheeses: Effects of Cheese Cultures, Processing, and Aging, *J. Agric. Food Chem. 40*, 1817–1821.
23. Lin, H., Boylston, T.D., Chang, M.J., Luedecke, L.O., and Shultz, T.D. (1995) Survey of the Conjugated Linoleic Acid Content of Dairy Products, *J. Dairy Sci. 78*, 2358–2365.
24. Banni, S., Carta, G., Contini, M.S., Angioni, E., Deiana, M., Dessi, M.A., Melis, M.P., and Corongiu, F.P. (1996) Characterization of Conjugated Diene Fatty Acids in Milk, Dairy Products, and Lamb Tissues, *J. Nutr. Biochem. 6*, 150–155.
25. Fritsche, J., and Steinhart, H. (1998) Amounts of Conjugated Linoleic Acid (CLA) in German Foods and Evaluation of Daily Intake, *Z. Lebensm. Unters. Forsch. A. 2065*, 77–82.
26. Kelly, M.L., and Bauman, D.E. (1996) Conjugated Linoleic Acid: A Potent Anticarcinogen Found in Milk Fat, in *Proceedings of the Cornell Nutrition Conference for Feed Manufacturers,* Cornell University Department of Animal Sciences, Ithaca, NY, pp. 68–74.
27. McGuire, M.A., McGuire, M.K., McGuire, M.S., and Griinari, J.M. (1997) Bovinic Acid: The Natural CLA, in *Proceedings of the Cornell Nutrition Conference for Feed Manufacturers,* Cornell University Department of Animal Sciences, Ithaca, NY, pp. 217–226.
28. Parodi, P.W. (1977) Conjugated Ocatadecadienoic Acids of Milk Fat, *J. Dairy Sci. 60*, 1550–1553.

29. Ha, Y.L., Grimm, N.K., and Pariza, M.W. (1989) Newly Recognized Anticarcinogenic Fatty Acids: Identification and Quantification in Natural and Processed Cheeses, *J. Agric. Food Chem. 37,* 75–81.
30. Aneja, R.P., and Murthi, T.N. (1990) Conjugated Linoleic Acid Contents of Indian Curds and Ghee, *Indian J. Dairy Sci. 43,* 231–238.
31. Chin, S.F., Liu, W., Storkson, J.M., Ha, Y.L., and Pariza, M.W. (1992) Dietary Sources of Conjugated Dienoic Isomers of Linoleic Acid, a Newly Recognized Class of Anticarcinogens, *J. Food Comp. Anal. 5,* 185–197.
32. Shantha, N.C., Decker, E.A., and Ustunol, Z. (1992) Conjugated Linoleic Acid Concentration in Processed Cheese, *J Am. Oil Chem. Soc. 69,* 425–428.
33. Shantha, N.C., and Decker, E.A. (1993) Conjugated Linoleic Acid Concentrations in Processed Cheese Containing Hydrogen Donors, Iron and Dairy-Based Additives, *Food Chem. 47,* 257–261.
34. Shantha, N.C., Crum, A., and Decker, E.A. (1994) Evaluation of Conjugated Linoleic Acid Concentrations in Cooked Beef, *J. Agric. Food Chem. 42,* 1757–1760.
35. Shantha, N.C., Ram, L.N., O'Leary, J., Hicks, L.C., and Decker, E.A. (1995) Conjugated Linoleic Acid Concentrations in Dairy Products as Affected by Processing and Storage, *J. Food Sci. 60,* 695–697.
36. McGuire, M.K., Park, Y., Behre, R.A., Harrison, L.Y., Shultz, T.D., and McGuire, M.A. (1997) Conjugated Linoleic Acid Concentrations of Human Milk and Infant Formula, *Nutr. Res. 17,* 1277–1283.
37. Hanson, T.W., and McGuire, M.A. (1998) Evaluation of Conjugated Linoleic Acid Content in the Meat of Wildlife from the Pacific Northwest, in *Proceedings of the Second Comparative Nutrition Society Symposium,* Comparative Nutrition Society, Silverspring, MD, pp. 68–73.
38. Jensen, R.G., Lammi-Keefe, C.J., Hill, D.W., Kind, A.J., and Henderson, R. (1998) The Anticarcinogenic Conjugated Fatty Acid, $9c,11t$-18:2, in Human Milk: Confirmation of Its Presence, *J. Hum. Lact. 14,* 23–27.
39. Ritzenthaler, K., McGuire, M.K., Falen, R., Shultz, T.D., and McGuire, M.A. (1998) Estimation of Conjugated Linoleic Acid (CLA) Intake, *FASEB J. 12,* A527.
40. Ip, C., Scimeca, J.A., and Thompson, H. (1994) Conjugated Linoleic Acid, a Powerful Anticarcinogen from Animal Fat Sources, *Cancer 74,* 1050–1054.
41. Parodi, P.W. (1994) Conjugated Linoleic Acid: An Anticarcinogenic Fatty Acid Present in Milk Fat, *Austr. J. Dairy Technol. 49,* 93–97.
42. Adolf, T., Eberhardt, W., Heseker, H., Hartmann, S., Herwig, A., Matiaske, B., Moch, K.J., Schneider, R., and Kuebler, W. (1994) *Lebensmittel und Naehrstoffaufnahmein der Bundesrepublik Deutschland,* (Kueble, W., Anders, H.J., und Heeschen, W., eds.) VERA Schriftenreihe, Band XII, Wissenschaftlicher Fachverlag Dr. Fleck, Niderkleen.
43. United States Department of Agriculture, Agriculture Research Service (1997) Data Tables: Intakes of 19 Individual Fatty Acids: Results from 1994–96 Continuing Survey of Food Intakes by Individuals. AGR Food Surveys Research Group. http://www.barc.usda.gov/bhnrc/foodsurvey/home.htm{11/04/98}
44. National Research Council (1996) *Carcinogens and Anticarcinogens in the Human Diet,* National Academy of Science, Washington, DC.
45. Ip, C., Scimeca, J.A., and Thompson, H. (1995) Effect of Timing and Duration of Dietary Conjugated Linoleic Acid on Mammary Cancer Prevention, *Nutr. Cancer 24,* 241–247.

Chapter 30

Formation, Contents, and Estimation of Daily Intake of Conjugated Linoleic Acid Isomers and *trans*-Fatty Acids in Foods

J. Fritsche[a], R. Rickert[b], and H. Steinhart[b]

[a]U.S. Food and Drug Administration, Center of Food Safety and Applied Nutrition, Office of Food Labeling, Washington, DC 20204
[b]Institute of Biochemistry and Food Chemistry, Department of Food Chemistry, University of Hamburg, 20146 Hamburg, Germany

Introduction

Interest in *trans* fatty acids (TFA) and conjugated linoleic acid isomers (CLA) has increased in the last decade. Although high dietary intakes of TFA (>10 g/d) have been linked to adverse physiologic effects [e.g., increased total cholesterol, lipoprotein(a), and low density lipoprotein levels, which are considered risk factors for coronary heart disease (1–5)], animal studies have suggested that CLA act as anticarcinogenic and antiatherogenic agents (6–10). The role of CLA in lipid metabolism is not yet fully understood, and many approaches have been undertaken to explore the unique physiologic properties of CLA.

Formation of CLA and TFA

Both CLA and TFA are formed in the rumen during the biohydrogenation of polyunsaturated fatty acids such as linoleic acid (18:2n-6, LA). The anaerobic rumen bacterium *Butyrivibrio fibrisolvens* converts LA to CLA, predominantly to the 18:2 *c*9*t*11 isomer (rumenic acid, RA; Fig. 30.1) (11). Numerous other geometrical and positional CLA isomers are also formed. These intermediates are converted to *cis* and *trans* monounsaturated fatty acids, predominantly oleic acid (18:1 *c*9), elaidic acid (18:1 *t*9) and *trans* vaccenic acid (18:1 *t*11). Finally, monounsaturated fatty acids are transformed to stearic acid (18:0, Fig.30.1) (11). Because of their origin, both TFA and CLA occur together in all ruminant products. More detailed information about the biosynthesis of CLA is provided in Chapter 13 of this book.

The ratio of precursors (e.g., 18:2n-6) to intermediates (TFA or CLA) or product (18:0) is influenced by rumen conditions (e.g., rumen pH or amounts of dietary polyunsaturated fatty acids) and can be affected by different factors, which are discussed in Chapter 16. In addition to the dietary factors, environmental and intrinsic factors (e.g., season or breed) affect the TFA and CLA content of dairy products.

RA can also be formed by Δ 9-desaturation of vaccenic acid (Fig. 30.1) (12). The autoxidation of LA in foods containing high amounts of polyunsaturated fat is another process by which trace amounts of CLA may be produced. However, this process

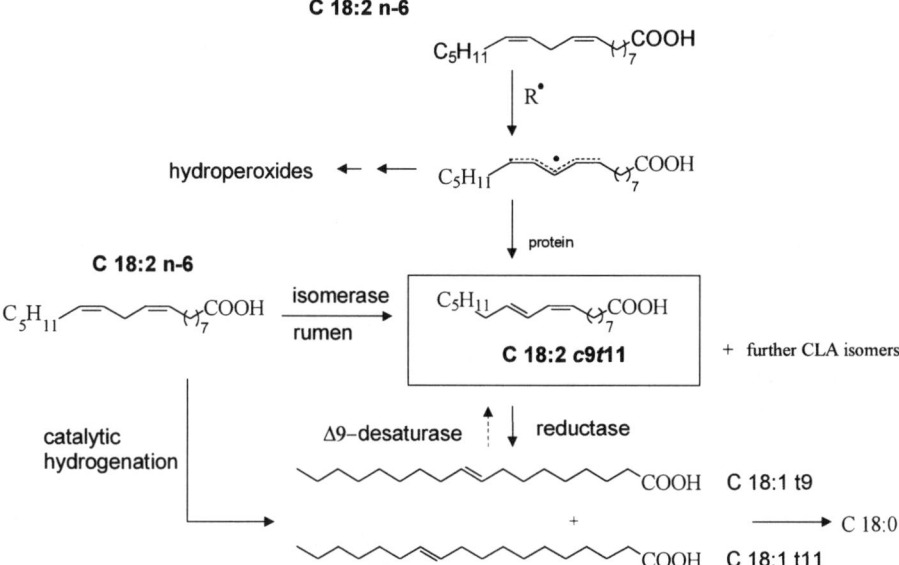

Fig. 30.1. Formation of conjugated linoleic acids (CLA) and *trans* fatty acids.

occurs only in the absence of air and the presence of proteins, which may act as proton donors (13).

Compared with the extent of biohydrogenation, TFA is formed to a much greater extent than CLA during industrial hydrogenation of vegetable oils. The amounts of TFA and CLA in partially hydrogenated vegetable oils (PHVO) depend on the deodorization conditions (e.g., temperature, duration, catalyst, or hydrogen pressure). Dry heating (e.g., 275°C, 3 h) led to an increased isomerization rate of 18:3n-3 in rapeseed oil (14). More recently, Schöne *et al.* (15) reported that deodorization temperatures >230°C for longer than 2 h increased the 18:3 $c9c12t15$ content up to sixfold [0.31 ± 0.02% of total fatty acid methyl esters (FAME)].

The triacylglycerol composition of the heated oil is also important for the isomerization of polyunsaturated fatty acids, especially 18:2n-6 or 18:3n-3. It has been shown that the *sn*-2 position of heated monoacid triacylglycerols (18:2,18:2,18:2 or 18:3,18:3,18:3) in canola oil is more sensitive to isomerization then are the *sn*-1 or *sn*-3 positions (16).

Contents in Food

Edible Oils and Margarines

Edible oils (e.g., refined or unrefined walnut, olive, sunflower, safflower, grapeseed, soybean, avocado, cashew, or peanut oil) and coconut oil are common in the human diet and in general contain only negligible amounts of CLA and TFA (Table 30.1). Extreme processing conditions may lead to detectable CLA and

TABLE 30.1 Mean Conjugated Linoleic Acid (CLA) and *trans* Fatty Acid (TFA) Contents of Edible Oils and Margarines[a]

Food	n	Origin/Year	Reference
Edible oils			
Unrefined oils	8	Germany/1973/4	17
Refined oils	7	UK/1984	18
Unrefined oils	11		
Soybean oil, hydrogenated	3	U.S./1991	19
Edible oils	6	Germany/1992	20
Coconut oil, olive oil	16	U.S./1992	21
Edible oils, refined	9	Germany/1997	22,23
Margarines			
Margarines	26	Italy/1967	24
Margarines	10	U.S./1973	25
Margarines	83	Germany/1973,1976	17
Margarines	9	U.S./1977	26
Margarines	7	U.S./1977	27
Margarines	9	Canada/1979	28
Margarines	40	U.S./1983	29
Margarines	84	U.S./1985	30
Margarines	10	Greece/1991	31
Margarine		U.S./1991	19
Margarines	16	Germany/1992	20
Margarines	9	Germany/1993	32
Margarines	46	Germany/1994	33
Margarines	9	Belgium/1994/5	34
	7	Hungary	
	3	GB	
Margarines	17	Turkey/1994	35
Shortenings	43	Denmark/1995	36
Hard margarines (>40% LA)	20		
Tub margarines	12	France/1995	37
Margarines	24	Germany/1995	38
Margarines	17	Germany/1997	22,23
Margarines	6	U.S./1997	39
Margarines	?	Canada/1998	40

[a]Values in parentheses are minimum and maximum levels.

conjugated octadecatriene (COT) concentrations in the range from <0.001 to 0.2% in edible oils (19,41).

Margarines contain varying amounts of PHVO and, therefore, TFA. PHVO provides desirable physical and sensory properties (e.g., increased melting point and better mouth feel) and prevents flavor reversion (14). CLA does not arise in considerable amounts under normal processing conditions (23). However, a few investigators have reported detectable amounts of CLA in margarines (Table 30.1).

18:1 t	18:2 c/t	Total TFA (% of total fat)	CLA
		0.1–0.2	
		1.6 (0.2–6.7)	
		0.5 (0.2–1.0)	
			0.3–1.8
0.25 (0–1.5)	0.05 (0–0.2)	0.30 (0–1.6)	
			0.01–0.07
<0.01	<0.01	<0.01	<0.01
		22.5 (0.9–34.3)	
		24.1 (0–36.0)	
		0.1–53.2	
18.0 (6.9–31.4)			
		20.4 (6.3–33.6)	
		21.3 (8.7–32.9)	
		18.4 (6.8–31.0)	
19.89 (10.74–30.06)			
5.4–9.5	0–3.65		
			0.6
7.9 (0.1–23.0)	0.5 (0.1–1.7)	8.5 (0.6–23.5)	
		5.0 (0–10.6)	
9.32 (0.17–25.90)			
5.7 (0.2–16.7)	0.4 (0–0.8)	6.4 (0.5–17.8)	
13.0 (2.0–22.3)	0.5 (0–0.8)	14.1 (2.0–24.5)	
9.3 (9.3–10.9)	0.5 (0–1.0)	9.77 (7.7–11.3)	
0–34.5			0.3–2.0
6.8 (0–13.7)	<0.2		<0.2
1.3 (0–5.6)	<0.2		<0.2
8.0 (0–17.6)	0.4 (0.06–1.40)		
4.6 (0–20.1)	0.5 (0–0.7)	5.1 (0.4–21.0)	
1.17 (0.05–4.35)	0.17 (0–0.43)	1.5 (0.15–4.88)	<0.01
0.04–13.29	0–0.26		
0.2–37.3	0.2–6.7	18.8 (0.9–46.4)	

During recent years, the TFA content in German and other European margarines has decreased significantly (22). This trend is probably attributable to increased public health interests linking TFA with higher risk of coronary heart disease (CHD) and atherosclerosis. One means of achieving lower TFA contents in PHVO is the partial replacement of TFA by palm, palm kernel, or coconut oil to obtain acceptable product consistency (37).

Milk and Dairy Products

Milk fat is a major dietary source of TFA and CLA in the Western diet because of its ruminant origin. The 18:1 *t*, 18:2 *cis*9,*trans*12 and 18:2 *trans*9,*cis*12, total TFA, and CLA contents of milk and dairy products are presented in Table 30.2.

The total TFA content in milk ranged from ~1.5 to 6.5% (48,51) and in dairy products from 2.0 to 6.1% (22). The predominant TFA isomers in dairy products are the octadecenoic acid isomers, especially *trans*-vaccenic acid. Other TFA such as myristelaidic acid (14:1 *t*9), palmitelaidic acid (16:1 *t*9), positional 18:1 *trans* isomers, linolelaidic acid (18:2 *t*9*t*12), and linolenelaidic acid isomers (18:3n-3 *t*) are also found in low amounts in dairy products (22,47). For example, the contents of palmitelaidic acid varied between 0.04 (goat cheese) and 0.21% (Lightdammer) (22). The wide variations in TFA content in dairy products may be explained by the

TABLE 30.2 Mean Conjugated Linoleic Acid (CLA) and *trans* Fatty Acid (TFA) Contents of Milk and Dairy Products[a]

Food	n	Origin/Year	Reference
Milk	116	Australia/1971	42
Milk	13	Canada/1983	43
Milk		Australia/1988	44
Milk	13	Austria/1994	45
Milk	5	Germany/1992	20
Milk	2	U.S./1994	
Milk	1756	Germany/1995	47
Milk	132	Germany/1996 Indoor Pasture Ecological	48
Milk	63	Sweden/1996	49
Milk	7	Germany/1997	22,23
Dairy products			
Cheese	27	Germany/1992	20
Butter	5	Germany/1992	20
Natural cheese	13	U.S./1992	21
Processed cheese	4		
Homogenized milk	3		
Butter	4		
Cheese	15	U.S./1994	50
Fermented dairy products	3	U.S./1995	46
Fluid milk products	4		
Cheese	15		
Cheese	25	Germany/1997	22,23
Butter	12		

[a]Values in parentheses are SD or minimum and maximum levels.

varying TFA content of milk fat. Irish butter (mean 6.5%, Fritsche *et al.*, unpublished results) tends to contain higher amounts of TFA than German butter (mean 3.5%). These variations in TFA in butters from different countries may be due to differences in breed of dairy cow or feeding regimens.

The wide variation of TFA and CLA contents in dairy products may also be due to different processing parameters such as different heat treatment procedures during pasteurization. An influence of fermentation on the TFA and CLA content in yogurt or cheese is also possible. For example, TFA contents in long ripened propionic acid–fermented cheeses (e.g., Old Emmentaler, 5.76% of total FAME) or yogurt with added probiotic cultures seem to be slightly elevated (22). Bacterial surface-ripened cheeses such as Münster cheese and Tilsiter did not show different CLA contents compared with wholly lactic acid–fermented cheeses (23). Additional factors that may affect the CLA/TFA content in dairy products, such as

18:1 t	18:2 c/t	Total TFA (% of total fat)	CLA
		6.01 (4.27–7.64)	
		4.0–5.7	
			0.9–1.2
		1.75–5.20 (0.99)	
1.6–4.0	0.3–0.8	2.4–5.5	
			0.45 (0.06)
		3.62 (1.29–6.75)	
1.53 (0.11)	0.24 (0,04)	3.32 (0.29)	0.34 (0.04)
2.54 (0.46)	0.21 (0,05)	4.73 (0.73)	0.61 (0.08)
3.04 (0.68)	0.43 (0.11)	5.78 (1.23)	0.80 (0.21)
			0.25–1.77
3.13 (0.24)	0.48 (0.28)	5.12 (1.02)	1.16 (0.10)
1.0–5.1	0.1–1.6	1.6–7.5	
1.5–6.3	0.5–0.8	2.5–7.9	
			0.29–0.71
			0.45–0.52
			0.55 (0.03)
			0.47 (0.04)
			0.32–0.89
			0.38–0.47
			0.40–0.64
			0.36–0.80
2.09 (0.89)	0.51 (0.14)	3.89 (1.92)	0.84 (0.38)
		4.43 (1.99)	0.94 (0.48)

the role of H-donators or the content of natural and synthetic antioxidants, are discussed in a recent review article (52).

More recently, Lavillonnière et al. (53) investigated the influence of geographic factors on the CLA content of French cheeses (Comté cheese, n = 12). These authors reported that cheese produced from milk of cows grazing in the plains in the Jura Mountains or at medium alititude contained more CLA than cheeses from milk of cows grazing on the high plateau (20.8 ± 4.0 and 20.6 ± 1.1 vs. 15.0 ± 0.7 mg CLA/g fat, respectively).

However, the influence of processing conditions on the CLA and TFA content in dairy products should be carefully evaluated. The natural variation of the raw material may lead to overestimation of the effects of food processing. Therefore, standardized raw material should be used to investigate possible changes in TFA and CLA contents in individual dairy products after each processing procedure (e.g., homogenization, pasteurization, or ripening).

Meat, Meat Products and Fish

Meat and Meat Products

The TFA and CLA contents in meat and meat products are shown in Table 30.3. Meat from nonruminants such as pork or poultry shows distinctly lower TFA contents than meat from ruminants. The TFA content in beef is similar to that in milk or other dairy products (20). *Trans* vaccenic acid is the predominant isomer in meat and accounts for ~80% of total TFA (22).

The CLA content in meat from ruminants is higher than that in meat from nonruminants (i.e., 1.20% in lamb and 0.12% in pork, respectively) (23). In the case of nonruminants, CLA may arise from dietary sources such as the feeding of meat meal and tallow. CLA may also arise by endogenous formation by intestinal bacteria, as has been shown in rats (54).

The TFA content in meat products varied from 0.2 to 3.4% of total fat (Table 30.3). These values fall between those for meat and those for ruminants and nonruminants, depending on the ingredients (e.g., meat or fat) used. The CLA contents in meat products ranged from 0.27 to 0.44%. The CLA content of the meat products also seems to reflect their ingredients, and it is influenced neither by processing conditions nor by fermentation (23).

Fish

Unprocessed fish oil has a negligible TFA content (up to 1.01%) compared with that of meat or dairy products (22). The predominant TFA isomers are the octadecenoic acid isomers (average, 60% of total TFA). Carp has the highest TFA content (1.01%), which might be caused by feeding TFA-containing feed to the fish in aquaculture.

The CLA levels ranged from 0.01% in pike-perch to 0.09% in carp (23). Currently, limited data dealing with CLA content in fish are available. Seafood such as shrimp (0.6 ± 0.10 mg total CLA/g fat) or mussels (0.4 ± 0.04 mg total CLA/g fat) is similar in content to fish (e.g., lake trout 0.5 ± 0.05 mg total CLA/g fat) (21). These authors could not find 18:2 *c*9*t*11 and therefore stated the CLA content in marine foods as total CLA content. The marine formation of CLA remains uncertain.

Cakes, Pastries, and Other Processed Foods

The TFA content in chocolates, cakes and pastries, snack food, and other foods ranged from <0.01% (e.g., plain chocolate) to ~34% (French fries, instant soups) and originates from both dairy fat and PHVO present in the foods (22).

These wide variations of TFA have also been observed in the same product category (e.g., potato chips 1.22–22.01% (22). A reason for this variability may be that these products often contain a blend of PHVO of different sources (e.g., soybean, canola, corn, or sunflower oil) with varying TFA contents. The CLA contents in these food items originate predominantly from dairy fat and reflect its CLA level. In composed foods such as milk chocolates or nut nougat cremes, which contain a blend of milk and cocoa fat, the percentage of CLA is, therefore, lower than in dairy products. On the other hand, French fries and frying fat, which can contain high amounts of PHVO, have negligible contents of CLA (<0.01%) (23).

Isomeric Distribution of CLA Isomers

In contrast to the well-reported distribution of *trans* octadecenoic acid isomers in ruminant products, only limited data are available on the isomeric distribution of CLA isomers in biological matrices. This lack of information is due to difficulties in the chromatographic separation of the individual minor isomers.

In addition to the major CLA isomer *c*9*t*11, seven minor CLA isomers, i.e., *c*10,*t*12; *t*10,*c*12; 11*c*,13*c*; 9*c*,11*c*; 10*c*,12*c*; 9*t*,11*t*; and 10*t*,12*t*, were reported in cheese on the basis of gas chromatography-mass spectrometry (GC-MS) analysis after acid-catalyzed methylation (56). These data have to be evaluated with caution, however, e.g., with attention to the influence of methylation procedures on the isomeric distribution, as discussed in Chapter 6. More recently, the identity of five additional minor CLA isomers, i.e., 8*c*,10*t*; 8*c*,10*c*; 8*t*,10*t*; 11*t*,13*t*; and 11?,13?, in cheese was reported (53). A total of 16 CLA isomers were determined by tandem-column silver-ion high-performance liquid chromatography (Ag$^+$-HPLC), gas chromatography-Fourier transform infrared spectroscopy (GC-FTIR), and GC-MS (57).

The isomeric distribution of CLA isomers in European cheeses and American beef fat is shown in Figure 30.2A (57, Fritsche *et al.*, unpublished data). The *c*9*t*11 amount in American beef is slightly lower than that in European cheeses, and the *t*7*c*9 content in American beef tends to be elevated. In addition, the CLA pattern of an alkali-isomerized CLA reference is presented in Figure 30.2B (57,58). The recognition of different CLA profiles of natural and synthetic sources is important because the isomers may have different physiologic effects.

TABLE 30.3 Mean Conjugated Linoleic Acid (CLA) and *trans* Fatty Acid (TFA) Contents in Meat and Meat Products[a]

Food	n	Origin/Year	Reference
Meat			
Round beef	4	U.S./1992	21
Fresh ground beef	4		
Lamb	4		
Pork	2		
Chicken	2		
Fresh ground turkey	2		
Calf	3	Germany/1992	20
Beef	5		
Lamb	3		
Sheep			
Mutton	3		
Pork	3		
Poultry	5		
Rabbit			
Horse	2		
Kangaroo	2		
Pork	4	Germany/1997	22,23
Lamb	2		
Turkey	2		
Beef steak	2		
Beef liver	2		
Rabbit	2		
Chicken	2		
Bulls	6	Germany/1998	55
Steers	6		
Meat products			
Meat products	25	Germany/1992	20
Meat prdoducts	18	U.S./1992	21
Meat products	4	Germany/1993	32
Meat products	10	Germany/1997	22,23
Meat products	3	U.S./1997	39

[a]Values in parentheses are SD.

The overall isomeric distribution in cheese is similar to that in beef fat and totally different from that of the alkali-isomerized LA, which contains a multiplicity of major CLA isomers [e.g., $c9t11$, $t8c10$, $t10c12$, and $c11t13$ in almost equal amounts (57,58)]. In contrast to earlier studies, which reported >90% $c9t11$ of total CLA (determined by GC only), more recent studies reported slightly lower $c9t11$ amounts [80.8 ± 1.9% total CLA (59); 83.5 ± 1.5% total CLA (57)] using GC and Ag+-HPLC techniques. These different CLA amounts are probably due to an overestimation based on coelutions of $c9t11$ with 7,9 c/t and 8,10 c/t when only GC was

18:1 t	18:2 c/t	Total TFA (% of total fat)	CLA
			0.29 (0.01)
			0.27 (0.02)
			0.56 (0.03)
			0.06 (0.01)
			0.09 (0.002)
			0.25 (0.004)
0.7–1.3	trace–0.2	0.9–1.7	
1.4–2.4	0–0.3	1.9–3.2	
5.2–7.0	0.6–0.9	6.6–8.8	
2.2	0.4	3.2	
6–8.9	0.9–1.2	8.2–10.6	
0.2–0.4	trace	0.2–0.4	
0.2–1.2	0–0.2	0.2–1.4	
0.3	trace	0.4	
0.2	trace	0.2	
7.9	0.8	9.8	
0.22–0.42	0.02–0.03	0.46–0.68	0.12–0.15
7.53	0.23	9.80	1.20
0.89	0.08	1.19	0.20
2.29	0.19	3.57	0.65
2.27	0.03	2.84	0.43
0.16	0.03	0.37	0.11
0.52	0.10	0.88	0.15
			0.76 (0.15)
			0.86 (0.15)
0.2–2.6	trace–0.3	0.2–3.4	
			0.08–0.33
		2.6 (2.6)	
0.14–0.53	0.02–0.12	0.48–1.09	0.27–0.44
			0.20–0.87

used. More details about the GC coelutions of CLA isomers are provided by John Roach in Chapter 9.

Estimation and Evaluation of Daily TFA and CLA Intake

Dietary CLA and TFA intake has been estimated from dietary questionnaires or recall data, analysis of adipose tissue or milk fat data, or food content analysis data. Commonly, the basis of TFA intake calculation varies, leading to varying intake estimations. Another reason for the variation is the differing eating habits of individuals

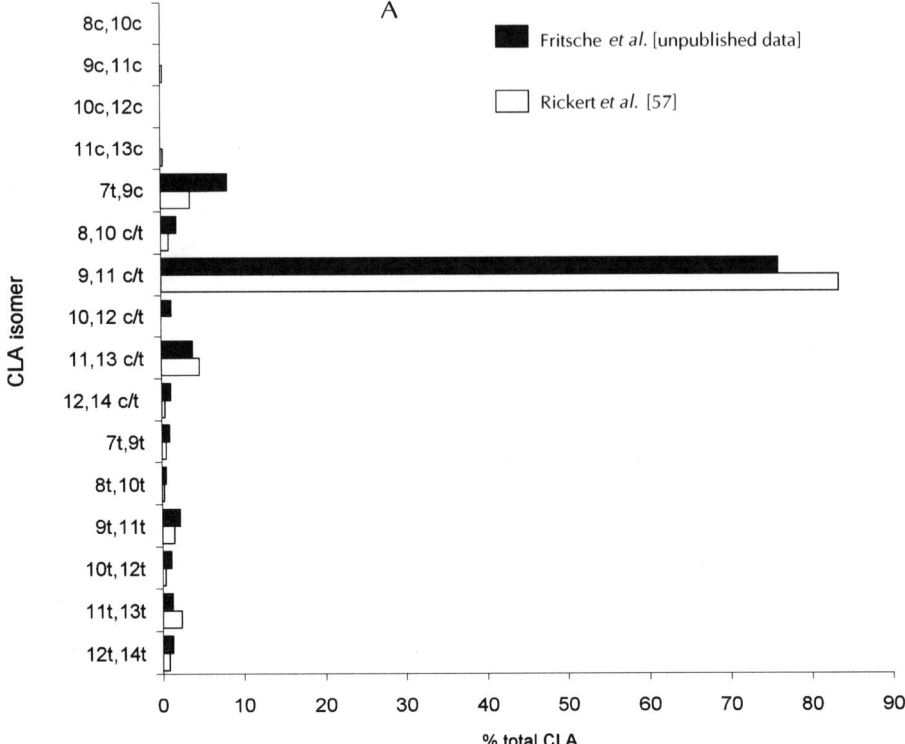

Fig. 30.2. (A) Conjugated linoleic acid (CLA) isomeric distribution in American beef fat [solid bars, n = 32, (Fritsche et al., unpublished data)] compared with European cheese [open bars, n = 16 (Ref. 57)].

as well as of cultures. Tables 30.4 and 30.5 present an overview of the daily TFA and CLA intake in different countries.

As can be seen from Table 30.4, individuals from Mediterranean countries such as Spain or Italy seem to have a lower TFA intake compared with individuals from other European countries. Pfalzgraf and Steinhart (64) estimated the daily TFA intake in Germany in 1992 to be 4.1 g/d for men. In view of decreased TFA contents in German margarines and lower meat consumption since 1992, the estimation of the daily TFA intake in Germany was updated in 1997 to 2.3 g/d (22). These results show a decrease of TFA intake by ~40% since 1992. The TFA intake reported by Fritsche and Steinhart (22) is in the same range as the estimate by Demmelmair et al. (32). Wolff reported elevated 18:1 t intake levels for Germany (3.7 g/d) in the range of the values published five years ago (66). This higher 18:1 t intake is probably caused by overestimation of TFA amounts for German margarines.

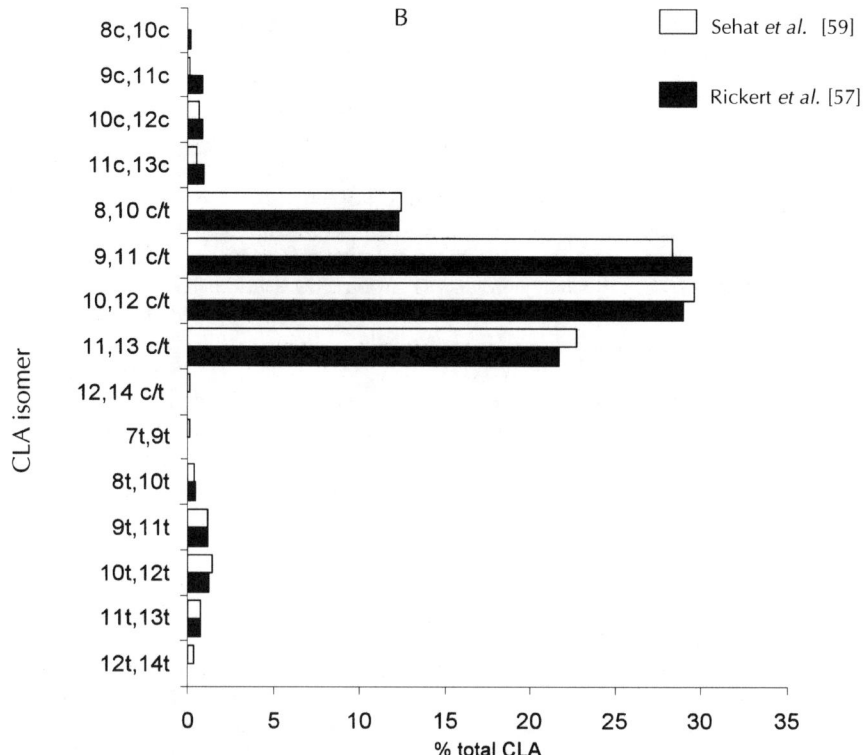

Fig. 30.2. (B) CLA isomeric distribution in commercial CLA mixtures produced by alkali isomerization of linoleic acid (open bars, Ref. 58; solid bars, Ref. 57).

Recently, the British Ministry of Agriculture, Fisheries and Food published average TFA intake data. IR and GC were used to determine the TFA levels from frying oils and foods cooked in frying oils. The average intake was estimated to be 2.0 and 1.7 g/d for men and women, respectively (68).

The TFA intake in the United States and Canada seems to be higher than that in European countries, although the per capita availability of TFA from household salad oils, cooking oils, margarines, and spreads has decreased since 1984 (62). The decreased TFA intake is probably due to the replacement by most manufacturers of unhydrogenated oil for household salads and cooking oils. The per capita TFA consumption from household shortenings decreased in a similar manner [from ~0.55 g/d in 1984 (60) to 0.31 g/d in 1989], probably caused by a decline in market size during the 1980s. Although margarine and spread production has increased in the United States since 1984, the per capita availability of TFA from margarines and spreads has decreased

TABLE 30.4 Estimation of Daily *trans* Fatty Acid (TFA) Intake in Different Countries

Country	Year	/Reference	TFA (g/d)	Basis
U.S.	1986	60	7.6	Market size and share data combined with TFA product data
GB	1984	61	7	
U.S.	1989	62	8.1	Market size and share data combined with TFA product data
U.S.	1990	63	12.8	Availability data
Germany	1992	64	3.4 (women)	German Nutrition Study and TFA survey-data
			4.1 (men)	
Spain	1994	65	2	
France	1995	66	2.8	Availability data from ruminant sources and margarines
Italy			1.7	
Germany			3.7	
Netherlands			4.7	
Belgium-Luxembourg			4.9	
Denmark			5.8	
Spain			1.5	
Portugal			2.1	
Greece			2.2	
Ireland			2.6	
GB			3.1	
Canada	1995	67	8.4	TFA data from human milk, dietary data
Germany	1996	32	1.5–3.1 (children)	Dietary plans
Germany	1997	22	1.9 (women)	German Nutrition Study and TFA-survey data (1996)
			2.3 (men)	
GB	1997	68	1.7 (women)	Analysis data from frying oils and fried foods
			2.0 (men)	
U.S.	1998	69	2.2	USDA data base, total daily fat and energy intake across all ages and
U.S.	1999	70	5.3	Both genders, USDA data base, food intake data from 1989 to 1991

slightly during this period. This may have been caused by the continuing popularity of tub margarines, which have lower TFA contents than stick margarines (62). Per capita TFA consumption from meat and dairy products has remained relatively constant and was estimated to be 1.34 g/(person·d) in 1989 for the United States (without edible tallow) (62).

Since 1989, many restaurants have replaced edible tallow by PHVO (containing ~30% TFA) in order to lower saturated fatty acid (SFA) levels in the products. Many of the fast-food chains are now considering changing from the use of solid PHVO to

TABLE 30.5 Estimation of Daily Conjugated Linoleic Acid (CLA) Intake in Different Countries[a]

Country	Year/Reference		CLA (g/d)	Basis
Australia	1994	44	0.5–1.5	Nutition data and CLA content of milk fat and CLA content of ruminant depot fat
Germany	1997	71	0.31	German Nutrition Study and CLA data
Germany	1997	23	0.35 (women)	German Nutrition Study and CLA-survey data
			0.43 (men)	
U.S.	1997	72	0.23 (women)	Semiquantitative food-frequency questionnaire and 3-d dietary record, lactating women
	1998	73	0.14 (0.08) (men)	3-d dietary record and nutrition CLA data
			0.05 (0.04) (women)	College-aged subjects
Finland	1998	74	0.04–0.31	Food duplicate method and 5-wk dietary record, experimental diet

[a]Values in parentheses are SD.

liquid oils, which have lower amounts of both SFA and TFA. Therefore, the contribution of TFA from fast food will probably return to an order of magnitude similar to that reported before 1989.

In summary, the trends mentioned above have resulted in relatively constant total TFA content in the U.S. diet since the 1980s (62). If zero *trans* fat or low-*trans* fat products become available to and accepted by the American consumer, the total TFA intake could be decreased in the future and may reach levels similar to those in Europe.

Currently, only limited data concerning the CLA intake are available (Table 30.5). For example, the daily CLA intake in Germany has been estimated to be 0.36 g/d for women and 0.44 g/d for men (23). These data are in the same range as results from Jahreis (71), who estimated a slightly lower daily CLA intake in Germany (Table 30.5). Compared with the estimated daily TFA intake, CLA intake in Germany is approximately one fifth that amount. More detailed information concerning CLA intake that is focused on the United States is provided by McGuire *et al.* in Chapter 29.

Conclusion

Conjugated linoleic acid isomers and *trans* fatty acids with isolated or nonconjugated double bonds are formed either by biohydrogenation of polyunsaturated fatty acids or by catalytic hydrogenation of vegetable oils in varying amounts. In ruminant products, CLA and TFA contents are affected by season and feeding regimen. Hydrogenation conditions (e.g., temperature or duration) have a major effect on the TFA content in foods, leading to extremely varying TFA levels in PHVO, whereas CLA amounts in vegetable oils or PHVO are negligible. It can therefore be concluded that foods traditionally high in TFA do not necessarily contain significant amounts of CLA. On the other hand, foods naturally rich in CLA always contain significant amounts of TFA. The CLA/TFA intake reflects individual, dietary, and ethnic consumption habits. Reported values are highly dependent on the limited and often equivocal data available. Therefore, further studies are necessary to provide a more reliable assessment of CLA/TFA contents in the diet, especially with regard to CLA supplementation or TFA reduction.

Alkali-isomerized linoleic acid mixtures, which are available as CLA references, have a distinctly different isomeric distribution compared with ruminant-produced CLA. For example, the latter contains ~10% of total CLA 18:2 $t7c9$, an isomer that is absent or present only in trace amounts (<0.05%) in alkali-isomerized CLA mixtures. To provide a final evaluation of individual CLA isomers, analytical techniques that provide comprehensive information on isomeric distribution should be used in further investigations.

References

1. Katan, M.B., Zock, P.L., and Mensink, R.P. (1995) *Trans*-Fatty Acids and Their Effects on Lipoproteins in Humans, *Annu. Rev. Nutr. 15,* 473–493.
2. Zock, P.L., and Katan, M.B. (1992) Hydrogenation Alternatives: Effects of *trans* Fatty Acids and Stearic Acid Versus Linoleic Acid on Serum Lipids and Lipoproteins in Humans, *J. Lipid Res. 33,* 439–445.
3. Lichtenstein, A.H., Ausman, L.M., Carrasco, W., Jenner, J.L., Ordovas, J.M., and Schaefer, E.J. (1993) Hydrogenation Impairs the Hydrogenation Effect of Corn Oil in Humans, *Arterioscler. Thromb. 13,* 154–161.
4. Judd, J.T., Clevidence, A., Muesing, R.A., Wittes, J., Sunkin, M.E., and Podczasy, J.J. (1994) Dietary *trans* Fatty Acids: Effects on Plasma Lipids and Lipoproteins in Healthy Men and Women, *Am. J. Clin. Nutr. 59,* 861–868.
5. Almendingen, K., Jordal, O., Kierulf, P., Sandstad, B., and Pedersen, J.I. (1995) Effects of Partially Hydrogenated Fish Oil, Partially Hydrogenated Soybean Oil and Butter on Serum Lipoproteins and Lp(a) in Men, *J. Lipid Res. 36,* 1370–1384.
6. Ip, C., and Scimeca, J.A. (1997) Conjugated Linoleic Acid and Linoleic Acid Are Distinctive Modulators of Mammary Carcinogenesis, *Nutr. Cancer 27,* 131–135.
7. Ip, C., Jiang, C., Thompson, H.J., and Scimeca, J.A. (1997) Retention of Conjugated Linoleic Acid in the Mammary Gland Is Associated with Tumor Inhibition During the Post-Initiation Phase of Carcinogenesis, *Carcinogenesis 18,* 755–759.
8. Belury, M.A., Bird, C., and Wu, B. (1995) Dietary Conjugated Dienoic Linoleate Modulation of Phorbol Ester-Elicited Tumor Promotion in Mouse Skin, *Proc. Am. Assoc. Cancer Res. 36,* 596–599.
9. Liew, C., Schut, H.A.J., Chin, S.F., Pariza, M.W., and Dashwood, R.H. (1995) Protection of Conjugated Linoleic Acids Against 2-Amino-3-methylimidazol [4,5-f]quinoline-Induced Colon Carcinogenesis in the F344 Rat: A Study of Inhibitory Mechanisms, *Carcinogenesis 16,* 3037–3043.
10. Shultz, T.D., Chew, B.P., Seaman, W.R., and Luedecke, L.O. (1992) Inhibitory Effect of Conjugated Dienoic Derivatives of Linoleic Acid and β-Carotene on the *in vitro* Growth of Human Cancer Cells, *Cancer Lett. 63,* 125–133.

11. Kepler, C.R., Hirons, K.P., McNeill, J.J., and Tove, S.B. (1966) Intermediates and Products of the Biohydrogenation of Linoleic Acid by *Butyrivibrio fibrisolvens, J. Biol Chem. 241,* 1350–1354.
12. Pollard, M.R., Gunstone, F.D., James, A.T., and Morris, L.J. (1980) Desaturation of Positional and Geometrical Isomers of Monoenoic Fatty Acids by Microsomal Preparations from Rat Liver, *Lipids 15,* 306–314.
13. Britton, M., Fong, C., Wickens, D., and Yudkin, J. (1992) Diet as a Source of Phospholipid Esterified 9,11-Octadecadienoic Acid in Humans, *Clin. Sci. 83,* 97–101.
14. Frankel, E.N. (1987) in *Handbook of Soy Oil Processing and Utilization,* (Ericson, D.R., Pryde, E.H., Brekke, O.L., Mounts, T.L., and Falb R.A., eds.) pp. 229–244, American Soybean Association, St. Louis, MO, and American Oil Chemists' Society, Champaign, IL.
15. Schöne, F., Fritsche, J., Bargholz, J., Leiterer, M., Jahreis, G., and Matthäus, B. (1998) Zu den Veränderungen von Rapsöl und Leinöl während der Verarbeitung, *Fett/Lipid 100,* 539–545.
16. Martin, J.C., Nour, M., Lavillonnière, F., and Sébédio, J.L. (1998) Effect of Fatty Acid Positional Distribution and Triacylglycerol Composition on Lipid By-Products Formation During Heat Treatment: II. *Trans* Isomers, *J. Am. Oil Chem. Soc. 75,* 1073–1078.
17. Heckers, H., and Melcher, F.W. (1978) *Trans*-Isomeric Fatty Acids Present in West German Margarines, Shortenings, Frying and Cooking Fats, *Am. J. Clin. Nutr. 31,* 1041–1049.
18. Kochhar, S.P., and Matsui, T. (1984) Essential Fatty Acid and *trans* Contents of Some Oils, Margarine and Other Food Fats, *Food Chem. 13,* 85–101.
19. Mossoba, M.M., McDonald, R.E., Armstrong, D.J., and Page, S.W. (1991) Identification of Minor C18 Triene and Conjugated Diene Isomers in Hydrogenated Soybean Oil and Margarine by GC-MI-FTIR Spectroscopy, *J. Chromatogr. Sci. 29,* 324–330.
20. Pfalzgraf, A., Timm, M., and Steinhart, H. (1994) Gehalte von *trans*-Fettsäuren in Lebensmitteln, *Z. Ernährungswiss 33,* 24–43.
21. Chin, S.F., Liu, W., Storkson, J.M., Ha, Y.L., and Pariza, M.W. (1992) Dietary Sources of Conjugated Dienoic Isomers of Linoleic Acid, a Newly Recognized Class of Anticarcinogens, *J. Food Comp. Anal. 5,* 185–197.
22. Fritsche, J., and Steinhart, H. (1997) Contents of *trans*-Fatty Acids (TFA) in German Foods and Estimation of Daily Intake, *Fett/Lipid 99,* 314–318.
23. Fritsche, J., and Steinhart, H. (1998) Amounts of Conjugated Linoleic Acid (CLA) in German Foods and Evaluation of Daily Intake, *Z. Lebensm. Unters. Forsch. A,* 77–82.
24. Galoppini, C., and Molino, M (1967) *Trans*-Fatty Acid Content in Margarines, *Ric. Sci. 37,* 262–264.
25. Carpenter, D.L., and Slover, H.T. (1973) Lipid Composition of Selected Margarines, *J. Am. Oil Chem. Soc. 50,* 372–376.
26. Perkins, E.G., McCarthy, T.P., O'Brien, M.A., and Kummerow, F.A. (1977) The Application of Packed Column Gas Chromatographic Analysis to the Determination of *trans* Unsaturation, *J. Am. Oil Chem. Soc. 54,* 279–281.
27. Ottenstein, D.M., Wittings, L.A., Walker, G., and Mahadevan, V. (1977) *Trans*-Fatty Acids in Commercial Margarine Samples Determined by Gas Liquid Chromatography on OV-275, *J. Am. Oil Chem. Soc. 54,* 207–209.

28. Sahasrabudhe, M.P., and Kurian, C.J. (1979) Fatty Acid Composition of Margarines in Canada, *J. Inst. Can. Sci. Technol. Aliment. 12,* 140–144.
29. Enig, M.G., Pallansch, L.A., Shampugna, J., and Keeney, M. (1983) Fatty Acid Composition of the Fat in Selected Food Items with Emphasis on *trans* Components, *J. Am. Oil Chem. Soc. 60,* 1788–1795.
30. Slover, H.T., Thompson, R.H., Davis, C.S., and Merola, G.V. (1985) Lipids in Margarines and Margarine-Like Foods, *J. Am. Oil Chem. Soc. 62,* 775–786.
31. Kafatos, A., Chrysafidis, D., and Peraki, E. (1994) Fatty Acids Composition of Greek Margarines. Margarine Consumption by the Population of Crete and Its Relationship to Adipose Tissue Analysis, *Int. J. Food Sci. Nutr. 45,* 107–114.
32. Demmelmair, H., Festl, B., Wolfram, G., and Koletzko, B. (1996) *Trans*-Fatty Acid Contents in Spreads and Cold Cuts Usually Consumed by Children, *Z. Ernährungswiss 35,* 235–240.
33. Molkentin, J., and Precht, D. (1995) Determination of *trans*-Octadecenoic Acids in German Margarines, Shortenings, Cooking and Dietary Fats by Ag-TLC/GC, *Z. Ernährungswiss 34,* 314–317.
34. De Greyt, W., Radanyi, O., Kellens, M., and Huyghebaert, A. (1996) Contribution of *trans*-Fatty Acids from Vegetable Oils and Margarines to the Belgian Diet, *Fett/Lipid 98,* 30–33.
35. Kayahan, M., and Tekin, A. (1994) Research on the Quantity of *trans*-Fatty Acids and Conjugated Fatty Acids in Margarines Produced in Turkey, *Gida 19,* 147–153.
36. Ovesen, L., Leth, T., and Hansen, K. (1996) Fatty Acid Composition of Danish Margarines and Shortenings with Special Emphasis on *trans*-Fatty Acids, *Lipids 31,* 971–975.
37. Bayard, C.C., and Wolff, R.L. (1995) *Trans*-18:1 Acids in French Tub Margarines and Shortenings: Recent Trends, *J. Am. Oil Chem. Soc. 72,* 1485–1489.
38. Pfalzgraf, A., and Steinhart, H. (1995) Gehalte von *trans*-Fettsäuren in Margarinen, *Dtsch. Lebensm. Rundsch. 91,* 113–114.
39. Ali, L.H., Angyal, G., Weaver, C.M, and Rader, J. (1997) Comparison of Capillary Column Gas Chromatographic and AOAC Gravimetric Procedures for Total Fat and Distribution of Fatty Acids in Foods, *Food Chem. 58,* 149–160.
40. Ratnayake, W.M.N., Pelletier, G., Hollywood, R., Bacler, S., and Leyte, D. (1998) *Trans* Fatty Acids in Canadian Margarines: Recent Trends, *J. Am. Oil Chem. Soc. 75,* 1587–1594.
41. Yurawecz, M.P., Molina, A.A., Mossoba, M.M., and Ku, Y. (1993) Estimation of Conjugated Octadecatrienes in Edible Fats and Oils, *J. Am. Oil Chem. Soc. 70,* 1093–1099.
42. Parodi, P.W., and Dunstan, R.J. (1971) The *trans* Unsaturated Content of Queensland Milk Fat, *Austr. J. Dairy Technol. 26,* 60–62.
43. DeMan, L., and DeMan, J.M. (1983) *Trans*-Fatty Acids in Milk Fat, *J. Am. Oil Chem Soc. 60,* 1095–1098.
44. Parodi, P.W. (1994) Conjugated Linoleic Acid: An Anticarcinogenic Fatty Acid Present in Milk Fat, *Austr. J. Dairy Technol. 49,* 93–97.
45. Henninger, M., and Ulberth, F. (1994) *Trans*-Fatty Acid Content of Bovine Milk Fat, *Milchwissenschaft 49,* 555–558.
46. Lin, H., Boylston, T.D., Luedecke, L.O., and Shultz, T.D. (1995) Survey of the Conjugated Linoleic Acid Contents of Dairy Products, *J. Dairy Sci. 78,* 2358–2365.

47. Precht, D., and Molkentin, J. (1995) *Trans*-Fatty Acids: Implications for Health, Analytical Methods, Incidence in Edible Fats and Intake, *Nahrung 39,* 343–374.
48. Jahreis, G., Fritsche, J., and Steinhart, H. (1996) Monthly Variations of Milk Composition with Special Regard to Fatty Acids Depending on Season and Farm Management Systems—Conventional Versus Ecological, *Fett/Lipid 98,* 356–359.
49. Jiang, J., Bjoerck, L., Fondén, R., and Emanuelson, M. (1996) Occurrence of Conjugated *cis*-9, *trans*-11-Octadecadienoic Acid in Bovine Milk: Effects of Feed and Dietary Regimen, *J. Dairy Sci. 79,* 438–445.
50. Shantha, N.C., Decker, E.A., and Ustunol, Z. (1992) Conjugated Linoleic Acid Concentration in Processed Chesse, *J. Am. Oil Chem. Soc. 69,* 425–428.
51. Precht, D. (1995) Variation of *trans*-Fatty Acids in Milk Fats, *Z. Ernährungswiss 34,* 27–29.
52. Fritsche, J., and Steinhart, H. (1998) Analysis, Occurrence, and Physiological Properties of *trans* Fatty Acids (TFA) with Particular Emphasis on Conjugated Linoleic Acid Isomers (CLA)—a Review, *Fett/Lipid 100,* 190–210.
53. Lavillonnière, F., Martin, J.C., Bougnoux, P., and Sébédio, J.L. (1998) Analysis of Conjugated Linoleic Acid Isomers and Content in French Cheeses, *J. Am. Oil Chem. Soc. 75,* 343–352.
54. Chin, S.F., Storkson, J.M., Albright, K.J., Cook, M.E., and Pariza, M.W. (1994) Conjugated Linoleic Acid Is a Growth Factor for Rats as Shown by Enhanced Weight Gain and Improved Feed Efficiency, *J. Nutr. 124,* 2344–2349.
55. Fritsche, S., and Fritsche, J. (1998) Occurrence of Conjugated Linoleic Acid Isomers in Beef, *J. Am. Oil Chem. Soc. 75,* 1449–1451.
56. Ha, Y.L., Grimm, N.K., and Pariza, M.W. (1989) Newly Recognized Anticancerogenic Fatty Acids: Identification and Quantitation in Natural and Processed Cheeses, *J. Agric. Food Chem. 37,* 75–81.
57. Rickert, R., Fritsche, J., Steinhart, H., Sehat, N., Kramer, J.K.G., Yurawecz, M.P., Mossoba, M.M., Roach, J.A.G., Eulitz, K., and Ku, Y. (1999) Enhanced Resolution of Conjugated Linoleic Acid Isomers by Tandem-Column Silver-Ion High Performance Liquid Chromatography, *J. High Resolu. Chromatogr., 22,* 144–148.
58. Sehat, N., Yurawecz, M.P., Roach, J.A.G., Mossoba, M.M., Kramer, J.K.G., and Ku, Y. (1998) Silver-Ion High-Performance Liquid Chromatographic Separation and Identification of Conjugated Linoleic Acid Isomers, *Lipids 33,* 217–221.
59. Sehat, N., Kramer, J.K.G., Mossoba, M.M., Yurawecz, M.P., Roach, J.A.G., Eulitz, K., Morehouse, K.M., and Ku, Y. (1998) Identification of Conjugated Linoleic Acid Isomers in Cheese by Gas Chromatography, Silver-Ion High-Performance Liquid Chromatography and Mass Spectral Reconstructed Ion Profiles. Comparison of Chromatographic Elution Sequences, *Lipids 33,* 963–971.
60. Hunter, J.E., and Appelwhite, T.H. (1986) Isomeric Fatty Acids in the US Diet: Levels and Health Perspectives, *Am. J. Clin. Nutr. 44,* 707–717.
61. Burt, R., and Buss, D.H. (1984) Dietary Fatty Acids in the UK, *Br. J. Clin. Pract. 31,* 20–23.
62. Hunter, J.E., and Applewhite, T.H. (1991) Reassessment of *trans*-Fatty Acid Availability in the US Diet, *Am. J. Clin. Nutr. 54,* 363–369.
63. Enig, M.G., Atal, S., Keeney, M., and Sampugna, J. (1990) Isomeric *trans*-Fatty Acids in the US Diet, *J. Am. Coll. Nutr. 9,* 471–486.
64. Pfalzgraf, A., and Steinhart, H. (1992) Aufnahme *trans*-isomerer Fettsäuren–Eine Abschätzung auf Basis der nationalen Verzehrsstudie 1991, *Z. Ernährwiss. 31,* 196–204.

65. Boatella, J., Rafecas, M., and Codony, R. (1993) Isomeric *trans*-Fatty Acids in the Spanish Diet and Their Relationships with Changes in Fat Intake Patterns, *Eur. J. Clin. Nutr. 47*, S62–S65.
66. Wolff, R.L. (1995) Content and Distribution of *trans*-18:1 Acids in Ruminant Milk and Meat Fats. Their Importance in European Diets and Their Effect on Human Milk, *J. Am. Oil Chem. Soc. 72*, 259–272.
67. Ratnayake, W.M.N., and Chen, Z.Y. (1995) in *Development and Processing of Vegetable Oils for Human Nutrition,* (Przybylski, R., and McDonald, B.E., eds.) pp. 20–35, AOCS Press, Champaign, IL.
68. British Ministry of Agriculture, Fisheries & Food (1997) *Trans*-Fatty Acids in Frying Oils and Fried Foods, Food Surveillance Information Sheet No. 112.
69. Lemaitre, R.N., King, I.B., Patterson, R.E., Psaty, B.M., Kestin, M., and Heckbert, S.R. (1998) Assessment of *trans*-Fatty Acid Intake with Food Frequency Questionnaire and Validation with Adipose Tissue Levels of *trans*-Fatty Acids, *Am. J. Epidemiol. 148*, 1085–1093.
70. Allison, D.B., Egan, S.K., Barraj, L.M., Caughman, C., Infant, M., and Heimbach, J.T. (1999) Estimated Intakes of *trans* Fatty and Other Fatty Acids in the US Population, *J. Am. Diet Assoc. 99*, 166–174.
71. Jahreis, G. (1997) Krebshemmende Fettsäuren in Milch und Rindfleisch, *Ernährungs Umschau 44*, 168–172.
72. Park, Y.S., Behre, R.A., McGuire, M.A., Shultz, T.D., and McGuire, M.K. (1997) Dietary Conjugated Linoleic Acid (CLA) and CLA in Human Milk, *FASEB J. 11*, Abstr. 1387.
73. Ritzenthaler, K., McGuire, M.K., Falen, M.K., Shultz, T.D., and McGuire, M.A. (1998) Estimation of Conjugated Linoleic Acid (CLA) Intake, *FASEB J. 12*, A527.
74. Salminen, I., Mutanen, M., Jauhiainen M., and Aro, A. (1998) Dietary *trans* Fatty Acids Increase Conjugated Linoleic Acid Levels in Human Serum, *J. Nutr. Biochem. 9*, 93–98.

Chapter 31

Conjugated Linoleic Acid and Experimental Atherosclerosis in Rabbits

David Kritchevsky

The Wistar Institute, Philadelphia, PA 19104

Introduction

Oxidized cholesterol has been found to be atherosclerotic (1) and to cause arterial injury in rabbits (2). Steinberg and his colleagues (3) showed that oxidized low density lipoprotein (LDL) plays a crucial role in atherogenesis. They also reported that a pharmaceutical agent with antioxidant properties, probucol [4,4'-(isopropylidenedithio)bis(2,6-di-*t*-butylphenol)] could prevent oxidation of LDL *in vitro* and *in vivo* and also prevent progression of atherosclerosis in the Watanabe heritable hyperlipidemic (WHHL) rabbit. The action is due to its antioxidant properties and is independent of its hypolipidemic properties (4,5). Earlier probucol had been found to inhibit cholesterol-induced atherosclerosis in rabbits (6). Reports that conjugated linoleic acid (CLA) exhibited antioxidant activity *in vivo* (7) and *in vitro* (8) prompted investigation of its effects on atherogenesis in cholesterol-fed rabbits. The CLA used in our studies was a mixture of conjugated dienoic derivatives of linoleic acid, primarily the *cis* 9,*trans* 11, and *trans* 10,*cis* 12 derivatives.

Experimental Studies

In the first study, control and test groups of six rabbits each (3 male, 3 female) were fed a semipurified diet containing 0.1% cholesterol for 22 wk (Table 31.1). The diet of the test rabbits was augmented with 0.5 g/d of CLA. At necropsy, plasma total cholesterol, LDL-cholesterol, and triglyceride levels were lower than those of controls but not significantly so. Total plasma cholesterol and triglyceride levels (mg/dL) for the CLA-fed group were 1000 and 140, and for the control group they were 1175 and 165, respectively. The plasma LDL/high density lipoprotein (HDL) cholesterol ratios were 16.5 for the control group and 10.9 for the CLA-fed group. Histological examination of the aortas showed maximal plaque thickness (mm) in the thoracic aorta to be the same in both groups; however, maximal plaque thickness in the abdominal aorta of CLA-fed rabbits was 27% smaller. Lipid deposition and connective tissue development were less severe in the aortas of the CLA-fed rabbits (9). On the basis of the thiobarbituric acid-reactive substances (TBARS) assay, levels of plasma peroxides were the same for both groups.

A second experiment was designed to confirm the apparent antiatherogenicity of CLA and also to examine its effects on progression or regression of established atherosclerotic lesions. Atheromata, once established in rabbits, do not regress after removal of the atherogenic stimulus and may become slightly more severe. Some hypolipidemic drugs, added to a cholesterol-free diet after lesions have been established, may cause a slight reduction in severity of atherosclerosis (10,11). A group of 30 rabbits was fed a semipurified atherogenic diet containing 0.2% cholesterol (control) (Table 31.2); 10 rabbits were fed the same diet augmented with 1% CLA. After 90 d, the rabbits receiving the control diet were bled and randomized into three groups of 10 rabbits each of equal serum cholesterol levels. One group of control rabbits and the CLA-fed rabbits were necropsied. The two remaining groups were placed on a cholesterol-free diet (Table 3) with or without 1% CLA. These rabbits were necropsied 90 d later.

The results of the atherogenesis experiment are detailed in Table 31.4. There were no significant differences in weight gain or serum lipids. However, the rabbits

TABLE 31.1 Atherogenic Diet (I)

Ingredient[a]	%	% Calories
Casein	25.0	26.2
L-Arginine-HCL	0.6	
DL-Methionine	0.2	
Sucrose	19.46	20.4
Cornstarch	19.3	20.3
Coconut oil	12.0	28.3
Corn oil	2.0	4.7
Cellulose	15.0	
Vitamin mix	1.0	
Mineral mix	5.34	
Cholesterol	0.1	

[a]Conjugated linoleic acid added at the expense of sucrose in the test diet.

TABLE 31.2 Atherogenic Diet (II)

Ingredient[a]	%	% Calories
Casein	25.0	26.0
DL-Methionine	0.2	
Sucrose	19.48	20.3
Starch	20.0	21.0
Coconut oil	12.0	28.0
Corn oil	2.0	4.7
Cellulose	15.0	
Mineral mix	5.0	
Vitamin mix	1.0	
Choline bitartrate	0.12	
Cholesterol	0.2	

[a]Conjugated linoleic acid added at the expense of starch in the test diet.

fed CLA exhibited 31% less severe atheromata in the aortic arch and a 40% lower severity in the thoracic aorta. After rabbits were on the regression regimen for 90 d, serum cholesterol had fallen by 83% on the control regimen and 67% on the CLA diet. The percentage of HDL-cholesterol had risen by 121 and 74%, respectively. Serum triglycerides were unchanged in the control group but were 58% lower in the CLA-fed animals. Examination of the level of atherosclerosis revealed that there had been practically no change in the control rabbits, but severity of atherosclerosis in the aortic arch and thoracic aorta of the CLA-fed rabbits had been reduced by 31 and 30%, respectively. Involved area had been reduced by 31%. The data are detailed in Table 31.5.

TABLE 31.3 Regression Diet

Ingredient[a]	%	% Calories
Casein	24.0	27.3
DL-Methionine	0.3	
Sucrose	30.58	34.8
Starch	20.0	22.7
Corn oil	6.0	15.3
Cellulose	14.0	
Mineral mix	4.0	
Vitamin mix	1.0	
Choline bitartrate	0.12	

[a]Conjugated linoleic acid added at expense of corn oil in the test diet.

TABLE 31.4 Necropsy Data for Rabbits Fed 0.2% Cholesterol With (Control) or Without 1% Conjugated Linoleic Acid (CLA) for 90 Days

	Group (n = 10)	
	Control	CLA
Weight gain (g)	822 ± 1.22	742 ± 87
Liver weight (g)	109 ± 5	111 ± 7
Liver (% body weight)	3.13 ± 0.14	3.24 ± 0.18
Serum (mg/mL)		
Cholesterol	430 ± 40	559 ± 53
% HDL-C[a]	6.1 ± 0.44	3.6 ± 0.51
Triglyceride	77 ± 6	135 ± 36
Aorta (0–4 scale)		
Arch	2.39 ± 0.47	1.65 ± 0.37
Thoracic	2.35 ± 0.36	1.40 ± 0.40
Area (%)	49 ± 10	30 ± 10

[a]HDL-C, high density lipoprotein cholesterol.

TABLE 31.5 Necropsy Data for Rabbits with Established Atherosclerosis Fed Corn Oil With (Control) or Without 1% Conjugated Linoleic Acid (CLA) for 90 Days

	Group (n = 10)	
	Control	CLA
Weight gain (g)	461 ± 65	246 ± 37
Liver weight (g)	99 ± 9	83 ± 6
Liver (% body weight)	2.33 ± 0.20	2.31 ± 0.13
Serum (mg/mL)		
Cholesterol	73 ± 10	140 ± 24
% HDL-C[a]	13.5 ± 1.02	10.6 ± 0.81
Triglyceride	77 ± 13	57 ± 5
Aorta (0–4 scale)		
Arch	2.35 ± 0.35	1.65 ± 0.26
Thoracic	2.30 ± 0.40	1.65 ± 0.22
Area (%)	51 ± 11	34 ± 6

[a]HDL-C, high density lipoprotein cholesterol.

The next study was designed to investigate different concentrations of CLA for their antiatherogenic potential as well as for their effects on preestablished lesions. Accordingly, one large group of 40 rabbits was fed the atherogenic diet described in Table 31.2. Groups of eight rabbits were fed the same diet augmented with 0.1, 0.5, or 1.0% CLA. The post–cholesterol–feeding phase of the study involved feeding rabbits with established lesions the cholesterol-free diet (Table 31.3) or the same diet containing 0.1, 0.5, or 1.0% CLA. The results of the first phase of the study are summarized in Table 31.6. As in the earlier study, there were no differences in serum lipids. The effects of CLA on atherogenesis were striking. Even at 0.1% of the diet, CLA reduced severity of atherosclerosis in the aortic arch and thoracic aorta by 28 and 41%, respectively. When fed as 0.5% or 1% of the diet, CLA significantly reduced atherosclerotic involvement in both the aortic arch and thoracic aorta. The percentage of esterified cholesterol present in the aortic tissue is a value that can also be used as a measure of severity of atherosclerosis. This parameter paralleled the other findings.

Rabbits with preestablished atherosclerosis were placed into four groups of seven rabbits each. One group was fed the control diet, and the others were fed the same diet plus 0.1, 0.5, or 1.0% cholesterol. As in the earlier study, after 90 d of consuming a cholesterol diet, the severity of lesions in the aortic arch and thoracic aorta was increased (by 12 and 4%, respectively). Addition of 0.1 or 0.5% CLA to the diet had virtually no effect on the severity of atherosclerosis (Table 31.7). However, consistent with the earlier study, at 1% of the diet, CLA reduced the severity of aortic arch atheromata by 27% and that in the thoracic aorta by 45% ($P < 0.01$). Analytical data were not available at the time of this writing.

Administration of cholesterol and saturated fat to hamsters leads to aortic sudanophilia (12). Nicolosi *et al.* (13) fed male hamsters (10 per group) an atherogenic

TABLE 31.6 Necropsy Data for Rabbits Fed 0.2% Cholesterol with 0.0 (Control), 0.1, 0.5, or 1.0% Conjugated Linoleic Acid (CLA) for 90 Days

	Group (n = 8)			
	Control	0.1% CLA	0.5% CLA	1.0% CLA
Weight gain (g)	104 ± 46	3 ± 108	67 ± 43	50 ± 79
Liver weight (g)	68 ± 5	77 ± 6	66 ± 4	78 ± 6
Liver (% body weight)	2.73 ± 0.16	3.22 ± 0.35	2.63 ± 0.12	3.35 ± 0.32
Serum (mg/dL)				
Cholesterol (%)	983 ± 118	1281 ± 116	1263 ± 104	1103 ± 134
% HDL-C[a]	5.0 ± 0.9	3.3 ± 0.54	3.3 ± 0.58	5.0 ± 1.14
Triglyceride	190 ± 32	246 ± 47	205 ± 48	216 ± 38
Aorta (0–4 scale)				
Arch	2.36 ± 0.39[bc]	1.69 ± 0.23[d]	0.88 ± 0.20[bd]	1.00 ± 0.28[c]
Thoracic	2.21 ± 0.42[ef]	1.31 ± 0.28	0.75 ± 0.21[e]	0.94 ± 0.27[f]
Area (%)	44 ± 12[g]	32 ± 7[h]	11 ± 4[gh]	18 ± 6
% Ester cholesterol	74.7	52.0	34.1	44.3

[a]HDL-C, high density lipoprotein cholesterol.
[b–h]Values in a horizontal row bearing the same letter are significantly different.

TABLE 31.7 Necropsy Data for Rabbits with Established Atherosclerosis Fed Corn Oil with 0.0 (Control), 0.1, 0.5, or 1% Conjugated Linoleic Acid (CLA) for 90 Days

	Group			
	Control (n = 7)	0.1% CLA (n = 6)	0.5% CLA (n = 7)	1.0% CLA (n = 6)
Weight gain (g)	312 ± 99	265 ± 104	298 ± 84	242 ± 73
Liver weight (g)	52 ± 3	58 ± 6	64 ± 5[a]	47 ± 3[a]
Liver (% body weight)	1.82 ± 0.07[b]	2.05 ± 0.16	2.28 ± 0.20[bc]	1.75 ± 0.13[a]
Aorta (0–4 scale)				
Arch	2.64 ± 0.28	2.25 ± 0.28	2.50 ± 0.29	1.95 ± 0.40
Thoracic	2.29 ± 0.36[d]	2.33 ± 0.44	2.00 ± 0.15[e]	1.25 ± 0.17[de]
Area (%)	53 ± 7	53 ± 10	49 ± 5	30 ± 10

[a–e]Values in horizontal row bearing the same letter are significantly different.

diet consisting of 88.9% commercial ration, 10% coconut oil, 1% safflower oil, and 0.12% cholesterol or the same diet augmented with CLA (0.25–5.0%) or linoleic acid (5%). Total cholesterol levels were reduced in all of the test groups. Fatty streak area was reduced by 19, 26, and 30%, respectively, in hamsters fed 0.25, 0.50, or 1.0% CLA. Linoleic acid feeding reduced fatty streak area by 25% (Table 31.8). It was determined in a subsequent study (Wilson et al., unpublished data) in which hamsters (12/group) were fed the atherogenic diet alone or the diet plus either 1% CLA or 1% linoleic acid that both CLA and linoleic acid significantly lowered total

TABLE 31.8 Plasma Lipids and Aorta Fatty Streak Area in Hamsters Fed 0.12% Cholesterol for 11 Weeks[a]

Group	Cholesterol (mg/dL)	Triglycerides (mg/dL)	Area ($\mu m^2/mm^2$ (100)
Control	690 ± 24[bc]	1099 ± 212[de]	53 ± 14
0.25% CLA	510 ± 99	467 ± 92[df]	43 ± 18
0.5% CLA	546 ± 41[b]	492 ± 44[eg]	39 ± 11
5.0% CLA	530 ± 60[c]	1003 ± 158[fg]	37 ± 15
5.0% LA	590 ± 42	791 ± 199	40 ± 18

[a]CLA, conjugated linoleic acid; LA, linoleic acid.
[b-g]Values in a column bearing the same letter are significantly different by t-test.

plasma cholesterol levels (control, −327 ± 16; CLA, −285 ± 11; LA, −264 ± 8 mg/dL). Aortic fatty streak area ($\mu m^2/mm^2$ (100) was 19.4 ± 2.25 in the controls, 10.3 ± 1.55 in the hamsters fed CLA, and 17.1 ± 2.62 in those fed linoleic acid. The difference between the CLA and control groups was significant. LDL oxidation was reduced significantly in the CLA-fed hamsters.

Conclusion

The data show that dietary CLA has the potential to inhibit cholesterol-induced atherosclerosis or sudanophilia in rabbits and hamsters, respectively. More striking is the apparent ability of CLA to reduce severity of preinduced atherosclerotic lesions in rabbits. The mechanism of action remains to be elucidated.

Acknowledgment

Supported, in part, by a Research Career Award (HL-00734) from the National Institutes of Health.

References

1. Imai, H., Werthessen, N.T., Taylor, C.B., and Lee, K.T. (1976) Angiotoxicity and Atherosclerosis Due to Contaminants of U.S.P. Grade Cholesterol, *Arch. Pathol. Lab. Med. 100*, 565–572.
2. Cook, R.P., and MacDougal, J.D.B. (1968) Experimental Atherosclerosis in Rabbits After Feeding Cholestanetriol, *Br. J. Exp. Pathol. 49*, 265–271.
3. Steinberg, D., Parthasarathy, S., Carew, T.E., Khoo, J.C., and Witztum, J.L. (1989) Beyond Cholesterol. Modifications of Low Density Lipoprotein That Increase Its Atherogenicity, *N. Engl. J. Med. 320*, 915–924.
4. Parthasarathy, S., Young, S.G., Witztum, J.L., Pittman, R.C., and Steinberg, D. (1986) Probucol Inhibits Oxidative Modification of Low Density Lipoproteins, *J. Clin. Investig. 77*, 641–644.

5. Carew, T.E., Schwenke, D.C., and Steinberg, D. (1987) Antiatherogenic Effect of Probucol Unrelated to Its Hypocholesterolemic Effect: Evidence That Antioxidants *in vivo* Can Selectively Inhibit Low Density Lipoprotein Degradation in Macrophage-Rich Fatty Streaks and Slow the Progression of Atherosclerosis in the Watanabe Heritable Hyperlipidemic Rabbit, *Proc. Natl. Acad. Sci. USA 84*, 7725–7729.
6. Kritchevsky, D., Kim, H.K., and Tepper, S.A. (1971) Influence of 4,4′- (isopropylidenedithio)bis(2,6-di-*t*-butylphenol) (DHS81) on Experimental Atherosclerosis in Rabbits, *Proc. Soc. Exp. Biol. Med. 136*, 1216–1221.
7. Ip, C., Chin, S.F., Scimeca, J.A., and Pariza, M.W. (1990) Mammary Cancer Prevention by Conjugated Dienoic Derivatives of Linoleic Acid, *Cancer Res. 51*, 6118–6124.
8. Ha, Y.L., Storkson, J., and Pariza, M.W. (1990) Inhibition of Benzo(a)pyrene-Induced Mouse Forestomach Neoplasia by Conjugated Dienoic Derivatives of Linoleic Acid, *Cancer Res. 50*, 1097–1101.
9. Lee, K.N., Kritchevsky, D., and Pariza, M.W. (1994) Conjugated Linoleic Acid and Atherosclerosis in Rabbits, *Atherosclerosis 108*, 19–25.
10. Kritchevsky, D., Moynihan, J., Langan, J., Tepper, S.A., and Sachs, M.L. (1961) Effects of D- and L-Thyroxine and D- and L-3,5,3′-Triiodothyronine on Development and Regression of Experimental Atherosclerosis in Rabbits, *J. Atheroscler. Res. 1*, 211–221.
11. Kritchevsky, D., Sallata, P., and Tepper, S.A. (1968). Influence of Ethyl *p*- Chlorophenoxyisobutyrate (CPIB) upon Establishment and Progression of Experimental Atherosclerosis in Rabbits, *J. Atheroscler. Res. 8*, 755–761.
12. Kowala, M.C., Nunnari, J.J., Durham, S.K., and Nicolosi, R.J. (1991) Doxazosin and Cholestyramine Similarly Decrease Fatty Streak Formation in the Aortic Arch of Hyperlipidemic Hamsters, *Atherosclerosis 91*, 35–49.
13. Nicolosi, R.J., Rogers, E.J., Kritchevsky, D., Scimeca, J.A., and Huth, P.J. (1997) Dietary Conjugated Linoleic Acid Reduces Plasma Lipoproteins and Early Aortic Atherosclerosis in Hypercholesterolemic Hamsters, *Artery 22*, 266–277.

Chapter 32

Modulation of Diabetes by Conjugated Linoleic Acid

Martha A. Belury[a] and John P. Vanden Heuvel[b]

[a]Department of Foods and Nutrition, Purdue University, West Lafayette, IN
[b]Department of Veterinary Sciences and Center for Molecular Toxicology, The Pennsylvania State University, University Park, PA

Introduction

Diabetes mellitus (DM) is a disease that affects ~10–12 million Americans (1). In the United States, Type 2 DM is the most prevalent form of diabetes, comprising 90–95% of the diagnosed cases. Individuals who have insulin resistance and a relative insulin deficiency exhibit the elevated postprandial blood glucose levels that typify this disease. Insulin resistance may be improved with weight reduction, dietary management, and/or pharmacological treatments. In some cases, exogenous insulin therapy is also used to improve insulin resistance in Type 2 DM patients. Even with combinatory therapies, hyperglycemia is seldom restored to normal.

As discussed in previous chapters, conjugated linoleic acid (CLA) has received considerable attention because of the plethora of beneficial effects seen in laboratory animals (2). It is important to note that although CLA has been extensively examined for its therapeutic prowess, a clearly defined mechanism by which this fatty acid exerts its effects has not emerged. Our laboratories have shown that CLA ameliorates the symptoms of Type 2 DM in the Zucker *fa/fa* diabetic rat and affects differentiation of adipose tissue (3). In addition, we have identified a possible mechanism of action of CLA, namely, binding to and activating a steroid hormone receptor called peroxisome proliferator-activated receptor (PPARα and γ). We will detail the effects of CLA on PPARγ and its possible utility as an antidiabetic agent in this chapter.

Thiazolidinediones and the Treatment of Type 2 DM

Development of insulin resistance is an early event in the progression of the disease, and reversing this insensitivity fulfills a critical need in the treatment of Type 2 DM. Thiazolidinediones are a class of oral insulin-sensitizing agents that improve glucose utilization without stimulating insulin release (4). Instead, *in vitro* tests with these chemicals demonstrate that thiazolidinediones promote adipocyte differentiation by causing a coordinate increase in adipocyte-specific gene expression (5). PPARγ was originally discovered as a member of adipocyte regulatory factor complex and is essential for adipocyte differentiation (6). PPARγ was shown to be expressed in an adipocyte-specific manner; its expression is induced early in the course of differentiation of preadipocyte cell lines and is pivotal in the adipogenic signaling pathways (6). In addition, the PPARγ ligand, BRL49653, has been shown to induce differentiation of cultured preadipocytes (3T3-L1 cells) (5). Therefore, a key regulatory pathway has emerged that is relevant to the treatment of type 2 DM. That is, PPARγ ligands such as the thiazolidinediones bind to and activate PPARγ. This steroid hormone receptor thereby regulates gene expression and enhances adipocyte differentiation. The increase in adipocyte differentiation correlates with an improvement in insulin sensitivity and glucose uptake. This mechanism of action has been attributed to the thiazolidinedione, troglitazone, a drug recently approved by the Food and Drug Administration for use in humans for type 2 DM therapy under the name Rezulin (Parke-Davis, Ann Arbor, MI). For reasons we will discuss subsequently, we believed CLA acts in a manner similar to thiazolidinediones.

CLA and Peroxisome Proliferator-Activated Receptors

Recently, we have shown that the biological effects of CLA are similar to those of a group of hypolipidemic drugs known as peroxisome proliferators (PP) (7). These exogenous chemicals resemble natural fatty acids in that they contain an acidic functional group and a large hydrophobic region. PP exert their biological responses as the result of altered gene expression subsequent to activation of a subgroup of steroid hormone receptors known as PPAR (peroxisome proliferator-activated receptors; PPARα, β/δ, and γ) (reviewed in Ref. 8). Activation of PPARα, the predominant PPAR subtype expressed in the liver, has been associated with hypolipidemia and hepatic peroxisome proliferation (9,10). This receptor has also been associated with cell cycle control, hepatocyte DNA synthesis, and cell division as evidenced by the fact that the relative ability to activate PPARα correlates with tumor cytostasis of phenylacetate and its derivatives (10). PP also have antiatherosclerotic properties (11), an effect that requires PPARα (12). Evidence to support CLA as a member of the PP family and as a PPARα activator includes induction of the peroxisomal enzyme acyl-CoA oxidase (ACO) at the mRNA and protein level in liver of mice fed this fatty acid (7). Recently, we have shown that CLA, in particular the 9Z,11E isomer is a potent PPARα activator and high affinity ligand (Belury et al., unpublished data). Furthermore, many other fatty acids including linoleic, palmitic, oleic and arachidonic acids are known PPARα activators (13).

In addition to the lipid-lowering effects of PP, many of these agents are modulators of glucose and insulin action (14). As described above, PPARγ was originally identified as part of a protein complex required for adipocyte differentiation (6). Ectopic expression of PPARγ and/or its activation converts NIH 3T3 fibroblasts to adipocytes, indicating that this receptor regulates genes required to obtain the adipocyte phenotype (15,16). Due in part to effects on adipogenesis, PPARγ activators decrease circulating glucose levels and also reverse the insulin resistance that is typical of noninsulin-dependent diabetes mellitus (17).

The similarity of biological effects of CLA and thiazolidinediones (see below) led us to speculate that CLA would bind to PPARγ, activate this receptor to a transcriptionally active form and subsequently regulate gene expression and enhance adipocyte differentiation. Each of these processes was examined with CLA isomers.

The binding of 9Z,11E CLA was examined with the use of a scintillation proximity assay (Vanden Heuvel et al., unpublished data) (Fig. 32.1) and is detailed for other isomers in Table 32.1. The affinity of CLA to PPARγ was similar to that of other fatty acids with a K_d of ~10 μmol/L. Activation of PPARγ was determined by using a reporter assay in which COS-1 or CV-1 cells were cotransfected with PPARγ and a reporter gene that responds to this receptor. As shown in Figure 32.2, various CLA isomers were able to activate PPARγ and drive expression of the luciferase reporter, although the amount of induction was quite low. Altered expression of ACO and in particular aP2 is indicative of activation of PPARγ and enhanced differentiation of adipocytes. As detailed in Figure 32.3, various CLA isomers were potent inducers of aP2 mRNA expression, indicating increased adipogenesis.

Taken together, these data suggest that CLA results in biological events similar to those seen with thiazolidinediones, such as increased aP2 expression and conversion of 3T3-L1 preadipocytes to adipocytes (Vanden Heuvel et al., unpublished

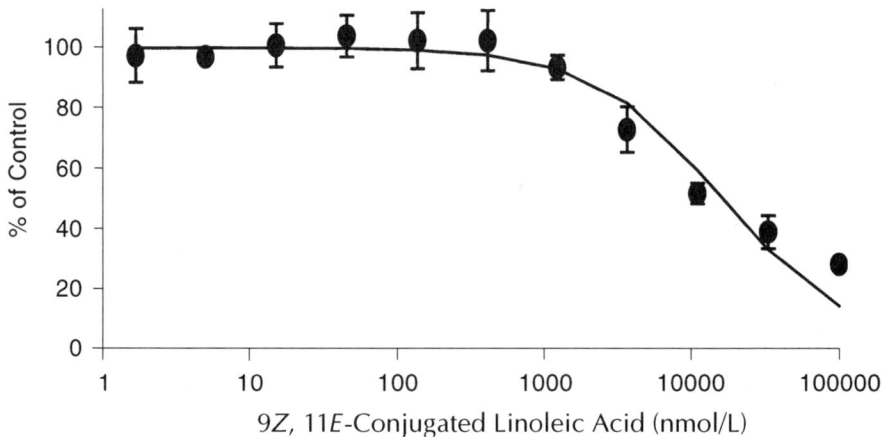

Fig. 32.1. Binding of conjugated linoleic acid to human peroxisome proliferator-activated receptor γ (PPARγ). Scintillation proximity assays were performed using human PPARγ. Each concentration was assayed in duplicate.

TABLE 32.1. K_i Values of Conjugated Linoleic Acid (CLA) Binding to Peroxisome Proliferator-Activated Receptor γ

Compound	K_i (μmol/L)
9Z,11E-CLA	6.45 ± 0.77
10E,12Z-CLA	7.10 ± 0.42
9E,11E-CLA	4.7 ± 0.71
Furan-CLA	5.5 ± 3.2
GW2331	0.34

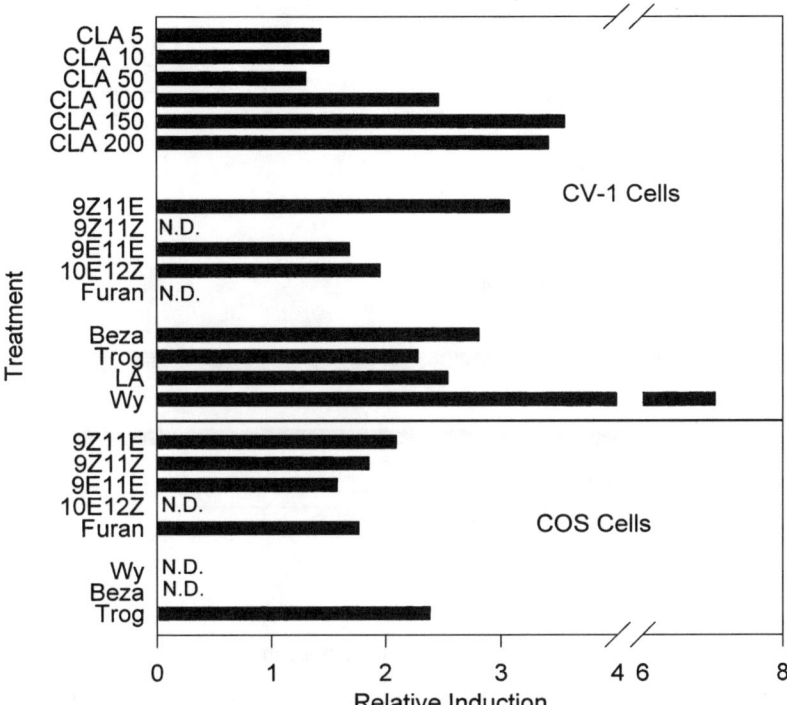

Fig. 32.2. Activation of full-length peroxisome proliferator-activated receptor (PPAR) in transient transfection assays. CV-1 (top) or COS-1 (bottom) cells were transfected with full length PPARγ plasmid along with a peroxisome proliterator response element (PPRE)-luciferase-reporter and β-gal vectors. Abbreviations: CLA, conjugated linoleic acid; LA, linoleic acid; mCLA, methylated CLA; 9Z11E, 9E11E, 9Z11Z and furan, individual CLA isomers at 100 μmol/L each; Wy, 100 μmol/L Wy 14,643; PFDA, 100 μmol/L perfluorodecanoic acid; Beza, 100 μmol/L bezafibrate; Trog, 100 μmol/L troglitazone; ND, not determined. Numbers after the abbreviations represent the concentration in micromoles. Data are corrected for transfection efficiency and are expressed relative to dimethyl sulfoxide (DMSO)-treated cells (mean of n = 3–9).

results). In addition, in the diabetic rat model described below, CLA resulted in increased expression of markers of adipocyte differentiation (i.e., induction of aP2 mRNA expression). However, the binding and activation of PPARγ by CLA were similar to the effect seen with fatty acids with no antidiabetic properties such as linoleic acid. A reasonable conclusion may be that CLA is metabolized to a more active form with increased ability to bind to and activate PPARγ compared with the parent compound. Nonetheless, the key result shown in these studies was that in certain model systems, CLA did indeed act as an antidiabetic agent through enhancement of adipocyte differentiation.

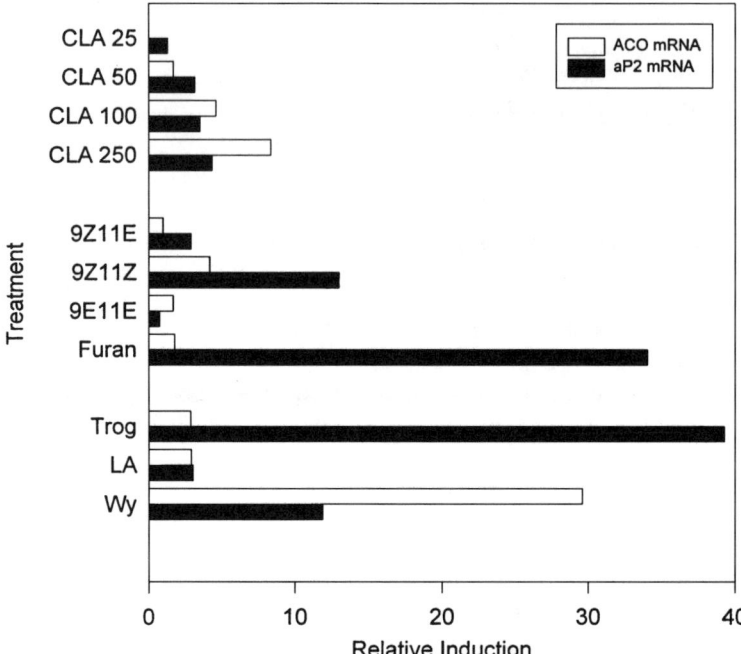

Fig. 32.3. Effects of conjugated linoleic acid (CLA) on gene expression in 3T3-L1 cells. Mouse preadipocyte cells were treated at confluence for 48 h with induction medium that contained the indicated treatments. Abbreviations are as in legend to Figure 32.2. Subsequently, induction medium plus insulin was added to the cells. Total RNA was extracted 2 d after the addition of induction medium plus insulin. Quantitative reverse transcriptase-polymerase chain reaction (RT-PCR) was performed using internal standards specific for ACO (open box) or aP2 (filled box). The data (average of three samples) are expressed as a percentage of the vehicle [dimethyl sulfoxide (DMSO)]-treated cells.

Effects of CLA Treatment on Indices of Diabetes in Rats

Because CLA acted as a PPARγ activator in 3T3-L1 cells, we hypothesized that this fatty acid would have beneficial effects on indices of diabetes. For the following studies, we utilized a rat model of type 2 DM, the Zucker diabetic fatty (*fa/fa*; ZDF) rat (3). ZDF rats fed a semipurified diet containing 1.5% CLA (by weight) exhibited normal glycemic responses in a glucose tolerance test compared with rats fed a control diet containing no CLA (3). In addition, the CLA-fed ZDF rats exhibited significantly reduced fasting glucose ($P < 0.05$), hyperinsulinemia ($P < 0.001$), hypertriglyceridemia ($P < 0.05$), free fatty acid levels ($P < 0.05$; Table 32.2; M.A. Belury, unpublished data), and hyperleptinemia ($P < 0.001$) (3,18) compared with ZDF rats fed control diet. Furthermore, dietary CLA increased steady-state levels of aP2 mRNA in adipose tissue compared with levels in rats fed control diets.

TABLE 32.2. Effect of Dietary Conjugated Linoleic Acid (CLA) on Diabetes in Zucker Diabetic Fatty (ZDF) Rats

	Control diet	CLA diet	Thiazolidinedione diet
Fasting glucose (mg/dL)	104 ± 19	79 ± 11^a	79 ± 6^a
Insulin (ng/mL)	38 ± 3	20 ± 3^a	7 ± 0.5^b
Triglyceride (mg/dL)	408 ± 148	200 ± 152^a	57 ± 12^b
Free fatty acids (mmol/L	1.969 ± 0.401	1.004 ± 0.262^a	0.778 ± 0.377^a
aP2 mRNA^c (%)	100	505 ± 155^a	600 ± 200^a

^{a}Significantly different from control diet group.
^{b}Significantly different from control and CLA diet groups.
^{c}Expressed as a percentage of the control diet (100%).

Summary

Many of the actions of CLA isomers are similar to those of PPAR activators such as fibrate hypolipidemic drugs and the insulin-sensitizing thiazolidinedione drugs. These studies demonstrate that CLA induces PPAR-responsive genes and that CLA display a degree of geometric and positional specificity of isomer activation of PPAR that has not been described previously. As described above, CLA is a term that describes a variety of positional and geometric dienoic fatty acid isomers found in meat and dairy products (19,20). Of the possible isomers, the *cis*-9, *trans*-11 (or 9Z,11*E*-) and *trans*-9, *cis*-11 (9*E*,11Z-) were proposed to be the most biologically relevant CLA molecules (21). Induction of expression of the luciferase reporter supported this idea, although other isomers were also effective activators of PPARγ. In many of our experiments, the antidiabetic drug troglitazone was used as a prototypical PPARγ activator to compare with CLA. In CV-1 cells, CLA and troglitazone resulted in a similar amount of PPARγ activation. From earlier experiments, it was observed that CLA and troglitazone reversed the symptoms of diabetes in the Zucker *fa/fa* diabetic rat, including decreased triglycerides, insulin and leptin, and improved glucose utilization (3,18). Therefore, CLA may represent not only an important agent for the treatment of type 2 DM, but may also be a tool for the examination of adipocyte differentiation and the role of PPARγ in this process. To our knowledge, this is the first demonstration of a molecular mechanism of action by which CLA exerts one of its many biological effects. Because various PPAR are targets for pharmaceutical intervention, in addition to showing that this important dietary fatty acid may be functioning through a receptor-mediated mechanism, the possibility of other uses of CLA may be entertained. Because CLA isomers share many attributes with the clinical thiazolidinedione drug, troglitazone, and may possibly exhibit fewer side effects, CLA may represent a promising new insulin-sensitizing agent for the treatment of type 2 DM.

Acknowledgments

Supported by the National Cattleman's Beef Association (J.P.V.H., M.A.B.), American Cancer Society (M.A.B.), and National Institutes of Health (J.P.V.H.).

References

1. American Diabetes Association (1997).
2. Belury, M.A., and Vanden Heuvel, J.P. (1997) Protection Against Cancer and Heart Disease by the Dietary Fat, Conjugated Linoleic Acid: Potential Mechanisms of Action. *Nutr. Dis. Update J. 1*, 58–63.
3. Houseknecht, K.L., Vanden Heuvel, J.P., Moya-Carnarena, S.Y., Portocarrero, C.P., Peck, L.W., Nickel, K.P., and Belury, M.A. (1998) Dietary Conjugated Linoleic Acid Normalizes Impaired Glucose Tolerance in the Zucker Diabetic Fatty *fa/fa* Rat, *Biochem. Biophys. Res. Commun. 244*, 678–682.
4. Willson, T.M., Lehmann, J.M., and Kliewer, S.A. (1996) Discovery of Ligands for the Nuclear Peroxisome Proliferator-Activated Receptors, *Ann. N.Y. Acad. Sci. 804*, 276–283.
5. Lehmann, J.M., Moore, L.B., Smith-Oliver, T.A., Wilkison, W.O., Willson, T.M., and Kliewer, S.A. (1995) An Antidiabetic Thiazolidinedione Is a High Affinity Ligand for Peroxisome Proliferator-Activated Receptor γ (PPAR γ), *J. Biol. Chem. 270*, 12953–12956.
6. Tontonoz, P., Graves, R.A., Budavari, A.I., Erdjument-Bromage, H., Lui, M., Hu, E., Tempst, P., and Spiegelman, B.M. (1994). Adipocyte-Specific Transcription Factor ARF6 Is a Heterodimeric Complex of Two Nuclear Hormone Receptors, PPARγ and RXRα, *Nucleic Acids Res. 22*, 5628–5634.
7. Belury, M.A., Moya-Camarena, S.Y., Liu, K.-L., and Vanden Heuvel, J.P. (1997) Dietary Conjugated Linoleic Acid Induces Peroxisome-Specific Enzyme Accumulation and Ornithine Decarboxylase Activity in Mouse Liver, *J. Nutr. Biochem. 8*, 579–584.
8. Latruffe, N., and Vamecq, J. (1997) Peroxisome Proliferators and Peroxisome Proliferator Activated Receptors (PPARs) as Regulators of Lipid Metabolism, *Biochimie 79*, 81–94.
9. Lee, S.S., Pineau, T., Drago, J., Lee, E.J., Owens, J.W., Kroetz, D.L., Fernandez-Salguero, P.M., Westphal, H., and Gonzalez, F.J. (1995) Targeted Disruption of the Alpha Isoform of the Peroxisome Proliferator-Activated Receptor Gene in Mice Results in Abolishment of the Pleiotropic Effects of Peroxisome Proliferators, *Mol. Cell Biol. 15*, 3012–3022.
10. Gonzalez, F.J. (1997) Recent Update on the PPAR Alpha-Null Mouse, *Biochimie 79*, 139–144.
11. Matsuoka, N., Jingami, H., Masuzaki, H., Mizono, M., Nakaishi, S., Suga, J., Tanaka, T. Yamamoto, T., and Nakao, K. (1996) Effects of Gemfibrozil Administration on Very Low Density Lipoprotein Receptor mRNA Levels in Rabbits, *Atherosclerosis 126*, 221–226.
12. Inoue, I., Noji, S., Shen, M.Z., Takahashi, K., and Katayama, S. (1997) The Peroxisome Proliferator-Activated Receptor α (PPARα) Regulates the Plasma Thiobarbituric Acid-Reactive Substance (TBARS) Level, *Biochem. Biophys. Res. Commun. 237*, 606–610.
13. Gottlicher, M., Demoz, A., Svensson, D., Tollet, P., Berge, R.K., and Gustafsson, J.A. (1993) Structural and Metabolic Requirements for Activators of the Peroxisome Proliferator-Activated Receptor, *Biochem. Pharmacol. 46*, 2177–2184.
14. Spiegelman, B.M., Hu, E., Kim, J.B., and Brun, R. (1997) PPAR gamma and the Control of Adipogenesis, *Biochimie 79*, 111–112.

15. Hirose, T., and Kurebayashi, S. (1997) Adipocyte Differentiation and Nuclear Receptor, *Nippon Rinsho 55*, 1596–1604.
16. MacDougald, O.A., and Lane, M.D. (1995) Transcriptional Regulation of Gene Expression During Adipocyte Differentiation, *Annu. Rev. Biochem. 64*, 345–373.
17. Saltiel, A.R., and Olefsky, J.M. (1996) Thiazolidinediones in the Treatment of Insulin Resistance and Type II Diabetes, *Diabetes 45*, 1661–1669.
18. Houseknecht, K., Vanden Heuvel, J., Moya-Camarena, S., Portocarrero, C., Peck, L., Nickel, K., and Belury, M. (1998) Amelioration of Impaired Glucose Tolerance by Conjugated Linoleic Acid (CLA) in ZDF (*fa/fa*) Rats, *Diabetes 47*, A245.
19. Jiang, J., Bjoerck, L., Fonden, R., and Emanuelson, M. (1996) Occurrence of Conjugated *cis*-9,*trans*-11-Octadecadienoic Acid in Bovine Milk: Effects of Feed and Dietary Regimen, *J. Dairy Sci. 79*, 438–445.
20. Lin, H., Boylston, T.D., Chang, M.J., Luedecke, L.O., and Shultz, T.D. (1995) Survey of the Conjugated Linoleic Acid Contents of Dairy Products, *J. Dairy Sci. 78*, 2358–2365.
21. Pariza, M.W., Ha, Y.L., Benjamin, H., Sword, J.T., Gruter, A., Chin, S.F., Storkson, J., Faith, N., and Albright, K. (1991) Formation and Action of Anticarcinogenic Fatty Acids, *Adv. Exp. Med. Biol. 289*, 269–272.

Chapter 33

Conjugated Linoleic Acid as a Nutraceutical: Observations in the Context of 15 Years of n-3 Polyunsaturated Fatty Acid Research

Howard R. Knapp

University of Iowa College of Medicine, Department of Internal Medicine, Iowa City, IA 52242–1081

The initial observations arousing interest in the biological effects of conjugated linoleic acids (CLA) centered on their ability to reduce chemically induced cancers in animal models. The present volume contains a summary of this work (Chapter 34), as well as contributions on potential antiatherosclerotic effects in cholesterol-fed animals (Chapter 31), and the amelioration of diabetes in rats (Chapter 32). It is interesting that although n-3 polyunsaturated fatty acids (PUFA) showed potent effects in animal models of cancer (1), retarded the progression of experimental vascular disease in many species (2), and improved diabetic control in rats (3), their use has not yielded impressive results in cancer patients (4), has not altered progression of vascular disease in one human model of atherosclerosis, angioplasty restenosis (5), and has worsened rather than improved glucose intolerance in human diabetics (6). Despite considerable governmental support and thousands of publications from many research groups, the only significant current clinical use in adults for n-3 fatty acids is for hypertriglyceridemia in patients who are intolerant of the first-line agents (niacin and the fibrate drugs) or who have an inadequate response to them. Considerable evidence has now accumulated that n-3 PUFA are also essential for optimal brain development in neonates (7), and they are added to infant formula in a number of countries (8). Other clinical uses of n-3 PUFA remain under investigation, but without the widespread enthusiasm of years past. One might ask, then, whether the current situation for CLA investigation appears at all promising in the context of the past 15 years of work on n-3 PUFA, and whether useful observations can be made that may increase the efficiency of its therapeutic development or promote realistic projections for the future for CLA.

In contrast to the suggestive epidemiologic data on beneficial effects of dietary n-3 PUFA in cardiovascular disease (2), there is little evidence for a cancer-preventing (primarily breast) effect of CLA in human populations if it is valid to extrapolate the consumption of dairy products to CLA intake (Chapter 35, this volume). The human diet is made up of food items, not individual fatty acids, and the many compounds co-ingested with CLA in dairy products could certainly obscure any effect this mixture of fatty acids might have in specific individuals. Although one cannot say that the dietary survey data rule out a useful preventive effect of CLA against breast cancer in women, it is equally true that no testable hypotheses have emerged as a result of the studies published thus far.

Beginning in the 1970s and early 1980s, it was noted that a number of human disease states existed in which there was an elevation in the detectable levels of plasma lipid conjugated dienes. It eventually became widely accepted that the predominant component of the "conjugated diene" in human blood and bile lipids was actually CLA (9–12), and much of the work done by this group was of a very high standard from the chemistry standpoint. Earlier workers (13) had shown that plasma dienes were reduced when vitamin E was taken as a dietary supplement. From the eicosanoid literature, however, it is known that plasma lipids are invariably contaminated with lipid hydroperoxides as a result of a small activation of platelets during blood sampling. For instance, when blood clots at 37°C in a test tube, ~250 ng of thromboxane/mL are formed in parallel with similar concentrations of the cyclooxygenase product HHT and the lipoxygenase product 12-hydroxyeicosatetranoic acid (12-HETE). Because of this enormous capacity for platelets to produce lipid peroxides, even the most fastidiously sampled plasma has on the order of 200–500 pg/mL thromboxane B_2 present, whereas metabolism studies have shown clearly that the true circulating concentration of thromboxane in human blood is on the order ≤0.5 pg/mL. Thus, blood samples invariably have some degree of contamination from ex vivo sampling artifact, and studies in humans must be interpreted in this regard. It is known, for instance, that plasma longer-chain polyunsaturates increased in a number of disease states, including rheumatoid arthritis (14) and toxemia of pregnancy (15). The latter syndrome is also associated with increased production of prostacyclin and thromboxane, whereas normal pregnancy is associated with a marked increase in prostacyclin synthesis alone. Clearly, if there are changes in the fatty acid profiles in blood and tissue lipids in various diseases, one could expect changes in their oxidation products whether due to biological conditions or sampling artifact.

It is frequently mentioned in articles dealing with changes in the fatty acid composition of blood and tissue lipids in human subjects that dietary modification has an unknown degree of influence over such findings, i.e., patients with different diseases could certainly be modifying their diets unconsciously to result in altered fatty acid profiles. It has been shown that increased trans fatty acids in the diet result in an increase in CLA levels in human serum (16), but that the consumption of linoleic acid as safflower oil does not result in such an increase (17). Other workers have indicated that CLA itself induces some peroxisomal processes in mouse liver (18); the interactions between such effects of CLA and the effects of disease processes themselves are difficult to understand without doing prospective trials in humans. It has been shown, for instance, that patients with pancreatic disease have an increase in lipid peroxidation products in bile, and these products include CLA (19,20). Could this relate to altered hepatic peroxisomal metabolism or cytochrome P-450 activities in such disease states? Certainly it remains an open question whether increased CLA in these patients is an adaptive response toward production of a beneficial substance or merely a biological by-product of the disease process itself. There have been conflicting reports on the association of plasma CLA with diseases in which lipid peroxidation is believed to occur. Alcoholics are known to have reductions in hepatic glutathione concentrations; this makes them particularly susceptible to a number of drug toxicities that involve free radical formation in the liver such as that with acetaminophen. Some authors have indicated an increase in conjugated dienes in alcoholic liver disease (21); others have noted that there was no difference between alcoholic and normal subjects but that vitamin E status played a critical role in determining the levels of conjugated dienes in plasma (10). On the basis of the observations presented in the literature to date, it would be difficult to draw firm conclusions related to the role of CLA production in the pathogenesis of any human disease. However, one would certainly expect to find increases in disease states associated with altered fatty acid metabolism that might yield increased levels of linoleic acid, or in cases in which lipid peroxidation was known to be increased either through the cytochrome P-450 system or peroxisomal oxidation. With the recent advances in analytical methodology presented in this textbook, further work will certainly be able to shed considerable light on metabolism and effects of CLA in numerous human disease states and determine whether it may be a useful substance with which to supplement patients' diets or is merely an index of a biological process that has no relevance in pathogenesis.

Until recently, experiments in the studies on CLA in animals have been conducted largely with ill-defined mixtures of CLA isomers (22). Recent analytical developments have enormously improved the ability to isolate and measure the different isomers, so that the various mixtures used can be characterized and the different CLA isomers can be quantified in tissue samples (Chapters 6–12, this volume). Until this is accomplished, mechanistic interpretation of any effects found is not possible. It is also difficult to assess in any quantitative way much of the work published in the past because the studies were conducted with CLA isomer mixtures of unknown proportions and with unknown degrees of incorporation into different tissues and organelles. New developments will now allow the stereospecific synthesis of large enough quantities of pure CLA isomers so that mechanistic studies can be conducted in the animal models previously described, allowing tentative selection of potential surrogate markers that may serve as a means of conducting future human studies. As with n-3 fatty acids, it would be most useful to measure some physiologic or biochemical parameter that has a well-understood relationship to the clinical process one hopes to benefit with CLA ingestion. Accumulating this sort of "preclinical" data will allow assessment of just what endpoints can be measured in human studies, as well as the potential utility of specific CLA isomers for applications in human subjects.

Although it will be highly advantageous to perform studies with single isomers of CLA, this by itself will not pave the way for its smooth development as a nutraceutical. In the case of n-3 PUFA, because many investigators were performing studies using different types of fish oils containing different proportions of fatty acids and having a wide variety of non-triglyceride components, a commitment was made by the U.S. Department of Commerce to set up a pilot plant in Charleston, SC to produce high-purity menhaden oil of well-defined composition, and eventually to produce ethyl esters of both eicosapentaenoic acid (EPA) and docosahexaenoic acid (DHA) for use in clinical trials. All of this effort (i.e., the Fish Oils Technical Materials program) produced an environment that was highly favorable for development of the undeniable effects in humans of n-3 fatty acid supplements into useful therapies for human disease. So, what happened? Despite considerable evidence of the potential for n-3 fatty acids as a useful therapy for several aspects of cardiovascular disease, this health claim was denied by the U.S. Food and Drug Administration (FDA) as being premature (23). The reasoning behind this decision is instructive for those interested in developing uses for CLA in any of the diseases in which this fatty acid mixture has shown potential in animal studies.

First, it was pointed out by the FDA reviewers that people eat foods, not specific fatty acids. Many studies had been performed using marine oils but, it was reasoned, it was impossible to determine which component or combination of components could have been responsible for the effects found. In fact, the mixtures used in various studies did sometimes yield different results, and it is possible that some of the components could have been exerting different, even antagonistic effects. Clearly, studies utilizing one pure fatty acid would be easier to interpret. There was no clear dose-response relationship found in either epidemiologic studies or, as originally published, in the widely discussed Diet And Reinfarction Trial (DART), which showed a striking benefit of ingesting a few helpings of oily fish per week in preventing death from a second myocardial infarction (24). A *post-hoc* analysis of DART did demonstrate a dose- response effect, but this has not been widely discussed because the therapeutic applications of fish oil have passed out of vogue. From the available literature, it was difficult to determine an "optimal" dose or dosing schedule to accomplish a particular therapeutic goal. Finally, there was concern that widespread increases in n-3 fatty acid intake posed unknown levels of hazard to subsets of the American public, such as diabetics, those with bleeding disorders or a tendency towards stroke, and children. Of course, each of these points could be debated in detail, but not rebutted in a manner sufficient to overcome FDA objections in the context of regulations in effect at that time.

In fairness to the FDA reviewers, it must be pointed out that a central problem in developing a number of therapies has been an incomplete understanding of their human pharmacology. An excellent example of this is aspirin, which is a very safe (by drug standards) and highly effective agent in the secondary prevention of stroke and myocardial infarction (25). It took many, many large trials and 20 years of work to convincingly demonstrate this useful therapeutic effect for the following reasons: There was no understanding of how aspirin worked to retard platelet aggregation; optimal doses to achieve this effect were not known; and it was unclear how this effect actually related to the thrombotic events themselves. Clearly, we did not have a good understanding of the optimal dose of specific n-3 fatty acids needed to exert a number of the particular effects desired, the mechanisms of the biochemical changes found, or how the biochemical changes and other surrogate endpoints (e.g., vascular control or angioplasty restenosis) being studied actually related to the vascular endpoints, such as thrombosis, infarction, and atherogenesis, that we were trying to prevent. Because fish oils and their components are natural products (as is the mixture of conjugated linoleic acids), the active compounds cannot be patented, although some limited protection could be obtained for their use in specific conditions, or for special processes developed for their purification or synthesis. This greatly discouraged large pharmaceutical firms from entering the fish oil arena because the considerable expenditure required to demonstrate safety and efficacy to the standards of regulatory bodies could benefit any vendor of marine oil products and not only the firm making the large investment of resources.

In the early human studies using fish oils, a number of biochemical and physiologic effects were observed by many investigators (e.g., bleeding time prolonged or blood triglycerides reduced) that had also been seen in animal studies (26). The few human studies on effects of CLA on body composition reported thus far have provided a disappointing contrast to the many animal studies that showed increased lean body mass, and did not show significant effects on body mass composition (Chapter 27, this volume). As noted above for fish oil studies, the few CLA clinical studies that have been performed with a mixture of isomers could bear repeating with a pure isomer showing unequivocal activity in pertinent animal models, now that the synthesis and measurement of specific isomers have become technically feasible. Also, it could be that the dose or dosing schedule of the CLA mixture used in studies thus far was suboptimal. From a practical standpoint, however, if 2–3 g/d of a fatty acid fail to alter body composition during the course of weight loss in obese individuals in a carefully conducted study run by experts, it seems unlikely that a similar mixture presented as a food item from the grocery store will exert beneficial effects in individuals who have a low likelihood of losing weight or exerting dietary modification for any prolonged period on their own. Clearly, a thorough understanding of the biochemistry and pharmacology of the individual CLA isomers will allow more definitive studies to address this issue.

The initial investigations on the effects of n-3 fatty acids in the Inuit people (26) were followed by clinical studies on small groups of subjects that showed beyond any reasonable doubt that marine oils taken for a short time in pharmacologic doses did produce biochemical and physiologic effects in humans, but were unlikely to exert strikingly harmful effects. For example, the highly unsaturated oil was easily peroxidizable; concern was raised whether it could affect fertility adversely, but Inuit have been consuming a diet of seal, whale, and fish over many generations. Fish oil supplements do cause a modest prolongation of bleeding time in normal volunteers, but women in the Faeroe Islands who consume large amounts of n-3 fatty acids have not been noted to bleed excessively during childbirth (27), and cardiac patients taking sizable doses did not experience excessive blood loss during coronary artery bypass operations (5). In addition, several generations of British and American children were exposed to widespread ingestion of a daily teaspoonful of cod-liver oil to prevent rickets, and this was not accompanied by any evidence of adverse effect except for the memory of a bad taste. Subsequent investigations have put aside many of the legitimate concerns about toxic effects of n-3 fatty acids, but the pace of new work in this area has clearly declined. Similar concerns no doubt await CLA research. The accumulation of specific CLA isomers in cardiac mitochondria (28), for example, will prompt additional studies to allay concerns about the possibility that chronic CLA ingestion could result in a myocarditis like that seen in animals fed large quantities of long-chain monoenes. Clearly, a focused approach of application of specific CLA isomers without theoretically objectionable properties in animal models or human volunteers will have the best chance of reaching the stage of serious clinical trials.

The failure of n-3 PUFA to become a useful therapy for human diseases in which they had potent activity in analogous animal models could be due to species differences, but it is possible that pharmacologic issues provided overwhelmingly confounding variables. Our incomplete understanding of PUFA kinetics in specific pools, of competing biological effects, of the potential mechanisms involved, or of the correct dose and dosage form could have also led to failure in critical clinical trials. Could the use of mixtures of fatty acids have resulted in effects that were antagonistic to or different from those anticipated from animal studies? As mentioned previously, one concern of clinicians about the work done so far in CLA and cancer is, that studies have by necessity been conducted with impure mixtures and inadequate analytical methods. In addition, all but a few of the published works in animal cancers have been chemoprevention studies, with uncertain extrapolation to spontaneous tumor development in humans. The few studies on using CLA as a therapy for tumors already present have not provided much encouragement for further work. Thus, the only human cancer studies for which there is presently a reasonable rationale to test CLA would be chemoprevention trials that are, by nature, large, expensive, and of long duration. To become a candidate compound for such a study requires compelling preliminary data in appropriate animal models, but so far there is little evidence that CLA prevents tumor appearance in strains of nonimmunosuppressed animals with a high incidence of spontaneous tumor development. If these data prove positive, studies in such types of animals would be most encouraging and provide the stimulus needed to gear up for the huge effort involved in human oncology trials.

The recent development of the ability to measure carefully the different CLA isomers in body fluids and tissues and to synthesize large quantities of specific isomers will now allow mechanistic studies to be conducted in animals. Once we achieve an understanding of the mechanisms of specific biological events caused by particular CLA isomers, it will then be possible to pick appropriate animal models to determine the potential utility of certain CLA isomers in defined types of human disease. This also will allow the development of an understanding of the animal and clinical pharmacology and toxicology of the compound of interest. Studies with an optimal design and dosing regimen will be required to demonstrate whether the compound of interest has real benefit for human patients. It is hoped that the many recent analytical advances discussed in this volume will contribute to this goal. Based on experiences with n-3 PUFA, significant and unique benefits must be demonstrated before an investment would be made in clinical testing of a specific CLA formulation.

References

1. Cave, W.T. (1992) n-3 Fatty Acid Effects on Tumorigenesis, in *Fish, Fish Oil and Human Health,* (Frölich, J.C., and von Schacky, C., eds.) pp. 179–189, W. Zuckschwerdt Verlag, Munich.
2. Leaf, A., and Kang, J.X. (1998) Omega 3 Fatty Acids and Cardiovascular Disease, *World Rev. Nutr. Diet. 83,* 24–37.
3. Storlien, L.H., Kraegen, E.W., Chisholm, D.J., Ford, G.L., Bruce, D.G., and Pascoe, W. (1987) Fish Oil Prevents Insulin Resistance Induced by High-Fat Feeding in Rats, *Science 237,* 885–888.
4. Karmali, R.A. (1996) Historical Perspective and Potential Use of n-3 Fatty Acids in Therapy of Cancer Cachexia, *Nutrition 12* (Suppl. 1), S2–S4.
5. Leaf, A., Jorgensen, M.B., Jacobs, A.K., Cote, G., Schoenfeld, D.A., Scheer, J., Weiner, B.H., Slack, J.D. Kellett, M.A., Raizner, A.E., et al. (1994) Do Fish Oils Prevent Restenosis after Coronary Angioplasty? *Circulation 90,* 2248–2257.
6. Borkman, M., Chisholm, D.J., Furler, S.M., Storlein, L.H., Kraegen, E.W., Simons, L.A., and Chesterman, C.N. (1989) Effects of Fish Oil Supplementation on Glucose and Lipid Metabolism in NIDDM, *Diabetes 38,* 1314–1319.
7. Innis, S.M., Sprecher, H., Hachey, D., Edmond, J., and Anderson, R.E. (1999) Neonatal Polyunsaturated Fatty Acid Metabolism, *Lipids 34,* 139–149.
8. Koletsko, B., and Sinclair, A. (1999) Long-Chain Polyunsaturated Fatty Acids in Diets for Infants: Choices for Recommending and Regulating Bodies and for Manufacturers of Dietary Products, *Lipids 34,* 215–220.
9. Iversen, S.A., Cawood, P., Madigan, M.J., Lawson, A.M., and Dormandy, T.L. (1984) Identification of Diene Conjugated Component of Human Lipid as Octadeca-9,11-dienoic Acid, FEBS Lett. 171, 320–324.
10. Cawood, P., Wickens, D.G., Iversen, S.A., Braganza, J.M., and Dormandy, T.L. (1983) The Nature of Diene Conjugation in Human Serum, Bile and Duodenal Juice, FEBS Lett. 162, 239–243.
11. Iversen, S.A., Cawood, P., and Dormandy, T.L. (1985) A Method for the Measurement of a Diene-Conjugated Derivative of Linoleic Acid, 18:2(9,11), in Serum Phospholipid, and Possible Origins, Ann. Clin. Biochem. 22, 137–140.
12. Dormandy, T.L., and Wickens, D.G. (1987) The Experimental and Clinical Pathology of Diene Conjugation, Chem. Phys. Lipids 45, 353–364.
13. Di Luzio, N.R. (1972) Protective Effect of Vitamin E on Plasma Lipid Dienes in Man, J. Agric. Food Chem. 20, 486–489.
14. Bruderlein, H., Daniel, R., Boismenu, D., Julien, N., and Couture, F. (1981) Fatty Acid Profiles of Serum Phospholipids in Patients Suffering Rheumatoid Arthritis, Prog. Lipid Res. 20, 625–631.
15. Wickens, D., Wilkins, M.H., Lunec, J., Ball, G., and Dormandy, T.L. (1981) Free-Radical Oxidation (Peroxidation) Products in Plasma in Normal and Abnormal Pregnancy, Ann. Clin. Biochem. 18, 158–162.

16. Salminen, I., Mutanen, M., Jauhiainen, M., and Aro, A. (1998) Dietary trans Fatty Acids Increase Conjugated Linoleic Acid Levels in Human Serum, J. Nutr. Biochem. 9, 93–98.
17. Herbel, B.K., McGuire, M.K., McGuire, M.A., and Shultz, T.D. (1998) Safflower Oil Consumption Does Not Increase Plasma Conjugated Linoleic Acid Concentrations in Humans, Am. J. Clin. Nutr. 67, 332–337.
18. Belury, M.A., Moya-Camarena, S.Y., Liu, K.-L., and Vanden Heuvel, J. P. (1997) Dietary Conjugated Linoleic Acid Induces Peroxisome-Specific Enzyme Accumulation and Ornithine Decarboxylase Activity in Mouse Liver, J. Nutr. Biochem. 8, 579–584.
19. Braganza, J.M., Wickens, D.G., Cawood, P., and Dormandy, T.L. (1983) Lipid-Peroxidation (Free-Radical-Oxidation) Products in Bile from Patients with Pancreatic Disease, Lancet, 375-378.
20. Guyan, P.M., Uden, S., and Braganza, J.M. (1990) Heightened Free Radical Activity in Pancreatitis, Free Radic. Biol. Med. 8, 347–354.
21. Shaw, S., Rubin, K.P., and Lieber, C.S. (1983) Depressed Hepatic Glutathione and Increased Diene Conjugates in Alcoholic Liver Disease: Evidence of Lipid Peroxidation, Dig. Dis. Sci. 28, 585–589.
22. Christie, W.W., Dobson, G., and Gunstone, F.D. (1997) Isomers in Commercial Samples of Conjugated Linoleic Acid, *Lipids 32,* 1231.
23. Wallingford, J.C., and Yetty, E.A. (1991) Development of the Health Claims Regulations: The Case of Omega-3 Fatty Acids and Heart Disease, *Nutr. Rev. 49,* 323–331.
24. Burr, M.L., Gilbert, J.F., Holliday, R.M., Elwood, P.C., Fehily, A.M., Rogers, S., Sweetnam, P.M., and Deadman, N.M. (1989) Effects of Changes in Fat, Fish, and Fibre Intakes on Death and Myocardial Reinfarction: Diet and Reinfarction Trial (DART), *Lancet,* 757–761.
25. Hennekens, C.H., Buring, J.E., Sandercock, P., *et al.* (1991) Aspirin and Other Antiplatelet Agents in the Secondary and Primary Prevention of Cardiovascular Diseases, *Circulation 80,* 749–756.
26. Bang, H.O., and Dyerberg, J. (1980) Lipid Metabolism and Ischemic Heart Disease in Greenland Eskimos, *Adv. Nutr. Res. 31,* 1–22.
27. Olsen, S.F., Hansen, H.S., Sommer, S., Jensen, B., Sørensen, T.I.A., Secher, N.J., and Zachariassen, P. (1991) Gestational Age in Relation to Marine n-3 Fatty Acids in Maternal Erythrocytes: A Study of Women in the Faroe Islands and Denmark, *Am. J. Obstet. Gynecol. 164,* 1203–1209.
28. Kramer, J.K.G., Sehat, N., Dugan, M.E.R., Mossoba, M.M., Yurawecz, M.P., Roach, J.A.G., Eulitz, K., Aalhus, J.L., Schaefer, A.L., and Ku, Y. (1998) Distributions of Conjugated Linoleic Acid (CLA) Isomers in Tissue Lipid Classes of Pigs Fed a Commercial CLA Mixture Determined by a Gas Chromatography and Silver-Ion High-Performance Liquid Chromatography, *Lipids 33,* 549–558.

Chapter 34
Cancer Inhibition in Animals

Joseph A. Scimeca

Kraft Foods, Incorporated, Glenview, IL 60025

Introduction

More than a decade has passed since a serendipitous discovery from Mike Pariza's laboratory revealed the existence of a substance from fried hamburger that actually *inhibited* rather than promoted mutagenesis. Soon after their research finding, these researchers isolated and identified the substance as conjugated linoleic acid (CLA) (1). In contrast to the many hundreds of phytochemicals that possess varying degrees of anticarcinogenic activity, CLA is unique in that it is a fatty acid found in highest amounts in animal products, specifically ruminant meat and dairy products (2). Furthermore, CLA was found to possess anticarcinogenic activity that turned out be quite potent. To illustrate this point, research conducted by Clement Ip and co-workers has determined that the lowest dietary level of CLA shown to inhibit tumors in a chemically induced rat model of carcinogenesis is 0.1% (w/w). By way of comparison, fish oil is an anticancer agent that is also not plant derived, but efficacious levels usually exceed 10% in the diet. Hence, at a dietary level of 0.1%, CLA has a 100-fold greater potency.

At present, CLA refers to a group of isomers of linoleic acid (9 *cis*,12 *cis*-octadecadienoic acid). The characteristic that binds the CLA isomers as a group is that, unlike linoleic acid, the two double bonds along the 18-carbon fatty acid chain are not separated by a methylene carbon, and hence are termed "conjugated." Of the many positional and geometric isomers of CLA, the 9 *cis*,11 *trans*-octadecadienoic acid (9 *cis*,11 *trans*-18:2) isomer predominates in food products (2).

Milk fat is the richest natural dietary source of CLA, and ~85% of milk fat is the 9 *cis*,11 *trans*-18:2 isomer (2–4). This isomer is produced by rumen bacteria as an intermediate in the biohydrogenation of dietary linoleic acid. Sequential reduction steps convert the CLA isomer to vaccenic acid (11 *trans*-18:1) and then to stearic acid. However, evidence is emerging that rumen production may not be the only source of 9 *cis*,11 *trans*-18:2. Bauman, Griinari and co-workers (5,6) have found evidence for an alternate endogenous pathway involving the desaturation of vaccenic acid (11 *trans*-18:1) by the Δ^9 desaturase enzyme, thus also resulting in the 9 *cis*,11 *trans*-18:2 isomer.

Typically, CLA is found in milk fat at concentrations of 4–6 mg/g lipid, although levels are strongly influenced by the bovine diet (7–9). Some treatments are able to increase CLA by as much as fivefold (7), thus offering the possibility of producing dairy products containing CLA-enriched milk fat. Moreover, an increase in the CLA level in milk fat would make testing of CLA-enriched dairy products in animal models of chemoprevention entirely feasible.

The following sections present the published experimental literature that indicates an effect of CLA on cancer inhibition, primarily in animal models of chemically induced cancer, but does include studies with human cancer cell lines. The studies are broadly organized by the type of study, either *in vivo* or *in vitro*. Within each type of study, the CLA research is subdivided by site of tumor inhibition for *in vivo* experiments and roughly by histological cell type for *in vitro* experiments. Because a considerable amount of research involved the investigation of the effect of CLA on a variety of endpoints related to carcinogenesis, which offers some insight into the potential mechanism(s) of action, a separate section briefly reviews these studies.

Every attempt was made to perform a thorough review of the published literature (and in a few instances, some unpublished literature) on the subject at hand, although peer-reviewed articles were given greater attention than research abstracts. All experimental results mentioned in the text are statistically significant as determined by the study author(s). Nonsignificant results were not included in the review of each of the studies. Little has been included in this review about the various isomers of CLA and potential differences in biological activity among them for the simple reason that there is an absence of data. Therefore, no attempt was made to identify or characterize the source of the CLA used or the isomer composition, for each of the studies. Nearly all of the studies reviewed utilized commercially obtained (e.g., Nu-Chek-Prep, Elysian, MN) synthetic CLA that was made by the isomerization of linoleic acid. Until recently, gas chromatography (GC) analysis has indicated that the bulk of the isomeric composition was equally split at ~43% 9 *cis*,11 *trans*-18:2 and/or 9 *trans*,11 *cis*-18:2, and ~45% 10 *trans*,12 *cis*-18:2. Recently, more sophisticated analysis by Yurawecz and colleagues (10) has determined that these isomers make up a lower percentage of CLA composition, i.e., ~29% for each positional/geometric combination. The remaining 40% included isomers not previously found, *cis,trans/trans,cis*-11,13-18:2 (22%) and *cis,trans/trans,cis*-8,10-18:2 (12%), as well as other minor isomers. This finding is notable and may have important bearing on the interpretation of research findings. The reader is referred to chapters elsewhere in this book for an understanding of the chemistry and analysis of CLA.

In Vivo Tumor Inhibition

Mammary Cancer

Inhibition of carcinogenesis in the rat mammary gland by CLA treatment has been studied more thoroughly than in any other tissue. In particular, Ip and co-workers have demonstrated the inhibition of chemically induced rat mammary tumors by CLA under a wide range of experimental conditions. In total, the dietary administration of CLA has been shown to inhibit chemically induced rat mammary gland tumors in

\>1000 animals, in a dozen separate experiments. Dietary CLA has been shown to be effective at various doses, at various stages of carcinogenesis, for varying dosing durations and regimens, regardless of level or type of fat in the diet, and independent of the presence of linoleate in the diet (see Table 34.1 for a summary of the chemically induced rat mammary cancer studies).

Ip and co-workers (11) were the first to show an inhibition of cancer by the *dietary* administration of CLA. Female Sprague-Dawley rats were fed an AIN-76A basal diet supplemented with 0.5, 1.0, or 1.5% synthetically prepared CLA. Controls received the same diet without CLA. Rats received their experimental diets (n = 30/group) 2 wk before gavage treatment with the carcinogen 7,12-dimethylbenz(a)anthracene (DMBA) (10 mg/rat) at ~50 d of age. Rats continued to receive experimental diets until study termination 24 wk after DMBA administration. Results revealed a dose-responsive reduction in the total number of mammary tumors by 32–60%. Tumor incidence was similarly reduced by 17–50% and tumor multiplicity by 33–60%. Median tumor latency was unaffected by CLA treatment. CLA feeding at dietary levels >1.0% did not seem to offer any additional protection.

In a follow-up paper, Ip and colleagues (12) further examined the dose-responsive inhibitory effect of CLA on mammary tumors. Similar to the first study, rats were given a modified basal AIN-76A diet supplemented with 0.05, 0.1, 0.25, or 0.5% CLA (n = 50/group) 2 wk before DMBA administration at 50 d of age. Controls (n = 50) received the same diet without CLA. Sample size was increased to ensure adequate statistical power due to the reduced tumor incidence produced by a lower dose of the DMBA (5 mg/rat). Rats consumed their experimental diets until study termination 36 wk after carcinogen treatment. Results showed a dose-responsive reduction in the total number of mammary tumors by 36–58% for CLA levels from 0.1 to 0.5%. Tumor incidence was reduced by ~37% for the 0.25 and 0.5% CLA groups. In a second experiment, the authors investigated the effect of short-term CLA feeding on mammary cancer inhibition. Female Sprague-Dawley rats were fed 1.0% CLA from weaning at 3 wk until 1 wk past carcinogen treatment, a period that corresponds to the time of peak mammary gland maturation. Rats were treated at d 50 with one of two different carcinogens, i.e., DMBA, which required metabolic activation, or methylnitrosourea (MNU), a direct-acting alkylating agent. CLA feeding reduced the total number of mammary tumors by 39 and 34% in the DMBA and MNU models, respectively. Similar results were found for tumor incidence. This experiment provided the following three important findings regarding the mode of action of CLA: (i) the anticancer effect of CLA was not limited only to DMBA; (ii) the timing of CLA exposure may be important; and (iii) CLA may be able to modulate the susceptibility of the target organ to carcinogenesis.

For their next study, these researchers continued to use the MNU model [50 mg/kg body weight (BW) given intraperintoneally (i.p.)] to investigate the effect of CLA feeding on the prevention of mammary gland carcinogenesis (13). This model was chosen over the DMBA model because MNU is a direct-acting carcinogen; hence its use precluded any potential confounding effect of CLA on carcinogen metabolism. In the first experiment, rats were fed a 1% CLA diet from weaning until MNU injection at 42 or 56 d of age (n = 30/group). Hence, the length of CLA feeding was 3 or 5 wk before MNU administration. CLA was not added to the diet after MNU treatment. Controls were injected with MNU at 42 or 56 d (n = 30/group) and were fed the same basal diet without CLA throughout the duration of the experiment. CLA feeding (3 or 5 wk) reduced the total number of mammary tumors by 47 and 42%, respectively. Similarly, 3 or 5 wk of CLA feeding reduced the tumor incidence by 32 and 39%, respectively. This experiment demonstrated that short-term CLA feeding during the critical window of mammary gland maturation was sufficient to confer some degree of lasting protection against a subsequent chemically initiated neoplastic transformation. A second experiment extended these findings by examining the effect of interrupted vs. continuous CLA feeding provided after MNU treatment on mammary carcinogenesis. As in the first experiment, CLA was added to the modified basal AIN-76A diet at a level of 1% by weight. Rats were maintained on the CLA diets (n = 30/group) for 4, 8, or 20 wk, starting immediately after MNU injection. Short-term post-initiation CLA feeding (i.e., 4 or 8 wk) was ineffective in mammary cancer inhibition. A reduction in tumor burden and incidence was observed only in rats that received an uninterrupted diet (i.e., 20 wk) of CLA throughout the post-initiation stages of carcinogenesis. A third experiment was conducted to compare the efficacy of CLA administered as a free fatty acid vs. esterification as a triglyceride. This experiment was important because all previous research employed CLA as a free fatty acid, which is unlike the triglyceride form naturally found in food. Both forms of CLA (n = 30 rats/group) were added to the basal diet at 1%, starting at weaning and continuing until MNU injection at 56 d of age. Both forms of CLA reduced both mammary tumor burden and incidence equally, thus indicating that CLA as a free fatty acid was absorbed as efficiently as the triglyceride form and was valid therefore for use in subsequent studies.

In a 1996 paper, Ip and co-workers (14) examined in three experiments the effect of various types and levels of fat in the diet on the efficacy of mammary cancer suppression by CLA. The DMBA model and basal diet as described in their 1991 paper were employed in these studies. The first experiment was designed to test whether the level of fat in the diet might affect the efficacy of cancer inhibition by CLA. A custom-blended vegetable fat that simulated the fatty acid profile of the average U.S. diet was added to the basal AIN-76A diet at 10, 13.3, 16.7, and 20% by weight. Plant oils were used to minimize the CLA content of the fat blend. Test diets at each fat level also included 1% CLA, whereas the corresponding controls did not (n = 32/group for all 8 groups). Rats consumed the experimental diets 1 wk before DMBA treatment and continued until study termination at 23 wk. Results indicated a similar magnitude of mammary tumor burden inhibition by CLA, ranging from 56 to 50% across the range of dietary fat intakes. Hence, the authors concluded that the efficacy of CLA prevention was independent of the level of fat in the diet. For their second experiment, the investigators examined the effect of the type of fat in the diet on CLA chemoprevention. Two diets were employed as follows: an exclusively 20%

TABLE 34.1. Inhibition of Rat Mammary Carcinogenesis by Conjugated Linoleic Acid (CLA)

Reference	Experimental model[b]
Ip et al. (11)	DMBA model
Ip et al. (12) (Exp. 1)	DMBA model (5mg)
Ip et al. (12) (Exp. 2)	DMBA model → or MNU model →
Ip et al. (13) (Exp. 1)	MNU model (at 42 d of age) MNU model (at 56 d of age)
Ip et al. (13) (Exp. 2)	MNU model (at 56 d of age)
Ip et al. (13) (Exp. 3)	MNU model (at 56 d of age)
Ip et al. (14) (Exp. 1)	DMBA model
Ip et al. (14) (Exp. 2)	DMBA model
Ip et al. (14) (Exp. 3)	DMBA model
Ip and Scimeca (15)	DMBA model (at 55 d of age)
Ip et al. (16)	DMBA model
Thompson et al. (17)	DMBA model

[a]Compared with controls without CLA.
[b]Carcinogen treatment at 50 d of age, unless otherwise noted. DMBA, dimethylbenz(a)anthracene; MNU, methylnitrosourea.
[c]CLA from 1–2 wk before carcinogen treatment and continued until study termination, unless otherwise noted.
[d]NS, nonsignificant determination.
[e]CLA started from weaning (21 d of age).

	Major findings[a]	
Dietary treatment[c]	% Reduction in total tumor number	% Reduction in tumor incidence
0.5% CLA	32	17
1.0% CLA	56	42
1.5% CLA	60	50
0.05% CLA	NS[d]	NS
0.10% CLA	36	NS
0.25% CLA	50	39
0.50% CLA	56	35
	39	35
1% CLA for 5 wk (started at weaning)	34	32
1% CLA for 3 wk[e]	47	32
1% CLA for 5 wk[e]	42	39
1% CLA[f] for 4 wk	NS	NS
1% CLA[f] for 8 wk	NS	NS
1% CLA[f] for 20 wk	59	48
1% CLA for 5 wk[e] as Free fatty acid	49	37
1% CLA for 5 wk[e] as Triglyceride	46	33
1% CLA/vegetable fat blend at 10%	56	41
1% CLA/vegetable fat blend at 13.3%	46	42
1% CLA/vegetable fat blend at 16.7%	51	32
1% CLA/vegetable fat blend at 20%	50	34
1% CLA + 20% corn oil	49	52
1% CLA + 8% corn oil/12% lard	47	50
20% corn oil + 0.5% CLA	39	25
20% corn oil + 1.0% CLA	57	46
20% corn oil + 1.5% CLA	61	50
2% linoleate[h] + 0.5% CLA[g]	35	32
2% linoleate[h] + 1.0% CLA	59	53
2% linoleate[h] + 1.5% CLA	51	53
2% linoleate[h] + 2.0% CLA	63	58
12% linoleate[i] + 0.5% CLA[g]	33	16
12% linoleate[i] + 1.0% CLA	52	48
12% linoleate[i] + 1.5% CLA	57	52
12% linoleate[i] + 2.0% CLA	57	52
1% CLA[g] for 4 wk	NS	NS
1% CLA[g] for 8 wk	NS	NS
1% CLA[g] for 20 wk	70	50
1% CLA weaning to 50 d	49	34
1% CLA from 55 d to term	54	46
1% CLA weaning to term	57	46

[f]CLA started immediately after MNU.
[g]CLA treatment started 4 d after DMBA.
[h]Formulated with 3.4 g corn oil and 16.6 g coconut oil per 100 g diet.
[i]Formulated with 20 g corn oil/100 g diet.

corn oil diet or a mixture of 8% corn oil and 12% lard. The lard provided a negligible amount of CLA compared with the 1% CLA added to the test diet. All diets, including the controls without CLA, were started 1 wk before DMBA and continued until study termination (n = 30/group). The magnitude of the tumor burden inhibition by CLA for the polyunsaturated fat diet (i.e., corn oil) was similar to that of the predominately saturated fat diet (i.e., corn oil and lard). Moreover, the magnitude of tumor suppression was similar to that found for the various fat levels in the first experiment. The third experiment involved feeding rats 0.5, 1.0, or 1.5% CLA added to a basal diet containing 20% corn oil (n = 30/group). Results from this study were compared with the investigators' first study (11) that used an identical design, but with a 5% corn oil diet. This experiment thus allowed the investigators to evaluate the effect of linoleate on CLA chemoprevention because the diets used in the two studies differed markedly in their linoleate content (~12 g vs. 3 g linoleate/100 g diet, respectively). As in the 1991 experiment, this experiment showed a dose-responsive inhibition in mammary cancer by CLA, with a similar magnitude of suppression of tumor burden and incidence for both experiments. Hence, these results suggested that the chemopreventative action by CLA was dissociated from linoleate. On the basis of these findings, it was concluded that the chemopreventative effect of CLA is independent of the level or type of fat in the diet.

Ip and Scimeca (15) pursued this apparent lack of interaction between CLA and linoleate. The investigators used the DMBA model in an experiment that involved 300 female Sprague-Dawley rats. One half of the rats received a basal diet containing 3.4% corn oil and 16.6% coconut oil (w/w), providing 2% linoleate in the diet. The other half received the basal diet containing 20% corn oil (w/w), providing 12% linoleate in the diet. Both diets maintained total fat at 20%. Four days after the animals were treated with DMBA (55 d of age), experimental diets were supplemented with CLA at 0.5, 1, 1.5, or 2%. Control rats continued to receive CLA-unsupplemented diets. Each of the 10 groups consisted of 30 rats. Results indicated a dose-responsive reduction in total tumor number regardless of dietary linoleate level. For rats consuming the 2% linoleate diet, increasing the CLA level from 0.5 to 2% reduced the total tumor number by 35–63%. Similarly, for rats consuming the 12% linoleate diet, increasing CLA in the diet from 0.5 to 2% reduced the total number by 33–57%. Tumor incidence data for the 2 and 12% linoleate-fed rats indicated that increasing CLA in the diet from 0.5 to 2% produced a 32–58% and 16–52% reduction, respectively. The authors concluded that CLA chemoprevention was maximized at 1% and was independent of the linoleate level in the diet.

In another paper published in 1997, Ip and colleagues (16) returned to their earlier investigation into the timing of CLA administration on the prevention of mammary cancer. Employing the DMBA model and a 20% corn oil diet, 1% CLA was added to the basal diet starting 4 d after carcinogen treatment for a duration of 4, 8, or 20 wk (n = 30/group). As in the 1995 experiment, short-term post-initiation CLA feeding (i.e., 4 or 8 wk) was ineffective in mammary cancer inhibition. A reduction in tumor burden and incidence, 70 and 50%, respectively, was observed only in rats that received an uninterrupted diet (i.e., 20 wk) of CLA throughout the post-initiation stages of carcinogenesis. Hence, based on the similarity of these results to the earlier study in which MNU was used and the animals were maintained on a 5% corn oil diet, the chemopreventative activity of CLA is apparently independent of the specific mutational event, the availability of linoleate in the diet, and the level of fat in the diet.

In their most recent paper, these investigators (17) returned to studying the morphologic effect of CLA on the rat mammary gland epithelium and the resultant susceptibility to carcinogenesis. Once again, the researchers used the DMBA model, administering the carcinogen to the rats at 50 d of age. One group of rats was fed a 1% CLA diet from weaning until ~50 d of age. The second group received the 1% CLA diet from 55 d of age until study termination at 21 wk post-DMBA. The third group consumed the 1% diet from weaning until the end of the experiment. CLA was not added the diet for the control group. Each of the four groups consisted of 30 rats. CLA feeding reduced total tumor number by about the same degree, 49–57%, regardless of the dosing regimen. Similarly, tumor incidence was reduced by relatively the same magnitude, 34–46%, regardless of the dosing regimen. It is instructive to point out that continuous CLA feeding (i.e., the third group) produced no greater reduction in cancer suppression than limited feeding. Hence, no additive effect was observed when CLA was administered during active mammary gland maturation and during the period of tumor progression. Finally, these results utilizing the DMBA model, combined with similar precarcinogen protection observed with MNU administration, suggest that the protective effect of CLA is independent of carcinogen-specific gene mutations.

Only two studies have investigated the effect of CLA on mammary carcinogenesis without employing the chemically induced model. In the first, Visonneau et al. (18) examined the effect of feeding CLA on transplanted human breast adenocarcinoma cells in severe combined immunodeficient mice (SCID). In this study, SCID mice were fed a 1% CLA diet for 2 wk before subcutaneous injection with MDA-MB468 cells and continued to consume the CLA-containing diet until study termination at either wk 9 or 14 (n = 2 for each time period). Controls received the same diet without CLA supplementation (total n = 5). Although only a few mice were employed in what should be considered a pilot study, the data were sufficiently robust to observe CLA-induced significant reductions in tumor weight and area at both time points. Additionally, the researchers found evidence of CLA inhibition of metastatic spread to the lung, bone marrow, and peripheral blood. These findings deserve additional follow-up studies to confirm the observations on tumor inhibition and to investigate further the effect of CLA on the histomorphologic features of human breast adenocarcinoma tumors.

The second study also employed transplanted mammary tumor cells in mice, but differed from the previous one in many aspects. Rather than using immune-compromised mice, Wong et al. (19) chose a normal strain of female mice (BALB-c) and implanted the murine mammary tumor cells, WAZ-2T (-SA). Female mice were fed a semipurified synthetic diet containing 0, 0.1, 0.3, or 0.9% CLA. Dietary CLA and safflower oil were maintained at 0.9% (w/w) by replacing the safflower oil with the added CLA. Corn oil was added at 4.1% to bring the total fat level up to 5%. Mice were fed their experimental diets (n = 20/group) for 2 wk before tumor cell injection. Daily tumor measurements began 21 d later and continued until study termination at 45 d post-injection. Results indicated that CLA at any dose did not affect tumor volume, final tumor weight, tumor incidence, or latency. The authors concluded that CLA did not reduce the growth of this transplanted mammary tumor cell line. This particular study does have two important considerations that bear attention. The WAZ-2T (-SA) cell line is an extremely aggressive metastatic cell line that normally produces tumors in 100% of challenged mice within 35 d of injection. Second, this is a hormone-nonresponsive cell line, which may explain the lack of an effect by CLA, also observed with *in vitro* experiments utilizing the estrogen-unresponsive MDA-MB-231 cell line (see section on *in vitro* tumor inhibition).

Skin Cancer

Although not nearly equal in extent to the investigations on mammary cancer inhibition, the effect of CLA on skin carcinogenesis has received a fair amount of examination and was the tumorigenic site of the first study to demonstrate an anticancer effect of CLA. From the many hundreds of chemicals in a partially purified extract from fried ground beef, Pariza and co-workers (1) were able to identify a substance that exhibited anticarcinogenic activity. Using the classic two-stage mouse epidermal carcinogenesis model, Pariza and co-workers were able to identify this substance as CLA. In this experiment, synthetically prepared CLA was topically applied 7 d (20 ng/mouse), 3 d (20 ng/mouse), and 5 min (10 ng/mouse) before dermal treatment with 50 nmol DMBA per mouse (n = 20/group). One week later, the female CD-1 mice received twice weekly topical treatments with 6 µg 12-*O*-tetradecanoyle-13-acetate (TPA) until study termination at 16 wk post- DMBA. Compared with control mice (n = 30), CLA-treated mice had an ~15% lower papilloma incidence and 50% fewer papillomas per mouse.

Using this same model of mouse skin carcinogenesis, in which the initiation and post-initiation stages of carcinogenesis are operationally and temporally separated, (20) confirmed and extended the findings of Pariza and colleagues by examining whether increasing the dietary levels of CLA produced a corresponding reduction in skin tumors. Using synthetic CLA, Belury *et al.* (20) supplemented the basal AIN-76 diet with 0, 0.5, 1.0, or 1.5% (w/w) CLA at the expense of dextrose. Female SENCAR mice were fed the 0% CLA (control) diet for 3 wk before a single dorsal application with DMBA (10 nmol in 100 µL acetone), and then continued on the diet for another week. Mice were then switched to experimental diets and maintained on their respective diets until study termination (n = 36/group). Four weeks after initiation with the carcinogen DMBA, the tumor promoter TPA (1 µg in 200 µL acetone) or 200 µL acetone vehicle (n = 16 mice/diet) was applied to the same dorsal area as the DMBA twice weekly for 25 wk. The study was terminated 35 wk after start of the TPA treatment. Results revealed that 24 wk after TPA treatment was begun, papilloma yield (average number of tumors/mouse) was significantly reduced for the 1.0 and 1.5% CLA groups by 28 and 39%, respectively. There was a modest reduction in papilloma incidence of ~15% for mice fed the 1.5% diet. There appeared to be no effect of CLA on tumor size. In addition, CLA feeding did not affect carcinoma incidence or number. Although CLA feeding reduced body weight by ~10%, starting at wk 12, the data did not support a direct correlation between CLA-induced body weight reduction and tumor inhibition. More importantly, this study confirmed the findings by Ip *et al.* (11) that dietary CLA protected against cancer, and that this effect was independent of any anti-initiator or blocking action of CLA. Additionally, these results would seem to suggest that CLA has less biopotency against mouse skin carcinogenesis and perhaps a less steep dose-response curve than for mammary cancer inhibition.

Forestomach Cancer

Although its been nearly 10 years since CLA was shown to be protective against mouse forestomach tumors by Ha *et al.* (21), this remains the sole study of this cancer site. As a follow-up to their initial study on CLA cancer inhibition using the mouse skin carcinogenesis model, Pariza and colleagues decided to test the anticancer activity of CLA by using the mouse benzo(a)pyrene-induced forestomach cancer model. Three groups (n = 25) of female ICR mice received the following intragastric (i.g.) treatments on Monday and Wednesday of each week: (i) 0.1 mL synthetic CLA plus 0.1 mL olive oil; (ii) 0.1 mL linoleate plus 0.1 mL olive oil; and (iii) 0.1 mL olive oil alone (Experiment 1) or plus 0.1 mL saline (Experiments 2 and 3). On Friday, all animals were given benzo(a)pyrene (2 mg; i.g.). This sequence was repeated for 4 wk, with the study terminated at 22 wk post-initiation with the first benzo(a)pyrene dose. Results from three replicate experiments found that CLA treatment, compared with linoleate controls, reduced the number of forestomach tumors by 51–60% and the tumor multiplicity by 39–56%. CLA reduced tumor incidence vs. linoleate and olive oil controls, in two of three experiments, by 21 and 22%, respectively.

Intestinal Cancer

Although there have been two studies that have addressed the effect of CLA on chemically induced intestinal carcinogenesis, only one study was of sufficiently long duration to observe direct effects on intestinal tumorigenesis. In the shorter study, Liew *et al.* (22) demonstrated the potential for dietary CLA to inhibit colon tumors in male Fisher 344 (F344) rats by measuring preneoplastic changes, i.e., aberrant crypt foci (ACF). To initiate colonic carcinogenesis, the investigators used the heterocyclic amine, 2-amino-3-methylimidazo[4,5-*f*]quinoline (IQ), which has been shown to produce tumors in F344 rats at several sites, including the liver, skin, small intestine, and colon. The rats were gavaged with either CLA or safflower oil (SFO) every other day for a 4-wk period (n = 10/group). To facilitate the uptake of CLA, an equal volume of olive oil was used as a carrier. The safflower group also received olive oil. The dose of CLA was calculated to provide an average daily intake equivalent to that consumed by rats consuming a diet containing 0.5% CLA [based on the study by Ip et al. (12), a 300-g rat fed a 0.5% CLA diet consumes ~75 mg CLA/d]. During wk 3 and 4, IQ was given on alternate days of CLA or safflower treatment. The carcinogen was given at a dose of 100 mg/kg BW in 0.1 mL vehicle composed of 55% methanol in saline. The study was terminated at the end of wk 16, and colons were excised and scored for aberrant crypt foci (ACF). Ten control rats were given the IQ without any dietary treatment, and all 10 developed ACF. Groups of rats given the CLA or SFO treatment, but no IQ, did not develop any ACF (n = 5/group). Correcting for the number of rats bearing ACF, the results indicated that CLA treatment vs. no treatment reduced the number of ACF per rat by 74%. However, CLA treatment was not significantly different from SFO treatment with regard to ACF/rat or aberrant crypts/rat, although SFO treatment was also not different from IQ alone. A larger number of rats per treatment group might have yielded detectable differences between the CLA and SFO treatments.

In the second study on intestinal tumorigenesis, which was not a peer-reviewed study, male F344 rats were fed an AIN-76A basal diet supplemented with 0.5, 1.0, or 1.5% CLA (23). Controls received the same diet without CLA (n = 33). The rats consumed their experimental diets (n = 32–35/group) 2 wk before three weekly treatments with azoxymethane (15 mg/kg BW; subcutaneous). Results indicated that the incidence and multiplicity of small, large, or total intestinal tumors were not affected by CLA. Tumor size and location within the intestinal tract were also not affected. It should be noted that the tumor distribution was unusual in this experiment in that there was a greater incidence of tumors in the small intestine than in the large intestine. Normally, this AOM treatment regimen produces the opposite effect, i.e., a greater incidence of tumors in the large intestine than in the small intestine.

Prostate Cancer

All three of the studies that examined the effect of CLA on prostate tumors utilized transplanted DU145 human prostate cancer cells; however, each study employed a different animal model. It should be noted that of the three studies, only the study by Cesano et al. (24) has been published and hence peer reviewed. In this study, Cesano and co-workers subcutaneously implanted SCID mice with the prostate cancer cells (5×10^6 cells/mouse) 4 d after pretreatment with etoposide (30 mg/kg BW) to inhibit their innate immunity. Two weeks before injection, the mice were given their experimental diets (n = 10/group). The control group received a commercial food pellet diet containing 1.7% linoleate and 4.4% total fat. The CLA group received the same diet supplemented with 1% CLA. A third group received an "extra" 1% linoleate added to the diet. All diets were autoclaved, and GC fatty acid analysis showed no consequential change in the CLA or linoleate. Mice were maintained on their diets until study termination 12 wk after tumor cell injection.

Tumors became visible starting 2 wk post-inoculation and quickly grew to the size of ~1500 mm^3 by the end of the study. Results indicated significant reductions in tumor growth in CLA-fed mice compared with controls or linoleate-fed mice. At study end, average tumor volume in the CLA group was reduced by ~70 and 75% compared with the control and linoleate-fed groups, respectively. Excised tumor weight in the CLA group, compared with the control and linoleate groups, similarly reflected these reductions.

Despite this encouraging result on the inhibition of prostate cancer cell growth by CLA, two other studies (Rose, D.P. and Scimeca, J.A., unpublished) failed to show inhibition of prostate cancer growth by CLA. As with the study of Cesano *et al.* (24), both of these studies employed transplanted DU145 human prostate cancer cells in immunocompromised athymic mice, although the site of implantation differed. The first study was an orthotopic model in which the cells were injected directly into the prostate gland, thus allowing for localized interactions with adjoining tissues. The second study was a subcutaneous model, and the tumor cells were injected with a supporting Matrigel matrix, which promotes new blood formation, thus allowing for an effect on angiogenesis. Both studies employed the same AIN-76A diet, containing 5% corn oil (w/w), with and without 1% CLA (w/w) (n = 25/group). Mice consumed experimental diets 1 wk before tumor cell implantation (10^6 cells/mouse) and continued for another 11 wk until study termination. Neither study found an effect of CLA on terminal tumor weight (Table 34.2).

In Vitro Tumor Inhibition

Breast Cancer Cells

The majority of studies investigating the *in vitro* effect of CLA on cancer cell lines have utilized the estrogen-responsive MCF-7 human breast cancer cell line. Shultz *et al.* (25) were the first to examine the *in vitro* effect of CLA on MCF-7 growth. They incubated

TABLE 34.2. Effect of Conjugated Linoleic Acid (CLA) on Transplanted DU145 Prostate Tumor Cells[a]

Diet	Mice[b] (n)	Body weight (g)	Body weight – Tumor weight (g)	Tumor weight (g)
Subcutaneous tumors in Matrigel				
5% Corn oil	19	31.2 ± 2.9[c]	30.0 ± 3.5	1.25 ± 1.23
1% CLA	22	31.0 ± 1.6	30.0 ± 1.9	1.00 ± 0.88
Intraprostatic tumors				
5% Corn oil	22	27.3 ± 3.8	23.6 ± 4.2	3.71 ± 1.55
1% CLA	22	26.7 ± 3.2	23.6 ± 3.7	3.09 ± 1.74

[a]Data are from Rose, D.P. and Scimeca, J.A., unpublished.
[b]Differences between groups reflect early deaths.
[c]Values are means ± SD.

MCF-7 cells with physiologic concentrations of CLA (1.78 to 7.14×10^{-5} mol/L; corresponding to a concentration from 5 to 20 µg/mL) for 12 d and found that CLA reduced cancer cell growth compared with controls by 54% at 1.78×10^{-5} mol/L and by 100% at both 3.57×10^{-5} mol/L and 7.14×10^{-5} mol/L.

In a follow-up study by Shultz and associates (26), the antiproliferative effect of CLA on MCF-7 cells was compared with the stimulatory effect of LA. The investigators replicated the results from their first study and concluded that CLA inhibited the growth of MCF-7 cells in a dose- and time-dependent manner and was cytostatic at 1.78×10^{-5} mol/L and cytotoxic at $3.57–7.14 \times 10^{-5}$ mol/L.

For their third study, Shultz and colleagues (27) compared the antiproliferative effect of CL on MCF-7 cells with normal human mammary epithelial cells (HMEC). Unlike the first two studies, the cell lines were incubated only with CLA (1.78×10^{-6} to 3.57×10^{-5} mol/L; corresponding to a concentration ranging from 0.5 to 10 µg/mL) for 3 d before cell counting. Once again, compared with controls, CLA inhibited MCF-7 cell growth (11–43%) in a dose-dependent manner. However, CLA also inhibited HMEC cell growth by 18–37% compared with controls over the same concentration range.

Research conducted by Durgam and Fernandes (28) also found that incubating MCF-7 cells for 4 d with CLA at a concentration of 3.5×10^{-5} mol/L inhibited growth by 65–80% when measured by viable cell count and thymidine incorporation assay. However, the investigators also determined that CLA was ineffective at the same concentrations (1.7 to 7.1×10^{-5} mol/L) in reducing the growth of the estrogen-unresponsive MDA-MB-231 human breast cancer line, possibly indicating some estrogen-related interaction. Finally, an experiment was conducted to determine if the inhibitory effect of CLA on MCF-7 cell growth could be reversed. Results indicated that cells grown with 3.5×10^{-5} mol/L CLA for 4 d and then switched to normal media began to increase their growth rate, perhaps hinting at some inhibitory interaction of the CLA with mitogenic factors in the media. Upon the removal of CLA, the normal mitogenic pathway continued unchecked, resulting in resumed cell growth.

The remaining study, by Desbordes and Lea (29), examined the effect of CLA on MCF-7 cells and found an antiproliferative effect as measured by thymidine incorporation; however, the effect was not nearly as potent. A reduction (100%) in cell growth was observed only at 5×10^{-4} but not at 1×10^{-4} mol/L. This impotent effect of CLA under the conditions of this experiment may be due in part to the short incubation time of only 1 d. The investigators also found very similar effects of CLA on another human breast cancer cell line, T47D.

Prostate Cancer Cells

In contrast to the research on breast cancer cell lines, the investigation of CLA on prostate cell lines is limited. Watkins and associates (30) established canine prostate cancer lines, which they have claimed are immortalized and tumorigenic in athymic mice. *In vitro* proliferation assays were performed on primary prostate tumor cells and lung metastatic tumor cells in two separate experiments. Cell lines were incubated for 4 d with various concentrations of CLA (20–200 μmol/L). The primary prostate tumor cell line showed a dose-responsive reduction in proliferation compared with cells incubated in normal media, with a 75% inhibition at 200 μmol/L CLA concentration. Cell lines derived from lung metastasis were more variable with only one of three showing responsiveness to CLA antiproliferative activity. Overall, these results are not very compelling, especially given the relatively high concentration of CLA required for inhibition of tumor cell growth.

Miscellaneous Tumor Cell Lines

CLA has shown to be effective in inhibiting the growth of a variety of tumor cell lines in addition to human breast and canine prostate. Shultz et al. (25) showed a 60% and 37% reduction by 7.1×10^{-5} mol/L CLA on human malignant melanoma and colorectal tumor cell growth, respectively. Yoon et al. (31) reported that CLA inhibited the growth of two human liver cancer cell lines, HepG2 and SNU-128. DesBordes and Lea (29) found a 35% reduction by 1×10^{-4} mol/L CLA in the rat hepatoma cell line, 7800NJ. Results from work by Schonberg and Krokan (32) indicated that CLA was effective in suppressing the growth of three different lung adenocarcinoma cell lines (A-427, SK-LU-1, and A549), but not a human glioblastoma cell line (A-172). Dose-dependent effects were in the range of 1 to 4×10^{-5} mol/L, with a >50% inhibition in growth at 20 μmol/L for the most sensitive cell line, A-427. Finally, Visonneau et al. (33) reported an inhibitory effect of CLA (at concentrations from 10^{-4} to 10^{-6} mol/L) on the growth of a variety of cell lines including the following: breast, prostate, colon, and ovary cancer cells; melanoma; leukemia; mesothelioma; and glioblastoma.

In summary, results from the majority of *in vitro* studies have indicated that it would be worthwhile to examine the anticancer efficacy of CLA at additional tissue sites not yet studied *in vivo*.

Mechanistic Studies

CLA has been associated with a variety of biological events that are involved in the three broad stages of carcinogenesis, i.e., initiation, promotion, and progression. A number of studies have investigated the effect of CLA on these biological events in an attempt to elucidate the mechanism of action. This is an evolving area of research that has identified to date a few pieces of the puzzle, although the picture is far from complete. For that reason, no attempt will be made to present a unified mechanistic theory, and therefore the various biological events will be treated separately. Research published before 1994 has been for the most part reviewed and summarized by Scimeca et al. (23) and Belury (34). Since that time, however, a number of studies that shed further light on the role CLA plays in carcinogenesis have been published; these will be the focus of this section.

Antioxidant Capacity

Although initial suggestions were that CLA possessed some antioxidant capacity that may play some role in its anticancer activity, recent data have been to the contrary. Strong evidence against CLA having antioxidant activity is found in two studies using defined *in vitro* experiments and precise analytical methods. With the use of a model membrane system, van den Berg *et al.* (35) examined the antioxidant properties of CLA and found that it was not a free radical scavenger and did not appear to be converted to a metal chelator. Chen *et al.* (36) examined the antioxidant capacity of CLA in three forms, i.e., free fatty acids (the form used in most biological studies), methyl esters (CLAME), and triacylglycerols. The investigators measured the ability of the various forms of CLA to affect the oxidation of heated canola oil. They found that CLA and CLAME exerted a dose-dependent prooxidant activity in the oil. The CLA-containing triacylglycerol had no effect, which is an important observation because this is the form in which CLA would naturally be found in dairy and meat products. Both sets of investigators concluded that a role for CLA as an antioxidant was not plausible.

Ip *et al.* (14) provided supporting *in vivo* data against a chemopreventative effect of CLA through an antioxidant mechanism. Although dietary treatment with CLA (1% w/w) resulted in lower levels of malondialdehyde, which is an end product of lipid peroxidation, it failed to change the levels of 8-hydroxydeoxyguanosine, a marker of oxidatively damaged DNA, in mammary tissue. An effect of CLA on lipid peroxidation as indicated by a reduction in malondialdehyde is nonspecific to carcinogenesis, unlike the absence of an effect on 8-hydroxydeoxyguanosine, which has been shown to be linked to errors in DNA replication.

DNA Adduct Formation

CLA exerts effects on *in vivo* carcinogen adduct formation in several organ sites in both mice and rats (see Table 34.3 for a summary of studies). Zu and Schut (37) were the first to show an effect of CLA on the formation of IQ-DNA adducts. CLA was administered by gavage to CDF_1 mice every other day for 45 d (initially 50 µL, then 100 µL); control mice received trioctanoin (n = 4 for all groups). On study d 46, all animals were gavaged with a 50 mg/kg BW bolus of IQ, and tissues were collected 24 h later. Tissue IQ-DNA adducts were determined by ^{32}P-post-labeling. Results indicated differential effects for female and male mice. CLA treatment inhibited IQ-DNA adducts compared with controls in the liver for both sexes, but was effective only in the lungs, large intestine, and kidney in female mice.

Liew *et al.* (22) also examined the effect of CLA on IQ-DNA adduct formation, but administered the CLA to male F344 rats, utilizing the same experimental protocol employed in their ACF study (see *in vivo* section). Briefly, this protocol entailed the gavage administration of either CLA or safflower every other day for a 4-wk period (n = 7/group). During the last 2 wk, the IQ was given by gavage at a dose of 100 mg/kg BW on alternating days with the treatment. Tissues were collected 6 h after the final IQ dose, and ^{32}P-post-labeling was used to quantify IQ- DNA adducts. CLA

TABLE 34.3. Inhibition of Carcinogen-DNA Adducts by Conjugated Linoleic Acid (CLA)

Reference	Carcinogen[a]	Species	Organ	% Reduction in RAL[b] Males	Females
Zu and Schut (37)	IQ	CDF_1 mice	Liver	43	32
			Lungs	NS[c]	74
			Stomach	NS	NS
			Small intestine	NS	NS
			Large intestine	NS	39
			Kidney	NS	95
Liew et al. (22)	IQ	F344 rats	Liver	NS	NT[d]
			Colon	40	NT
Schut et al. (38)	PhIP	F344 rats	Liver	NT	58
			Mammary gland	NT	68

[a]IQ, 2-amino-3-methylimidazo[4,5-f]quinoline; PhIP, 2-amino-1-methyl-6-phenylimidazo[4,5-b]pyridine.
[b]RAL, relative adduct labeling.
[c]NS, nonsignificant determination.
[d]NT, not tested.

treatment reduced the relative adduct labeling in the colon (which supported the data indicating an inhibition of ACF; see *in vivo* section), but not in the liver, which conflicted with the finding by Zu and Schut (37).

In a second study from Schut's laboratory on DNA adduct formation in the liver and mammary gland, the effect of CLA was determined through dietary administration (38). As with the study of Liew et al. (22), DNA adduct formation was studied in F344 rats, but in females rather than males. In addition, a different carcinogen, 2-amino-1-methyl-6-phenylimidazo[4,5-b]pyridine (PhIP), was utilitized. In F344 rats, PhIP has been shown to produce mammary tumors in females and colon tumors in males. Rats received AIN-76A diets with 0, 0.1, 0.5, or 1.0% CLA for 2 wk before addition of PhIP (0.04% w/w) to the diet for four additional weeks. Rats then received a PhIP-free diet until study termination at 1, 8 or 15 d after PhIP removal [n = 4 or 2/(group time period) for liver or mammary gland DNA-adduct determination, respectively]. CLA at dietary level of 1% was effective in inhibiting PhIP-DNA adduct formation in both the liver and mammary gland, but only on d 1 after removal of the carcinogen.

Taken together, these studies provide suggestive evidence that CLA administered during the initiation stage of carcinogenesis can act as a blocking agent. The anticancer effect of CLA on certain organs such as the liver and large intestine, in addition to the mammary gland, deserves attention. In addition, there is a need to understand more completely the mechanism of the blocking action of CLA.

Carcinogen Activation and Detoxification

Although there is evidence that CLA has chemopreventative effects in animal models, the mechanism by which it exerts this action is unknown. ˍ et al. (11) investigated the effect of CLA on two detoxifying enzymes, glutathione-S-transferase and

UDP-glucuronyl transferase. One month of feeding CLA at dietary levels from 0.25 to 1.5% failed to affect liver or mammary gland UDP-glucuronyl transferase activities or liver UDP-glucuronyl transferase activity. Liew et al. (22) examined the mechanistic aspects of CLA and concluded that the inhibitory action cannot be attributed to simple substrate-ligand interaction or to electrophile scavenging and antioxidant activity. They also studied the *in vitro* and *in vivo* ability of CLA to modulate the enzymatic activation of a procarcinogen (i.e., IQ) and concluded that CLA can exert favorable activity.

Estrogen-Related Activity

Despite the well-established connection between mammary gland carcinogenesis and estrogen regulation, little research has pursued this line of investigation with regard to CLA and cancer inhibition. This is largely because the available data are weak and lack compelling support for further inquiry into the effect of CLA on estrogen regulation. What little evidence exists for a possible role lies with *in vitro* work by Durgam and Fernandez (28) that compared the effect of CLA on estrogen-responsive and estrogen-unresponsive mammary tumor cells (see *in vivo* section). These investigators proposed that CLA inhibited the expression of the protooncogene c-*myc* in estrogen-responsive MCF-7 cells. Because expression of c-*myc* is known to be modulated by estrogens among other factors, the authors suggested that the reduction in c-*myc* expression observed in their study may relate to a CLA involvement in an estrogen-, or some other growth factor-regulated mechanism. However, supporting evidence for an effect of CLA on any specific growth factor related to carcinogenesis has yet to appear.

Eicosanoid-Related Activity

Given the well-established metabolic pathway of eicosanoid biosynthesis from linoleate and other polyunsaturated fatty acids (PUFA) plus the biological role eicosanoids play in carcinogenesis, the similarity in structure between the CLA isomers and the PUFA linoleate made this an obvious avenue to investigate as a possible mechanism for the anticarcinogenic action of CLA. Belury and colleagues have especially pursued this line of research, investigating the hypothesis that CLA inhibits cancer, and more specifically skin carcinogenesis, *via* an eicosanoid-mediated mechanism. The metabolism of CLA is beyond the scope of this review, but it is sufficient to note that CLA has been shown to be incorporated into membrane phospholipids and neutral lipids in a number of tissues (1,14,22,39–41). Furthermore, CLA can serve as a substrate for fatty acid enzymes, such as Δ^6 desaturase, elongase, and Δ^5 desaturase (42,43).

It is instructive to note that the incorporation of CLA into the membrane phospholipid vs. the neutral lipid fraction does not occur to the same relative degree in different tissues. In the rat mammary gland in which CLA treatment caused a relatively slight perturbation of the phospholipid fraction, including an inability to displace linoleate or arachidonate, cancer inhibition by CLA does not appear to be tightly linked to modifications in the eicosanoid pathway (14,15). The same cannot be said for the CLA inhibition of skin carcinogenesis in which research by Belury and colleagues (40) has shown that in cultured murine keratinocytes, CLA treatment resulted in reduced prostaglandin E2 synthesis and ornithine decarboxylase activity (ODC) induced by the tumor promoter TPA, compared with linoleate (LA) and arachidonate (AA) treatment. Because prostaglandin E is known to modulate ODC activity, which in turn is an established biomarker for tumor promotion, these investigators concluded that CLA inhibits skin carcinogenesis at least in part through an eicosanoid-mediated mechanism. *In vivo* data are still needed to establish whether an effect of CLA on eicosanoid biosynthesis occurs through direct competition with LA or *via* the formation of CLA–metabolites.

Mammary Gland Development and Maturation

Recently, there has been evidence that CLA can exert an effect on mammary gland development and maturation, with apparent long-lasting beneficial consequences. Ip, Thompson and colleagues (13,17) showed that feeding female Sprague-Dawley rats an AIN-76A diet containing 1% CLA (w/w), for a short time period (3–5 wk) from weaning until carcinogen injection, conferred long-lasting protection against mammary gland tumorigenesis (see *in vivo* section). This short-term feeding during the critical time of mammary gland morphogenesis and development appears to make the mammary gland epithelial cells less susceptible to neoplastic transformation. Potential protective changes by CLA in the mammary gland include the following: (i) a modification in the rate of epithelial cell growth and death; (ii) a morphogenic alteration in the mammary gland composition; and (iii) an indirect effect on the mammary gland through an action on the surrounding adipose tissue. All three questions were addressed in the paper by Thompson *et al.* (17). The proliferative activity of the rat mammary epithelium was measured by bromodeoxyuridine (BrdUrd) labeling. Weanling rats were fed for 1 mo a basal diet with or without 1% CLA (n = 12/group). Mammary epithelial cells were categorized into one of three compartments of the gland, i.e., duct, terminal end bud, or lobulalveolar bud. Results indicated that the percentage of cells that reacted positively to the labeling was reduced by 29% for both the terminal end bud and the lobulalveolar compartments. The terminal end bud cells are the progenitors for the lobulalveolar bud cells, and the former serve as the primary target site for carcinogens that induce mammary tumors. The morphology of the mammary gland after CLA treatment was also studied by using whole-mount densitometric analysis of digitalized images. Once again, weanling rats were fed for 1 mo a basal diet with or without 1% CLA (n = 10/group). Results indicated that CLA treatment caused a 21% reduction in density of the mammary gland epithelium, without any apparent effect on the area of the mammary fat pad occupied by the mammary epithelium. The lipid content (neutral, phospholipid, and total) of the mammary fat pad was unaffected by CLA feeding. The investigators concluded that CLA had no apparent effect on the surrounding fat tissue of the mammary epithelium but was able to suppress the lateral branching, as well as the proliferative rate, of the terminal end bud cells, thus decreasing the risk from chemical initiation of carcinogenesis. Whether CLA is influencing mammary epithelial cell growth and function *via* a direct action or is doing so indirectly through an action of mammary fat pad adipocyte remains to be investigated.

Promotion

Data from Belury's laboratory (40,44,45) indicated that CLA treatment inhibited skin tumorigenesis, at least in part, by alteration of the promotional effect of TPA. Furthermore, this antipromotional effect of CLA may be exerted *via* a reduction in ODC activity and appears to be an eicosanoid-mediated activity (see above). ODC activity, which is a classical biomarker for tumor promotion, has been shown to be reduced by CLA treatment in three murine tissues, i.e., forestomach, keratinocytes, and hepatocytes. Finally, studies by Ip and colleagues have also demonstrated that CLA can exert an antipromotional effect on mammary tumorigenesis based on evidence of cancer inhibition obtained through the post-initiation feeding of CLA (13,16; see *in vivo* section). However, compared with skin carcinogenesis, much less attention has been paid to the mechanism of this antipromotional effect of CLA in mammary carcinogenesis.

Peroxisome Proliferator-Activated Receptors

Peroxisome proliferator-activated receptors (PPAR) belong to a superfamily of nuclear hormone receptors that are ligand-dependent transcription factors. Ligands for PPAR include the hypolipidemic drug, clofibrate, perfluorodecanoate, and some PUFAS like arachidonate and linoleate. Evidence is emerging that CLA may also belong to this lipid class of molecules that can activate PPAR. Belury, Vanden Heuval, and co-workers have shown that dietary CLA treatment can apparently activate the PPARγ subtype in the male Zucker diabetic fatty *fa/fa* rat (46) and activate the PPARα subtype in the female SENCAR mouse (45). The apparent effects of CLA on PPARα are especially interesting because they involve transcription of genes encoding enzymes involved with lipid metabolism. Specifically, Belury *et al.* (45) demonstrated that feeding female SENCAR mice diets supplemented with 1.0 or 1.5% CLA for 6 wk dramatically increased liver mRNA levels of acyl-CoA oxidase, cytochrome P4504A1, and liver fatty acid binding protein, compared with controls. Measurement of acyl-CoA oxidase protein accumulation confirmed that the CLA-associated changes in mRNA were carried through to protein expression. These tantalizing findings provide impetus for further research into understanding the role of CLA activation of PPAR in cancer inhibition.

Conclusion

The evidence that CLA is an anticarcinogen is abundant and certain. In more than 10 years of research on the *in vivo* inhibition of tumor development and growth, involving nearly two dozen experiments and >1000 laboratory animals, only three experiments have failed to show positive results. CLA treatment is efficacious in several different models of carcinogenesis in several strains of both mice and rats. Tumorigenesis originating from five different tissues responds favorably to CLA treatment. No tissue site of carcinogenesis has been studied more extensively than the mammary gland. Dietary CLA treatment has been shown to be effective against mammary gland carcinogenesis under a wide range of experimental regimens, such as duration and timing of feeding. CLA effectiveness is independent of varied dietary conditions, such as level or type of fat. Dose-response studies have demonstrated that CLA is effective against DMBA-induced mammary cancer at levels as low as 0.1% in the diet. *In vitro* research has been fairly extensive, with most of the effort utilizing the MCF-7 human breast cancer line. Many other cancer cell lines were also tested *in vitro* and found responsive to CLA treatment.

In his principles of chemoprevention, Wattenberg (47) divided inhibitors of carcinogenesis into the following three categories on the basis of the time in the carcinogenic process in which they exert their effect: (i) compounds that prevent the formation of carcinogens from precursor compounds; (ii) blocking agents; and (iii) suppressing agents. Evidence from mechanistic and tumorigenesis studies seems to indicate that CLA possesses the latter two types of chemopreventative action. Data are lacking concerning the ability of CLA to affect the formation of carcinogens from precursor compounds; hence, no conclusions can be drawn. Studies on the inhibition of DNA adduct formation provide some evidence that CLA treatment can act as a blocking agent. Wattenburg (47) characterizes suppressing agents as those compounds that inhibit carcinogenesis when administered after carcinogen treatment. Evidence for classifying CLA as a suppressing agent comes from many different studies, but the studies by Ip and colleagues on the effect of postcarcinogen CLA treatment on mammary tumorigenesis stand out. Finally, CLA exerts an effect on mammary gland development and maturation that cannot be pigeon-holed into one of Wattenburg's categories, and therefore belongs to a fourth category of anticancer agents whose action is on "target tissue susceptibility." It is tempting to speculate that these effects of CLA on the developing mammary gland, as well as other anticarcinogenic effects, are related to the activation of PPAR. However, multiple parallel mechanistic pathways are a more likely explanation, given the multifaceted properties of CLA, the fact the CLA is more than a single molecule, and that carcinogenesis is a multistage process. Nonetheless, investigation of the relationship between CLA and PPAR activation offers to shed light on some of the biological action of CLA.

The role of dietary CLA in human cancer risk has received little attention. As indicated above, there are some data showing a favorable effect of CLA on human cancer cells. More extensive evidence from human studies is generally lacking, although there are a few epidemiologic studies that have focused on evaluating the role of dairy products in relation to various cancers, and at least one has found an inverse relationship between milk intake and breast cancer (48).

Among anticarcinogens, CLA is unique for the following reasons: (i) It is an animal-derived product with its highest levels found in ruminant meat and dairy products; (ii) it exerts a potent action through dietary intervention under a wide range of conditions; and (iii) it possesses multiple modes of action. In addition, evidence indicates that the dietary administration of CLA at elevated levels is safe (49). The time has come to conduct clinical studies with CLA in order to judge its potential as a chemopreventative agent for humans.

Acknowledgments

The author is indebted to Dr. Kreutler for her helpful advice and enduring support for his involvement in CLA research. The author also thanks Ms. Sue Eickhoff for her professional assistance in the preparation of this manuscript. Finally, the author is grateful for the support and understanding of his family, Diana, Maggie, and Danny, during the preparation of this review.

References

1. Ha, Y.L., Grimm, N.K, and Pariza, M.W. (1989) Newly Recognized Anticarcinogenic Fatty Acids: Identification and Quantification in Natural and Processed Cheeses, *J. Agric. Food Chem. 37,* 75–81.
2. Chin, S.F., Liu, W., Storkson, J.M., Ha, Y.L., and Pariza, M.W. (1992) Dietary Sources of Conjugated Dienoic Isomers of Linoleic Acid, a Newly Recognized Class of Anticarcinogens, *J. Food Compos. Anal. 5,* 185–197.
3. Shantha, N.C., Ram, L.N., O'Leary, J., Hicks, C.L., and Decker, E.A. (1995) Conjugated Linoleic Acid Concentrations in Dairy Products as Affected by Processing and Storage, *J. Food Sci. 60,* 695–697.
4. Yurawecz, M.P., Roach, J.A.G., Sehat, N., Mossoba, M.M., Kramer, J.K.G., Fritsche, J., Steinhart, H., and Ku, Y. (1998) A New Conjugated Linoleic Acid Isomer, 7 *trans*, 9 *cis*-Octadecadienoic Acid, in Cow Milk, Cheese, Beef and Human Milk and Adipose Tissue, *Lipids 33,* 803–809.
5. Griinari, J.M., Nurmela, K.V.V., Corl, B.A., Chouinard, Y.P., and Bauman, D.E. (1998) The Endogenous Synthesis of Milkfat Conjugated Linoleic Acid (CLA) from Absorbed Vaccenic Acid in Dairy Cows, 89th AOCS Annual Meeting (abstr.), May 10–13.
6. Corl, B.A., Chouinard, Y.P., Bauman, D.E., Dwyer, D.A., Griinari, J.M., and Nurmela, K.V. (1998) Conjugated Linoleic Acid in Milk Fat of Dairy Cows Originates in Part by Endogenous Synthesis from *trans*-11 Octadecenoic Acid, *J. Dairy Sci. 81* (Suppl. 1), 233.
7. Kelly, M.L., Berry, J.R., Dwyer, D.A., Griinari, J.M., Chouinard, P.Y., Vanamburgh, M.E., and Bauman, D.E. (1998) Dietary Fatty Acid Sources Affect Conjugated Linoleic Acid Concentrations in Milk from Lactating Dairy Cows, *J. Nutr. 128,* 881–885.
8. Kelly, M.L., Kolver, E.S., Bauman, D.E., Vanamburgh, M.E., and Muller, L.D. (1998) Effect of Intake of Pasture on Concentrations of Conjugated Linoleic Acid in Milk of Lactating Cows, *J. Dairy Sci. 81,* 1630–1636.
9. Stanton, C., Lawless, F., Murphy, J., and Connolly, B. (1997) Conjugated Linoleic Acid (CLA)—A Health Promoting Component of Dairy Fats, I. Biological Properties of CLA, *Farm Food 7,* 19–20.
10. Sehat, N., Yurawecz, M.P., Roach, J.A.G., Mossoba, M.M., Kramer, J.K.G., and Ku, Y. (1998) Silver-Ion High-Performance Liquid Chromatographic Separation and Identification of Conjugated Linoleic Acid Isomers, *Lipids 33,* 217–221.

11. Ip, C., Chin, S.F., Scimeca, J.A., and Pariza, M.W. (1991) Mammary Cancer Prevention by Conjugated Dienoic Derivative of Linoleic Acid, *Cancer Res. 51,* 6118–6124.
12. Ip, C., Singh, M., Thompson, H.J., and Scimeca, J.A. (1994) Conjugated Linoleic Acid Suppresses Mammary Carcinogenesis and Proliferative Activity of the Mammary Gland in the Rat, *Cancer Res. 54,* 1212–1215.
13. Ip, C., Scimeca, J.A., and Thompson, H. (1995) Effect of Timing and Duration of Dietary Conjugated Linoleic Acid on Mammary Cancer Prevention, *Nutr. Cancer 24,* 241–247.
14. Ip, C., Briggs, S.P., Haegele, A.D., Thompson, H.J., Storkson J., and Scimeca, J.A. (1996) The Efficacy of Conjugated Linoleic Acid in Mammary Cancer Prevention Is Independent of the Level or Type of Fat in the Diet, *Carcinogenesis 17,* 1045–1050.
15. Ip, C., and Scimeca, J.A. (1997) Conjugated Linoleic Acid and Linoleic Acid Are Distinctive Modulators of Mammary Carcinogenesis, *Nutr. Cancer 27,* 131–135.
16. Ip, C., Jiang, C., Thompson, H.J., and Scimeca, J.A. (1997) Retention of Conjugated Linoleic Acid in the Mammary Gland Is Associated with Tumor Inhibition During the Post-Initiation Phase of Carcinogenesis, *Carcinogenesis 18,* 755–759.
17. Thompson, H., Zhu, Z.J., Banni, S., Darcy, K., Loftus, T., and Ip, C. (1997) Morphological and Biochemical Status of the Mammary Gland as Influenced by Conjugated Linoleic Acid—Implication for a Reduction in Mammary Cancer Risk, *Cancer Res. 57,* 5067–5072.
18. Visonneau, S., Cesano, A., Tepper, S.A., Scimeca, J.A., Santoli, D., and Kritchevsky, D. (1997) Conjugated Linoleic Acid Suppresses the Growth of Human Breast Adenocarcinoma Cells in SCID Mice, *Anticancer Res. 17,* 969–973.
19. Wong, M.W., Chew, B.P., Wong, T.S., Hosick, H.L., Boylston, T.D., and Shultz, T.D. (1997) Effects of Dietary Conjugated Linoleic Acid on Lymphocyte Function and Growth of Mammary Tumors in Mice, *Anticancer Res. 17,* 987–993.
20. Belury, M.A., Nickel, K.P., Bird, C.E., and Wu, Y.M. (1996) Dietary Conjugated Linoleic Acid Modulation of Phorbol Ester Skin Tumor Promotion, *Nutr. Cancer 26,* 149–157.
21. Ha, Y.L., Storkson, J., and Pariza, M.W. (1990) Inhibition of Benzo(a)Pyrene-Induced Mouse Forestomach Neoplasia by Conjugated Dienoic Derivatives of Linoleic Acid, *Cancer Res. 50,* 1097–1101.
22. Liew, C., Schut, H.A.J., Chin, S.F., Pariza, M.W., and Dashwood, R.H. (1995) Protection of Conjugated Linoleic Acids Against 2-Amino-3-methylimidazo[4,5-*f*] quinoline-Induced Colon Carcinogenesis in the F344 Rat—A Study of Inhibitory Mechanisms, *Carcinogenesis 16,* 3037–3043.
23. Scimeca, J.A., Thompson, H.J., and Ip, C. (1994) Effect of Conjugated Linoleic Acid on Carcinogenesis, *Adv. Exp. Med. Biol. 364,* 59–65.
24. Cesano, A., Visonneau, S., Scimeca, J.A., Kritchevsky, D., and Santoli, D. (1998) Opposite Effects of Linoleic Acid and Conjugated Linoleic Acid on Human Prostatic Cancer in SCID Mice, *Anticancer Res. 18,* 1429–1434.
25. Shultz, T.D., Chew, B.P., Seaman, W.R., and Luedecke, L.O. (1992) Inhibitory Effect of Conjugated Dienoic Derivatives of Linoleic Acid and Beta-Carotene on the *in vitro* Growth of Human Cancer Cells, *Cancer Lett. 63,* 125–133.
26. Shultz, T.D., Chew, B.P., and Seaman, W.R. (1992) Differential Stimulatory and Inhibitory Responses of Human MCF-7 Breast Cancer Cells to Linoleic Acid and Conjugated Linoleic Acid in Culture, *Anticancer Res. 12,* 2143–2145.

27. Cunningham, D.C., Harrison, L.Y., and Shultz, T.D. (1997) Proliferative Responses of Normal Human Mammary and MCF-7 Breast Cancer Cells to Linoleic Acid, Conjugated Linoleic Acid and Eicosanoid Synthesis Inhibitors in Culture, *Anticancer Res. 17*, 197–203.
28. Durgam, V.R., and Fernandes, G. (1997) The Growth Inhibitory Effect of Conjugated Linoleic Acid on MCF-7 Cells Is Related to Estrogen Response System, *Cancer Lett. 116*, 121–130.
29. Desbordes, C., and Lea, M.A. (1995) Effects of C18 Fatty Acid Isomers on DNA Synthesis in Hepatoma and Breast Cancer Cells, *Anticancer Res. 15*, 2017–2021.
30. Cornell, K., Waters, D.J., Watkins, B., and Robinson, J.P. (1998) Conjugated Linoleic Acid Exhibits Differential Inhibition of Canine Prostate Cells *in vitro*, *Proc. Am. Assoc. Cancer Res. 39*, 590.
31. Yoon, C.S., Ha, T.Y., Rho, J.H., Sung, K.S., and Cho, I.J. (1997) Inhibitory Effect of Conjugated Linoleic Acid on *in vitro* Growth of Human Hepatoma, *FASEB J. 11*, A578.
32. Schonberg, S., and Krokan, H.E. (1995) The Inhibitory Effect of Conjugated Dienoic Derivatives (CLA) of Linoleic Acid on the Growth of Human Tumor Cell Lines Is in Part Due to Increased Lipid Peroxidation, *Anticancer Res. 15*, 1241–1246.
33. Visonneau, S., Cesano, A., Tepper, S.A., Scimeca, J., Santoli, D., and Kritchevsky, D. (1996) Effect of Different Concentrations of Conjugated Linoleic Acid (CLA) on Tumor Cell Growth *in vitro*, *FASEB J. 9*, A869.
34. Belury, M.A. (1995) Conjugated Dienoic Linoleate: A Polyunsaturated Fatty Acid with Unique Chemoprotective Properties, *Nutr. Rev. 53*, 83–89.
35. van den Berg, J.J., Cook, N.E., and Tribble, D.L. (1995) Reinvestigation of the Antioxidant Properties of Conjugated Linoleic Acid, *Lipids 30*, 599–605.
36. Chen, Z.Y., Chan, P.T., Kwan, K.Y., and Zhang, A. (1997) Reassessment of the Antioxidant Activity of Conjugated Linoleic Acids, *J. Am. Oil Chem. Soc. 74*, 719–753.
37. Zu, H.X., Schut, H.A. Inhibition of 2-Amino-3-methylimidazo[4,5-*f*]quinoline-DNA Adduct Formation in CDF1 Mice by Heat-Altered Derivatives of Linoleic Acid, *Food Chem. Toxicol. 30*, 9–16.
38. Schut, H.A.J., Cummings, D.A., and Smale, M.H.E., Josyula, S., and Friesen, M.D. (1997) Adducts of Heterocyclic Amines: Formation, Removal and Inhibition by Dietary Components, *Mutat. Res. Fund. Mol. Microbiol. 376*, 185–194.
39. Belury, M.A., and Kempasteczko, A. (1997) Conjugated Linoleic Acid Modulates Hepatic Lipid Composition in Mice, *Lipids 32*, 199–204.
40. Liu, K.L., and Belury, M.A. (1997) Conjugated Linoleic Acid Modulation of Phorbol Ester-Induced Events in Murine Keratinocytes, *Lipids 32*, 725–730.
41. Kramer, J.K.G., Sehat, N., Dugan, M.E.R., Mossoba, M.M., Yurawecz, M.P., Roach, J.A.G., Eulitz, K., Aalhus, J.L., Schaefer, A.L., and Ku, Y. (1998) Distributions of Conjugated Linoleic Acid (CLA) Isomers in Tissue Lipid Classes of Pigs Fed a Commercial CLA Mixture Determined by Gas Chromatography and Silver-Ion High-Performance Liquid Chromatography, *Lipids 33*, 549–558.
42. Banni, S., Day, B.W., Evans, R.W., Corongin, F.P., and Lombardi, B. (1995) Detection of Conjugated Diene Isomers of Linoleic Acid in Liver Lipids of Rats Fed a Choline-Devoid Diet Indicates That the Diet Does Not Cause Lipoperoxidation, *Nutr. Biochem. 6*, 281–289.
43. Sébédio, J.L., Juaneda, P., Dobson, G., Ramilison, I., Martin, J.D., and Chardigny, J.M. (1997) Metabolites of Conjugated Isomers of Linoleic Acid (CLA) in the Rat, *Biochim. Biophys. Acta 1345*, 5–10.

44. Benjamin, H., Storkson, J.M., Albright, K., and Pariza, M.W. (1990) TPA-Mediated Induction of Ornithine Decarboxylase Activity in Mouse Forestomach and Its Inhibition by Conjugated Dienoic Derivatives of Linoleic Acid, *FASEB J. 4,* A508.
45. Belury, M.A., Moyacamarena, S.Y., Liu, K.L., and Heuvel, J.P.V. (1997) Dietary Conjugated Linoleic Acid Induces Peroxisome-Specific Enzyme Accumulation and Ornithine Decarboxylase Activity in Mouse Liver, *J. Nutr. Biochem. 8,* 579–584.
46. Houseknecht, K.L., Vanden Heuvel, P., Moya-Camarena, S.Y., Portocarrero, C.P., Peck, L.W., Nickel, K.P., and Belury, M.A. (1998) Dietary Conjugated Linoleic Acid Normalizes Impaired Glucose Tolerance in the Zucker Diabetic Fatty *fa/fa* Rat, *Biochem. Biophys. Res. Commun. 244,* 678–682.
47. Wattenburg, L.W. (1985) Chemoprevention of Cancer, *Cancer Res. 45,* 1–8.
48. Knekt, P., Järvinen, R., Seppänen, R., Pukkala, E., and Aromaa, A. (1996) Intake of Dairy Products and the Risk of Breast Cancer, *Br. J. Cancer 73,* 687–691.
49. Scimeca, J.A. (1998) Toxicological Evaluation of Dietary Conjugated Linoleic Acid in Male Fischer 344 Rats, *Food Chem. Toxicol. 36,* 391–395.

Chapter 35
Intake of Dairy Products and Breast Cancer Risk

Paul Knekt[a] and Ritva Järvinen[b]

[a]National Public Health Institute, Helsinki, Finland
[b]Department of Clinical Nutrition, University of Kuopio, Kuopio, Finland

Introduction

Conjugated linoleic acid (CLA) has been hypothesized to provide protection against breast cancer by means of several mechanisms (1). CLA has been shown to inhibit proliferation of human malignant breast cancer cell lines in *in vitro* studies (2). In animal studies, CLA inhibited mammary tumorigenesis (3). As far as we know, no results on the predictive value of tissue concentration or dietary intake of CLA and occurrence of breast cancer in humans have thus far been published. Milk lipid is the richest natural source of CLA (4); accordingly, its intake is directly correlated with the intake of milk and other dairy products. Consumption of milk and dairy products may therefore represent a proxy measure for intake of this fatty acid.

Several reviews have been published, including the associations between intake of dairy products and breast cancer risk (5,6). This study updates the information on the relation between intake of milk and other dairy products and occurrence of breast cancer in analytical epidemiologic studies in humans.

Populations and Methods

Study Population and Study Design

To date, analytical observational epidemiologic studies from ~30 different populations have presented results on consumption of dairy products and breast cancer risk. In this type of study, the investigator looks for associations existing between intake of dairy products and breast cancer occurrence, but makes no changes in the intake of the population. The observational studies considered are of cohort, nested case-control, or case-control design.

In a cohort study, a group of persons whose intake of dairy products has been established is followed up over a period of time with respect to breast cancer occurrence. When a sufficient number of breast cancer cases have occurred, the breast cancer incidence among individuals with high intake of dairy products is compared with that among those with low intake. To date, two cohort studies conducted in Norway (7,8), one in the United States (9), and one in Finland (10) have been completed on the predictive value of dairy product intake in breast cancer occurrence (Table 35.1).

A nested case-control study is a modification of a cohort study in which intake of dairy products for breast cancer cases and controls is determined prediagnostically, and the controls are selected from the entire cohort at risk. Such a design is particularly effective when information has been collected at the baseline examination from the entire cohort in qualitative form; its formalization is resource consuming. This design was applied in two studies (11,12) because key variables were not available (Table 35.1). This allowed the missing data to be abstracted only for cases and their matched controls rather than for the total population. The study by Mills et al. (11) was based on 142 cases and 852 controls (6 controls per case, matched for sex, age, county, and ethnicity); that of Toniolo et al. (12) was based on 180 cases and 829 controls (5 controls per case, matched for age, menopausal status, and date of enrollment). A total of 68,910 individuals were monitored in the cohort studies or in the nested case-control studies, and the study populations varied from 2679 to 24,897 persons. During 4- to 25-y follow-up periods, a total of 740 breast cancer cases were diagnosed.

In a case-control study, a group of breast cancer patients is compared with a group of controls free of the disease with respect to intake of dairy products. In all, 27 case-control studies comparing the intake of some dairy products among breast cancer patients and controls were considered (Table 35.2). Results exist for populations from 14 different countries. The number of cancer cases in different studies varied from 68 to 2569. In ~50% of the studies, the controls were hospital patients; in the majority of studies, the controls were selected by matching for sex, age, and possibly for geographic area.

Assessment of Dairy Product Intake

Consumption data on milk and other dairy products were collected as a part of a food-frequency questionnaire or by a dietary history method (Table 35.1). One cohort study (10) and four case-control studies (21,22,29,30) reported that they utilized a dietary history interview method to estimate habitual diet of the participants. The majority of food-frequency questionnaires were also administered by interview (Table 35.1). In some studies relatively few details were given about the questionnaire used to estimate food consumption (7,11,14,17,18,20,25,32,40).

The number of food items in the food-frequency questionnaires varied greatly (Table 35.1). Approximately one third of the questionnaires included <30 food items or did not specify the number of food items. One cohort study (7) and two case-control studies (17,25) described consumption specifically of milk and/or dairy products. On the other hand, several studies applied comprehensive food-frequency questionnaires, covering the majority of important food items used by the study population. A wide variety of different foods are usually covered by a dietary history interview; however, the exact number of food items may be difficult to specify from open-ended questions. In 50% of the studies, the dietary survey method was sufficiently comprehensive to enable calculation of total intake of dietary energy (Tables 35.3 and 35.4), but only in six studies was dietary intake allowed for in the statistical analyses (70,12,21,33,36,38).

With the exception of three studies, dietary investigations referred to the usual intake of milk and other foods during some period of adult life (Table 35.1). The time interval covered by the dietary survey varied from current diet to the 20 preceding years.

TABLE 35.1. Description of Cohort and Nested Case-Control Studies of the Association Between Intake of Dairy Products and Breast Cancer Occurrence

Country/(Reference)	Design	Cohort size	Age (y)	Starting year	Length of follow-up (y)
Norway/(7)	Cohort	2679	≤ 75	1967–69	11
Norway/(8)	Cohort	24,897	20–54	1977–83	7–13
U.S./(9)	Cohort	6156	32–86	1982–84	4
Finland/(10)	Cohort	4697	15–90	1966–72	25
U.S./(11)	Nested[c]	16,190	30–85	1960	20
U.S./(12)	Nested[d]	14,291	35–65	1985–90	6

[a]FFQ, food frequency; DH, dietary history; Q, questionnaire.
[b]I, interview; S, self-administered.
[c]852 controls matched for age, county and ethnicity.
[d]829 controls matched for age, menopausal status and date of enrollment.
NR, not reported

In one study (15), dietary habits were investigated in four different periods of life, whereas two other studies (20,41) inquired specifically about dietary habits during adolescence.

In most of the case-control studies, dietary data were collected for a period preceding the symptoms or diagnosis of the disease to avoid the potential effect of the disease on the responses (Table 35.2). Investigations of breast cancer cases were carried out during the patients' stay in the hospital or afterward at home, but not later than 1 y after the diagnosis. In one study based on a breast cancer screening examination (34), dietary data were collected before the mammographic examination.

Only a few studies gave estimates on the reliability of the dietary survey method used. Ursin et al. (7) reported that their questionnaire, repeated 3–4 mo apart, yielded a high coefficient of reproducibility (0.68) for questions on milk consumption. In the Finnish Mobile Clinic Health Examination Survey, the coefficient of reproducibility for consumption of milk products was 0.68 over an interval of 4–8 mo, and 0.54 over an interval of 4–7 y (10). In the Finnish study, the short-term repeatability for cheese and butter was fairly good (the coefficients were 0.51 and 0.59, respectively), whereas the long-term repeatability was low (the coefficients were 0.30 and 0.25, respectively) (42). Toniolo et al. (12) reported that coefficients for short-term (2–3 mo) and long-term (2–4 y) reproducibility were in the range of 0.50–0.60 for the food items considered. Four studies reported that their food-frequency questionnaires were validated (8,30,36,38).

Cases (n)	Dietary assessment				Dairy product
	Method[a]	Food items (n)	Type[b]	Time period	
29	Q	NR	S	NR	Milk
248	FFQ	80	S	Usual	Milk
53	FFQ	93	I	Usual	Milk
88	DH	>100	I	1-y period before interview	Dairy, milk, cheese, cream, butter, fermented milk, ice cream, milk fat
142	FFQ	21	S	NR	Milk, cheese
180	FFQ	71	S	Current	Dairy

Statistical Methods

The strength of association between consumption of dairy products and breast cancer risk was commonly expressed either as the relative risk (or odds ratio) of the cancer among different levels of intake or as a percentage of mean intake differences between cases and controls. The relative risks were calculated by comparing the risk among persons in different categories of the intake distribution. In this review, the lowest category is used as the reference category, and the risk of this level is settled unity. The relative risk can be interpreted as the risk of developing breast cancer for exposed persons in comparison to unexposed individuals. A relative risk <1.0 thus implies that exposure to dairy products is associated with a reduced risk of cancer. In ~50% of the studies, testing for a gradient between intake of dairy products and breast cancer risk also took place. The percentage mean differences were estimated as [(case mean − control mean)/control mean] × 100.

Results

The main results from analytical observational epidemiologic studies on the association between consumption of dairy products and occurrence of breast cancer published not earlier than 1980 are presented by study design. The relative risk between high- and low-level milk and other dairy product consumption and the mean difference in consumption of dairy products between cases and controls are shown in Tables 35.3 and 35.4.

TABLE 35.2. Description of Case-Control Studies of the Association Between Intake of Dairy Products and Breast Cancer Occurrence

Country/(Reference)	Time period	Age (y)	Cases (n)	Controls (n)	Type of control[a]
Canada/(13)	1976–77	30–80	577	826	P
Italy/(14)	1980–83	27–79	368	373	H
Canada/(15)	1980–82	<70	846	862	N
Greece/(16)	1983–84	NR	120	120	H
France/(17)	1976–80	58 (13)[d]	1010	1950	H
Italy/(18)	1983–85	26–74	1108	1281	H
Argentina/(19)	1984–85	56 (-)[d]	150	300	H, N
U.S./(20)	1980–83	21–54	172	190	P
Italy/(21)	1983–84	< 75	250	499	P
The Netherlands/ (22,23)	1985–87	25–44 55–64	133	289	P
Denmark/(24)	1983–84	< 70	1474	1322	P
U.S./(25)	1982	19–97	848	850	H
Canada/(26)	1985–87	40–59	68	343	P
Australia/(27)	1985–87	56 (1)	99	209	P
Argentina/(28)	1979–81	≤75	196	205	N
France/(29)	1983–87	28–66	409	515	H
U.S./(30)	1975–80	45–74	272	296	H, N
Japan/(31)	1990–91	≥20	908	908	H, S
Poland/(32)	1987	≥35	127	250	P
Switzerland/(33)	1990	32–75	107	318	H
Sweden/(34)	1987–90	40–74	265	432	P
Spain/(35)	1987–91	59 (-)[d]	100	100	H
Italy/(36) (37)	1991–94	23–74	2569	2588	H
Greece/(38)	1989–91	NR	820	1548	H, N
China/(39)	1984–85	20–69	834	834	P
U.S./(40)	1986–91	41–85	154	192	P
U.S./(41)	1990–92	<45	1647	1501	P

[a]H, hospital; P, population; N, neighborhood; S, screening.
[b]FFQ, food frequency; DH, dietary history; Q, questionnaire.
[c]I, interview; S, self-administered.
[d]Mean (SD).
[e]24-h recall for controls.
NR, not reported.

	Dietary assessment			
Method[b]	Food items (n)	Type[c]	Time period	Dairy product
FFQ	8	I	Usual	Milk, cheese, cream, butter
FFQ	NR	I	NR	Dairy
FFQ	31	S	4 periods	Milk
FFQ	120	I	Before disease	Dairy
Q	NR	S	NR	Milk, cheese, butter, yogurt
FFQ	NR	S	NR	Milk, cheese, butter/margarine/oil
FFQ	147	I	5 y to 6 mo before interview	Milk, cheese, butter/cream
FFQ	NR	I	Adolescence	Fat from milk/cheese/yogurt
DH	>70	I	Usual	Dairy, milk, cheese, butter
DH	236	I	1-y period before diagnosis	Milk, cheese, fermented milk, fat from milk products/cheese
FFQ	21	S	1-y period before diagnosis	Milk, butter
Q	NR	S	Before diagnosis	Milk, milk fat
FFQ 24-h[e]	333	NR	NR	Milk, cream, butter, milk dessert
FFQ	179	S	Current	Dairy, milk, butter/margarine
FFQ	40	I	20-y period before diagnosis	Milk, cheese, cream, butter, yogurt
DH	55	I	Usual	Milk, cheese, cream/butter, yogurt
DH	43	I	Before diagnosis	Dairy, ice cream
FFQ	8	S	NR	Dairy
FFQ	44	S	20 y ago	Butter
FFQ	50	I	Before symptoms	Milk, cheese, butter, yogurt
FFQ	60	S	Recent 6 mo	Dairy
FFQ	99	I	Before disease	Dairy
FFQ	79	I	2-y period before diagnosis	Milk, cheese, butter
FFQ	115	I	1-y period before disease	Milk products, butter
FFQ	63/68	I	Usual	Milk
FFQ	NR	I	2 y before interview	Dairy
FFQ	29	I	Adolescence	Dairy

TABLE 35.3. Results of Cohort and Nested Case-Control Studies of the Association Between Intake of Dairy Products and Breast Cancer Occurrence

Reference	Dairy product	Exposure Categories compared[a]	Unit	Relative risk
Ursin et al. (7)	Milk	≥2/<1	glasses/d	1.48
Gaard et al. (8)	Milk	≥5/1	glasses/d	1.71
	Whole milk	≥5/1	glasses/d	2.91
Byrne et al. (9)	Whole milk	>7/≤7	servings/wk	0.5
Knekt et al. (10)	Dairy	T3/T1	g/d	0.42
	Milk	T3/T1	g/d	0.42
	Cheese	T3/T1	g/d	1.25
	Cream	T3/T1	g/d	0.84
	Butter	T3/T1	g/d	0.59
	Fermented milk	T3/T1	g/d	1.37
	Ice-cream	T3/T1	g/d	0.63
	Milk fat	T3/T1	g/d	0.64
Mills et al. (11)	Milk	≥3/0	drinks/d	1.03
	Cheese	≥3/0	d/wk	1.04
Toniolo et al. (12)	Dairy	Q5/Q1	g/d	0.59

[a]T3/T1, highest vs. lowest tertile; Q5/Q1, highest vs. lowest quintile.
[b]AGE, age; ALC, alcohol consumption; ANT, antropologic measures; DEM, demographic factors; DIE, dietary factors; ENG, energy intake; EXE, exercise; HER, heredity; HOR, hormone use; REP, reproductive factors; SMO, smoking.
NR, not reported.

Cohort Studies

The population of a Norwegian study, conducted during 1967–1969 comprised 2679 female spouses and siblings (≤75 y of age) of individuals interviewed in a case-control study on gastrointestinal cancer (7) (Table 35.1). Milk intake was estimated through a mailed questionnaire with precoded alternatives as follows: no use, <2 glasses per week, ≥2 glasses per week but <1 glass per day, 1 glass per day, 2–3 glasses per day, 4–5 glasses per day, and ≥6 glasses per day. Milk consumption showed a coefficient of repeatability of 0.68 between two sets of replies collected 3–4 mo apart. During an 11-y follow-up, 29 individuals developed breast cancer according to the Cancer Registry of Norway. A nonsignificant positive association was observed between milk intake and breast cancer incidence (Table 35.3). The age- and residence-adjusted relative risk of breast cancer between those consuming ≥2 glasses per day in comparison to women consuming <1 glass per day was 1.48 (P-value for trend = 0.40). No interaction between milk intake and alcohol consumption was observed.

The Norwegian National Health Screening Service, conducted during 1977–1983, included 24,897 women from four counties, 20–54 y of age and free of cancer, who attended a health survey (8) (Table 35.1). A food-frequency questionnaire on the usual consumption of 80 food items was filled in at home and returned

95% Confidence interval	P value for trend	Controlled variables[b]
NR	0.40	AGE, DEM
0.86–3.38	0.30	AGE
1.38–6.14	0.08	
0.1–2.1	NR	AGE
0.23–0.78	0.02	AGE, ANT, DEM, DIE, ENG, REP, SMO
0.24–0.74	0.003	
0.75–2.08	0.66	
0.53–1.34	0.67	
0.35–0.99	0.17	
0.80–2.37	0.47	
0.35–1.15	0.32	
0.39–1.06	NR	
0.56–1.90	0.95	AGE, ANT, DEM, DIE, REP
0.61–1.75	0.98	
0.35–0.99	0.10	AGE, ANT, DEM, ENG, HER, REP

by mail. During a follow-up of 7–13 y, 248 breast cancer cases were identified through the Norwegian Cancer Registry. A positive association occurred between baseline whole-milk consumption and breast cancer incidence (Table 35.3). The age-adjusted relative risk between women consuming at least 5 glasses (0.75 L) per day in comparison to women consuming 1 glass per day was 2.91 [95% confidence interval (CI) = 1.38–6.14]. No association between intake of skim milk or low-fat milk or butter and breast cancer incidence was observed.

During 1982–1984, 6156 women 32–86 y of age with no history of breast cancer, from the U.S. National Health and Nutrition Examination Survey I, completed an interview including a food-frequency questionnaire (9) (Table 35.1). During a 4-y follow-up, 53 women reported breast cancer diagnosis. The age-adjusted relative risk of breast cancer for women who drank >7 servings of whole milk each week was 0.5 (CI = 0.1–2.1) compared with women who drank ≤7 servings each week.

The Finnish Mobile Clinic undertook health examinations in 25 cohorts from various parts of Finland during 1966–1972 (10) (Table 35.1). Food consumption data were obtained from 4697 women, 15–90 y of age and free of cancer. A modified dietary history interview method was used to survey the total habitual diet of examinees during the previous year. The amount of each individual food item per day was calculated by combining the amount of food directly reported in the interview with

TABLE 35.4. Results of Case-Control Studies of the Association Between Intake of Dairy Products and Breast Cancer Occurrence

Reference	Dairy product	Exposure Categories compared[a]	Unit	Relative Risk
Lubin et al. (13)	Skim milk	≥1/0	glasses/d	1.04
	Whole + 2% milk	≥3/0	glasses/d	0.77
	Cheese	7/<4	d/wk	1.11
	Cream	≥4/≤1	d/wk	0.92
	Butter	yes/no	Use at table	1.47
		yes/no	Use for frying	2.33
Talamini et al. (14)	Dairy	≥5/≤2	d/wk	3.2
Hislop et al. (15)	Whole milk; recent	30/<1	d/mo	1.55
	Whole milk; childhood	30/<1	d/mo	0.73
Katsouyanni et al. (16)	Dairy	—	times/wk	NR
Lé et al. (17)	Milk	Full cream/none		1.8
	Cheese	7/0	d/wk	1.5
	Butter	7/<7	d/wk	0.9
	Yogurt	7/0	d/wk	0.8
La Vecchia et al. (18)	Milk	—	portions/wk	NR
	Cheese	—	portions/wk	NR
	Butter/margarine/oil	high/low	Consumption score	1.27
Iscovich et al. (19)	Whole milk	Q4/Q1	g/d	0.38
	Skim milk	T3/T1	g/d	1.0
	Cheese	Q4/Q1	g/d	1.7
	Butter/cream	Q4/Q1	g/d	2.5
Pryor et al. (20)	Fat from milk/	Q4/Q1	g/d	0.4[c]
	cheese/yogurt	Q4/Q1	g/d	0.2[d]
Toniolo et al. (21)	Dairy	Q4/Q1	g/d	2.5
	Whole milk	Q4/Q1	g/d	1.8
	Skim milk	T3/T1	g/d	1.5
	Low-fat cheese	Q4/Q1	g/d	0.9
	High-fat cheese	Q4/Q1	g/d	2.4
	Butter	Q4/Q1	g/d	1.5
van't Veer et al. (22)	Milk	per 225 g		0.81
	Gouda cheese	per 60 g		0.56
	Fermented milk	per 225 g		0.63
van't Veer et al. (23)	Fat from dairy products			
Ewertz et al. (24)	Milk, standard	≥3/0	L/wk	1.55
	Milk, low-fat	≥3/0	L/wk	1.36
	Butter for frying	≥4/0	times/wk	0.82
	Butter on bread	Q4/Q1	amount	1.12
Mettlin et al. (25)	Whole milk	7/0	d/wk	1.5
	2% milk	7/0	d/wk	0.9
	Skim milk	7/0	d/wk	0.8
	Milk fat	Q4/Q1	index	1.1
Simard et al. (26)	Milk	% regularly consuming		NR
	Cream	% regularly consuming		NR
	Butter	% regularly consuming		NR
	Milk dessert	% regularly consuming		NR

95% Confidence interval	P-value for trend	Mean case-control difference (%)	Controlled variables[b]
0.7–1.6	NR	—	AGE
0.5–1.3	NR	—	
0.9–1.4	NR	—	
0.7–1.2	NR	—	
1.2–1.9	NR	—	
1.4–4.0	NR	—	
1.8–5.8	<0.001	—	AGE, ALC, ANT, DEM, DIE, HOR, REP, SMO
1.18–2.05	NR	—	AGE
0.53–1.02	NR	—	
NR	<0.10	—	AGE, DEM
1.3–2.4	<0.0001	—	AGE, ALC, DEM, DIE, HER,
1.0–2.3	0.01	—	REP
0.7–1.1	0.21	—	
0.6–1.0	0.02	—	
NR	NR	−5	AGE, ANT, DEM, HER,
NR	NR	+4	HOR, REP
0.98–1.64	0.01	—	
NR	<0.05	—	AGE, ANT, DEM, DIE, REP
NR	NS	—	
NR	NS	—	
NR	<0.05	—	
0.1–1.1	0.28	—	AGE, ANT, DEM, DIE, REP
0.0–0.8	0.01	—	
NR	0.001	+15	AGE, ENG
NR	0.006	+17	
NR	0.011	−3	
NR	NS	0	
NR	0.001	+41	
NR	NS	+8	
0.59–1.12	NR	−7	AGE, ALC, ANT, DEM,
0.33–0.95	NR	−9	HER, HOR, REP, SMO
0.41–0.96	NR	−26[e]	
		0	AGE
1.03–2.33	0.005	—	AGE, DEM
1.01–1.84	NS	—	
0.54–1.24	NR	—	
0.94–1.34	NR	—	
1.1–2.0	NR	—	AGE, DEM, SMO
0.7–1.1	NR	—	
0.6–1.0	NR	—	
0.8–1.5	NR	—	
NR	NR	−9	AGE
NR	NR	−60[e]	
NR	NR	−29[e]	
NR	NR	−16[e]	

Study	Food	Categories	Units	RR
Ingram et al. (27)	Dairy	M2/M1	g/d	1.3
	Milk/milk products	M2/M1	g/d	0.9
	Butter/margarine	M2/M1	g/d	1.2
Matos et al. (28)	Soft cheese	NR	times/wk	0.6
Richardson et al. (29)	Milk		L/wk	NR
	Cheese, low-fat		g/wk	NR
	Cheese, high-fat	>210/≤90	g/wk	1.4
	Butter/cream		g/wk	NR
	Yogurt		g/wk	NR
Goodman et al. (30)	Dairy	M2/M1	g/wk	1.6
	Ice cream	M2/M1	g/wk	1.0
Kato et al. (31)	Dairy	7/≤2	times/wk	0.71
Pawlega et al. (32)	Butter	daily/rare		0.5
Levi et al. (33)	Milk	T3/T1	freq. of cons.	1.0
	Cheese	T3/T1	freq. of cons.	3.0
	Butter	T3/T1	freq. of cons.	2.2
	Yogurt	T3/T1	freq. of cons.	0.9
Holmberg et al. (34)	Dairy	Q4/Q1	g/d	1.2
Landa et al. (35)	Dairy	T3/T1	times/mo	1.76
Franceschi et al. (36)	Milk	Q5/Q1	servings/wk	0.81
	Cheese	Q5/Q1	servings/wk	0.98
La Vecchia et al. (37)	Butter	Q5/Q1	g/d	0.95
Trichopoulou et al. (38)	Milk products	Q5/Q1	times/mo	1.00
	Butter	per 4 times/mo		1.01
Yuan et al. (39)	Milk	—	g/d	NR
Moysich et al. (40)	Dairy	—	g/mo	NR
Potischman et al. (41)	Dairy	Q4/Q1	servings/mo	0.96

[a]M2/M1, values above vs. under the median; T3/T1, highest vs. lowest tertile; Q4/Q1, highest vs. lowest quartile; Q5/Q1, highest vs. lowest quintile.
[b]AGE, age; ALC, alcohol consumption; ANT, anthropologic measures; DEM, demographic factors; DIE, dietary factors; ENG, energy intake; EXE, exercise; HER, heredity; HOR, hormone use; REP, reproductive factors; SMO, smoking.
[c]Premenopausal.
[d]Postmenopausal.
[e]Differs statistically significantly from zero.
NR, not reported.

that derived from mixed dishes. Energy intake was calculated. Short- and long-term repeatabilities of the daily milk product consumption were estimated by repeating the dietary interviews 4–8 mo and 4–7 y after the initial interviews. The intraclass correlation coefficients were 0.68 and 0.54, respectively. During a 25-y follow-up, 88 incident breast cancer cases occurred. An inverse association was suggested between intake of dairy products and occurrence of breast cancer (Table 35.3). The age-adjusted relative risk of breast cancer between the highest and lowest tertiles of intake of all dairy products was 0.42 (CI = 0.23–0.78; P-value for trend = 0.02). The association was due mainly to milk and butter intakes; the relative risks were 0.42 (CI = 0.24–0.74) and 0.59 (CI = 0.35–0.99), respectively. Exclusion of those breast cancer cases that occurred during the first 5 y of follow-up resulted in a relative risk of 0.49 (CI = 0.28–0.87) for milk intake. Adjustment for smoking, body mass index,

0.7–2.2	NR	−1	AGE, DEM
0.5–1.6	NR	−3	
0.7–2.0	NR	+5	
0.4–0.9	NR	—	AGE, DEM, DIE, REP
NR	NR	−2	AGE, ALC, DEM, HER, REP
NR	NR	−9	
1.0–1.9	0.02	+13[e]	
NR	NR	0	
NR	NR	0	
1.1–2.3	0.45	—	AGE, ANT, DEM, REP
0.7–1.4	0.96	—	
0.56–0.89	0.006	—	AGE, ANT, DIE, HER, REP
0.1–2.0	NR	—	AGE, ALC, ANT, DEM, SMO
NR	NS	—	AGE, ANT, ENG
NR	<0.01	—	
NR	<0.05	—	
NR	NS	—	
0.8–1.2	NR	—	AGE, DEM
NR	NS	—	AGE
0.67–0.98	<0.05	—	AGE, ALC, DEM, ENG, REP
0.81–1.18	NS	—	
0.8–1.1	0.08	—	AGE, ALC, DEM, DIE, ENG, REP
0.93–1.08	NS	—	AGE, ANT, DEM, DIE,
0.95–1.08	NS	—	ENG, REP
NR	NR	+46[e]	AGE
NR	NR	+8	NR
0.8–1.2	NR	—	AGE, ALC, DEM, EXE, HOR, REP

number of childbirths, occupation, and geographic area did not notably alter the association; the relative risk was 0.49 (CI = 0.27–0.86). A strong association between intake of milk and other foodstuffs (cereals, potatoes, vegetables, fruits and berries, meat, fish, and eggs) was observed. Adjustment for these and for energy intake, however, altered the milk intake/breast cancer relationship only slightly; the relative risk was 0.57 (CI = 0.28–1.13). Adjustment for intake of carbohydrates, protein, and fat also altered the association only slightly; the relative risk was 0.40 (CI = 0.21–0.76). The total amount of milk fat derived from dairy products was nonsignificantly inversely associated with breast cancer, with a relative risk of 0.64 (CI = 0.39–1.06). Adjustment for calcium and milk fat intake did not materially alter the association between milk and breast cancer occurrence. No interactions between milk intake and nondietary factors were observed.

Nested Case-Control Studies

A total of 16,190 California Seventh-Day Adventist women, >30 y of age, completed a 21-item food-frequency questionnaire in 1960 that included information on food and beverage consumption (11) (Table 35.1). During a 20-y follow-up, 142 fatal breast cancer cases were ascertained by California death certificates among the white females. Six controls were selected for each cancer case by individual matching for age, county, and ethnicity. Neither milk intake nor cheese intake was associated with breast cancer risk (Table 35.3). The relative risk of breast cancer among women consuming ≥3 glasses of milk per day in comparison to women with no or occasional milk consumption was 1.03 (CI = 0.56–1.90) in a model that included age at menarche, age at first pregnancy, age at menopause, percentage of desirable weight, education, and consumption of meat, cheese, milk, and egg. The corresponding value for cheese consumption on ≥3 d/wk in comparison to no or occasional cheese consumption was 1.04 (CI = 0.61–1.75). No interaction was found between consumption of these foods and five nondietary risk factors. The results of a subgroup of women who were menopausal at baseline were equivalent to those of the entire group.

Toniolo *et al.* (12) investigated the effect of dietary factors on breast cancer occurrence in the New York University Women's Health Study population, consisting of 14,291 individuals 35–65 y of age (Table 35.1). A nested case-control study design was selected, based on 180 breast cancer cases and 829 controls individually matched for age, menopausal status, phase of menstrual cycle, and date of enrollment. Consumption of dairy products was assessed with a self-administered food-frequency questionnaire that included 71 food items. Coefficients of short-term (2–3 mo) and long-term (2–4 y) repeatability varied from 0.50 to 0.60 for the food items. A significant inverse association was observed between intake of dairy products and breast cancer risk. The relative risk between the highest and lowest quintiles of intake was 0.59 (CI = 0.35–0.99) after adjustment for age, menopausal status, menstrual cycle, and energy intake (Table 35.3). Further adjustment for several potential confounding factors, including demographic, anthropologic, reproductive, and hereditary factors, did not notably alter the results.

Case-Control Studies

Lubin *et al.* (13) carried out a population-based case-control study in northern Alberta, Canada, based on 577 breast cancer cases and 826 controls (Table 35.2). The controls were an age-stratified random sample from the general population. A food-frequency questionnaire with eight items included cheese and cream. Questions were also asked about the amount and type of milk and use of butter. Of the intake of dairy products, only butter intake was significantly associated with breast cancer risk (Table 35.4). The age-adjusted relative risk of breast cancer among individuals using butter at the table in comparison to nonusers was 1.47 (CI = 1.2–1.9), whereas the relative risk of butter used for frying was 2.33 (CI = 1.4–4.0). The relative risks for the other dairy products varied from 0.77 to 1.11.

Talamini *et al.* (14) compared dietary factors of 368 breast cancer patients and 373 age-matched hospital controls in northern Italy (Table 35.2). The frequency of consumption per week of milk, cheese, and other dairy products was assessed. A significantly elevated risk of breast cancer was observed among women consuming milk, cheese, and other dairy products for ≥5 d/wk in comparison to women consuming them on ≤2 d (Table 35.4). The relative risk adjusted for marital status, was 3.4 (CI = 2.0–5.8). Further adjustment for potential confounding factors (alcohol and food intake, education, occupation, body mass index, parity, age at first childbirth, age at menarche and at menopause, oral contraceptive and other female hormone use, smoking, and methylaxanthine consumption) did not notably alter the results. The relative risk was 3.2 (CI = 1.8–5.8; P-value for trend < 0.001).

Hislop *et al.* (15) studied the association between childhood and recent diet and occurrence of breast cancer in British Columbia, Canada (Table 35.2). The study population consisted of 846 breast cancer patients, <70 years of age and 862 controls of similar age from neighbors or acquaintances. The intake of whole milk and other dairy products was assessed with a mailed self-administered food-frequency questionnaire. Recent intake of whole milk was significantly positively associated with breast cancer risk, and intake during childhood was inversely associated with the disease (Table 35.4). The relative risks of breast cancer among women using whole milk daily in comparison to those using it very rarely were 1.55 (CI = 1.18–2.05) and 0.73 (CI = 0.53–1.02), respectively. No notable differences were observed for preor postmenopausal breast cancer.

Katsouyanni *et al.* (16) studied 120 women in Athens, Greece who suffered from breast cancer and 120 hospital controls matched for age (Table 35.2). Dairy and other dietary data were obtained with a food-frequency questionnaire relating to the period preceding onset of the disease. A nonsignificant inverse trend ($P < 0.10$) was observed between intake of dairy products and breast cancer occurrence after adjustment for age and education. The association was totally eliminated after further adjustment for intake of other food groups.

Lé *et al.* (17) investigated the relationship between different dairy products and breast cancer risk in a case-control study in Paris, France (Table 35.2). A total of 1010 breast cancer cases diagnosed <1 y before the study and 1950 hospital controls matched for age and clinic were included. The intake of dairy products was assessed with a food-frequency questionnaire. The questionnaire included daily quantities of different milks, names of regularly consumed cheeses, and frequency or consumption of cheese, yogurt, and butter. The risk of breast cancer was increased among women consuming full cream milk or with daily intake of cheese (Table 35.4). After adjustment for occupation, reproductive and hereditary factors, and consumption of other dairy products and alcohol, the relative risks were 1.8 (CI = 1.3–2.4) and 1.5 (1.0–2.3), respectively. Butter intake was not associated with breast cancer occurrence and yogurt intake was inversely associated. The relative risk for butter was 0.9 (CI = 0.7–1.1) and for yogurt 0.8 (CI = 0.6–1.0). A significant inverse interaction between milk intake and alcohol consumption was observed, implying that the excess risk related to increasing cream content in milk was reduced for women who consumed an alcoholic beverage with meals.

La Vecchia et al. (18) studied dietary factors in relation to breast cancer among 26- to 74- y-old women in northeastern Italy (Table 35.2). The study population comprised 1108 women with histologically confirmed breast cancer diagnosed within the year preceding the study and 1281 hospital controls, comparable in age and geographic area. Consumption of milk and dairy products was assessed, using a food-frequency questionnaire. Women with a high fat score based on intake of butter, margarine, and oil combined had nonsignificantly elevated risk of breast cancer (Table 35.4). The relative risk was 1.27 (CI = 0.98–1.64) after adjustment for several potential nondietary risk factors and dietary factors. The number of portions of milk or cheese per week was not associated with breast cancer occurrence.

Iscovich et al. (19) investigated the association between different dietary factors and breast cancer in a case-control study in La Plata, Argentina based on 150 cases and 300 controls matched for age and residence block (Table 35.2). The controls consisted of two separate series, i.e., hospital controls and neighborhood controls. Information on the intake of dairy products, which was assessed by a food-frequency questionnaire, described the consumption during a 5-y period up to 6 mo before the interview. After adjustment for different potential confounding factors, intake of whole milk was inversely and intake of butter or cream was positively associated with breast cancer risk (Table 35.4). The relative risk between the highest and lowest quintiles of the intakes was 0.38 (P for trend < 0.05) and 2.5 (P for trend < 0.05), respectively, in the study referring to the neighborhood controls after adjustment for age, education, husband's occupation, age at first pregnancy, parity, and body mass index. The corresponding values in the study referring to the hospital controls were 0.85 for whole milk and 2.6 for butter or cream. Further adjustment for other food groups did not alter the inverse association between whole milk intake and breast cancer risk. The relative risk for milk-based desserts was 3.8 (P for trend < 0.05). Intake of skim milk or cheese was not significantly associated with occurrence of breast cancer.

Pryor et al. (20) carried out a population-based case-control study in Utah that was based on 172 Caucasian breast cancer patients identified through the Utah Cancer Registry and a pathology laboratory reporting system, and 190 age-matched controls (Table 35.2). Fat from milk, cheese, and yogurt eaten during adolescence was assessed based on a food-frequency questionnaire, administered by telephone interview. The relative risk of premenopausal breast cancer between the highest and lowest quintiles of dairy fat intake was 0.4 (CI = 0.1–1.1) and of postmenopausal breast cancer 0.2 (CI = 0.0–0.8) after adjustment for age, education, body mass index, age at first pregnancy, age at menarche, and fiber intake (Table 35.4).

Toniolo et al. (21) investigated the role of diet in breast cancer risk in Vercelli, Italy. The study population consisted of 250 breast cancer cases identified through records of all public and private hospitals of the area and 499 age-matched population controls (Table 35.2). The usual intake of whole milk, skim milk, low- and high-fat cheese, butter, and all dairy products combined was assessed with a dietary history questionnaire structured by meals. All interviews were conducted in the homes of the respondents for cases not earlier than 2 mo after surgery or other medical treatment. A strong positive association between intake of dairy products and breast cancer risk was noted (Table 35.4). The age- and energy-adjusted relative risk between the highest and lowest quartiles (or tertiles) of intake of all dairy products combined was 2.5 (P value for trend = 0.001). The corresponding value for whole milk was 1.8 (P value for trend = 0.006), for skim milk 1.5 (P value for trend = 0.011), high-fat cheese 2.4 (P value for trend = 0.001), and for butter 1.5. Only the intake of low-fat cheese showed no sign of any association with the disease.

In the central and southern Netherlands, van't Veer et al. (22) investigated the association between intake of fermented milk, Gouda cheese, and milk and breast cancer risk, in a case-control study based on 133 breast cancer patients and 289 population controls from the same areas and with age distribution similar to that of the cases (Table 35.2). A structured dietary history interview was completed to estimate usual consumption of all major dairy products used in The Netherlands. Interviews were conducted for cases during a home visit 3–6 mo after hospital discharge. An inverse association was observed between intake of fermented milk products and breast cancer occurrence (Table 35.4). The relative risk expressed per consumption of 225 g was 0.63 (CI = 0.41–0.96) after adjustment for several potential confounding factors. The stronger association was observed for curds and buttermilk; weaker associations were noted for yogurt and kefir. Intake of Gouda cheese also showed an inverse relation with a relative risk of 0.56 (CI = 0.33–0.95) per consumption of 60 g. The corresponding value for milk intake was 0.81 (CI = 0.59–1.12) per consumption of 225 g. A significant interaction between intakes of fermented milk and fiber was also reported (43). In all women, joint exposure to high intake of fermented milk products and high-fiber intake was 0.58 (CI = 0.32–1.05) and in younger age-groups 0.38 (CI = 0.16–0.89). No association was observed for estimated fat intake from dairy products (23).

Ewertz and Gill (24) studied Danish women, <70 y of age, in a case-control study of 1486 breast cancer cases and 1336 age-stratified population controls (Table 35.2). Milk intake was assessed with a self-administered food-frequency questionnaire that inquired about the subject's usual intake of 21 food items in the year before diagnosis. The age- and residence-adjusted relative risk of breast cancer was 1.55 (CI = 1.03–2.33) between individuals consuming at least 3 L standard milk per week in comparison with those consuming no milk (Table 35.4). The corresponding value for low-fat milk was 1.36 (CI = 1.01–1.84). There was no significant difference between cases and controls in frequency of frying with butter or amount of butter used on bread.

Mettlin et al. (25) carried out a case-control study in Buffalo, New York, of 848 breast cancer patients and 850 female hospital controls selected randomly within strata of residence (Table 35.2). Milk consumption was assessed based on a self-administered questionnaire. The questionnaire prompted the patient to consider general habits immediately preceding onset of the disease. Frequent use of whole milk was associated with elevated breast cancer risk. The relative risk between daily users of whole milk and nonusers was 1.5 (CI = 1.1–2.0) after adjustment for age, smoking, education, and county of residence (Table 35.4). Intake of 2% milk, skim milk, or milk fat was not associated with breast cancer risk. The relative risks varied from 0.8 to 1.1.

Simard et al. (26) carried out a case-control study in Montreal, Canada, based on 68 breast cancer cases and 343 controls matched for age (Table 35.2). Both cases and controls were drawn from the cohort of women attending the Canadian National Breast Screening Study. A semiquantitative food-frequency questionnaire was used to obtain data on food habits of cases, whereas a 24-h dietary recall served to assess the diet of controls. The proportions of individuals consuming cream, milk dessert, and butter were significantly lower in breast cancer cases than in controls (Table 35.4). The percentage differences were 60, 16, and 29%, respectively. A similar but nonsignificant difference for milk intake was observed among women ≥50 y of age.

Ingram et al. (27) investigated the relation between dietary factors and breast cancer risk in a case-control study carried out in Perth, Australia (Table 35.2). A total of 99 breast cancer patients and 209 controls, matched for age and area, were included. Current consumption of milk, butter or margarine, or all dairy products combined was estimated from a comprehensive food-frequency questionnaire (Table 35.4). Each subject was visited at home and requested to complete the questionnaire and return it by mail. No significant associations were observed between intake of dairy products and breast cancer occurrence. The relative risks varied within 0.9–1.3. When subjects who had changed their diet during the preceding 3 mo were excluded, consumption of butter and margarine was significantly positively associated with breast cancer risk; the relative risk was 2.3 (CI = 1.1–5.1).

Matos et al. (28) carried out a case-control study on eating habits and breast cancer in Buenos Aires, Argentina (Table 35.2). The cases included 196 women ≤75 y of age with newly diagnosed, histologically confirmed breast cancer and two controls per case, i.e., a friend and a consanguineous family member. Both groups of controls combined totaled 205 women. All participants were interviewed in the hospital and the cases after the surgery. The usual consumption of dairy products was assessed from a food-frequency questionnaire relating to the 20 y preceding the diagnosis of cancer cases or interview of the controls. No significant associations were observed for consumption of milk, yogurt, cream, butter, or hard cheese. The consumption of soft cheese, however, was negatively associated with risk after controlling for hard cheese intake. The relative risk was 0.6 (CI = 0.4–0.9).

Richardson et al. (29) studied the association between dietary factors and breast cancer risk in a case-control study in Montpellier, France, based on 409 cases, 28–66 y of age, and 515 hospital controls in the same age-group (Table 35.2). Intake of milk, butter and cream, low-fat and high-fat cheese, and yogurt was assessed with a dietary history-type questionnaire administered by interview. Persons with a high intake of high-fat cheese had an elevated risk of breast cancer (Table 35.4). The relative risk between high and low intake level was 1.4 (CI = 1.0–1.9) after adjustment for age, menopausal status, alcohol consumption, and several dietary variables.

Goodman et al. (30) investigated the association of diet and breast cancer risk in a case-control study in Hawaii (Table 35.2). The study population comprised 272 postmenopausal cases and 296 neighborhood controls matched for age, ethnic background, and residency. Intake of 43 different food items was assessed with a dietary history questionnaire. The frequency and amount of foods consumed during the usual week before diagnosis were indicated. Women with intake of dairy products over the median had an elevated breast cancer risk in comparison to those under the median (Table 35.2); the relative risk was 1.6 (CI = 1.1–2.3). The association was stronger in obese than in nonobese individuals; the relative risks were 2.2 (P value for trend = 0.04) and 1.4 (P value for trend = 0.11), respectively. Similar differences were also observed in categories of other variables. The association was stronger in Caucasians than in Japanese, in women ≥60 y of age compared with younger women, and in nulliparous women compared with women who had borne children. No association was revealed for ice-cream consumption.

Kato et al. (31) studied the relationship between dietary factors and breast cancer in eight prefectures of Japan (Table 35.2). A total of 908 cases and 908 hospital controls matched for age were included. Consumption of dairy products was assessed with a self-administered food-frequency questionnaire. The relative risk of breast cancer was 0.71 (CI = 0.56–0.89) among women consuming dairy products daily in comparison to those consuming dairy products <3 times per week. Adjustment for body mass index, green/yellow vegetables, age at first birth, and family history of breast cancer did not materially alter the results. The relationship was stronger in postmenopausal than in premenopausal women. The relative risks were 0.54 (CI = 0.38–0.76, P-value for trend < 0.001) and 0.89 (CI = 0.65–1.21, P-value for trend = 0.66), respectively.

Pawlega (32) conducted a case-control study in Cracow, Poland based on 127 breast cancer cases and 250 age-matched population controls (Table 35.2). All participants completed a mailed self-administered questionnaire about frequency of past (~20 y) consumption of 44 food items including cheese, milk, and butter. The relative risk of breast cancer among women <50 y of age, using butter daily, in comparison with those using it rarely, was 0.5 (CI = 0.1–2.0) after adjustment for age, education, social class, marital status, number of persons in household, smoking, body mass index, and drinking of vodka 20 y previously (Table 35.4).

Levi et al. (33) investigated the relation between dietary factors, including various dairy products and breast cancer risk in the Canton of Vaud, Switzerland in a case-control study based on 107 cases and 318 hospital controls (Table 35.2). A food-frequency questionnaire considering 50 indicator foods was assessed by interview. Weekly consumption of milk, yogurt, cheese, and butter before the occurrence of disease symptoms was elicited. Cheese and butter showed a positive association with breast cancer risk (Table 35.4). The age-adjusted relative risks were 3.0 (P-value for trend < 0.01) and 2.2 (P-value for trend < 0.05), respectively. After further adjustment for education and energy intake, the association between butter and breast cancer disappeared; the relative risk was 1.3.

Holmberg et al. (34) conducted a case-control study on diet and breast cancer in two Swedish counties among women 40–74 y of age (Table 35.2). A total of 265 breast cancer cases were recruited from a mammography-screening program and

432 controls, frequency-matched for age, month of mammography, and county of residence. The usual intake of dairy products during the previous 6 mo was questioned in a mailed food-frequency questionnaire that was returned before the screening examination. No significant association between total intake of dairy products and breast cancer occurrence was observed (Table 35.4). The relative risk between the highest and lowest quartiles of intake was 1.2 (CI = 0.8–1.8) after adjustment for age, county of residence, and month of mammography. The corresponding value for women >50 y of age was 1.3 (CI = 0.8–2.2) and for younger women 0.9 (CI = 0.4–2.0).

Landa et al. (35) carried out a case-control study in northern Spain based on 100 breast cancer cases and 100 hospital controls on the relation between diet and breast cancer risk (Table 35.2). Average frequency of dairy product consumption, including milk, skimmed milk, yogurt, and cheese was assessed as part of a 99-item food-frequency questionnaire related to the period preceding onset of the disease. A nonsignificant positive association was observed for total consumption of dairy products (Table 35.4). The relative risk of breast cancer between the highest and lowest tertiles of consumption times per month was 1.76.

A multicenter case-control study (36,37) on dietary factors and breast cancer risk was carried out in six Italian areas (Table 35.2). A total of 2569 cases and 2588 hospital controls were included. The cases were comparable to the controls in age and area of residence. An interviewer-administered food-frequency questionnaire was used to collect information on the subjects' habitual diet. Among different food groups, milk intake was significantly inversely related to breast cancer (Table 35.4). The relative risk between the highest and lowest quintiles of servings per week was 0.81 (CI = 0.67–0.98) after adjustment for age, study center, education, parity, alcohol consumption, and energy intake. Among different types of milk, only high intake of skim milk was inversely associated. Cheese intake was not significantly related to the disease. The relative risk was 0.98 (CI = 0.81–1.18); however, a positive association was present for intake of cheese at the end of a meal or as a snack. The relative risk between highest and lowest quintiles was 1.2 (CI = 1.0–1.4). The relative risk for butter intake after adjustment for several factors was 0.95 (CI = 0.8–1.1, P-value for trend = 0.08) (37). No notable differences were observed in strata of menopausal status. The relative risk among premenopausal women was 1.03 (CI = 0.94–1.12) and among postmenopausal 0.98 (CI = 0.91–1.06).

Trichopoulou et al. (38) conducted a case-control study based on 820 histologically confirmed breast cancer cases diagnosed in four major hospitals in Athens (Table 35.2). A total of 1548 controls were selected, one half among hospital visitors and the other half among orthopedic patients. The average frequency of consumption of milk and milk products during a period of 1 y before the onset of the disease was estimated from a validated semiquantitative food-frequency questionnaire. No association between intake of milk and milk products and breast cancer was observed (Table 35.4). The relative risk of breast cancer between the highest and lowest quintiles of intake was 1.00 (CI = 0.93–1.08) after adjustment for age, place of birth, parity, age at first pregnancy, age at menarche, menopausal status, body mass index, and energy intake. Butter intake similarly showed no association.

Yuan et al. (39) studied 534 breast cancer cases 20–69 y of age from Shanghai, and 300 cases 20–55 y of age from Tianjin, China, provided by the regional Cancer Registries (Table 35.2). Community controls were selected by individual matching for age. The usual frequencies and amounts of consumption of >60 food items were obtained by an interview conducted in the home. In both areas, statistically significantly higher intakes of cow's milk were observed in breast cancer patients than in controls (Table 35.4). The mean differences were 45 and 46% in Shanghai and Tianjin, respectively.

Moysich et al. (40) carried out a study in Erie and Niagara in western New York, based on 154 Caucasian postmenopausal breast cancer patients identified from most area hospitals (Table 35.2). A total of 192 community controls were selected by frequency matching for age and date of blood drawn. A food-frequency questionnaire reflecting the usual intake for 2 y before the interview was used. The monthly intake of dairy products was nonsignificantly (8%) higher in breast cancer patients than in controls (Table 35.4).

Potischman et al. (41) studied 1647 newly diagnosed breast cancer cases <45 y of age from cancer registries in three geographic areas (Atlanta, Seattle/Puget Sound, New Jersey) in the United States (Table 35.2). A total of 1501 controls were selected with frequency matching by age and geographic area. Diet during adolescence was assessed by asking subjects about their consumption of 29 food items at the ages of 12–13 y. The results were validated through use of questionnaires completed by the mothers. Intake of dairy products was defined as the intake of whole milk, ice cream, milk shakes, cheese, or butter. No differences were observed between the number of servings of dairy products consumed and breast cancer incidence (Table 35.4). The relative risk of breast cancer between individuals consuming >82 servings per month in comparison to those consuming <30 servings per month was 0.96 (CI = 0.8–1.2).

Summary of Findings

Overall, the studies on intake of dairy products combined, of milk, and of cheese presented both statistically significant inverse associations and statistically significant positive associations with breast cancer risk (Table 35.5). A considerable number of the studies found no association. The median relative risk between high and low intake of dairy products among the studies was 1.3 and the range of all studies was 0.4–3.2. The corresponding values for milk intake were 1.0 and 0.4–2.9, respectively, and for cheese intake 1.4 and 0.6–3.0, respectively. The few studies on intake of fermented milk showed all types of associations; the median relative risk was 1.0 and the range 0.6–1.4. With one exception, the studies on butter intake showed either a positive or no association. The median and range were 1.2 and 0.5–2.5, respectively. The few studies on cream (range = 0.8–0.9) and ice cream intake (range 0.6–1.0) in general gave no associations. Finally, of the three studies on fat intake from dairy products (range = 0.2–1.1), two gave an inverse association and one no association.

TABLE 35.5. Summary of the Strength of Association Between Intake of Dairy Products and Breast Cancer Occurrence

Dairy product	Study by relative risk		
	<0.8[a]	0.8–1.3	≥1.4[b]
All	Kato et al. (31)[c] Toniolo et al. (12)[c] Knekt et al. (10)[c]	Ingram et al. (27) Holmberg et al. (34) Potischman et al. (41)	Talamini et al. (14)[c] Toniolo et al. (21)[c] Goodman et al. (30)[c] Landa et al. (35)
Milk	Hislop et al. (15) Iscovich et al. (19)[c] Franceschi et al. (36)[c] Byrne et al. (9) Knekt et al. (10)[c]	Lubin et al. (13) Mills et al. (11) Van't Veer et al. (22) Ingram et al. (27) Levi et al. (33) Trichopoulou et al. (38)	Hislop et al. (15)[c] Lé et al. (17)[c] Toniolo et al. (21)[c] Ewertz and Gill (24)[c] Mettlin et al. (25)[c] Ursin et al. (7) Gaard et al. (8)[c]
Cheese	van't Veer et al. (22)[c] Matos et al. (28)[c]	Lubin et al. (13) Mills et al. (11) Franceschi et al. (36) Knekt et al. (10)	Lé et al. (17)[c] Iscovich et al. (19) Toniolo et al. (21)[c] Richardson et al. (29)[c] Levi et al. (33)[c]
Fermented milk	Lé et al. (17)[c] van't Veer et al. (22)[c]	Levi et al. (33)	Knekt et al. (10)
Butter	Pawlega et al. (32) Knekt et al. (10)[c]	Lé et al. (17) Ewertz and Gill (24) Ingram et al. (27) La Vecchia et al. (37) Trichopoulou et al. (38)	Lubin et al. (13)[c] La Vecchia et al. (18)[c] Iscovich et al. (19)[c] Toniolo et al. (21) Levi et al. (33)[c]
Cream		Lubin et al. (13) Knekt et al. (10)	
Ice cream	Knekt et al. (10)	Goodman et al. (30)	
Milk fat	Pryor et al. (20)[c] Knekt et al. (10)	Mettlin et al. (25)	

[a]Studies giving a relative risk significantly smaller than unity are included in this group.
[b]Studies giving a relative risk significantly greater than unity are included in this group.
[c]Differs significantly from unity.

Discussion

The studies considered gave no consistent evidence of an inverse association between intake of different dairy products and breast cancer occurrence. In fact, some of the studies found a significant inverse association, some a significant positive association, and some no association. Milk fat is a rich natural dietary source of conjugated linoleic acids (4). The use of milk and other dairy products as a proxy measure for intake of conjugated linoleic acid is problematic, however, because conjugated linoleic acid is also detected in other foods, especially in red meats (4). In addition, several other constituents of milk may either increase or reduce the risk of breast cancer. On the basis of results from ecological studies, it has been suggested that high intake of saturated fatty acids may be a risk factor for breast cancer (44,45). Although the confirming evidence from analytic epidemiologic cohort studies is scarce (46), it cannot be excluded that the positive associations observed between milk intake and breast cancer risk are due to such mechanisms. In contrast, however, none of the studies that reported on intake of milk fat and breast cancer risk found any positive association (10,20,23,35). No systematical differences in the strengths of association were observed for various concentrations of fat in the milk. In addition to the suggestion that intake of conjugated fatty acids reduces breast cancer risk, there are also hypotheses concerning several other substances in milk that may provide protection against breast cancer. Such substances include calcium and vitamin D (47), lactic acid bacteria (48), and milk proteins (49). Although the results from epidemiologic studies on these hypotheses are controversial, the possibility remains that the inverse associations reported are due to some substance in the milk other than conjugated linoleic acid.

In principle, a well-designed observational study could give reliable information concerning the existence of an association between milk intake and breast cancer risk, but several methodological factors may bias the results. Such factors arise from the study design, sample size, validity and reliability of the dietary method, range of dairy product intake reported, confounding factors, and effect modification. A total of six of the studies considered were of cohort or nested case-control design (Table 35.1), whereas all others were of case-control design (Table 35.2). Potentially, two types of bias may occur in case-control studies, i.e., selection bias and information bias. The possibility of selection bias was pronounced in those case-control studies (Table 35.2) in which hospital patients were used as controls, because such controls were not necessarily representative of the populations from which the cancer cases originated. Case-control studies are subject to information bias because knowledge about the disease may affect the dietary habits or the responses of cancer cases. In cohort studies, information on intake of dairy products is collected before diagnosis of cancer, and thus there is less chance that the disease has affected dietary habits. To minimize this possible bias, the breast cancer cases diagnosed during the first 5 y of follow-up were excluded in one study (10). However, the exclusion did not notably change the results. A small number of cancer cases, i.e., the low power of the study, may be one reason for the failure to detect significant associations in some of the studies published (7,9).

The choice and use of the dietary assessment method may bias the results. Assessment of intake may be inaccurate, because the food lists may be incomplete and there may be errors in the subject's estimates of the frequency or size of the portion eaten. The precision by which information on food items is collected may be inadequate. Milk and other dairy products are consumed as such and as constituents of mixed dishes. Apparently, only those studies utilizing comprehensive food-frequency questionnaires or dietary history methods have been able to account for the total consumption of these foods. The precision of the dietary transformation tables, which constitute the basis of estimating intake of dairy products, may also be inadequate because the dairy product content in a particular foodstuff under different circumstances may vary considerably. Although coefficients of short-term reproducibility of milk and other dairy products were in general adequately high (7,10,50), the method may give an inaccurate picture of long-term intake because changes in dietary habits may weaken the representativeness of the estimated consumption. In the Finnish Mobile Clinic Health Examination Survey, long-term repeatability was relatively high for consumption of milk but much lower for consumption of cheese and butter (42). Because the intake of dairy products in all studies was estimated on the basis of a single measurement, a conservative estimate of the strength of association was obtained.

One potential reason for the lack of association in some studies may be that they were conducted in populations with intake of dairy products at levels having no effect or constant effects on breast cancer development. The majority of studies quoted presented a direct estimate in grams, liters, glasses, portions, or servings on the intake of different dairy products consumed by study subjects (Tables 35.3 and 35.4). Because of differences in the accuracy and specificity of the dietary questionnaires applied in different studies, however, comparisons of the intake levels of dairy products among the studies may not be straightforward. According to national average consumption levels (51), approximately two thirds of the studies were drawn from populations that can be characterized by a high average consumption level of fat-containing dairy products. Only two of the studies were carried out among populations with a very low average consumption level of milk and dairy products (31,39).

With few exceptions, inquiries concerning the consumption of milk and dairy products referred to the usual dietary intakes during some period of adult life. Age at the time of exposure, however, has been shown to be pivotal in determining the subsequent risk of breast cancer such that the younger the age at exposure to a risk factor, the greater the risk (52). Accordingly, it may be possible that the most important period for effects of dairy products and conjugated linoleic acid in breast cancer development is early in life during maturation of the mammary gland (3). Two of the three studies that reported on consumption of dairy products in childhood or adolescence suggested that high consumption of whole milk in childhood (15) or high intake of fat from dairy products in adolesecence (20) were associated with reduced breast cancer risk. The third study investigating premenopausal breast cancer found no association for consumption during adolescence (41).

A confounding factor, i.e., a risk factor for breast cancer associated with intake of milk and dairy products, may affect the association between intake of these foods and breast cancer risk, possibly by hiding the true association or by resulting in an artefactual association. Intake of milk was observed to be associated with several lifestyle and other nondietary factors (age, smoking status, alcohol consumption, occupation, and geographical area) potentially related to the risk of breast cancer (7,10). Milk consumption was also associated with the intake of several potentially protective or harmful foods (vegetables, fruits, cereals, potatoes, eggs, and meat) and nutrients (carbohydrates, protein, and fat) (7,10). With few exceptions, practically all studies adjusted the results for age (Tables 35.3 and 35.4). Only some of the studies, however, were adjusted for reproductive behavior and alcohol consumption. Energy intake was adjusted for in seven studies (10,12,21,33,36–38) and other single dietary factors were allowed in nine studies (Tables 35.3 and 35.4). Because control for confounding was incomplete in several studies, the possibility remains that the associations or the lack of association observed was due to other components in the diet or to dietary patterns or other lifestyle habits related to high intake of milk and dairy products.

It is possible that the effect of conjugated linoleic acid and other components of milk and dairy products on breast cancer occurrence is present only at specific levels of other effect-modifying factors. In such situations, the study of total populations will give conservative estimates of the association. It has been suggested that pre- and postmenopausal breast cancer have different risk factors. However, studies investigating the importance of dairy products separately in pre- and postmenopausal breast cancer did not find notable differences in the results between the two groups (11,15,18,20,31,34,36,37). To date, a small number of studies have described the associations in other subgroups of the study populations (7,10,17,30). A significant inverse interaction was observed between milk consumption and use of alcohol in one study (17), whereas no similar interaction was observed in another study (7). An interaction between intake of fermented milk and fiber was detected in one study (43). In another study, no interactions with various nondietary factors (age, occupation, geographic area, smoking, body mass index, or number of childbirths) were found (10). One study reported a stronger positive association among nonobese than among obese persons, and the association was also stronger in Caucasians than in Japanese, among women ≥ 60 y of age, and among nulliparous women (30).

In summary, the results of analytic epidemiologic studies are controversial, giving both reduced and elevated breast cancer risk associated with higher consumption of milk and other dairy products. In addition, a considerable number of the studies failed to find any association. These results may be due to the fact that there are potential mechanisms of components in milk and other dairy products leading to both reduction in and excess risk of breast cancer. The discrepant results may also be due to several methodological issues. Because of several competing biological mechanisms, the reduced risks observed cannot be attributed to conjugated linoleic acid intake. The next step in studying the effect of this fatty acid on breast cancer risk is to carry out epidemiologic studies on serum concentration of conjugated linoleic fatty acid and breast cancer risk.

References

1. Parodi, P.W. (1997) Cows' Milk Fat Components as Potential Anticarcinogenic Agents, *J. Nutr. 127,* 1055–1060.
2. Durgam, V.R., and Fernandes, G. (1997) The Growth Inhibitory Effect of Conjugated Linoleic Acid on MCF-7 Cells Is Related to Estrogen Response System, *Cancer Lett. 116,* 121–130.

3. Thompson, H., Zhu, Z., Banni, S., Darcy, K., Loftus, T., and Ip, C. (1997) Morphological and Biochemical Status of the Mammary Gland as Influenced by Conjugated Linoleic Acid: Implication for a Reduction in Mammary Cancer Risk, *Cancer Res. 57,* 5067–5072.
4. Chin, S.F., Liu, W., Storkson, J.M., Ha, Y.L., and Pariza, M.W. (1992) Dietary Sources of Conjugated Dienoic Isomers of Linoleic Acid, a Newly Recognized Class of Anticarcinogens, *J. Food Compos. Anal. 5,* 185–197.
5. Boyd, N.F., Martin, L.J., Noffel, M., Lockwood, G.A., and Trichler, D.L. (1993) A Meta-Analysis of Studies of Dietary Fat and Breast Cancer Risk, *Br. J. Cancer 68,* 627–636.
6. Jain, M. (1998) Dairy Foods, Dairy Fats, and Cancer: A Review of Epidemiological Evidence, *Nutr. Res. 18,* 905–937.
7. Ursin, G., Bjelke, E., Heuch, I., and Vollset, S.E. (1990) Milk Consumption and Cancer Incidence: A Norwegian Prospective Study, *Br. J. Cancer 61,* 454–459.
8. Gaard, M., Tretli, S., and Loken, E.B. (1995) Dietary Fat and the Risk of Breast Cancer: A Prospective Study of 25,892 Norwegian Women, *Int. J. Cancer 63,* 13–17.
9. Byrne, C., Ursin, G., and Ziegler, R.G. (1996) A Comparison of Food Habit and Food Frequency Data as Predictors of Breast Cancer in the NHANES I/NHEFS Cohort, *J. Nutr. 126,* 2757–2764.
10. Knekt, P., Järvinen, R., Seppänen, R., Pukkala, E., and Aromaa, A. (1996) Intake of Dairy Products and the Risk of Breast Cancer, *Br. J. Cancer 73,* 687–691.
11. Mills, P.K., Annegers, J.F., and Phillips, R.L. (1988) Animal Product Consumption and Subsequent Fatal Breast Cancer Risk Among Seventh-Day Adventists, *Am. J. Epidemiol. 127,* 440–453.
12. Toniolo, P., Riboli, E., Shore, R.E., and Pasternack, B.S. (1994) Consumption of Meat, Animal Products, Protein, and Fat and Risk of Breast Cancer: A Prospective Cohort Study in New York. *Epidemiology 5,* 391–397.
13. Lubin, J.H., Burns, P.E., Blot, W.J., Ziegler, R.G., Lees, A.W., and Fraumeni, J.F., Jr. (1981) Dietary Factors and Breast Cancer Risk, *Int. J. Cancer 28,* 685–689.
14. Talamini, R., La Vecchia, C., Decarli, A., Franceschi, S., Grattoni, E., Grigoletto, E., Liberati, A., and Tognoni, G. (1984) Social Factors, Diet and Breast Cancer in a Northern Italian Population, *Br. J. Cancer 49,* 723–729.
15. Hislop, T.G., Coldman, A.J., Elwood, J.M., Brauer, G., and Kan, L. (1986) Childhood and Recent Eating Patterns and Risk of Breast Cancer, *Cancer Detect. Prev. 9,* 47–58.
16. Katsouyanni, K., Trichopoulos, D., Boyle, P., Xirouchaki, E., Trichopoulou, A., Lisseos, B., Vasilaros, S., and MacMahon, B. (1986) Diet and Breast Cancer: A Case-Control Study in Greece, *Int. J. Cancer 38,* 815–820.
17. Lé, M.G., Moulton, L.H., Hill, C., and Kramar, A. (1986) Consumption of Dairy Produce and Alcohol in a Case-Control Study of Breast Cancer, *J. Natl. Cancer Inst. 77,* 633–636.
18. La Vecchia, C., Decarli, A., Franceschi, S., Gentile, A., Negri, E., and Parazzini, F. (1987) Dietary Factors and the Risk of Breast Cancer, *Nutr. Cancer 10,* 205–214.
19. Iscovich, J.M., Iscovich, R.B., Howe, G., Shiboski, S., and Kaldor, J.M. (1989) A Case-Control Study of Diet and Breast Cancer in Argentina, *Int. J. Cancer 44,* 770–776.
20. Pryor, M., Slattery, M.L., Robison, L.M., and Egger, M. (1989) Adolescent Diet and Breast Cancer in Utah, *Cancer Res. 49,* 2161–2167.
21. Toniolo, P., Riboli, E., Protta, F., Charrel, M., and Cappa, A.P. (1989) Calorie-Providing Nutrients and Risk of Breast Cancer, *J. Natl. Cancer Inst. 81,* 278–286.

22. van't Veer, P., Dekker, J.M., Lamers, J.W., Kok, F.J., Schouten, E.G., Brants, H.A., Sturmans, F., and Hermus, R.J. (1989) Consumption of Fermented Milk Products and Breast Cancer: A Case-Control Study in the Netherlands, *Cancer Res. 49*, 4020–4023.
23. van't Veer, P., Kok, F.J., Brants, H.A., Ockhuizen, T., Sturmans, F., and Hermus, R.J. (1990) Dietary Fat and the Risk of Breast Cancer, *Int. J. Epidemiol. 19*, 12–18.
24. Ewertz, M., and Gill, C. (1990) Dietary Factors and Breast-Cancer Risk in Denmark, *Int. J. Cancer 46*, 779–784.
25. Mettlin, C.J., Schoenfeld, E.R., and Natarajan, N. (1990) Patterns of Milk Consumption and Risk of Cancer, *Nutr. Cancer 13*, 89–99.
26. Simard, A., Vobecky, J., and Vobecky, J.S. (1990) Nutrition and Lifestyle Factors in Fibrocystic Disease and Cancer of the Breast, *Cancer Detect. Prev. 14*, 567–572.
27. Ingram, D.M., Nottage, E., and Roberts, T. (1991) The Role of Diet in the Development of Breast Cancer: A Case-Control Study of Patients with Breast Cancer, Benign Epithelial Hyperplasia and Fibrocystic Disease of the Breast, *Br. J. Cancer 64*, 187–191.
28. Matos, E.L., Thomas, D.B., Sobel, N., and Vuoto, D. (1991) Breast Cancer in Argentina: Case-Control Study with Special Reference to Meat Eating Habits, *Neoplasma 38*, 357–366.
29. Richardson, S., Gerber, M., and Cenee, S. (1991) The Role of Fat, Animal Protein and Some Vitamin Consumption in Breast Cancer: A Case Control Study in Southern France, *Int. J. Cancer 48*, 1–9.
30. Goodman, M.T., Nomura, A.M., Wilkens, L.R., and Hankin, J. (1992) The Association of Diet, Obesity, and Breast Cancer in Hawaii, *Cancer Epidemiol. Biomark. Prev. 1*, 269–275.
31. Kato, I., Miura, S., Kasumi, F., Iwase, T., Tashiro, H., Fujita, Y., Koyama, H., Ikeda, T., Fujiwara, K., Saotome, K., Asaishi, K., Abe, R., Nihei, M., Ishida, T., Yokoe, T., Yamamoto, H., and Murata, M. (1992) A Case-Control Study of Breast Cancer Among Japanese Women: With Special Reference to Family History and Reproductive and Dietary Factors, *Breast Cancer Res. Treat. 24*, 51–59.
32. Pawlega, J. (1992) Breast Cancer and Smoking, Vodka Drinking and Dietary Habits. A Case-Control Study, *Acta Oncol. 31*, 387–392.
33. Levi, F., La Vecchia, C., Gulie, C., and Negri, E. (1993) Dietary Factors and Breast Cancer Risk in Vaud, Switzerland, *Nutr. Cancer 19*, 327–335.
34. Holmberg, L., Ohlander, E.M., Byers, T., Zack, M., Wolk, A., Bergström, R., Bergkvist, L., Thurfjell, E., Bruce, A., and Adami, H.O. (1994) Diet and Breast Cancer Risk. Results from a Population-Based, Case-Control Study in Sweden, *Arch. Intern. Med. 154*, 1805–1811.
35. Landa, M.C., Frago, N., and Tres, A. (1994) Diet and the Risk of Breast Cancer in Spain, *Eur. J. Cancer Prev. 3*, 313–320.
36. Franceschi, S., Favero, A., La Vecchia, C., Negri, E., Dal Maso, L., Salvini, S., Decarli, A., and Giacosa, A. (1995) Influence of Food Groups and Food Diversity on Breast Cancer Risk in Italy, *Int. J. Cancer 63*, 785–789.
37. La Vecchia, C., Negri, E., Franceschi, S., Decarli, A., Giacosa, A., and Lipworth, L. (1995) Olive Oil, Other Dietary Fats, and the Risk of Breast Cancer (Italy), *Cancer Causes Control 6*, 545–550.
38. Trichopoulou, A., Katsouyanni, K., Stuver, S., Tzala, L., Gnardellis, C., Rimm, E., and Trichopoulos, D. (1995) Consumption of Olive Oil and Specific Food Groups in Relation to Breast Cancer Risk in Greece, *J. Natl. Cancer Inst. 87*, 110–116.

39. Yuan, J.-M., Wang, Q.-S., Ross, R.K., Henderson, B.E., and Yu, M.C. (1995) Diet and Breast Cancer in Shanghai and Tianjin, China. *Br. J. Cancer 71*, 1353–1358.
40. Moysich, K.B., Ambrosone, C.B., Vena, J.E., Shields, P.G., Mendola, P., Kostyniak, P., Greizerstein, H., Graham, S., Marshall, J.R., Schisterman, E.F., and Freudenheim, J.L. (1998) Environmental Organochlorine Exposure and Postmenopausal Breast Cancer Risk, *Cancer Epidemiol. Biomark. Prev. 7*, 181–188.
41. Potischman, N., Weiss, H.A., Swanson, C.A., Coates, R.J., Gammon, M.D., Malone, K.E., Brogan, D., Stanford, J.L., Hoover, R.N., and Brinton, L.A. (1998) Diet During Adolescence and Risk of Breast Cancer Among Young Women, *J. Natl. Cancer Inst. 90*, 226–233.
42. Järvinen, R., Seppänen, R., and Knekt, P. (1993) Short-Term and Long-Term Reproducibility of Dietary History Interview Data, *Int. J. Epidemiol. 22*, 520–527.
43. van't Veer, P., van Leer, E.M., Rietdijk, A., Kok, F.J., Schouten, E.G., Hermus, R.J., and Sturmans, F. (1991) Combination of Dietary Factors in Relation to Breast-Cancer Occurrence, *Int. J. Cancer 47*, 649–653.
44. Gaskill, S.P., McGuire, W.L., Osborne, C.K., and Stern, M.P. (1979) Breast Cancer Mortality and Diet in the United States, *Cancer Res. 39*, 3628–3637.
45. Carroll, K.K. (1992) Dietary Fat and Breast Cancer, *Lipids 27*, 793–797.
46. Hunter, D.J., Spiegelman, D., Adami, H.-O., Beeson, L., van den Brandt, P.A., Folsom, A.R., Fraser, G.E., Goldbohm, R.A., Graham, S., Howe, G.R., Kushi, L.H., Marshall, J.R., McDermott, A., Miller, A.B., Speizer, F.E., Wolk, A., Yaun, S.-S., and Willett, W. (1996) Cohort Studies of Fat Intake and the Risk of Breast Cancer—A Pooled Analysis, *N. Engl. J. Med. 334*, 356–361.
47. Newmark, H.L. (1994) Vitamin D Adequacy: A Possible Relationship to Breast Cancer, *Adv. Exp. Med. Biol. 364*, 109–114.
48. Biffi, A., Coradini, D., Larsen, R., Riva, L., and Di Fronzo, G. (1997) Antiproliferative Effect of Fermented Milk on the Growth of a Human Breast Cancer Cell Line, *Nutr. Cancer 28*, 93–99.
49. McIntosh, G.H., Regester, G.O., Le Leu, R.K., Royle, P.J., and Smithers, G.W. (1995) Dairy Proteins Protect Against Dimethylhydrazine-Induced Intestinal Cancers in Rats, *J. Nutr. 125*, 809–816.
50. Salvini, S., Hunter, D.J., Sampson, L., Stampfer, M.J., Colditz, G.A., Rosner, B., and Willett, W.C. (1989) Food-Based Validation of a Dietary Questionnaire: The Effects of Week-to-Week Variation in Food Consumption, *Int. J. Epidemiol. 18*, 858–867.
51. Kesteloot, H., Lesaffre, E., and Joossens, J.V. (1991) Dairy Fat, Saturated Animal Fat, and Cancer Risk, *Prev. Med. 20*, 226–236.
52. Colditz, G.A., and Frazier, A.L. (1995) Models of Breast Cancer Show That Risk Is Set by Events of Early Life: Prevention Efforts Must Shift Focus, *Cancer Epidemiol. Biomark. Prev. 4*, 567–571.

Index

Absorption of CLA, 307, 327–328
Acid catalyzed methylation
    Confirmation of methoxy artifacts by GC-MS, 134
    Isomerization of CLA, 69–71, 74–78, 85–86, 88–89
    Theory of artifact formation, 71–75
Adipose tissue
    CLA in female breast, 277–278, 280
    CLA in human, 76, 239, 283, 350–352
    CLA in nonruminant, 7, 86, 244–245, 283, 307–310, 329, 350
    CLA in ruminant, 7, 76
    Δ9 Desaturase, 187–188
    Intramuscular fat, 15, 363–364
    Metabolites of CLA, 307–309
    Site of synthesis of CLA from *trans*-11-18:1 in rodents, 205–206
Adipocyte differentiation
    Trans-10,*cis*-12-CLA, the active isomer, 16–17, 344–345
    Effect of CLA, 15–16, 344–346, 404–407
Ag^+-TLC (*see* silver ion thin-layer chromatography)
Ag^+-HPLC (*see* silver ion high performance liquid chromatography)
Alcoholism, effect on CLA levels, 414
Arachidonic acid (20:4n-6)
    Levels affected by CLA, 13, 309–310, 329–331
Atherosclerosis
    Effect of CLA, 14, 397–402

Beef
    Antimutagenicity of, 12
    CLA isomers, 76, 88, 240–241, 384, 386–388
Bile
    CLA content in phosphatidylcholine, 6, 86, 241
Biohydrogenation
    Effect of ionophores, 209–213
    Effect of low-fiber on rumen biohydrogenation, 184
    cis-9, *trans*-10 isomerase, 181–183
    cis-12, *trans*-11 isomerase, 183–184
    Cis/trans isomerization in rumen, 182–184
    Of C_{18} unsaturated FA, 180–185

Origin of different positional CLA isomers, 182–184, 186–187
Biosynthesis of CLA, 29–30, 47–48, 180–190, 379
Body composition
    Effect of CLA, 15–17, 288–289, 350–352, 359–364
    CLA content, 205–206, 289–290
    See also repartitioning (fat to lean)
Bone
    CLA content, 243
    Modeling or remodeling, 255–257
    Regulation of bone metabolism by cytokines, 257–258
    Regulation of bone metabolism by eicosanoids, 258–259
    Regulation of bone metabolism by growth factors, 258
    Structure and growth, 253–257
Bovinic acid, 5
Breast cancer
    Breast cancer cells, 432–433
    Chemically induced mammary tumors in rats, 278–280, 421–428
    Mammary development and risk of cancer, 300
    Relationship to CLA, 277–280, 301–304, 313, 421–428
    Relationship to lactation, 299–302
Butyrivibrio fibrisolvens, 7, 26, 30, 85, 183, 193, 327, 378

Cancer
    Breast cancer; *see* breast cancer
    CLA content in cancer tissue, 241–242, 277–278, 280
    Forestomach cancer, 429–430
    Intestinal/colon cancer, 13, 430–431
    Mammary cancer, chemically induced, 278–280, 421–428
    Mammary cancer, non-chemically induced, 428
    Possible mechanism of CLA action in cancer inhibition, 434–439
    Prostate cancer, 431
    Skin cancer, 428–429
Cheese
    CLA content, 382–387

CLA isomer distribution, 93, 95, 98–100, 240, 385–387
  Selected ion recording of CLA-FAME from cheese, 134
  EI spectra of DMOX derivatives of CLA isomers, 128–129
  Separation of CLA isomers by Ag+-HPLC, 99
  Separation of CLA isomers by GC, 134
Chicks
  Immune response, 14, 227–234
Cis-9, trans-11-CLA
  Adipocyte differentiation, 15–17, 344–345
  Anticarcinogenic response, 16, 101–102, 277–280
  Activation of peroxisomal proliferation, 16, 405–408, 439
  FTIR assignments, 144–146
  1H-NMR and 13C-NMR assignments, 155–161, 165–167, 173–178
  Mass spectrum of its DMOX derivative, 129
  Metabolites in rats, 307–311, 319–324
  Relationship to trans-11-18:1 in milk fat, 189–190, 222–223
  Separation by Ag+-HPLC, 94–100, 243–248
  Separation by GC, 85, 89, 91–93, 243–248, 283, 324
  Synthesis, 21–33,42–48,102–104
Cis-11, trans-13-CLA
  Formation during alkali-isomerization, 102–104
  FTIR assignments, 144–146
  1H-NMR and 13C-NMR assignments, 169–178
  Mass spectrum of its DMOX derivative, 128
  No metabolites detected in rats fed CLA, 321
  Separation by Ag+-HPLC, 94–100, 243–248
  Separation by GC, 89, 91–93, 243–248
Clinical studies
  Using CLA, 350–352, 413–414
  Using n-3 PUFA, 412–418

Clofibrate, 202, 204
Commercial preparation of CLA using alkali catalysts
  Analysis of CLA soaps, 52
  Choice of catalyst, 46–47
  Choice of vessel, 47
  Effect of solvents, 45–46
  Effect of crystallization, 103–104
  Formation of monoglycerides of ethylene glycol, 48, 50
  Metal content, 48–50
  Minor by-products of the reaction, 48
  Quality characteristics, 50–53
  Preparation of 9c,11t- and 10t,12c-18:2 isomers, 43–45, 102–104
  Preparation of 8t,10c- and 11c,13t-18:2 isomers, 44–45, 102–104
  Products of linoleic acid, 43–44, 102–104
  Products of linolenic acid, 49
  Reaction kinetics, 43–45
  Separation of the total CLA preparation, 51–52, 66
  Separation of CLA isomers by Ag+-HPLC, 95, 102–104, 389
  Separation of CLA isomers by GC, 91–92, 324
  Synthetic approaches, 42–43
Conjugated linoleic acid (CLA)
  Absorption, 307, 327–328
  Analysis of CLA isomers by GC-EIMS, 126–139
  Biochemical synthesis of CLA, 29–30, 47–48
  Biosynthesis of CLA in rumen, 6–7, 180–185, 379
  Biosynthesis of CLA in rodents, 203–206, 319–324
  13C-NMR assignments, 51, 152–161, 164–178
  Clinical studies, 350–352, 413–414
  Common name, 5, 201–206, 215, 238, 260, 296, 369, 378
  Dietary sources, 18, 348–349, 370, 379–388
  Differences in CLA isomer responses, 16–18, 344–345, 405–408
  Early history, 1–8

Effect of dietary fiber on CLA content in cow milk, 190–194
Effect of dietary ionophores on CLA content in cow milk, 191, 193, 209–213
Effect on atherosclerosis, 14, 397–402
Effect on body composition, 15–16, 348–352, 359–361
Effect on bone metabolism, 253–268
Effect on carcinogenesis, 12–14, 16, 420–440
Effect on diabetes, 404–409
Effect on growth, 14, 226–231, 288–290, 302–304, 350–352, 359, 365–366, 398–401, 431–432
Effect on immunoglobulin production, 332–334
Effect on mammary cancer, 277–280, 421–428, 432–433
Effect on pork quality, 359–367
Effect on retinol, 13, 310–312
Effect on tumor number and size, 278–279, 421–434
Estimated intake in humans, 348, 352, 369–375
Epidemiological studies: dairy products vs breast cancer, 444–467
Factors affecting CLA content in rumen milk and meat, 190–194, 215–223
FTIR assignments, 3–4, 141–146
^{1}H-NMR assignment, 155–161
In animal depot fat, 7, 86, 244–245, 283, 350, 359–361
In bile lipids, 6, 86, 241
In cheese and dairy products, 76, 93, 98–100, 240, 370, 382–384
In cow milk, 3, 5, 76, 93, 98–100, 212–213, 215–233, 382–384
Incorporation into neutral vs phospholipids, 205–207, 241–249, 289, 307–309, 313, 329–331
Incorporation into tissue lipids, 205–207, 239–249, 307–314, 328–331
Incorporation into tissue phospholipids, 205–207, 239, 241–249, 289, 309, 328–331
In fish lipids, 285–291, 384–385

Inhibits cholesterol-induced atherosclerosis in rabbits, 397–402
In humans, 76, 239–242, 277–278, 280, 312–314, 350–352
In human milk, 296–304
In margarine, 8, 379–381
In meat and meat products, 384–387
In milk phospholipids, 5–6
In nonruminants, 7, 205–207, 242–249
In oilseeds, 26, 379–381
In ruminants, 6–7, 85, 180–194, 210–213, 215–223, 240–241
In vegetable oils, 7, 285–286, 290–291, 379–381
Inverse association of CLA levels to breast cancer, 278–279
Metabolites, 100–101, 307–311, 312–314, 319–324
Methylation, 64–78, 290
Oxidation of CLA, 55–61
Reduces severity of preinduced atherosclerotic lesions, 400–402
Regulation of stearoyl-CoA desaturase mRNA, 340–346
Regulatory aspects, 415, 417–418
Role in immunity, 14–15, 229–234, 310, 332–336
Separation by Ag$^+$-TLC, 4, 93–94, 288–289
Separation by Ag$^+$-HPLC, 94–104, 244–248, 322
Separation by GC, 4, 85–93, 134, 244–246, 324
Separation by reversed phase HPLC, 52, 100–101, 307–309, 320–321
Site of CLA biosynthesis in rumen, 184–188
Sources of CLA intake for humans, 369–375, 379–387
Structure, 13, 43, 44, 65
Synthesis of CLA isomers, 21–33
Synthesis of commercial CLA products, 39–53, 102–104
Tissue synthesis of CLA from *trans*-11-18:1, 186–190, 201–207, 222, 379
Uncertainty of CLA isomer composition of diets used, 101–102
UV absorption, 1–2, 52

Conjugated octadecatrienoic acids
  Content in edible oils, 380
  FTIR, 146
  Intermediates in biohydrogenation of 18:3, 181, 184–185
  Metabolites of CLA, 307–311, 319–324
  Products of alkali isomerization of linolenic acid, 49
  See also - and -eleostearic acid
Cytokines
  Effect of CLA, 231–233, 334–335
  Role in immunity, 15, 227–234
  Role in bone formation, 253, 257–258, 260–268

Dairy products
  CLA composition, 240, 297, 385–387
  Correlation of risk of breast cancer and intake of, 444–467
  GC-EIMS spectra of CLA isomers, 128–129
  Separation of CLA isomers by Ag^+-HPLC, 76, 98–99
  Separation of CLA isomers by GC, 88–89, 93, 134–136
  Source of CLA, 370, 382–384, 418
Desaturation
  $\Delta 6$ and $\Delta 5$ desaturation in rats, 321
  $\Delta 9$ desaturation in humans, 202–203, 206–207
  $\Delta 9$ desaturation in rodents, 203–207
  $\Delta 9$ desaturation in rumen tissue, 185–190
  Of *trans* isomers of linoleic acid, 319, 324
  Of *trans* isomers of linolenic acid, 319
Diabetes
  Effect of CLA, 404–409
  Relation to CLA intake during breast feeding, 303
DMOX; *see* derivatization of CLA for GC-EIMS
Derivatization of CLA for GC-EIMS
  DMOX (4,4-dimethyloxazoline) derivatives, 111–113, 117, 118, 320–323
  GC conditions for analysis of CLA-DMOX derivatives, 126–127
  Hydroxylation followed by trimethylsilation, 113–114

Methyl esters; *see* methylation of CLA
MTAD (4-methyl-1,2,4-triazoline-3,5-dione) derivative, 114, 277
On-site derivatization (conversion of double bonds), 112–114
Picolinyl ester, 112–113, 117, 118
Pyrrolidine derivatives, 111–112, 116, 118
Remote-site derivatization (modifying carboxyl group), 111–113
Separation by GC, 114–116
Diacetylenic fatty acids, 25–26

Eicosanoids
  Effect of CLA, 15, 231–233, 309–310, 314, 324, 331–332, 435
  Role in bone formation, 253, 257–260
  Role in bone metabolism, 261–268
  Role in immune response, 227–234
Electron spin resonance
  Measuring effect of CLA-PC in membranes, 248–249
Eleosteric (elaeostearic) acid ($9c,11t,13t$-18:3), 2, 25, 146
Eleosteric acid ($9t,11t,13t$-18:3), 146
Epidemiological studies
  Breast cancer in women, 14, 277–278, 312–314
  Case-control studies: CLA vs breast cancer, 277–278, 446–448, 451–454, 456–463
  Cohort studies: dairy intake vs breast cancer, 445, 449, 453–456
  Intake of dairy products and risk of breast cancer, 444–467
  n-3 PUFA and cardiovascular disease, 412, 415–416
  Nested case-control studies: dairy intake vs breast cancer, 445, 449–450, 456
  Significance of epidemiological studies, 463–467
  Type of epidemiological studies, 444–446
Estimated consumption of CLA in humans
  Comparison to levels of CLA used in animal studies, 352
  For the average person, 348, 370–375, 387–391, 420–421

Ethylene glycol
  As solvent in alkali-isomerization of linoleic acid, 42, 43, 46, 48, 50
  Formation of monoester with CLA, 48, 50
Extraction of lipids from
  Commercial CLA preparations, 67–68
  Tissues, 203, 243–244

Fish oil
  CLA content, 285–291, 370, 384–385
  CLA content in fish fed CLA mixtures, 288–291
  CLA content in fish fed vegetable oils, 290–291
  Effect on CLA content, 191–193
  Estimated consumption of fish foods, 285
  Inhibition of rumen biohydrogenation, 192–193
Free fatty acids
  Industrial preparation from vegetable oils, 40–41
  Derivatization, 67–69, 74–78, 110–113
Fourier transform infrared (FTIR)
  Characteristic FTIR absorption of CLA isomers, 3–4, 84–85, 141–142, 144–146
  FTIR of CLA DMOX derivatives, 142, 144–146
  FTIR of conjugated octadecatrienoic acids, 146
  FTIR of furan fatty acids, 147–149
  FTIR of methoxy artifacts, 146–149
  GC-direct deposition (DD)-FTIR, 144–149
  GC-light pipe (LP)-FTIR, 143
  GC-matrix isolation (MI)-FTIR, 143
Furan fatty acids
  Conversion to dioxoenes, 59–60
  GC-FTIR analysis, 147–149
  GC separation, 147–149
  Natural occurrence, 58
  Oxidation, 59–60
  Product of CLA oxidation, 57–58
  Rate of oxidation compared to CLA, 60

Gas chromatography (GC) of CLA
  Complementary technique to Ag$^+$-HPLC, 98, 100, 320
  Columns used, 76–77, 87–91, 283
  Elution order, 87, 93
  Equivalent chain length, 87
  Identification of methoxy artifacts, 71, 75, 147–149
  Interfering fatty acid methyl esters, 86–88, 285–291
  Separation of CLA isomers, 4, 87–93, 242–246, 277–278, 286–289, 324
  Separation of positional and geometric CLA isomers, 91–93
  Using capillary columns, 87–93, 283–289, 324
  Using packed columns, 4, 85–87
GC-electron impact mass spectrometry (GC-EIMS)
  Derivatization of the double bonds (on-site), 113–114
  Derivatization of the carboxyl group (remote-site), 111–113
  Diagnostic ions for DMOX derivatives of CLA isomers, 130–132, 135–137
  Fragmentation pattern of CLA derivatives, 116–122, 128–132, 320–324
  GC conditions for analysis of CLA derivatives, 114–116, 126–127
  Identification of CLA metabolites, 120–122, 320–324
  Ion trap MS detection, 136–139
  Limitation and benefits of GC-MS of FAME, 110, 134
  Magnetic sector MS detection of CLA, 133–136
  MS of remote-site derivatives of CLA, 116–121
  MS of on-site derivatives of CLA, 121–122
  Quadrupole MS detection of CLA, 133
  Reconstructed ion chromatogram, 135–137
  Selected EI spectra of CLA isomers present in cheese, 128–129
  Selected ion recording (SIR), high resolution, 133–135
Growth
  Effect of CLA in chicken, 14, 226–231, 350
  Effect of CLA in fish, 288–290

Effect of CLA in humans, 302–304, 350–352
Effect of CLA in pigs, 350, 359, 365–366
Effect of CLA in rabbits, 398–401
Effect of CLA in rodents, 350, 352, 431–432
Growth hormones, 349–350, 354

Heart
  CLA content, 329
  CLA isomer distribution in, 243–248
Humans
  CLA content in adipose, 76, 239, 283, 313
  CLA content in foods, 370, 379–391
  CLA content in blood, 239–240, 312–314
  CLA metabolites, 312–314
  CLA produced during disease conditions, 413–414
  Effect of CLA on body composition, 350–352
  Estimated CLA intake, 348, 370–375, 387–392
  Infant formula, 297–298, 302, 374

Infrared (*see* Fourier transform infrared)
Iodine catalyzed isomerization of CLA isomers, 25, 30, 91, 92, 95–96
Immunity
  Role of CLA in immune-related disorders, 233–234, 303
  Role of CLA, 231–233, 335–336
  Role of CLA in preventing immune-induced wasting, 14, 230–231
  Role of cytokines, 227–229, 334–335
  Role of eicosanoids, 229–234, 331–332
  Role of immune response in growth, 227–228
Immunoglobulin, effect of CLA, 332–334
Interleukin, 227–229, 231–234, 257–258, 261–266, 334–335
Ionophores
  Decreased methane production in dairy cows fed, 209, 212
  Effect on animal performance, 209, 212
  Effect on CLA content, 191, 193, 210–213
  Effect on fatty acid metabolism, 209–213

Effect on milk fat content, 212
Tests with continuous rumen cultures, 210–211
Isomerization of CLA isomers by
  Acid catalysts, 71–78
  Iodine and light, 25, 30, 91–93, 95–97
  p-Toluenesulfinic acid, 92

Linoleic acid
  Biohydrogenation, 180–186, 210–211
  Content in vegetable oils, 39–40
  Effect on chemopreventative action of CLA, 425–427
  Enrichment using molecular sieves, 42
  Enrichment using solvent crystallization, 41
  Enrichment using solvent-free crystallization, 41–42
  Enrichment using two-phase solvent systems, 42
  Enrichment of using urea adducts, 42
  Metabolism of *trans* linoleic acid isomers, 319, 324
  Metabolism affected by CLA, 309–311
  Used in synthesis of CLA, 22, 26–27, 39, 42–48, 102–104
Linolenic acid
  Biohydrogenation, 180–185
  Evaluation of n-3 PUFA research findings, 412–418
  Metabolism of *trans* linolenic acid isomers, 319
Liver
  CLA content, 329
  CLA isomer distribution, 242–247
  Fatty acid changes due to CLA, 329–331
  Identification of CLA metabolites in rat, 319–324

Mammary cancer
  CLA content in breast adipose, control vs cancer patients, 277–278, 280
  Case-control study in women, CLA vs breast cancer, 277–278
  Effect of CLA on chemically induced mammary tumors, 12, 278–280, 310, 421–428

Effect of CLA on mammary tumor cells in mice, 428
Effect of free vs esterified CLA, 425
Proliferation (metastasis) suppressed by CLA, 12, 428
Mammary gland
  Site of CLA biosynthesis in the rumen, 188
Margarine, CLA content, 8, 379–381
Membrane
  Artificial membrane containing CLA-PC, 248–249
  Significance of CLA incorporation, 244–249
Methoxy artifacts
  GC-FTIR analysis, 146–149
  GC separation, 71, 75, 147–149
  Produced from allylic hydroxy fatty acids, 72–75
  Produced from acid catalyzed methylation of CLA, 69–72, 74–78
Methylation of CLA
  Acid catalyzed, 69–78, 88–89, 243, 290
  Base catalyzed, 68–69, 72, 75–78, 243–244, 277
  Formation of methoxy artifacts, 69–72, 74–78, 146–149
  Methylation of allylic hydroxy monoenes, 72–75
  Prior alkali hydrolysis of unknown CLA preparations, 68
  Review of methods used, 74–78, 90–91
  Testing completion of methylation, 75–78
  Using BF_3/methanol, 69–77, 88–89, 239–243, 290
  Using HCl/methanol, 69–78, 240–243
  Using sodium methoxide/methanol, 69, 77–78, 243–244, 277
  Using tetramethylguanidine (TMG), 72, 77
  Using TMS-diazomethane for free fatty acids, 68–69, 78, 244
Mice
  Immune-deficient, 13, 230–231
  Stearoyl-CoA desaturase mRNA, 16, 342–344
Milk
  CLA content in cow milk, 3, 5, 76, 93, 98–100, 212–213, 215–233, 382–384

CLA content in human milk, 297–299
CLA in milk phospholipids, 5–6
CLA isomer distribution, 76, 86, 98–100, 190–194, 212–213, 240
Composition of human milk, 296–297
Decrease in milk fat synthesis, 183, 192–194, 212
Effect of dietary oil/fat on CLA content, 191–192, 217–219
Effect of dietary fiber on CLA content, 190–194
Effect of pasture feeding on CLA content, 191, 194, 220–221
Effect of season and management, 221–223
Species, age and breed differences in cow CLA content, 188, 215–217
Variation in CLA content, 191, 216–217
Modified tall oil (source of CLA)
  Fed to pigs, 366
Monensin; see ionophores

n-3 Polyunsaturated fatty acids (PUFA)
  Evaluation of n-3 PUFA results, 412–418
  FDA assessment of beneficial effects, 415
NMR
  ^{13}C-NMR assignments of geometric CLA isomers, 51, 152–161, 164–178
  Effect of concentration on ^{13}C-NMR chemical shifts, 174–175
  ^{1}H-NMR assignments of geometric CLA isomers, 155–161
  Identification of CLA isomers in mixtures by ^{13}C-NMR, 173–175
  Previous NMR assignments, 153–154
  Quantitation of CLA isomers by ^{13}C-NMR, 175–178

Oilseeds
  CLA content, 26
  Conjugated octadecatrienoic acids, 26, 146
  Santalbic acid in; see also santalbic acid, 26
Oxidation of CLA
  Autoxidation of CLA, 58
  By singlet oxygen, 56–58
  Formation of furan fatty acids, 57–58
  Formation of dioxoenes, 59–61

Identification of CLA products by GC-FTIR, 147–149
Photo-oxidation of CLA, 57
Rate comparison to other fatty acids, 55, 58

Patents on CLA synthesis, 42–48
Partial hydrazine reduction, 5, 25, 86, 321
Peroxisomal-oxidation
  Activation by PPAR, 310, 314, 336, 405–409, 439
  Metabolites of CLA by, 308, 324
Phospholipids
  CLA incorporation into individual phospholipids, 86, 243–248, 328–331
  CLA incorporation into total phospholipids, 5–6, 239, 241–242, 289–291, 319–324
  Effect of synthetic 1-stearoyl-2-CLA-PC in membranes, 248–249
Pig
  CLA isomer distribution in tissues, 243–248
  Effect of CLA on carcass composition, 350, 359–366
  Effect of CLA on growth performance, 359, 365–366
  Effect of CLA on pork loin characteristics, 363–366
  Effect of CLA on pork quality, 361–366
  Growth promoters used, 354
  Pig feeding studies, 350, 355–356
  Pork quality measurements, 356–358
  Sensory evaluation of pork loins, 358, 364–365
Positional distribution of CLA
  In chylomicron triacylglycerol, 328
  In cow and sheep bile phosphatidylcholine, 6
  In women breast adipose tissue triacylglycerol, 279
Peroxisome proliferator activated receptor (PPAP)
  Activation by CLA, 16, 310, 314, 336, 404–409, 439
  Involvement in adipocyte differentiation, 344, 349, 404–408
Prostaglandins; *see* eicosanoids

Rabbit
  Atherogenic diets, 398–399
  CLA inhibits cholesterol-induced atherosclerosis, 14, 397–402
  CLA reduces severity of preinduced atherosclerosis, 400–402
Rat
  Accumulation of CLA, 307–309, 320, 328–331
  Chemically induced mammary cancer, 12, 278–280, 421–428
  Metabolic studies of CLA, 319–324
  Growth, 14, 350
Reconstructed ion chromatogram for DMOX derivatives, 135–137
Reductive ozonolysis, 5, 85, 86
Reduction of acetylenic bond(s), 25, 28–29
Regulatory aspects
  Of CLA, 417–418
  Of fish oils (linolenic acid), 415
Repartitioning (fat to lean)
  Effect of different CLA isomers, 16–18, 344–345
  In animals, 15, 350, 352, 359
  In humans, 303, 350–352
  Intramuscular fat in pigs, 359–360, 363–364
  Potential mechanism of action, 349–350
Retinol (vitamin A)
  Effect of CLA, 13, 311–312
Reversed-phase high performance liquid chromatography (RP-HPLC)
  Commercial CLA preparations, 52, 66–67
  Concentration of CLA metabolites, 67, 100–101, 320–321
  Identification of CLA metabolites, 100–101, 307–311, 320–321
Ricinoleic acid (12-OH, 9 *cis*-18:1), 22, 30, 48, 86, 153
Rumenic acid (9 *cis*-11 *trans*-18:2), 5, 201–206, 215, 238, 260, 296, 369, 378

Santalbic (ximenynic) acid (11*t*-octadec-9-ynoic), 25, 26, 27, 28, 153
Selected ion recording (mass spectrometry), 133–135
Serum, CLA content, 239–240, 243, 308, 313, 329

Silver ion high performance liquid chromatography (Ag+-HPLC)
  Complementary technique to GC, 98, 100, 320
  Effect of acetonitrile content in hexane, 96–97
  Elution order of CLA isomers using, 94
  Improved separation using up to six columns in series, 94–100
  Interfering fatty acids, 98–99
  Purification of CLA isomers/synthetic intermediate, 30
  Separation by chain length, 98, 100
  Separation of CLA isomers in tissue lipids, 243–248
  Separation of CLA metabolites, 320, 322
  Separation of positional and geometric isomers, 94–96
Silver ion thin-layer chromatography (Ag+-TLC)
  Elution order of CLA isomers, 93–94, 288
  Separation of CLA, 4, 85, 86, 288–290
Size exclusion chromatography, 51
Skin cancer, 426–427
Stearoyl-CoA desaturase mRNA
  Trans-10,*cis*-12-CLA, the active isomer, 16–17, 343–345
  Decrease in gene expression by CLA, 342–344
  Effect of CLA on enzyme activity, 340–342
  Regulation during adipocyte differentiation, 344–346
Supercritical fluid chromatography (SFC)
  Separation of total CLA preparations, 66–67
Synthesis of CLA by
  Alkali isomerization resulting in two CLA isomers, 21, 25, 39, 43–45, 102–104, 287–288
  Alkali isomerization resulting in four CLA isomers, 21, 25, 39, 43–45, 102–104
  Biochemical synthesis, 26–27, 29–30, 47–48
  Chemical synthesis, 24–26, 28–31
  CLA isomers synthesized, 31–33
  Dehydration of unsaturated hydroxy fatty acids, 22, 24, 48

Fate of minor components on, 48
Isotopically labeled CLA isomers, 27–30
Synthesis of isotopically labeled CLA
  Carbon (^{13}C), 27, 29
  Carbon (^{14}C), 27, 29, 33
  Deuterium (^{2}H), 27–29, 33
  Tritium (^{3}H), 27, 33
Synthesized conjugated fatty acid isomers; listing
  CLA isomers, 31
  Isotopically labelled isomers, 33
  PUFA containing conjugated double bonds, 32

Tetramethylguanidine (TMG), 72, 75, 77
Trans fatty acids
  Content in edible oils and margarine, 379–381
  Content in dairy products, 382–384
  Content in meat and meat products, 384–388
  Estimation of intake in different countries, 387–392
  Formation by ruminants, 180–185, 219, 378–379
  Formation during partial hydrogenation of vegetable oils, 379–380
Trans-7, *cis*-9-CLA
  Identification by ^{13}C-NMR, 173–176
  Identification by GC-EIMS, 135–137
  FTIR spectrum, 142
  Mass spectrum of its DMOX derivative, 129, 142
  Separation by Ag+-HPLC, 76, 94–99
  Separation by GC, 93, 137
Trans-8, *cis*-10-CLA
  ^{13}C-NMR assignments, 169–176, 178
  Formation during alkali-isomerization, 102–104
  Mass spectrum of its DMOX derivative, 129
  No metabolites detected in rats fed CLA, 321
  Separation of by Ag+-HPLC, 94–99
  Separation of by GC, 89, 91–93, 324
Trans-10, *cis*-12-CLA
  Anticarcinogenic response, 16

Biosynthesis, 190–194
^{13}C-NMR assignments, 153–154, 165–169
Effect on eicosenoids, 230
Formation during alkali-isomerization, 102–104
Involvement in changes of body composition, 16–17
Mass spectrum of its DMOX derivative, 128
Metabolism, 16, 229–230, 320–324
Regulation of stearoyl-CoA desaturase mRNA, 340–346
Separation by Ag$^+$-HPLC, 94–99
Separation by GC, 89, 91–93, 288, 324
Suppression of milk fat synthesis, 16, 183, 193–194
Trans-10-octadecenoic acid, 183–184, 192, 222–223
Trans-11-octadecenoic acid (vaccenic acid)
Intermediate in biohydrogenation of PUFA in ruminants, 181, 378–379
Precursor of CLA in tissues and milk lipids, 186–194, 205–206, 328
Relationship to *cis*-9, *trans*-11-CLA in milk fat, 189–190, 222
Results of feeding deuterated vaccenic acid, 202, 206–207
Tumor inhibition by CLA in cell lines
In breast cancer, 432–433
In prostate cancer, 433
In tumor, 433–434

UV absorption
Of CLA,
Of commercial CLA preparations, 52
Of polyunsaturated fatty acids, 2–3
Used in the detection of CLA, 1–2, 84–85, 91, 94–104, 285, 321–322

Vaccenic acid; see *trans*-11-octadecenoic acid
Vegetable oils
CLA content in, 7, 285, 379–381
Effect on CLA synthesis of minor components, 48
Effect of processing, 6–8, 379

Web site, 12